Instructor's Solutions Manual

for
McKeague and Turner's

Trigonometry

Fifth Edition

Ross Rueger
College of the Sequoias

THOMSON
™
BROOKS/COLE

Australia • Canada • Mexico • Singapore • Spain • United Kingdom • United States

Printed in the United States of America
1 2 3 4 5 6 7 07 06 05 04 03

Printer: Von Hoffmann Corporation

ISBN: 0-534-40394-8

For more information about our products,
contact us at:
Thomson Learning Academic Resource Center
1-800-423-0563

For permission to use material from this text,
contact us by:
Phone: 1-800-730-2214
Fax: 1-800-730-2215
Web: http://www.thomsonrights.com

Brooks/Cole—Thomson Learning
10 Davis Drive
Belmont, CA 94002-3098
USA

Asia
Thomson Learning
5 Shenton Way #01-01
UIC Building
Singapore 068808

Australia/New Zealand
Thomson Learning
102 Dodds Street
Southbank, Victoria 3006
Australia

Canada
Nelson
1120 Birchmount Road
Toronto, Ontario M1K 5G4
Canada

Europe/Middle East/South Africa
Thomson Learning
High Holborn House
50/51 Bedford Row
London WC1R 4LR
United Kingdom

Latin America
Thomson Learning
Seneca, 53
Colonia Polanco
11560 Mexico D.F.
Mexico

Spain/Portugal
Paraninfo
Calle/Magallanes, 25
28015 Madrid, Spain

Contents

Preface

This *Instructor's Solutions Manual* contains complete solutions to every even-numbered problem in the exercise sets for *Trigonometry, Fifth Edition* by Charles P. McKeague and Mark Turner. It also contains solutions to every problem in the chapter tests. I have attempted to format solutions for readability and accuracy, and apologize to you for any errors that you may encounter. If you have any comments, suggestions, error corrections, or alternative solutions please feel free to drop me a note. If you prefer, you can send me e-mail with corrections or comments (address at the bottom of this page).

I would like to thank a number of people for their assistance in preparing this manual. Thanks go to Rachael Sturgeon and Lisa Chow at Brooks/Cole Publishing for their valuable assistance and support. Special thanks go to Matt Bourez of College of the Sequoias for his meticulous error-checking of my solutions, and prompt return of my manuscript under tight deadlines.

I wish to express my appreciation to Pat McKeague and Mark Turner for asking me to be involved with this textbook. This book provides a complete course in trigonometry, and you will find the text very easy for your students to read.

Ross Rueger
College of the Sequoias
915 South Mooney Boulevard
Visalia, CA 93277
matmanross@aol.com

October, 2003

Chapter 1
The Six Trigonometric Functions

1.1 Angles, Degrees, and Special Triangles

2. 50° is an acute angle. The complement is $90° - 50° = 40°$, and the supplement is $180° - 50° = 130°$.

4. 90° is neither an acute or obtuse angle (it is a right angle). The complement is $90° - 90° = 0°$, and the supplement is $180° - 90° = 90°$.

6. 160° is an obtuse angle. The complement is $90° - 160° = -70°$, and the supplement is $180° - 160° = 20°$.

8. y is neither an acute or obtuse angle (its measure is unknown). The complement is $90° - y$, and the supplement is $180° - y$.

10. Adding the angles in triangle BDC:

$$B + 90° + 45° = 180°$$
$$B + 135° = 180°$$
$$B = 45°$$

12. Adding the angles in triangle ADC:

$$2\alpha + 90° + \alpha = 180°$$
$$3\alpha + 90° = 180°$$
$$3\alpha = 90°$$
$$\alpha = 30°$$

14. Since $A = 80°$, $\alpha = 180° - 80° - 90° = 10°$. Since $\alpha + \beta = 80°$:

$$10° + \beta = 80°$$
$$\beta = 70°$$

Since $\beta = 70°$, $B = 180° - 70° - 90° = 20°$.

16. No. The posts must be perpendicular to the ground for α and β to be complementary.

18. Since α and β are complementary, $\alpha = 90° - 52° = 38°$.

20. Since 1 revolution = 360°, the searchlight rotates 360° every 4 seconds. Thus it rotates 90° in 1 second.

22. Since 1 revolution = 360°, the earth rotates 360° in 24 hours. Thus it rotates 180° in 12 hours.

24. Let x represent the measure of each base angle. Since the angles must add to 180°:

$$x + x + 40° = 180°$$
$$2x + 40° = 180°$$
$$2x = 140°$$
$$x = 70°$$

26. Using the Pythagorean Theorem:

$$6^2 + 8^2 = c^2$$
$$36 + 64 = c^2$$
$$c^2 = 100$$
$$c = 10$$

28. Using the Pythagorean Theorem:

$$2^2 + b^2 = 6^2$$
$$4 + b^2 = 36$$
$$b^2 = 32$$
$$b = \sqrt{32} = 4\sqrt{2}$$

30. Using the Pythagorean Theorem:
$$a^2 + 10^2 = 26^2$$
$$a^2 + 100 = 676$$
$$a^2 = 576$$
$$a = 24$$

32. Using the Pythagorean Theorem:
$$5^2 + x^2 = \left(5\sqrt{2}\right)^2$$
$$25 + x^2 = 50$$
$$x^2 = 25$$
$$x = 5$$

34. Using the Pythagorean Theorem:
$$1^2 + x^2 = 2^2$$
$$1 + x^2 = 4$$
$$x^2 = 3$$
$$x = \sqrt{3}$$

36. Using the Pythagorean Theorem:
$$(x-1)^2 + x^2 = 5^2$$
$$x^2 - 2x + 1 + x^2 = 25$$
$$2x^2 - 2x - 24 = 0$$
$$x^2 - x - 12 = 0$$
$$(x-4)(x+3) = 0$$
$$x = 4, -3$$
Since x represents the side of a triangle, it cannot be a negative number. So the only solution is $x = 4$.

38. First find AC using the Pythagorean Theorem:
$$(AC)^2 + 5^2 = 13^2$$
$$(AC)^2 + 25 = 169$$
$$(AC)^2 = 144$$
$$AC = 12$$
Thus $DC = AC - AD = 12 - 4 = 8$. Now find BD using the Pythagorean Theorem:
$$8^2 + 5^2 = (BD)^2$$
$$64 + 25 = (BD)^2$$
$$(BD)^2 = 89$$
$$BD = \sqrt{89}$$

40. Since $AC = 8 + r$, using the Pythagorean Theorem:
$$12^2 + r^2 = (8+r)^2$$
$$144 + r^2 = 64 + 16r + r^2$$
$$144 = 64 + 16r$$
$$16r = 80$$
$$r = 5$$

42. Let x represent the distance across the pond. Using the Pythagorean Theorem:
$$25^2 + 60^2 = x^2$$
$$625 + 3600 = x^2$$
$$x^2 = 4225$$
$$x = 65$$
The distance across the pond is 65 yards.

44. If the shortest side is 3, the side opposite the 60° angle is $\sqrt{3} \cdot 3 = 3\sqrt{3}$, and the longest side is $2 \cdot 3 = 6$.

46. If the longest side is 5, the shortest side is $\frac{1}{2} \cdot 5 = \frac{5}{2}$, and the side opposite the 60° angle is $\sqrt{3} \cdot \frac{5}{2} = \frac{5}{2}\sqrt{3}$.

48. If the side opposite the 60° angle is 4, the shortest side is $\frac{4}{\sqrt{3}} \cdot \frac{\sqrt{3}}{\sqrt{3}} = \frac{4}{3}\sqrt{3}$, and the longest side is $2 \cdot \frac{4}{3}\sqrt{3} = \frac{8}{3}\sqrt{3}$.

50. Since 20 ft is now opposite the 60° angle, the length of the shortest side is $\dfrac{20}{\sqrt{3}} \cdot \dfrac{\sqrt{3}}{\sqrt{3}} = \dfrac{20}{3}\sqrt{3}$. The length of the escalator is the longest side, which is $2 \cdot \dfrac{20}{3}\sqrt{3} = \dfrac{40}{3}\sqrt{3} \approx 23.09$ ft.

52. To solve this problem we need to find the widths of the base and sides of the tent. Since 3 ft is the side opposite the 60° angle at the end, the shortest side is $\dfrac{3}{\sqrt{3}} \cdot \dfrac{\sqrt{3}}{\sqrt{3}} = \sqrt{3}$ ft and the longest side is $2\sqrt{3}$ ft. The width of the base and sides are therefore $2\sqrt{3}$ ft. The sides and floor of the tent have a total area of $3 \cdot 2\sqrt{3} \cdot 6 = 36\sqrt{3}$ ft, and the ends (which are triangles) have a total area of $2 \cdot \frac{1}{2} \cdot 2\sqrt{3} \cdot 3 = 6\sqrt{3}$ ft^2. Thus the total amount of material needed is $36\sqrt{3} + 6\sqrt{3} = 42\sqrt{3} \approx 72.7$ ft^2.

54. Since the shorter sides are each $\frac{1}{2}$, the longer side is $\frac{1}{2}\sqrt{2}$.

56. Since the longest side is $5\sqrt{2}$, the shorter sides are $\dfrac{5\sqrt{2}}{\sqrt{2}} = 5$.

58. Since the longest side is 12, the shorter sides are $\dfrac{12}{\sqrt{2}} \cdot \dfrac{\sqrt{2}}{\sqrt{2}} = 6\sqrt{2}$.

60. Since the height is 1,000 ft, which represents the shorter side, the longer side is $1000\sqrt{2}$ ft. Thus, if the bullet is traveling at 2,828 ft/sec, the total time is $\dfrac{1000\sqrt{2}\ \text{ft}}{2828\ \text{ft/sec}} \approx \frac{1}{2}$ sec.

62.   **a.**   Since $GC = 5$ cm and $CD = 5$ cm, and $\triangle GCD$ is a 45°-45°-90° triangle, the length of diagonal GD (which is the longest side) is $5\sqrt{2}$ cm.

    b.   Since $GD = 5\sqrt{2}$ cm from part (a) and $BD = 5$ cm, using the Pythagorean Theorem:
$$(GD)^2 + (BD)^2 = (GB)^2$$
$$\left(5\sqrt{2}\right)^2 + 5^2 = (GB)^2$$
$$50 + 25 = (GB)^2$$
$$(GB)^2 = 75$$
$$GB = \sqrt{75} = 5\sqrt{3}\ \text{cm}$$

64. The measure of $\angle GDH$ is 45°, since $\triangle GHD$ is a 45°-45°-90° triangle.

66. Let x represent the length of edge CD. Since $\triangle CDH$ is a 45°-45°-90° triangle, the length of diagonal CH is $x\sqrt{2}$. Now FH has a length of x, so using the Pythagorean Theorem with $\triangle CHF$:
$$(CH)^2 + (HF)^2 = (CF)^2$$
$$\left(x\sqrt{2}\right)^2 + x^2 = \left(5\sqrt{3}\right)^2$$
$$2x^2 + x^2 = 75$$
$$3x^2 = 75$$
$$x^2 = 25$$
$$x = 5$$
The length of edge CD is 5 ft.

68.   **a.**   Since $\triangle ODB$ is a right triangle, using the Pythagorean Theorem:
$$(OD)^2 + (DB)^2 = (OB)^2$$
$$1^2 + 2^2 = (OB)^2$$
$$(OB)^2 = 5$$
$$OB = \sqrt{5}$$

    b.   Since $OB = OE$, $OE = \sqrt{5}$.

    c.   Since $CE = CO + OE$, $CE = 1 + \sqrt{5}$.

    d.   Finding the ratio: $\dfrac{CE}{EF} = \dfrac{1 + \sqrt{5}}{2}$. This is called the golden ratio.

1.2 The Rectangular Coordinate System

2.   Graphing the ordered pair $(2,-4)$:

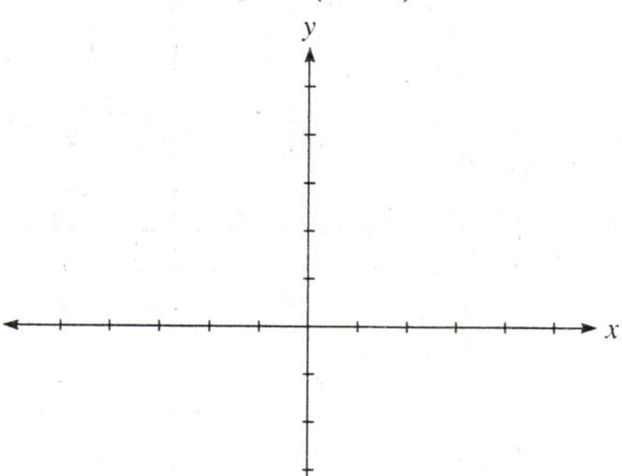

4.   Graphing the ordered pair $(-2,-4)$:

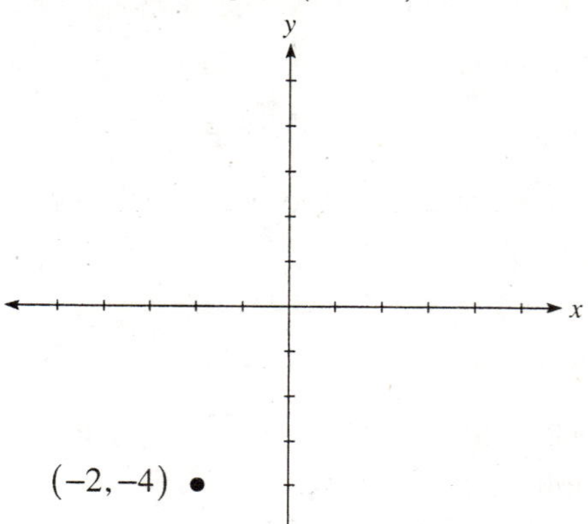

6.   Graphing the ordered pair $(-4,-2)$:

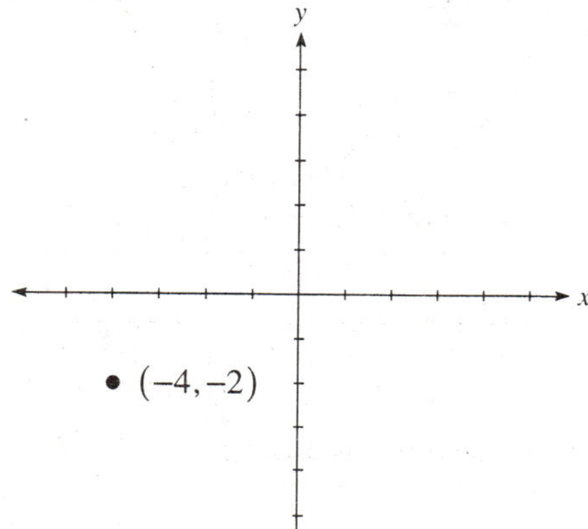

8.   Graphing the ordered pair $(-3,0)$:

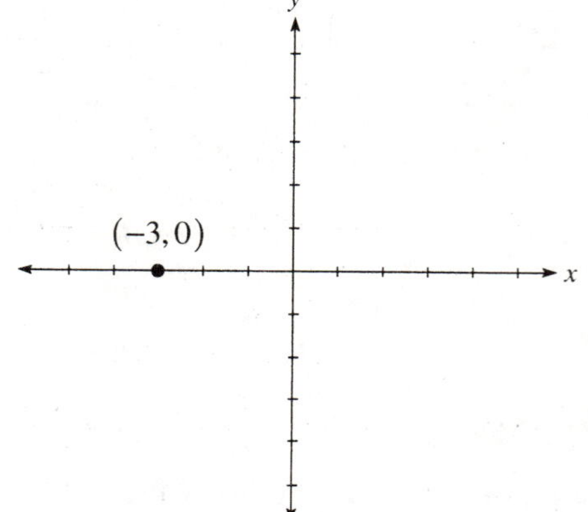

10.   Graphing the ordered pair $(0,2)$:

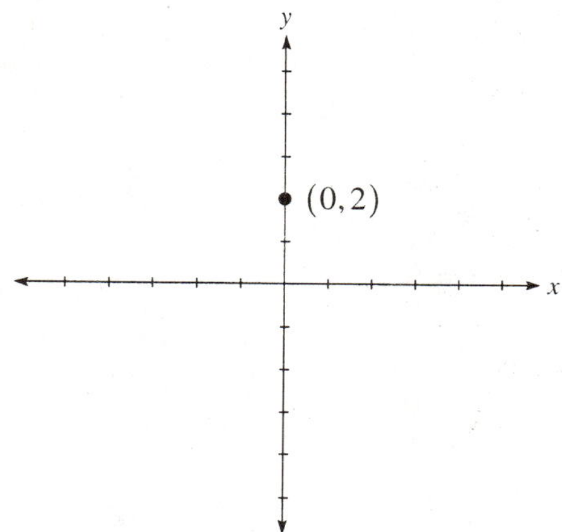

12.   Graphing the ordered pair $\left(5,\frac{1}{2}\right)$:

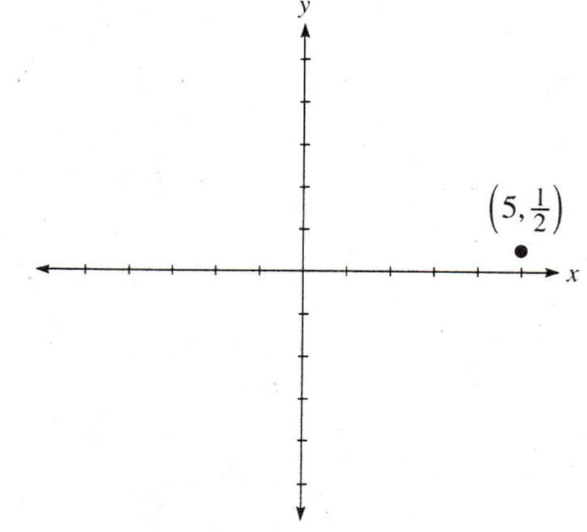

14. Graphing the line:

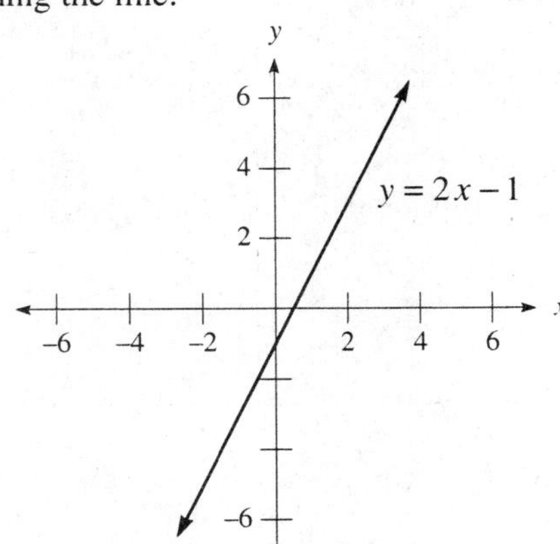

16. Graphing the line:

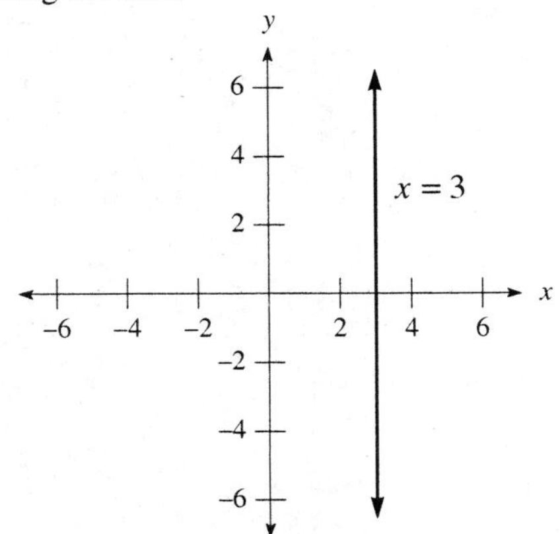

18. Graphing the parabola:

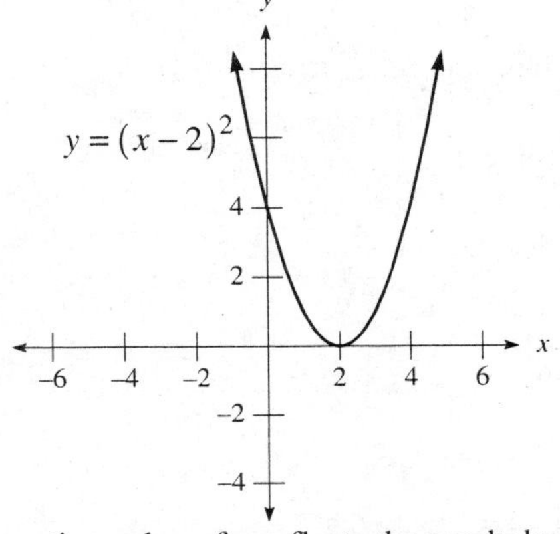

20. Graphing the parabola:

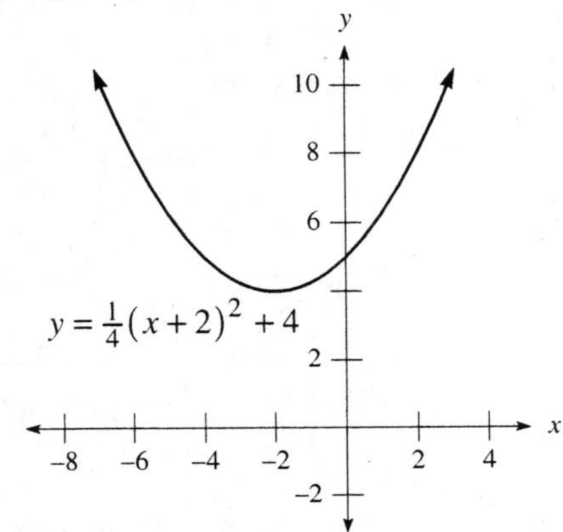

22. The negative value of a reflects the parabola across the x-axis (so that it now points down):

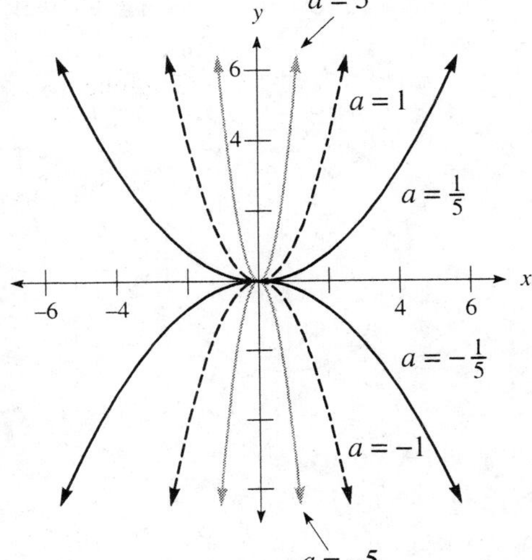

24.   When $k < 0$, the parabola is shifted down and when $k > 0$, the parabola is shifted up:

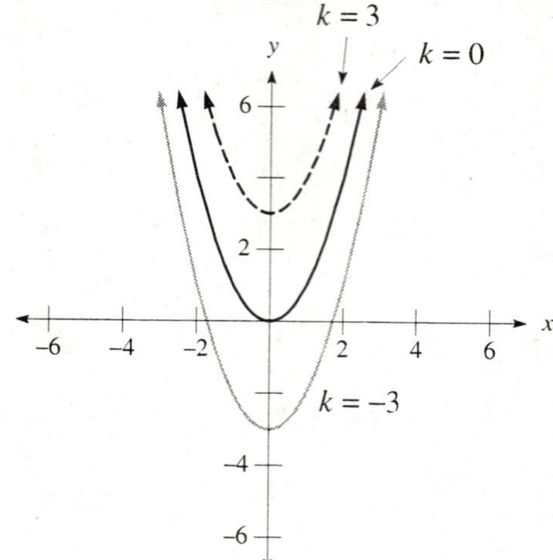

26.   The circle with have a center of $(0,0)$ and a radius of 6. Graphing the circle:

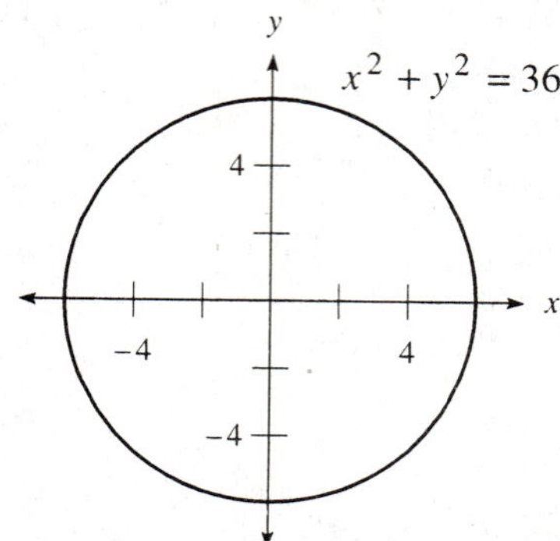

28.   The circle with have a center of $(0,0)$ and a radius of $\sqrt{6}$. Graphing the circle:

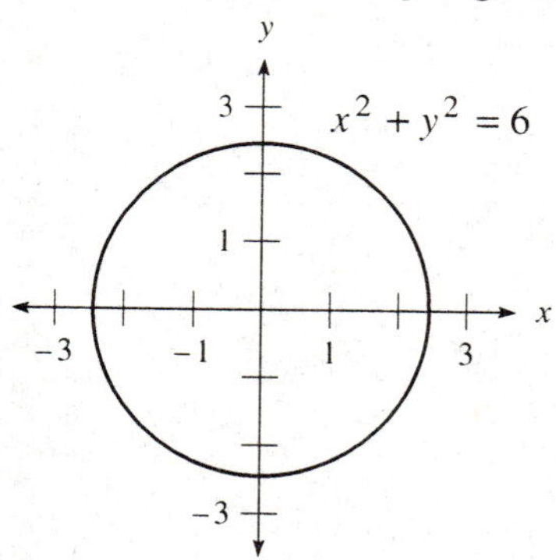

30.   The two points are $\left(-\frac{1}{2}, 0.8660\right)$ and $\left(-\frac{1}{2}, -0.8660\right)$.

32.   The two points are $(-0.7071, 0.7071)$ and $(-0.7071, -0.7071)$.

34.   The two points are $(0.8660, 0.5)$ and $(0.8660, -0.5)$.

36. Note that graphing the line and circle together results in two intersection points:

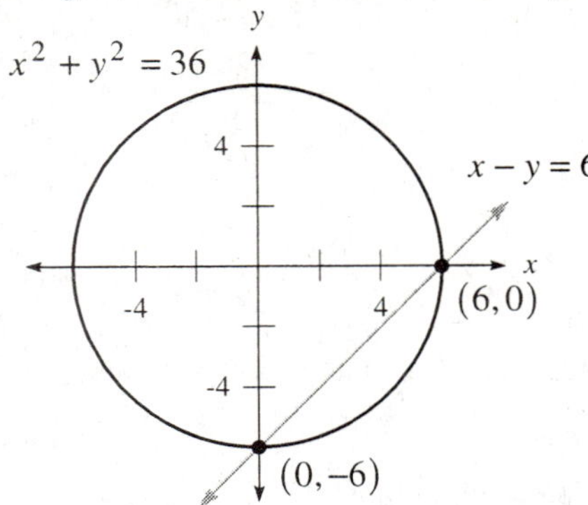

The intersection points are $(0,-6)$ and $(6,0)$.

38. Note that graphing the line and circle together results in two intersection points:

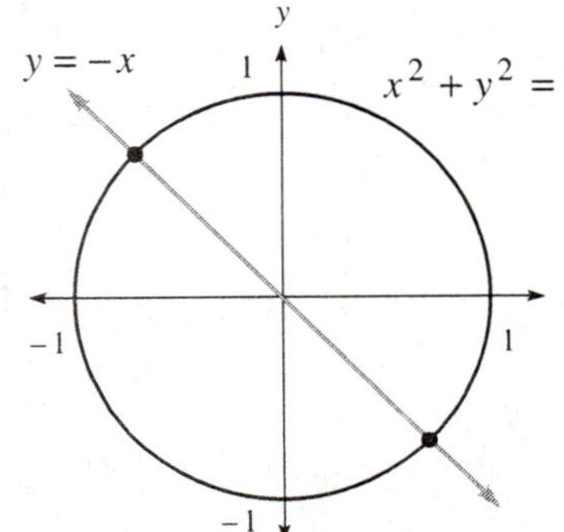

The points, however, may not be readily apparent. Substituting $y = -x$ into the equation for the circle:

$$x^2 + (-x)^2 = 1$$
$$2x^2 = 1$$
$$x^2 = \tfrac{1}{2}$$
$$x = \pm\frac{1}{\sqrt{2}} = \pm\frac{\sqrt{2}}{2}$$

Since $y = -x$, the two intersection points are $\left(\dfrac{\sqrt{2}}{2}, -\dfrac{\sqrt{2}}{2}\right)$ and $\left(-\dfrac{\sqrt{2}}{2}, \dfrac{\sqrt{2}}{2}\right)$.

40. Using the distance formula: $r = \sqrt{(8-4)^2 + (1-7)^2} = \sqrt{4^2 + (-6)^2} = \sqrt{16+36} = \sqrt{52} = 2\sqrt{13}$

42. Using the distance formula: $r = \sqrt{(0+3)^2 + (4-0)^2} = \sqrt{3^2 + 4^2} = \sqrt{9+16} = \sqrt{25} = 5$

44. Using the distance formula: $r = \sqrt{(-1+3)^2 + (6-8)^2} = \sqrt{2^2 + (-2)^2} = \sqrt{4+4} = \sqrt{8} = 2\sqrt{2}$

46. Using the points $(0,0)$ and $(12,-5)$ in the distance formula:

$$r = \sqrt{(12-0)^2 + (-5-0)^2} = \sqrt{12^2 + (-5)^2} = \sqrt{144+25} = \sqrt{169} = 13$$

48. Using the points $(7,y)$ and $(3,3)$ in the distance formula:

$$\sqrt{(3-7)^2 + (3-y)^2} = 5$$
$$\sqrt{16 + (3-y)^2} = 5$$
$$16 + (3-y)^2 = 25$$
$$(3-y)^2 = 9$$

$$3 - y = 3 \qquad \text{or} \qquad 3 - y = -3$$
$$y = 0 \qquad \text{or} \qquad y = 6$$

50.   Since the triangle from home plate to first base to second base is a right triangle, use the Pythagorean Theorem. Let x represent the distance from home plate to second base, so:

$$60^2 + 60^2 = x^2$$
$$3600 + 3600 = x^2$$
$$x^2 = 7200$$
$$x = \sqrt{7200} = 60\sqrt{2}$$

The distance is $60\sqrt{2} \approx 84.85$ ft.

52.   From Exercise 50, the distance from home plate to the pitcher's mound is $30\sqrt{2}$ ft, so the coordinates of the pitcher's mound are $\left(30\sqrt{2},0\right)$. Since the distance from the pitcher's mound to either first base or third base is also $30\sqrt{2}$ ft, the coordinates of first base are $\left(30\sqrt{2},-30\sqrt{2}\right)$ and the coordinates of third base are $\left(30\sqrt{2},30\sqrt{2}\right)$.

54.   All the points have negative y-coordinates in quadrants III and IV.

56.   The ratio $\dfrac{x}{y}$ is also negative in quadrant IV, since x is positive and y is negative in that quadrant.

58.   The vertex of the parabolic path of the human cannonball is $(80,60)$, so the equation of the path is $y = a(x-80)^2 + 60$. Substituting the point $(160,0)$:

$$0 = a(160-80)^2 + 60$$
$$-60 = 6400a$$
$$a = -\frac{60}{6400} = -\frac{3}{320}$$

Thus the equation is $y = -\frac{3}{320}(x-80)^2 + 60$. Substituting $x = 30$ and $x = 150$:

$$x = 30:\ y = -\tfrac{3}{320}(30-80)^2 + 60 = 36.5625 \text{ feet}$$
$$x = 150:\ y = -\tfrac{3}{320}(150-80)^2 + 60 = 14.0625 \text{ feet}$$

These values are verified using a graphing calculator.

60.   The complement is $90° - 30° = 60°$.   **62.**   The complement is $90° - 0° = 90°$.

64.   The supplement is $180° - 150° = 30°$.   **66.**   The supplement is $180° - 45° = 135°$.

68.   The angle between $0°$ and $360°$ which is coterminal with $-135°$ is $225°$.

70.   The angle between $0°$ and $360°$ which is coterminal with $-300°$ is $60°$.

72.   Drawing $45°$ in standard position:

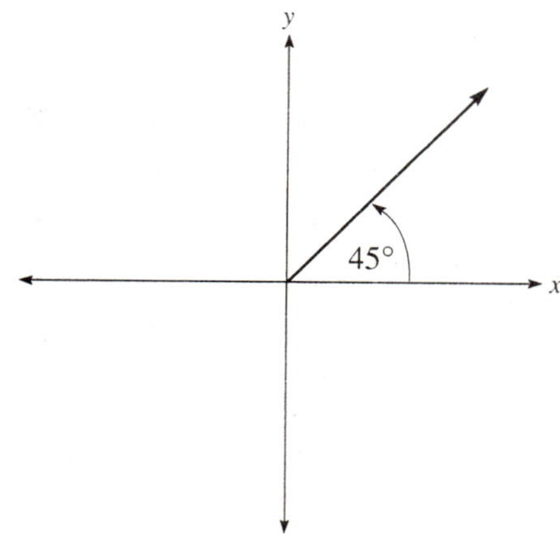

a.   A point on the terminal side of $45°$ is $(1,1)$.

b.   The distance to that point is $r = \sqrt{(1-0)^2 + (1-0)^2} = \sqrt{1+1} = \sqrt{2}$.

c.   Another angle which is coterminal with $45°$ is $-315°$.

74. Drawing 315° in standard position:

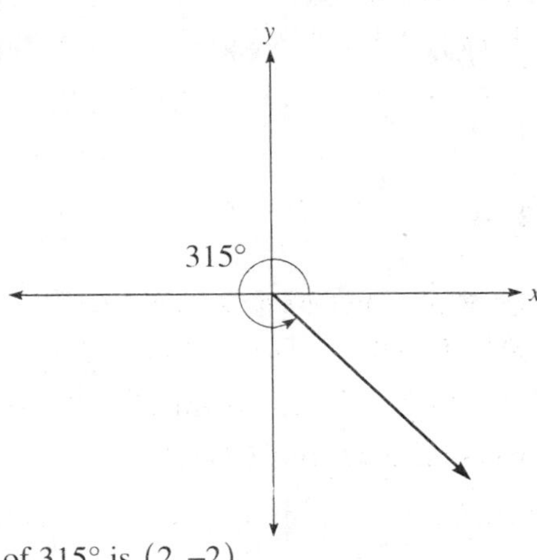

a. A point on the terminal side of 315° is $(2, -2)$.

b. The distance to that point is $r = \sqrt{(2-0)^2 + (-2-0)^2} = \sqrt{4+4} = \sqrt{8} = 2\sqrt{2}$.

c. Another angle which is coterminal with 315° is −45°.

76. Drawing 360° in standard position:

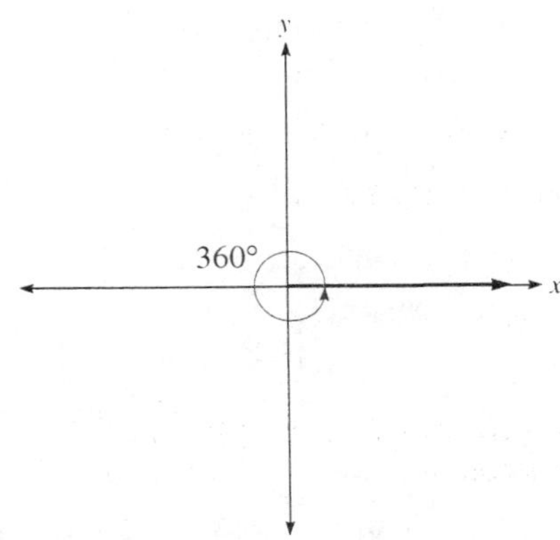

a. A point on the terminal side of 360° is $(5, 0)$.

b. The distance to that point is 5.

c. Another angle which is coterminal with 360° is 0°.

78. Drawing −90° in standard position:

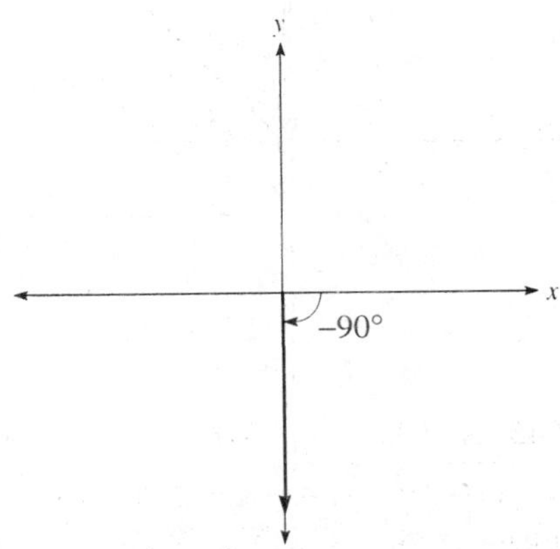

a. A point on the terminal side of −90° is $(0, -3)$.

b. The distance to that point is 3.

c. Another angle which is coterminal with −90° is 270°.

80.   Drawing $60°$ in standard position and labeling the point $(2, b)$:

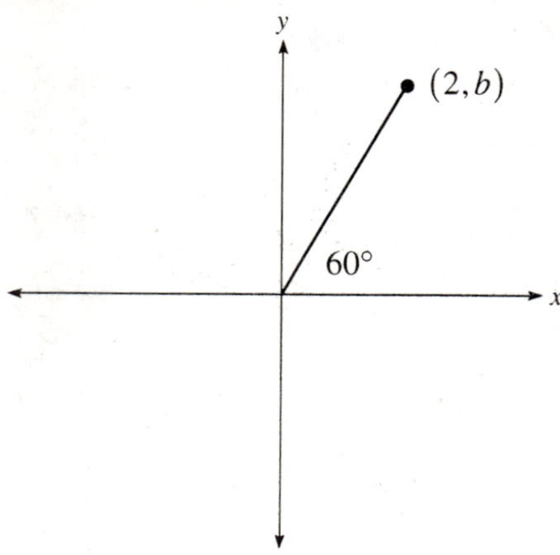

Since 2 is the shorter side of the triangle (opposite the $30°$ angle), the value of b must be $b = 2\sqrt{3}$.

82.   Drawing an angle whose terminal side contains the point $(2, -3)$:

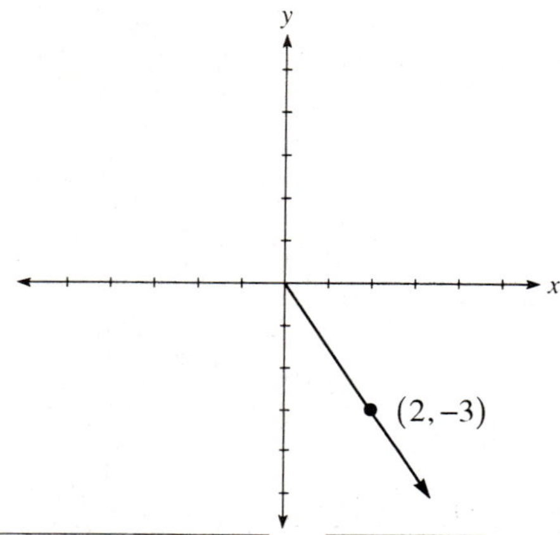

Using the distance formula: $r = \sqrt{(2-0)^2 + (-3-0)^2} = \sqrt{2^2 + (-3)^2} = \sqrt{4+9} = \sqrt{13}$

84.   Plotting the points:

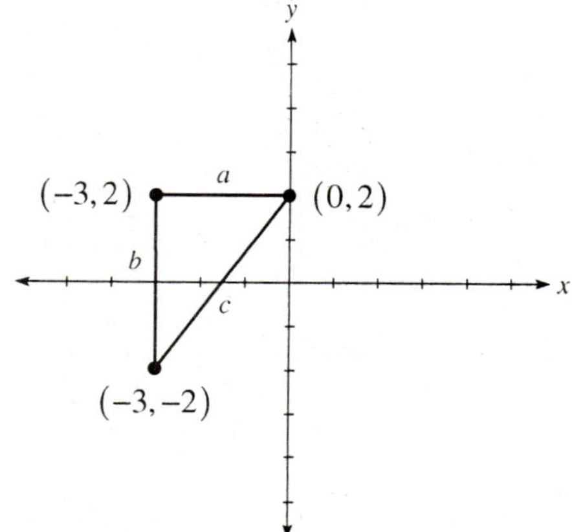

Note that $a = 3$, $b = 4$, and $c = \sqrt{(-3-0)^2 + (-2-2)^2} = \sqrt{(-3)^2 + (-4)^2} = \sqrt{9+16} = \sqrt{25} = 5$. Since $a^2 + b^2 = c^2$, by the Pythagorean Theorem these points form a right triangle.

86. Pascal's triangle appears as:

$$1$$
$$1 \quad 1$$
$$1 \quad 2 \quad 1$$
$$1 \quad 3 \quad 3 \quad 1$$
$$1 \quad 4 \quad 6 \quad 4 \quad 1$$
$$1 \quad 5 \quad 10 \quad 10 \quad 5 \quad 1$$
$$1 \quad 6 \quad 15 \quad 20 \quad 15 \quad 6 \quad 1$$

These numbers are the coefficients of the expansion of $(a+b)^n$, where n represents the line number of the triangle. The first few expansions are:

$$(a+b)^2 = a^2 + 2ab + b^2$$
$$(a+b)^3 = a^3 + 3a^2b + 3ab^2 + b^3$$
$$(a+b)^4 = a^4 + 4a^3b + 6a^2b^2 + 4ab^3 + b^4$$
$$(a+b)^5 = a^5 + 5a^4b + 10a^3b^2 + 10a^2b^3 + 5ab^4 + b^5$$
$$(a+b)^6 = a^6 + 6a^5b + 15a^4b^2 + 20a^3b^3 + 15a^2b^4 + 6ab^5 + b^6$$

This pattern continues for all natural numbers n.

1.3 Definition 1: Trigonometric Functions

2. Begin by finding the distance r from the origin to $(-3, -4)$:

$$r = \sqrt{(-3-0)^2 + (-4-0)^2} = \sqrt{9+16} = \sqrt{25} = 5$$

Now applying the definitions for the six trigonometric functions using the values $x = -3$, $y = -4$, and $r = 5$:

$$\sin\theta = \frac{y}{r} = -\frac{4}{5} \qquad\qquad \csc\theta = \frac{r}{y} = -\frac{5}{4}$$

$$\cos\theta = \frac{x}{r} = -\frac{3}{5} \qquad\qquad \sec\theta = \frac{r}{x} = -\frac{5}{3}$$

$$\tan\theta = \frac{y}{x} = \frac{-4}{-3} = \frac{4}{3} \qquad\qquad \cot\theta = \frac{x}{y} = \frac{-3}{-4} = \frac{3}{4}$$

4. Begin by finding $r = \sqrt{(3)^2 + (-4)^2} = \sqrt{9+16} = \sqrt{25} = 5$. Now applying the definitions for the six trigonometric functions using the values $x = 3$, $y = -4$, and $r = 5$:

$$\sin\theta = \frac{y}{r} = -\frac{4}{5} \qquad\qquad \csc\theta = \frac{r}{y} = -\frac{5}{4}$$

$$\cos\theta = \frac{x}{r} = \frac{3}{5} \qquad\qquad \sec\theta = \frac{r}{x} = \frac{5}{3}$$

$$\tan\theta = \frac{y}{x} = -\frac{4}{3} \qquad\qquad \cot\theta = \frac{x}{y} = -\frac{3}{4}$$

6. Begin by finding $r = \sqrt{(-12)^2 + (5)^2} = \sqrt{144+25} = \sqrt{169} = 13$. Now applying the definitions for the six trigonometric functions using the values $x = -12$, $y = 5$, and $r = 13$:

$$\sin\theta = \frac{y}{r} = \frac{5}{13} \qquad\qquad \csc\theta = \frac{r}{y} = \frac{13}{5}$$

$$\cos\theta = \frac{x}{r} = -\frac{12}{13} \qquad\qquad \sec\theta = \frac{r}{x} = -\frac{13}{12}$$

$$\tan\theta = \frac{y}{x} = -\frac{5}{12} \qquad\qquad \cot\theta = \frac{x}{y} = -\frac{12}{5}$$

8. Begin by finding $r = \sqrt{1^2 + 2^2} = \sqrt{1+4} = \sqrt{5}$. Now applying the definitions for the six trigonometric functions using the values $x = 1$, $y = 2$, and $r = \sqrt{5}$:

$$\sin\theta = \frac{y}{r} = \frac{2}{\sqrt{5}} \qquad\qquad \csc\theta = \frac{r}{y} = \frac{\sqrt{5}}{2}$$

$$\cos\theta = \frac{x}{r} = \frac{1}{\sqrt{5}} \qquad\qquad \sec\theta = \frac{r}{x} = \sqrt{5}$$

$$\tan\theta = \frac{y}{x} = 2 \qquad\qquad \cot\theta = \frac{x}{y} = \tfrac{1}{2}$$

10. Begin by finding $r = \sqrt{m^2 + n^2}$. Now applying the definitions for the six trigonometric functions using the values $x = m$, $y = n$, and $r = \sqrt{m^2 + n^2}$:

$$\sin\theta = \frac{y}{r} = \frac{n}{\sqrt{m^2+n^2}} \qquad\qquad \csc\theta = \frac{r}{y} = \frac{\sqrt{m^2+n^2}}{n}$$

$$\cos\theta = \frac{x}{r} = \frac{m}{\sqrt{m^2+n^2}} \qquad\qquad \sec\theta = \frac{r}{x} = \frac{\sqrt{m^2+n^2}}{m}$$

$$\tan\theta = \frac{y}{x} = \frac{n}{m} \qquad\qquad \cot\theta = \frac{x}{y} = \frac{m}{n}$$

12. Begin by finding $r = \sqrt{0^2 + (-5)^2} = \sqrt{25} = 5$. Now applying the definitions for the six trigonometric functions using the values $x = 0$, $y = -5$, and $r = 5$:

$$\sin\theta = \frac{y}{r} = \frac{-5}{5} = -1 \qquad\qquad \csc\theta = \frac{r}{y} = \frac{5}{-5} = -1$$

$$\cos\theta = \frac{x}{r} = \frac{0}{5} = 0 \qquad\qquad \sec\theta = \frac{r}{x} = \frac{5}{0}, \text{ which is undefined}$$

$$\tan\theta = \frac{y}{x} = \frac{-5}{0}, \text{ which is undefined} \qquad\qquad \cot\theta = \frac{x}{y} = \frac{0}{-5} = 0$$

14. Begin by finding $r = \sqrt{\left(\sqrt{2}\right)^2 + \left(\sqrt{2}\right)^2} = \sqrt{2+2} = \sqrt{4} = 2$. Now applying the definitions for the six trigonometric functions using the values $x = \sqrt{2}$, $y = \sqrt{2}$, and $r = 2$:

$$\sin\theta = \frac{y}{r} = \frac{\sqrt{2}}{2} \qquad\qquad \csc\theta = \frac{r}{y} = \frac{2}{\sqrt{2}} = \sqrt{2}$$

$$\cos\theta = \frac{x}{r} = \frac{\sqrt{2}}{2} \qquad\qquad \sec\theta = \frac{r}{x} = \frac{2}{\sqrt{2}} = \sqrt{2}$$

$$\tan\theta = \frac{y}{x} = \frac{\sqrt{2}}{\sqrt{2}} = 1 \qquad\qquad \cot\theta = \frac{x}{y} = \frac{\sqrt{2}}{\sqrt{2}} = 1$$

16. Begin by finding $r = \sqrt{(-3)^2 + \left(\sqrt{7}\right)^2} = \sqrt{9+7} = \sqrt{16} = 4$. Now applying the definitions for the six trigonometric functions using the values $x = -3$, $y = \sqrt{7}$, and $r = 4$:

$$\sin\theta = \frac{y}{r} = \frac{\sqrt{7}}{4} \qquad\qquad \csc\theta = \frac{r}{y} = \frac{4}{\sqrt{7}}$$

$$\cos\theta = \frac{x}{r} = -\frac{3}{4} \qquad\qquad \sec\theta = \frac{r}{x} = -\frac{4}{3}$$

$$\tan\theta = \frac{y}{x} = -\frac{\sqrt{7}}{3} \qquad\qquad \cot\theta = \frac{x}{y} = -\frac{3}{\sqrt{7}}$$

18. Begin by finding $r = \sqrt{(-80)^2 + (60)^2} = \sqrt{6400 + 3600} = \sqrt{10000} = 100$. Now applying the definitions for the six trigonometric functions using the values $x = -80$, $y = 60$, and $r = 100$:

$$\sin\theta = \frac{y}{r} = \frac{60}{100} = \frac{3}{5} \qquad\qquad \csc\theta = \frac{r}{y} = \frac{100}{60} = \frac{5}{3}$$

$$\cos\theta = \frac{x}{r} = -\frac{80}{100} = -\frac{4}{5} \qquad\qquad \sec\theta = \frac{r}{x} = -\frac{100}{80} = -\frac{5}{4}$$

$$\tan\theta = \frac{y}{x} = -\frac{60}{80} = -\frac{3}{4} \qquad\qquad \cot\theta = \frac{x}{y} = -\frac{80}{60} = -\frac{4}{3}$$

20. Begin by finding $r = \sqrt{(-9a)^2 + (12a)^2} = \sqrt{81a^2 + 144a^2} = \sqrt{225a^2} = 15a$. Now applying the definitions for the six trigonometric functions using the values $x = -9a$, $y = 12a$, and $r = 15a$:

$$\sin\theta = \frac{y}{r} = \frac{12a}{15a} = \frac{4}{5} \qquad\qquad \csc\theta = \frac{r}{y} = \frac{15a}{12a} = \frac{5}{4}$$

$$\cos\theta = \frac{x}{r} = -\frac{9a}{15a} = -\frac{3}{5} \qquad\qquad \sec\theta = \frac{r}{x} = -\frac{15a}{9a} = -\frac{5}{3}$$

$$\tan\theta = \frac{y}{x} = -\frac{12a}{9a} = -\frac{4}{3} \qquad\qquad \cot\theta = \frac{x}{y} = -\frac{9a}{12a} = -\frac{3}{4}$$

22. Begin by finding $r = \sqrt{4^2 + 3^2} = \sqrt{16 + 9} = \sqrt{25} = 5$. Now applying the definitions for $\sin\theta$, $\cos\theta$, and $\tan\theta$ using $x = 4$, $y = 3$, and $r = 5$:

$$\sin\theta = \frac{y}{r} = \frac{3}{5} \qquad\qquad \cos\theta = \frac{x}{r} = \frac{4}{5} \qquad\qquad \tan\theta = \frac{y}{x} = \frac{3}{4}$$

24. Begin by finding $r = \sqrt{5^2 + 3^2} = \sqrt{25 + 9} = \sqrt{34}$. Now applying the definitions for $\sin\theta$, $\cos\theta$, and $\tan\theta$ using $x = 5$, $y = 3$, and $r = \sqrt{34}$:

$$\sin\theta = \frac{y}{r} = \frac{3}{\sqrt{34}} \qquad\qquad \cos\theta = \frac{x}{r} = \frac{5}{\sqrt{34}} \qquad\qquad \tan\theta = \frac{y}{x} = \frac{3}{5}$$

26. Begin by finding r: $r = \sqrt{(6.36)^2 + (2.65)^2} = \sqrt{40.4496 + 7.0225} = \sqrt{47.4721} = 6.89$
Now applying the definitions for $\sin\theta$ and $\cos\theta$ using $x = 6.36$, $y = 2.65$, and $r = 6.89$:

$$\sin\theta = \frac{y}{r} = \frac{2.65}{6.89} \approx 0.385 \qquad\qquad \cos\theta = \frac{x}{r} = \frac{6.36}{6.89} \approx 0.923$$

28. Drawing $225°$ in standard position:

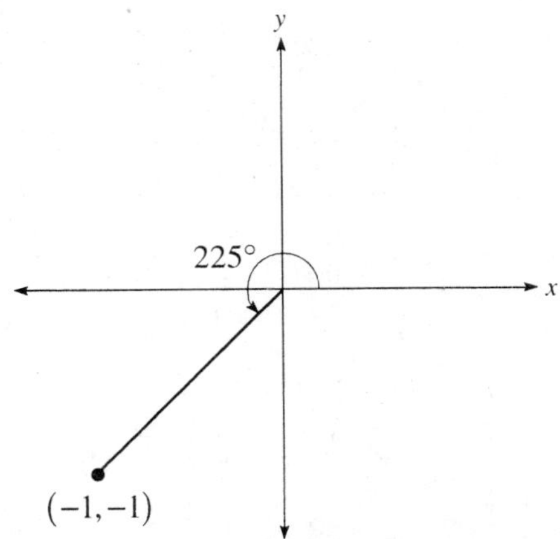

A point on the terminal side is $(-1,-1)$. Now find $r = \sqrt{(-1)^2 + (-1)^2} = \sqrt{1 + 1} = \sqrt{2}$. Appling the definitions for $\sin 225°$, $\cos 225°$, and $\tan 225°$:

$$\sin 225° = \frac{y}{r} = -\frac{1}{\sqrt{2}} = -\frac{\sqrt{2}}{2} \qquad \cos 225° = \frac{x}{r} = -\frac{1}{\sqrt{2}} = -\frac{\sqrt{2}}{2} \qquad \tan 225° = \frac{y}{x} = \frac{-1}{-1} = 1$$

30.   Drawing 180° in standard position:

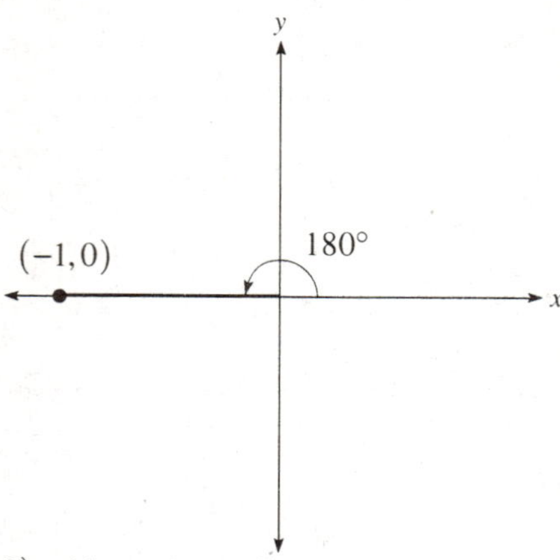

A point on the terminal side is $(-1,0)$ with $r = 1$. Appling the definitions for sin 180°, cos 180°, and tan 180°:

$$\sin 180° = \frac{y}{r} = \frac{0}{1} = 0 \qquad \cos 180° = \frac{x}{r} = \frac{-1}{1} = -1 \qquad \tan 180° = \frac{y}{x} = \frac{0}{-1} = 0$$

32.   Drawing −90° in standard position:

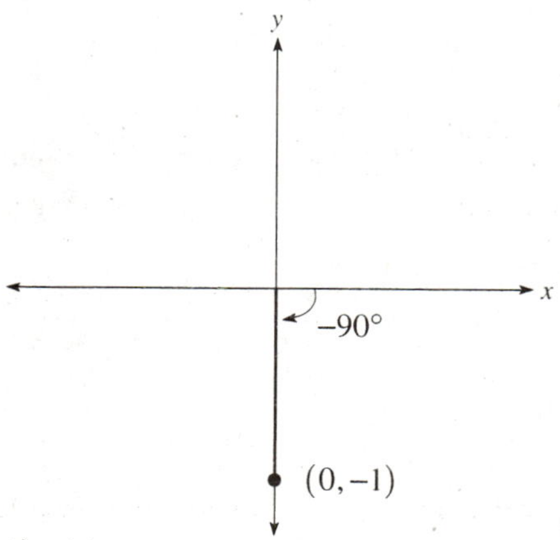

A point on the terminal side is $(0,-1)$ with $r = 1$. Appling the definitions for sin (−90°), cos (−90°), and tan (−90°):

$$\sin(-90°) = \frac{y}{r} = \frac{-1}{1} = -1 \quad \cos(-90°) = \frac{x}{r} = \frac{0}{1} = 0 \quad \tan(-90°) = \frac{y}{x} = \frac{-1}{0}, \text{ which is undefined}$$

34. Drawing $-135°$ in standard position:

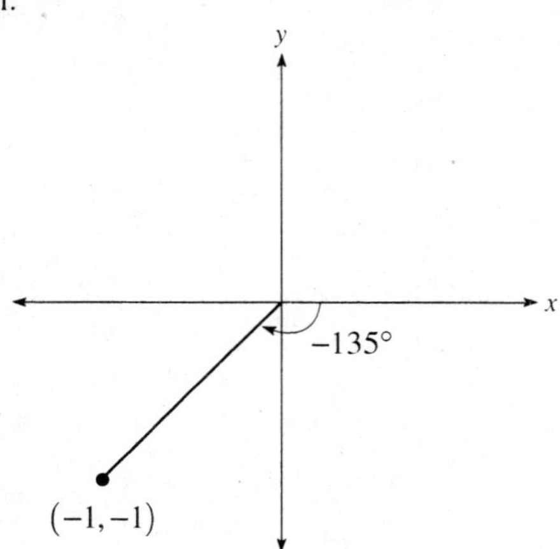

A point on the terminal side is $(-1,-1)$. Now find $r = \sqrt{(-1)^2 + (-1)^2} = \sqrt{1+1} = \sqrt{2}$. Appling the definitions for sin $(-135°)$, cos $(-135°)$, and tan $(-135°)$:

$$\sin(-135°) = \frac{y}{r} = -\frac{1}{\sqrt{2}} = -\frac{\sqrt{2}}{2} \qquad \cos(-135°) = \frac{x}{r} = -\frac{1}{\sqrt{2}} = -\frac{\sqrt{2}}{2} \qquad \tan(-135°) = \frac{y}{x} = \frac{-1}{-1} = 1$$

36. If $\cos\theta$ is negative, $x < 0$, and the terminal side of θ must lie in quadrants II or III.

38. If $\tan\theta$ is negative, $\dfrac{y}{x} < 0$, and the terminal side of θ must lie in quadrants II or IV.

40. If $\sin\theta$ is positive, $y > 0$, and if $\cos\theta$ is negative, $x < 0$, so the terminal side of θ must lie in quadrant II.

42. $\cos\theta$ and $\cot\theta$ are both positive in quadrant I, and $\cos\theta$ and $\cot\theta$ are both negative in quadrant II. Thus $\cos\theta$ and $\cot\theta$ have the same sign in quadrants I and II.

44. If $\sin\theta = \frac{12}{13}$, we can choose $y = 12$ and $r = 13$. To find x, use $x^2 + y^2 = r^2$:

$$x^2 + 12^2 = 13^2$$
$$x^2 + 144 = 169$$
$$x^2 = 25$$
$$x = \pm 5$$

Since θ terminates in quadrant II, $x < 0$ and thus $x = -5$. Using $x = -5$, $y = 12$, and $r = 13$ in the definitions:

$$\cos\theta = \frac{x}{r} = -\frac{5}{13} \qquad\qquad \sec\theta = \frac{r}{x} = -\frac{13}{5} \qquad\qquad \csc\theta = \frac{r}{y} = \frac{13}{12}$$

$$\tan\theta = \frac{y}{x} = -\frac{12}{5} \qquad\qquad \cot\theta = \frac{x}{y} = -\frac{5}{12}$$

46. If $\cos\theta = \frac{7}{25}$, we can choose $x = 7$ and $r = 25$. To find y, use $x^2 + y^2 = r^2$:

$$7^2 + y^2 = 25^2$$
$$49 + y^2 = 225$$
$$y^2 = 576$$
$$y = \pm 24$$

Since θ terminates in quadrant IV, $y < 0$ and thus $y = -24$. Using $x = 7$, $y = -24$, and $r = 25$ in the definitions:

$$\sin\theta = \frac{y}{r} = -\frac{24}{25} \qquad\qquad \csc\theta = \frac{r}{y} = -\frac{25}{24} \qquad\qquad \sec\theta = \frac{r}{x} = \frac{25}{7}$$

$$\tan\theta = \frac{y}{x} = -\frac{24}{7} \qquad\qquad \cot\theta = \frac{x}{y} = -\frac{7}{24}$$

48.   If $\tan\theta = \frac{12}{5}$, we can choose $y = 12$ and $x = 5$. Note that choosing $y = -12$ and $x = -5$ would also result in $\tan\theta = \frac{12}{5}$, but θ terminating in quadrant I requires $x > 0$ and $y > 0$. Finding r using $x^2 + y^2 = r^2$:

$$5^2 + 12^2 = r^2$$
$$25 + 144 = r^2$$
$$r^2 = 169$$
$$r = 13$$

Using $x = 5$, $y = 12$, and $r = 13$ in the definitions:

$$\sin\theta = \frac{y}{r} = \frac{12}{13} \qquad \csc\theta = \frac{r}{y} = \frac{13}{12} \qquad \cot\theta = \frac{x}{y} = \frac{5}{12}$$

$$\cos\theta = \frac{x}{r} = \frac{5}{13} \qquad \sec\theta = \frac{r}{x} = \frac{13}{5}$$

50.   If $\cos\theta = -\frac{20}{29}$, we can choose $x = -20$ and $r = 29$. To find y, use $x^2 + y^2 = r^2$:

$$(-20)^2 + y^2 = 29^2$$
$$400 + y^2 = 841$$
$$y^2 = 441$$
$$y = \pm 21$$

Since θ terminates in quadrant II, $y > 0$ and thus $y = 21$. Using $x = -20$, $y = 21$, and $r = 29$ in the definitions:

$$\sin\theta = \frac{y}{r} = \frac{21}{29} \qquad \csc\theta = \frac{r}{y} = \frac{29}{21} \qquad \sec\theta = \frac{r}{x} = -\frac{29}{20}$$

$$\tan\theta = \frac{y}{x} = -\frac{21}{20} \qquad \cot\theta = \frac{x}{y} = -\frac{20}{21}$$

52.   If $\csc\theta = \frac{13}{5}$, we can choose $y = 5$ and $r = 13$. To find x, use $x^2 + y^2 = r^2$:

$$x^2 + 5^2 = 13^2$$
$$x^2 + 25 = 169$$
$$x^2 = 144$$
$$x = \pm 12$$

Since $\cos\theta < 0$, $x < 0$ and thus $x = -12$. Using $x = -12$, $y = 5$, and $r = 13$ in the definitions:

$$\cos\theta = \frac{x}{r} = -\frac{12}{13} \qquad \sec\theta = \frac{r}{x} = -\frac{13}{12} \qquad \sin\theta = \frac{y}{r} = \frac{5}{13}$$

$$\tan\theta = \frac{y}{x} = -\frac{5}{12} \qquad \cot\theta = \frac{x}{y} = -\frac{12}{5}$$

54.   If $\tan\theta = -\frac{1}{2}$ and $\sin\theta > 0$, $y > 0$, so choose $y = 1$ and $x = -2$. Finding r using $x^2 + y^2 = r^2$:

$$(-2)^2 + 1^2 = r^2$$
$$4 + 1 = r^2$$
$$r^2 = 5$$
$$r = \sqrt{5}$$

Using $x = -2$, $y = 1$, and $r = \sqrt{5}$ in the definitions:

$$\sin\theta = \frac{y}{r} = \frac{1}{\sqrt{5}} \qquad \csc\theta = \frac{r}{y} = \sqrt{5} \qquad \cot\theta = \frac{x}{y} = -2$$

$$\cos\theta = \frac{x}{r} = -\frac{2}{\sqrt{5}} \qquad \sec\theta = \frac{r}{x} = -\frac{\sqrt{5}}{2}$$

56.   If $\cos\theta = \dfrac{\sqrt{2}}{2}$, we can choose $x = \sqrt{2}$ and $r = 2$. To find y, use $x^2 + y^2 = r^2$:

$$\left(\sqrt{2}\right)^2 + y^2 = 2^2$$
$$2 + y^2 = 4$$
$$y^2 = 2$$
$$y = \pm\sqrt{2}$$

Since θ terminates in quadrant I, $y > 0$ and thus $y = \sqrt{2}$. Using $x = \sqrt{2}$, $y = \sqrt{2}$, and $r = 2$ in the definitions:

$$\sin\theta = \frac{y}{r} = \frac{\sqrt{2}}{2} \qquad\qquad \csc\theta = \frac{r}{y} = \frac{2}{\sqrt{2}} = \sqrt{2} \qquad\qquad \sec\theta = \frac{r}{x} = \frac{2}{\sqrt{2}} = \sqrt{2}$$

$$\tan\theta = \frac{y}{x} = \frac{\sqrt{2}}{\sqrt{2}} = 1 \qquad\qquad \cot\theta = \frac{x}{y} = \frac{\sqrt{2}}{\sqrt{2}} = 1$$

58.   If $\tan\theta = 2$ and θ terminates in quadrant I, we can choose $y = 2$ and $x = 1$. Thus $r = \sqrt{2^2 + 1^2} = \sqrt{5}$. Using $x = 1$, $y = 2$, and $r = \sqrt{5}$ in the definitions:

$$\sin\theta = \frac{y}{r} = \frac{2}{\sqrt{5}} \qquad\qquad \csc\theta = \frac{r}{y} = \frac{\sqrt{5}}{2} \qquad\qquad \cot\theta = \frac{x}{y} = \frac{1}{2}$$

$$\cos\theta = \frac{x}{r} = \frac{1}{\sqrt{5}} \qquad\qquad \sec\theta = \frac{r}{x} = \sqrt{5}$$

60.   If $\cot\theta = \dfrac{m}{n}$, we can choose $x = m$ and $y = n$, so $r = \sqrt{m^2 + n^2}$. Using the definitions:

$$\sin\theta = \frac{y}{r} = \frac{n}{\sqrt{m^2 + n^2}} \qquad\qquad \csc\theta = \frac{r}{y} = \frac{\sqrt{m^2 + n^2}}{n} \qquad\qquad \tan\theta = \frac{y}{x} = \frac{n}{m}$$

$$\cos\theta = \frac{x}{r} = \frac{m}{\sqrt{m^2 + n^2}} \qquad\qquad \sec\theta = \frac{r}{x} = \frac{\sqrt{m^2 + n^2}}{m}$$

62.   If $\tan\theta = 1$, then $y = x$ is the line representing the terminal side of angle θ. Thus the corresponding angle is $\theta = 45°$.

64.   If $\cos\theta = -1$, then $x = -1$ and $y = 0$. This is the point $(-1, 0)$, thus the corresponding angle is $\theta = 180°$.

66.   A point on $y = 2x$ in quadrant III is $(-1, -2)$, so choose $x = -1$ and $y = -2$. Thus $r = \sqrt{(-1)^2 + (-2)^2} = \sqrt{1 + 4} = \sqrt{5}$. Using the definitions:

$$\sin\theta = \frac{y}{r} = -\frac{2}{\sqrt{5}} \qquad\qquad \cos\theta = \frac{x}{r} = -\frac{1}{\sqrt{5}}$$

68.   A point on $y = -3x$ in quadrant IV is $(1, -3)$, so choose $x = 1$ and $y = -3$. Thus $r = \sqrt{1^2 + (-3)^2} = \sqrt{1 + 9} = \sqrt{10}$. Using the definitions:

$$\sin\theta = \frac{y}{r} = -\frac{3}{\sqrt{10}} \qquad\qquad \tan\theta = \frac{y}{x} = \frac{-3}{1} = -3$$

70.   Drawing 45° and −45° in standard position:

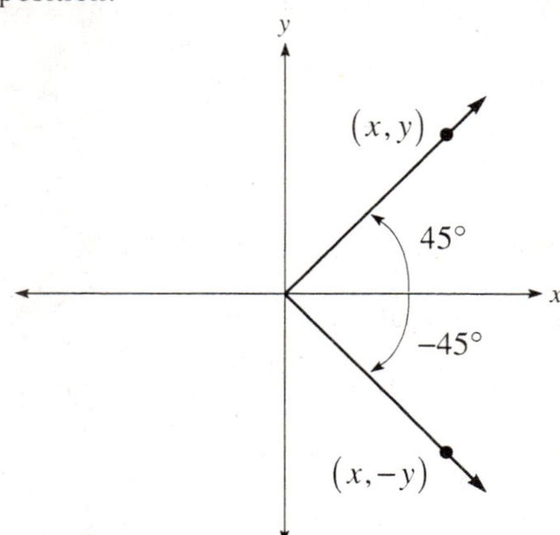

Let (x, y) represent a point on the terminal side of 45°. Then $(x, -y)$ will represent a point on the terminal side of −45°. Therefore: $\sin(-45°) = \dfrac{-y}{r} = -\dfrac{y}{r} = -\sin 45°$

72.   If $\cos\theta = \dfrac{5}{13}$, we can choose $x = 5$ and $r = 13$. To find y, use $x^2 + y^2 = r^2$:

$$5^2 + y^2 = 13^2$$
$$25 + y^2 = 169$$
$$y^2 = 144$$
$$y = \pm 12$$

Note that both values of y are possible, since the points $(5, 12)$ and $(5, -12)$ both satisfy the condition $\cos\theta = \dfrac{5}{13}$.

1.4 Introduction to Identities

2.   The reciprocal is $\dfrac{1}{4}$.

4.   The reciprocal is $-\dfrac{13}{5}$.

6.   The reciprocal is $-\dfrac{2}{\sqrt{3}}$.

8.   The reciprocal is a.

10.   Using the reciprocal identity: $\sec\theta = \dfrac{1}{\cos\theta} = \dfrac{1}{\sqrt{3}/2} = \dfrac{2}{\sqrt{3}}$

12.   Using the reciprocal identity: $\sin\theta = \dfrac{1}{\csc\theta} = \dfrac{1}{-13/12} = -\dfrac{12}{13}$

14.   Using the reciprocal identity: $\tan\theta = \dfrac{1}{\cot\theta} = \dfrac{1}{-b} = -\dfrac{1}{b}$

16.   Using the ratio identity: $\tan\theta = \dfrac{\sin\theta}{\cos\theta} = \dfrac{2/\sqrt{5}}{1/\sqrt{5}} = 2$

18.   Using the ratio identity: $\cot\theta = \dfrac{\cos\theta}{\sin\theta} = \dfrac{3/\sqrt{13}}{2/\sqrt{13}} = \dfrac{3}{2}$

20.   Finding $\cos^2\theta$: $\cos^2\theta = (\cos\theta)^2 = \left(\dfrac{1}{3}\right)^2 = \dfrac{1}{9}$

22.   Finding $\sec^3\theta$: $\sec^3\theta = (\sec\theta)^3 = (-3)^3 = -27$

24.   Using the ratio identity: $\cot\theta = \dfrac{\cos\theta}{\sin\theta} = \dfrac{-5/13}{-12/13} = \dfrac{5}{12}$

26.   Using the reciprocal identity: $\csc\theta = \dfrac{1}{\sin\theta} = \dfrac{1}{-12/13} = -\dfrac{13}{12}$

28. First find $\sin^2\theta$: $\sin^2\theta = 1 - \cos^2\theta = 1 - \left(\frac{5}{13}\right)^2 = 1 - \frac{25}{169} = \frac{144}{169}$

Since θ terminates in quadrant I, $\sin\theta > 0$, and thus $\sin\theta = \sqrt{\frac{144}{169}} = \frac{12}{13}$.

30. First find $\cos^2\theta$: $\cos^2\theta = 1 - \sin^2\theta = 1 - \left(\frac{\sqrt{3}}{2}\right)^2 = 1 - \frac{3}{4} = \frac{1}{4}$

Since θ terminates in quadrant II, $\cos\theta < 0$, and thus $\cos\theta = -\sqrt{\frac{1}{4}} = -\frac{1}{2}$.

32. First find $\cos^2\theta$: $\cos^2\theta = 1 - \sin^2\theta = 1 - \left(-\frac{4}{5}\right)^2 = 1 - \frac{16}{25} = \frac{9}{25}$

Since θ terminates in quadrant IV, $\cos\theta > 0$, and thus $\cos\theta = \sqrt{\frac{9}{25}} = \frac{3}{5}$.

34. First find $\sin^2\theta$: $\sin^2\theta = 1 - \cos^2\theta = 1 - \left(-\frac{1}{2}\right)^2 = 1 - \frac{1}{4} = \frac{3}{4}$

Since θ terminates in quadrant II, $\sin\theta > 0$, and thus $\sin\theta = \sqrt{\frac{3}{4}} = \frac{\sqrt{3}}{2}$.

36. First find $\sin^2\theta$: $\sin^2\theta = 1 - \cos^2\theta = 1 - \left(-\frac{1}{\sqrt{10}}\right)^2 = 1 - \frac{1}{10} = \frac{9}{10}$

Since θ terminates in quadrant III, $\sin\theta < 0$, and thus $\sin\theta = -\sqrt{\frac{9}{10}} = -\frac{3}{\sqrt{10}}$.

38. First find $\cos^2\theta$: $\cos^2\theta = 1 - \sin^2\theta = 1 - \left(\frac{2}{3}\right)^2 = 1 - \frac{4}{9} = \frac{5}{9}$

Since θ terminates in quadrant II, $\cos\theta < 0$, and thus $\cos\theta = -\sqrt{\frac{5}{9}} = -\frac{\sqrt{5}}{3}$. Now using the ratio identity:

$$\cot\theta = \frac{\cos\theta}{\sin\theta} = \frac{-\sqrt{5}/3}{2/3} = -\frac{\sqrt{5}}{2}$$

40. First find $\csc^2\theta$: $\csc^2\theta = 1 + \cot^2\theta = 1 + \left(-\frac{24}{7}\right)^2 = 1 + \frac{576}{49} = \frac{625}{49}$

Since θ terminates in quadrant IV, $\csc\theta < 0$, and thus $\csc\theta = -\sqrt{\frac{625}{49}} = -\frac{25}{7}$.

42. First find $\sec^2\theta$: $\sec^2\theta = 1 + \tan^2\theta = 1 + \left(\frac{7}{24}\right)^2 = 1 + \frac{49}{576} = \frac{625}{576}$

Since $\cos\theta < 0$, $\sec\theta < 0$, and thus $\sec\theta = -\sqrt{\frac{625}{576}} = -\frac{25}{24}$.

44. First find $\cos^2\theta$: $\cos^2\theta = 1 - \sin^2\theta = 1 - \left(\frac{12}{13}\right)^2 = 1 - \frac{144}{169} = \frac{25}{169}$

Since θ terminates in quadrant I, $\cos\theta > 0$, and thus $\cos\theta = \sqrt{\frac{25}{169}} = \frac{5}{13}$. Now use the ratio identities to find the remaining four trigonometric ratios:

$$\sec\theta = \frac{1}{\cos\theta} = \frac{1}{5/13} = \frac{13}{5} \qquad\qquad \csc\theta = \frac{1}{\sin\theta} = \frac{1}{12/13} = \frac{13}{12}$$

$$\tan\theta = \frac{\sin\theta}{\cos\theta} = \frac{12/13}{5/13} = \frac{12}{5} \qquad\qquad \cot\theta = \frac{\cos\theta}{\sin\theta} = \frac{5/13}{12/13} = \frac{5}{12}$$

46. First find $\sin^2\theta$: $\sin^2\theta = 1 - \cos^2\theta = 1 - \left(-\frac{1}{3}\right)^2 = 1 - \frac{1}{9} = \frac{8}{9}$

Since $\cos\theta < 0$ and θ does not terminate in quadrant II, it must terminate in quadrant III and thus $\sin\theta < 0$.

Therefore $\sin\theta = -\sqrt{\frac{8}{9}} = -\frac{2\sqrt{2}}{3}$. Now use the ratio identities to find the remaining four trigonometric ratios:

$$\sec\theta = \frac{1}{\cos\theta} = \frac{1}{-\frac{1}{3}} = -3 \qquad\qquad \csc\theta = \frac{1}{\sin\theta} = \frac{1}{-\frac{2\sqrt{2}}{3}} = -\frac{3}{2\sqrt{2}}$$

$$\tan\theta = \frac{\sin\theta}{\cos\theta} = \frac{-\frac{2\sqrt{2}}{3}}{-\frac{1}{3}} = 2\sqrt{2} \qquad\qquad \cot\theta = \frac{\cos\theta}{\sin\theta} = \frac{-\frac{1}{3}}{-\frac{2\sqrt{2}}{3}} = \frac{1}{2\sqrt{2}}$$

48. Since $\csc\theta = 2$, $\sin\theta = \frac{1}{\csc\theta} = \frac{1}{2}$. Now find $\cos^2\theta$: $\cos^2\theta = 1 - \sin^2\theta = 1 - \left(\frac{1}{2}\right)^2 = 1 - \frac{1}{4} = \frac{3}{4}$

Since $\cos\theta < 0$, $\cos\theta = -\sqrt{\frac{3}{4}} = -\frac{\sqrt{3}}{2}$. Now use the ratio identities to find the remaining three trigonometric ratios:

$$\sec\theta = \frac{1}{\cos\theta} = \frac{1}{-\frac{\sqrt{3}}{2}} = -\frac{2}{\sqrt{3}} \qquad \tan\theta = \frac{\sin\theta}{\cos\theta} = \frac{\frac{1}{2}}{-\frac{\sqrt{3}}{2}} = -\frac{1}{\sqrt{3}} \qquad \cot\theta = \frac{\cos\theta}{\sin\theta} = \frac{-\frac{\sqrt{3}}{2}}{\frac{1}{2}} = -\sqrt{3}$$

50. First find $\cos^2\theta$: $\cos^2\theta = 1 - \sin^2\theta = 1 - \left(\frac{3}{\sqrt{10}}\right)^2 = 1 - \frac{9}{10} = \frac{1}{10}$

Since θ terminates in quadrant II, $\cos\theta < 0$, and thus $\cos\theta = -\sqrt{\frac{1}{10}} = -\frac{1}{\sqrt{10}}$. Now use the ratio identities to find the remaining four trigonometric ratios:

$$\sec\theta = \frac{1}{\cos\theta} = \frac{1}{-\frac{1}{\sqrt{10}}} = -\sqrt{10} \qquad\qquad \csc\theta = \frac{1}{\sin\theta} = \frac{1}{\frac{3}{\sqrt{10}}} = \frac{\sqrt{10}}{3}$$

$$\tan\theta = \frac{\sin\theta}{\cos\theta} = \frac{\frac{3}{\sqrt{10}}}{-\frac{1}{\sqrt{10}}} = -3 \qquad\qquad \cot\theta = \frac{\cos\theta}{\sin\theta} = \frac{-\frac{1}{\sqrt{10}}}{\frac{3}{\sqrt{10}}} = -\frac{1}{3}$$

52. Since $\sec\theta = -4$, $\cos\theta = \frac{1}{\sec\theta} = -\frac{1}{4}$. Now find $\sin^2\theta$: $\sin^2\theta = 1 - \cos^2\theta = 1 - \left(-\frac{1}{4}\right)^2 = 1 - \frac{1}{16} = \frac{15}{16}$

Since θ terminates in quadrant II, $\sin\theta > 0$, and thus $\sin\theta = \sqrt{\frac{15}{16}} = \frac{\sqrt{15}}{4}$. Now use the ratio identities to find the remaining three trigonometric ratios:

$$\csc\theta = \frac{1}{\sin\theta} = \frac{1}{\frac{\sqrt{15}}{4}} = \frac{4}{\sqrt{15}} \qquad \tan\theta = \frac{\sin\theta}{\cos\theta} = \frac{\frac{\sqrt{15}}{4}}{-\frac{1}{4}} = -\sqrt{15} \qquad \cot\theta = \frac{\cos\theta}{\sin\theta} = \frac{-\frac{1}{4}}{\frac{\sqrt{15}}{4}} = -\frac{1}{\sqrt{15}}$$

54. Since $\sec\theta = b$, $\cos\theta = \dfrac{1}{\sec\theta} = \dfrac{1}{b}$. Now find $\sin^2\theta$: $\sin^2\theta = 1 - \cos^2\theta = 1 - \left(\dfrac{1}{b}\right)^2 = 1 - \dfrac{1}{b^2} = \dfrac{b^2 - 1}{b^2}$

Since θ terminates in quadrant I, $\sin\theta > 0$, and thus $\sin\theta = \sqrt{\dfrac{b^2 - 1}{b^2}} = \dfrac{\sqrt{b^2 - 1}}{b}$. Now use the ratio identities to find the remaining three trigonometric ratios:

$$\csc\theta = \dfrac{1}{\sin\theta} = \dfrac{1}{\sqrt{b^2 - 1}\Big/ b} = \dfrac{b}{\sqrt{b^2 - 1}} \qquad\qquad \tan\theta = \dfrac{\sin\theta}{\cos\theta} = \dfrac{\sqrt{b^2 - 1}\Big/ b}{1\big/ b} = \sqrt{b^2 - 1}$$

$$\cot\theta = \dfrac{\cos\theta}{\sin\theta} = \dfrac{1\big/ b}{\sqrt{b^2 - 1}\Big/ b} = \dfrac{1}{\sqrt{b^2 - 1}}$$

56. First find $\sin^2\theta$: $\sin^2\theta = 1 - \cos^2\theta = 1 - (0.51)^2 = 1 - 0.2601 = 0.7399$

Since θ terminates in quadrant I, $\sin\theta > 0$, and thus $\sin\theta = \sqrt{0.7399} \approx 0.86$. Now use the ratio identities to find the remaining four trigonometric ratios:

$$\sec\theta = \dfrac{1}{\cos\theta} = \dfrac{1}{0.51} \approx 1.96 \qquad\qquad \csc\theta = \dfrac{1}{\sin\theta} = \dfrac{1}{0.86} \approx 1.16$$

$$\tan\theta = \dfrac{\sin\theta}{\cos\theta} = \dfrac{0.86}{0.51} \approx 1.69 \qquad\qquad \cot\theta = \dfrac{\cos\theta}{\sin\theta} = \dfrac{0.51}{0.86} \approx 0.59$$

58. Since $\csc\theta = -2.45$, $\sin\theta = \dfrac{1}{-2.45} \approx -0.41$. Now find $\cos^2\theta$:

$$\cos^2\theta = 1 - \sin^2\theta = 1 - (-0.41)^2 = 1 - 0.1666 = 0.8334$$

Since θ terminates in quadrant III, $\cos\theta < 0$, and thus $\cos\theta = -\sqrt{0.8334} \approx -0.91$. Now use the ratio identities to find the remaining three trigonometric ratios:

$$\sec\theta = \dfrac{1}{\cos\theta} = \dfrac{1}{-0.91} \approx -1.10 \qquad \tan\theta = \dfrac{\sin\theta}{\cos\theta} = \dfrac{-0.41}{-0.91} \approx 0.45 \qquad \cot\theta = \dfrac{\cos\theta}{\sin\theta} = \dfrac{-0.91}{-0.41} \approx 2.22$$

60. Forming the right triangle to the point $(1, 3)$, note that $\tan\theta = \dfrac{y}{x} = \dfrac{3}{1} = 3$.

62. Forming the right triangle to the point $(1, m)$, note that $\tan\theta = \dfrac{y}{x} = \dfrac{m}{1} = m$.

1.5 More on Identities

2. Writing in terms of $\sin\theta$: $\csc\theta = \dfrac{1}{\sin\theta}$

4. Writing in terms of $\sin\theta$: $\sec\theta = \dfrac{1}{\cos\theta} = \pm\dfrac{1}{\sqrt{1 - \sin^2\theta}}$

6. Writing in terms of $\cos\theta$: $\sin\theta = \pm\sqrt{1 - \cos^2\theta}$

8. Writing in terms of $\cos\theta$: $\csc\theta = \dfrac{1}{\sin\theta} = \pm\dfrac{1}{\sqrt{1 - \cos^2\theta}}$

10. Writing in terms of $\sin\theta$ and $\cos\theta$, then simplifying: $\sec\theta\cot\theta = \dfrac{1}{\cos\theta} \cdot \dfrac{\cos\theta}{\sin\theta} = \dfrac{1}{\sin\theta}$

12. Writing in terms of $\sin\theta$ and $\cos\theta$, then simplifying: $\sec\theta\tan\theta\csc\theta = \dfrac{1}{\cos\theta} \cdot \dfrac{\sin\theta}{\cos\theta} \cdot \dfrac{1}{\sin\theta} = \dfrac{1}{\cos^2\theta}$

14. Writing in terms of $\sin\theta$ and $\cos\theta$, then simplifying: $\dfrac{\csc\theta}{\sec\theta} = \dfrac{1\big/\sin\theta}{1\big/\cos\theta} = \dfrac{1}{\sin\theta} \cdot \dfrac{\cos\theta}{1} = \dfrac{\cos\theta}{\sin\theta}$

16. Writing in terms of $\sin\theta$ and $\cos\theta$, then simplifying: $\dfrac{\csc\theta}{\cot\theta} = \dfrac{1/\sin\theta}{\cos\theta/\sin\theta} = \dfrac{1}{\sin\theta}\cdot\dfrac{\sin\theta}{\cos\theta} = \dfrac{1}{\cos\theta}$

18. Writing in terms of $\sin\theta$ and $\cos\theta$, then simplifying: $\dfrac{\cot\theta}{\tan\theta} = \dfrac{\cos\theta/\sin\theta}{\sin\theta/\cos\theta} = \dfrac{\cos\theta}{\sin\theta}\cdot\dfrac{\cos\theta}{\sin\theta} = \dfrac{\cos^2\theta}{\sin^2\theta}$

20. Writing in terms of $\sin\theta$ and $\cos\theta$, then simplifying: $\dfrac{\cos\theta}{\sec\theta} = \dfrac{\cos\theta}{1/\cos\theta} = \cos\theta\cdot\dfrac{\cos\theta}{1} = \cos^2\theta$

22. Writing in terms of $\sin\theta$ and $\cos\theta$, then simplifying: $\cot\theta - \csc\theta = \dfrac{\cos\theta}{\sin\theta} - \dfrac{1}{\sin\theta} = \dfrac{\cos\theta - 1}{\sin\theta}$

24. Writing in terms of $\sin\theta$ and $\cos\theta$, then simplifying: $\cos\theta\tan\theta + \sin\theta = \cos\theta\cdot\dfrac{\sin\theta}{\cos\theta} + \sin\theta = \sin\theta + \sin\theta = 2\sin\theta$

26. Writing in terms of $\sin\theta$ and $\cos\theta$, then simplifying:
$$\csc\theta - \cot\theta\cos\theta = \dfrac{1}{\sin\theta} - \dfrac{\cos\theta}{\sin\theta}\cdot\cos\theta = \dfrac{1}{\sin\theta} - \dfrac{\cos^2\theta}{\sin\theta} = \dfrac{1-\cos^2\theta}{\sin\theta} = \dfrac{\sin^2\theta}{\sin\theta} = \sin\theta$$

28. Adding the fractions: $\dfrac{\cos\theta}{\sin\theta} + \dfrac{\sin\theta}{\cos\theta} = \dfrac{\cos\theta}{\sin\theta}\cdot\dfrac{\cos\theta}{\cos\theta} + \dfrac{\sin\theta}{\cos\theta}\cdot\dfrac{\sin\theta}{\sin\theta} = \dfrac{\cos^2\theta + \sin^2\theta}{\sin\theta\cos\theta} = \dfrac{1}{\sin\theta\cos\theta}$

30. Subtracting the fractions: $\dfrac{1}{\cos\theta} - \dfrac{1}{\sin\theta} = \dfrac{1}{\cos\theta}\cdot\dfrac{\sin\theta}{\sin\theta} - \dfrac{1}{\sin\theta}\cdot\dfrac{\cos\theta}{\cos\theta} = \dfrac{\sin\theta - \cos\theta}{\sin\theta\cos\theta}$

32. Adding the fractions: $\cos\theta + \dfrac{1}{\sin\theta} = \dfrac{\cos\theta}{1}\cdot\dfrac{\sin\theta}{\sin\theta} + \dfrac{1}{\sin\theta} = \dfrac{\cos\theta\sin\theta + 1}{\sin\theta}$

34. Subtracting the fractions: $\dfrac{1}{\cos\theta} - \cos\theta = \dfrac{1}{\cos\theta} - \dfrac{\cos\theta}{1}\cdot\dfrac{\cos\theta}{\cos\theta} = \dfrac{1-\cos^2\theta}{\cos\theta} = \dfrac{\sin^2\theta}{\cos\theta}$

36. Multiplying the expressions: $(\cos\theta + 2)(\cos\theta - 5) = \cos^2\theta + 2\cos\theta - 5\cos\theta - 10 = \cos^2\theta - 3\cos\theta - 10$

38. Multiplying the expressions: $(3\sin\theta - 2)(5\cos\theta - 4) = 15\sin\theta\cos\theta - 10\cos\theta - 12\sin\theta + 8$

40. Multiplying the expressions: $(1 - \cos\theta)(1 + \cos\theta) = 1 - \cos\theta + \cos\theta - \cos^2\theta = 1 - \cos^2\theta = \sin^2\theta$

42. Multiplying the expressions: $(1 - \cot\theta)(1 + \cot\theta) = 1 - \cot\theta + \cot\theta - \cot^2\theta = 1 - \cot^2\theta$

44. Multiplying the expressions: $(\cos\theta + \sin\theta)(\cos\theta + \sin\theta) = \cos^2\theta + \sin\theta\cos\theta + \sin\theta\cos\theta + \sin^2\theta = 1 + 2\sin\theta\cos\theta$

46. Multiplying the expressions: $(\cos\theta - 2)(\cos\theta - 2) = \cos^2\theta - 2\cos\theta - 2\cos\theta + 4 = \cos^2\theta - 4\cos\theta + 4$

48. Substituting $x = \tan\theta$: $\sqrt{x^2 + 1} = \sqrt{\tan^2\theta + 1} = \sqrt{\sec^2\theta} = |\sec\theta|$

50. Substituting $x = 5\sin\theta$: $\sqrt{25 - x^2} = \sqrt{25 - (5\sin\theta)^2} = \sqrt{25 - 25\sin^2\theta} = \sqrt{25(1 - \sin^2\theta)} = \sqrt{25\cos^2\theta} = 5|\cos\theta|$

52. Substituting $x = 8\sec\theta$: $\sqrt{x^2 - 64} = \sqrt{(8\sec\theta)^2 - 64} = \sqrt{64\sec^2\theta - 64} = \sqrt{64(\sec^2\theta - 1)} = \sqrt{64\tan^2\theta} = 8|\tan\theta|$

54. Substituting $x = 5\tan\theta$:
$$\sqrt{4x^2 + 100} = \sqrt{4(5\tan\theta)^2 + 100} = \sqrt{100\tan^2\theta + 100} = \sqrt{100(\tan^2\theta + 1)} = \sqrt{100\sec^2\theta} = 10|\sec\theta|$$

56. Substituting $x = 4\sin\theta$: $\sqrt{64 - 4x^2} = \sqrt{64 - 4(4\sin\theta)^2} = \sqrt{64 - 64\sin^2\theta} = \sqrt{64(1 - \sin^2\theta)} = \sqrt{64\cos^2\theta} = 8|\cos\theta|$

58. Substituting $x = 6\sec\theta$:
$$\sqrt{4x^2 - 144} = \sqrt{4(6\sec\theta)^2 - 144} = \sqrt{144\sec^2\theta - 144} = \sqrt{144(\sec^2\theta - 1)} = \sqrt{144\tan^2\theta} = 12|\tan\theta|$$

60. Working from the left side: $\sin\theta\cot\theta = \sin\theta\cdot\dfrac{\cos\theta}{\sin\theta} = \cos\theta$

62. Working from the left side: $\cos\theta\csc\theta\tan\theta = \cos\theta\cdot\dfrac{1}{\sin\theta}\cdot\dfrac{\sin\theta}{\cos\theta} = \dfrac{\sin\theta\cos\theta}{\sin\theta\cos\theta} = 1$

64. Working from the left side: $\dfrac{\cos\theta}{\sec\theta} = \dfrac{\cos\theta}{1/\cos\theta} = \cos\theta\cdot\cos\theta = \cos^2\theta$

66. Working from the left side: $\dfrac{\sec\theta}{\tan\theta} = \dfrac{\frac{1}{\cos\theta}}{\frac{\sin\theta}{\cos\theta}} = \dfrac{1}{\cos\theta} \cdot \dfrac{\cos\theta}{\sin\theta} = \dfrac{1}{\sin\theta} = \csc\theta$

68. Working from the left side: $\dfrac{\csc\theta}{\sec\theta} = \dfrac{\frac{1}{\sin\theta}}{\frac{1}{\cos\theta}} = \dfrac{1}{\sin\theta} \cdot \dfrac{\cos\theta}{1} = \dfrac{\cos\theta}{\sin\theta} = \cot\theta$

70. Working from the left side: $\dfrac{\csc\theta\tan\theta}{\sec\theta} = \dfrac{\frac{1}{\sin\theta} \cdot \frac{\sin\theta}{\cos\theta}}{\frac{1}{\cos\theta}} = \dfrac{1}{\sin\theta} \cdot \dfrac{\sin\theta}{\cos\theta} \cdot \dfrac{\cos\theta}{1} = \dfrac{\sin\theta\cos\theta}{\sin\theta\cos\theta} = 1$

72. Working from the left side:

$$\cos\theta\cot\theta + \sin\theta = \cos\theta \cdot \dfrac{\cos\theta}{\sin\theta} + \sin\theta = \dfrac{\cos^2\theta}{\sin\theta} + \dfrac{\sin\theta}{1} \cdot \dfrac{\sin\theta}{\sin\theta} = \dfrac{\cos^2\theta + \sin^2\theta}{\sin\theta} = \dfrac{1}{\sin\theta} = \csc\theta$$

74. Working from the left side: $\tan^2\theta + 1 = \dfrac{\sin^2\theta}{\cos^2\theta} + 1 \cdot \dfrac{\cos^2\theta}{\cos^2\theta} = \dfrac{\sin^2\theta + \cos^2\theta}{\cos^2\theta} = \dfrac{1}{\cos^2\theta} = \sec^2\theta$

76. Working from the left side: $\sec\theta - \cos\theta = \dfrac{1}{\cos\theta} - \dfrac{\cos\theta}{1} \cdot \dfrac{\cos\theta}{\cos\theta} = \dfrac{1 - \cos^2\theta}{\cos\theta} = \dfrac{\sin^2\theta}{\cos\theta}$

78. Working from the left side: $\sec\theta\cot\theta - \sin\theta = \dfrac{1}{\cos\theta} \cdot \dfrac{\cos\theta}{\sin\theta} - \sin\theta = \dfrac{1}{\sin\theta} - \dfrac{\sin\theta}{1} \cdot \dfrac{\sin\theta}{\sin\theta} = \dfrac{1 - \sin^2\theta}{\sin\theta} = \dfrac{\cos^2\theta}{\sin\theta}$

80. Working from the left side: $(1 + \sin\theta)(1 - \sin\theta) = 1 + \sin\theta - \sin\theta - \sin^2\theta = 1 - \sin^2\theta = \cos^2\theta$

82. Working from the left side: $(\cos\theta + 1)(\cos\theta - 1) = \cos^2\theta + \cos\theta - \cos\theta - 1 = \cos^2\theta - 1 = -\left(1 - \cos^2\theta\right) = -\sin^2\theta$

84. Working from the left side: $1 - \dfrac{\sin\theta}{\csc\theta} = 1 - \dfrac{\sin\theta}{\frac{1}{\sin\theta}} = 1 - \sin\theta \cdot \sin\theta = 1 - \sin^2\theta = \cos^2\theta$

86. Working from the left side:

$$\begin{aligned}
(\cos\theta + \sin\theta)^2 - 1 &= (\cos\theta + \sin\theta)(\cos\theta + \sin\theta) - 1 \\
&= \cos^2\theta + \sin\theta\cos\theta + \sin\theta\cos\theta + \sin^2\theta - 1 \\
&= 1 + 2\sin\theta\cos\theta - 1 \\
&= 2\sin\theta\cos\theta
\end{aligned}$$

88. Working from the left side: $\sec\theta(\sin\theta + \cos\theta) = \dfrac{1}{\cos\theta}(\sin\theta + \cos\theta) = \dfrac{\sin\theta}{\cos\theta} + \dfrac{\cos\theta}{\cos\theta} = \tan\theta + 1$

90. Working from the left side: $\cos\theta(\csc\theta + \tan\theta) = \cos\theta\left(\dfrac{1}{\sin\theta} + \dfrac{\sin\theta}{\cos\theta}\right) = \dfrac{\cos\theta}{\sin\theta} + \sin\theta = \cot\theta + \sin\theta$

92. Working from the left side: $\cos\theta(\sec\theta - \cos\theta) = \cos\theta\left(\dfrac{1}{\cos\theta} - \cos\theta\right) = 1 - \cos^2\theta = \sin^2\theta$

Chapter 1 Test

1. The complement is $90° - 70° = 20°$ and the supplement is $180° - 70° = 110°$.

2. Using the Pythagorean Theorem:

$$3^2 + x^2 = 6^2$$
$$9 + x^2 = 36$$
$$x^2 = 27$$
$$x = \sqrt{27} = 3\sqrt{3}$$

3. Using the Pythagorean Theorem:

$$4^2 + x^2 = (x+2)^2$$
$$16 + x^2 = x^2 + 4x + 4$$
$$16 = 4x + 4$$
$$4x = 12$$
$$x = 3$$

4. Since $h = s$, $h = 5\sqrt{3}$. Since $r = \sqrt{2}\,h$, we have $r = \sqrt{2} \cdot 5\sqrt{3} = 5\sqrt{6}$. Since ΔBDC is a 30°-60°-90° triangle, $h = \sqrt{3}\,y$ and thus $y = \dfrac{h}{\sqrt{3}}$. So $y = \dfrac{5\sqrt{3}}{\sqrt{3}} = 5$. Finally, $x = 2y$, so $x = 10$.

5. Since $x = 2y$, $x = 2 \cdot 3 = 6$. Since $h = \sqrt{3}\,y$, $h = 3\sqrt{3}$. Since $s = h$, $s = 3\sqrt{3}$. Finally, since $r = \sqrt{2}\,s$, $r = \sqrt{2} \cdot 3\sqrt{3} = 3\sqrt{6}$.

6. Since ΔABC is a right triangle, using the Pythagorean Theorem:

$$(AB)^2 + 3^2 = 5^2$$
$$(AB)^2 + 9 = 25$$
$$(AB)^2 = 16$$
$$AB = 4$$

Since ΔDAB is a right triangle, using the Pythagorean Theorem:

$$(AB)^2 + (AD)^2 = (DB)^2$$
$$4^2 + 6^2 = (DB)^2$$
$$(DB)^2 = 16 + 36$$
$$(DB)^2 = 52$$
$$DB = \sqrt{52} = 2\sqrt{13}$$

7. Since the hour hand moves $360°$ in 12 hours, in 3 hours it must move $\dfrac{360°}{4} = 90°$.

8. If the longest side is 5, the shortest side is $\frac{1}{2}(5) = \frac{5}{2}$ and the middle side is $\frac{5}{2}\sqrt{3}$.

9. Since this is a 45°-45°-90° triangle, the hypotenuse has a length of $15\sqrt{2} \approx 21.2$ feet.

10. Constructing the triangles from the center of the pentagon, there are 5 congruent triangles with a total sum of angles of $5 \cdot 180° = 900°$. However, this sum includes the center angles, which add to $360°$, so the sum of the interior angles is $900° - 360° = 540°$. Since there are 5 of these interior angles, the measure of each is $\dfrac{540°}{5} = 108°$.

11. Constructing the triangles from the center of the hexagon, there are 6 congruent triangles with a total sum of angles of $6 \cdot 180° = 1080°$. However, this sum includes the center angles, which add to $360°$, so the sum of the interior angles is $1080° - 360° = 720°$. Since there are 6 of these interior angles, the measure of each is $\dfrac{720°}{6} = 120°$.

12. First find the x- and y-intercepts. For the x-intercept, let $y = 0$:

$$2x + 3(0) = 6$$
$$2x = 6$$
$$x = 3$$

The x-intercept is $(3,0)$. For the y-intercept, let $x = 0$:

$$2(0) + 3y = 6$$
$$3y = 6$$
$$y = 2$$

The y-intercept is $(0,2)$. Graphing the line:

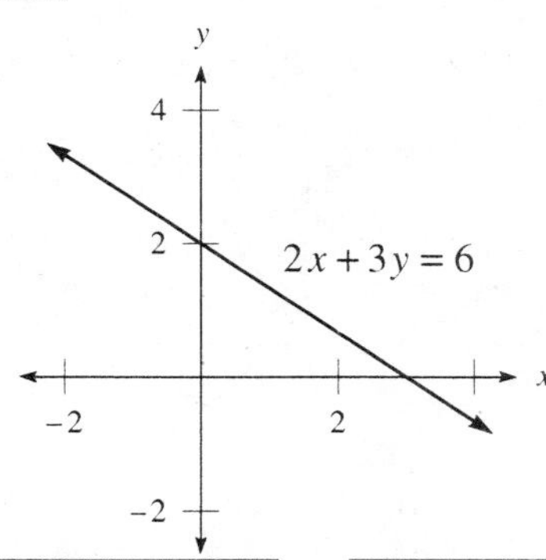

13. Using the distance formula: $r = \sqrt{(-1-4)^2 + (10+2)^2} = \sqrt{(-5)^2 + 12^2} = \sqrt{25 + 144} = \sqrt{169} = 13$

14. Using the distance formula: $r = \sqrt{(a-0)^2 + (b-0)^2} = \sqrt{a^2 + b^2}$

15. Using the distance formula:

$$\sqrt{(x+2)^2 + (1-3)^2} = \sqrt{13}$$
$$\sqrt{(x+2)^2 + 4} = \sqrt{13}$$
$$(x+2)^2 + 4 = 13$$
$$(x+2)^2 = 9$$
$$x + 2 = 3 \quad \text{or} \quad x + 2 = -3$$
$$x = 1 \quad \text{or} \quad x = -5$$

16. The vertex of the parabola is $(70,50)$, so the equation has the form $y = a(x - 70)^2 + 50$. Since the point $(140,0)$ is on this parabola:

$$0 = a(140 - 70)^2 + 50$$
$$-50 = 4900a$$
$$a = -\frac{1}{98}$$

Thus the equation is $y = -\frac{1}{98}(x - 70)^2 + 50$. Graphing the parabola:

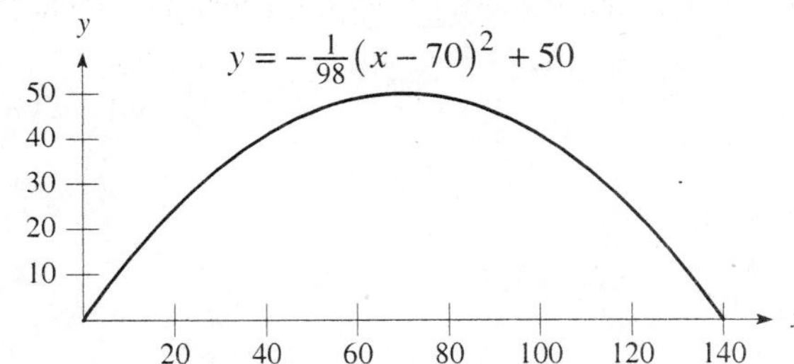

17. Choose $(0,1)$ as a point on the terminal side of $90°$. Therefore:

$$\sin 90° = \frac{y}{r} = \frac{1}{1} = 1 \qquad \cos 90° = \frac{x}{r} = \frac{0}{1} = 0 \qquad \tan 90° = \frac{y}{x} = \frac{1}{0}, \text{ which is undefined}$$

18. Choose $(1,-1)$ as a point on the terminal side of $-45°$. First find r: $r = \sqrt{1^2 + (-1)^2} = \sqrt{1+1} = \sqrt{2}$
Therefore:

$$\sin(-45°) = \frac{y}{r} = -\frac{1}{\sqrt{2}} \qquad \cos(-45°) = \frac{x}{r} = \frac{1}{\sqrt{2}} \qquad \tan(-45°) = \frac{y}{x} = \frac{-1}{1} = -1$$

19. If $\sin\theta < 0$ and $\cos\theta > 0$, $y < 0$ and $x > 0$. Thus θ lies in quadrant IV.

20. If $\csc\theta > 0$ and $\cos\theta < 0$, $y > 0$ and $x < 0$. Thus θ lies in quadrant II.

21. Begin by finding $r = \sqrt{(-6)^2 + 8^2} = \sqrt{36+64} = \sqrt{100} = 10$. Now applying the definitions for the six trigonometric functions using the values $x = -6$, $y = 8$, and $r = 10$:

$$\sin\theta = \frac{y}{r} = \frac{8}{10} = \frac{4}{5} \qquad \cos\theta = \frac{x}{r} = -\frac{6}{10} = -\frac{3}{5} \qquad \tan\theta = \frac{y}{x} = -\frac{8}{6} = -\frac{4}{3}$$

$$\csc\theta = \frac{r}{y} = \frac{10}{8} = \frac{5}{4} \qquad \sec\theta = \frac{r}{x} = -\frac{10}{6} = -\frac{5}{3} \qquad \cot\theta = \frac{x}{y} = -\frac{6}{8} = -\frac{3}{4}$$

22. Begin by finding $r = \sqrt{(-3)^2 + (-1)^2} = \sqrt{9+1} = \sqrt{10}$. Now applying the definitions for the six trigonometric functions using the values $x = -3$, $y = -1$, and $r = \sqrt{10}$:

$$\sin\theta = \frac{y}{r} = -\frac{1}{\sqrt{10}} \qquad \cos\theta = \frac{x}{r} = -\frac{3}{\sqrt{10}} \qquad \tan\theta = \frac{y}{x} = \frac{1}{3}$$

$$\csc\theta = \frac{r}{y} = -\sqrt{10} \qquad \sec\theta = \frac{r}{x} = -\frac{\sqrt{10}}{3} \qquad \cot\theta = \frac{x}{y} = 3$$

23. If $\sin\theta = \frac{1}{2}$, we can choose $y = 1$ and $r = 2$. To find x, use $x^2 + y^2 = r^2$:

$$x^2 + 1^2 = 2^2$$
$$x^2 + 1 = 4$$
$$x^2 = 3$$
$$x = \pm\sqrt{3}$$

Since θ terminates in quadrant II, $x < 0$ and thus $x = -\sqrt{3}$. Using $x = -\sqrt{3}$, $y = 1$, and $r = 2$ in the definitions:

$$\cos\theta = \frac{x}{r} = -\frac{\sqrt{3}}{2} \qquad \sec\theta = \frac{r}{x} = -\frac{2}{\sqrt{3}} \qquad \csc\theta = \frac{r}{y} = 2$$

$$\tan\theta = \frac{y}{x} = -\frac{1}{\sqrt{3}} \qquad \cot\theta = \frac{x}{y} = -\sqrt{3}$$

24. Since θ terminates in quadrant III, $x < 0$ and $y < 0$. If $\tan\theta = \frac{12}{5}$, we can choose $y = -12$ and $x = -5$. Then

$r = \sqrt{(-5)^2 + (-12)^2} = \sqrt{25+144} = \sqrt{169} = 13$. Now applying the definitions for the remaining trigonometric functions using the values $x = -5$, $y = -12$, and $r = 13$:

$$\sin\theta = \frac{y}{r} = -\frac{12}{13} \qquad \csc\theta = \frac{r}{y} = -\frac{13}{12} \qquad \cot\theta = \frac{x}{y} = \frac{5}{12}$$

$$\cos\theta = \frac{x}{r} = -\frac{5}{13} \qquad \sec\theta = \frac{r}{x} = -\frac{13}{5}$$

25. Let $x = 1$ and $y = -2$. Now find $r = \sqrt{1^2 + (-2)^2} = \sqrt{1+4} = \sqrt{5}$. Now applying the definitions for $\sin\theta$ and $\cos\theta$:

$$\sin\theta = \frac{y}{r} = -\frac{2}{\sqrt{5}} \qquad \cos\theta = \frac{x}{r} = \frac{1}{\sqrt{5}}$$

26. Using the ratio identity: $\csc\theta = \dfrac{1}{\sin\theta} = \dfrac{1}{-\frac{3}{4}} = -\frac{4}{3}$

27. Using the ratio identity: $\cos\theta = \dfrac{1}{\sec\theta} = -\frac{1}{2}$

28. Simplifying: $\sin^3\theta = (\sin\theta)^3 = \left(\frac{1}{3}\right)^3 = \frac{1}{27}$

29. Since $\sec\theta = 3$, $\cos\theta = \frac{1}{3}$. We can choose $x = 1$ and $r = 3$. To find y, use $x^2 + y^2 = r^2$:

$$1^2 + y^2 = 3^2$$
$$1 + y^2 = 9$$
$$y^2 = 8$$
$$y = \pm\sqrt{8} = \pm 2\sqrt{2}$$

Since θ terminates in quadrant IV, $y < 0$ and thus $y = -2\sqrt{2}$. Therefore:

$$\sin\theta = \frac{y}{r} = -\frac{2\sqrt{2}}{3} \qquad\qquad \tan\theta = \frac{y}{x} = -\frac{2\sqrt{2}}{1} = -2\sqrt{2}$$

30. Since $\sin\theta = \frac{1}{a}$, we can choose $y = 1$ and $r = a$. To find x, use $x^2 + y^2 = r^2$:

$$x^2 + 1^2 = a^2$$
$$x^2 = a^2 - 1$$
$$x = \pm\sqrt{a^2 - 1}$$

Since θ terminates in quadrant I, $x > 0$ and thus $x = \sqrt{a^2 - 1}$. Therefore:

$$\cos\theta = \frac{x}{r} = \frac{\sqrt{a^2 - 1}}{a} \qquad \csc\theta = \frac{r}{y} = a \qquad \cot\theta = \frac{x}{y} = \sqrt{a^2 - 1}$$

31. Multiplying the expressions: $(\sin\theta + 3)(\sin\theta - 7) = \sin^2\theta + 3\sin\theta - 7\sin\theta - 21 = \sin^2\theta - 4\sin\theta - 21$

32. Multiplying the expressions: $(\cos\theta - \sin\theta)(\cos\theta - \sin\theta) = \cos^2\theta - \sin\theta\cos\theta - \sin\theta\cos\theta + \sin^2\theta = 1 - 2\sin\theta\cos\theta$

33. Subtracting the fractions: $\dfrac{1}{\sin\theta} - \sin\theta = \dfrac{1}{\sin\theta} - \dfrac{\sin\theta}{1}\cdot\dfrac{\sin\theta}{\sin\theta} = \dfrac{1 - \sin^2\theta}{\sin\theta} = \dfrac{\cos^2\theta}{\sin\theta}$

34. Substituting $x = 2\sin\theta$: $\sqrt{4 - x^2} = \sqrt{4 - (2\sin\theta)^2} = \sqrt{4 - 4\sin^2\theta} = \sqrt{4\left(1 - \sin^2\theta\right)} = \sqrt{4\cos^2\theta} = 2|\cos\theta|$

35. Working from the left side: $\dfrac{\cot\theta}{\csc\theta} = \dfrac{\cos\theta/\sin\theta}{1/\sin\theta} = \dfrac{\cos\theta}{\sin\theta}\cdot\dfrac{\sin\theta}{1} = \cos\theta$

36. Working from the left side:

$$\cot\theta + \tan\theta = \frac{\cos\theta}{\sin\theta} + \frac{\sin\theta}{\cos\theta}$$
$$= \frac{\cos\theta}{\sin\theta}\cdot\frac{\cos\theta}{\cos\theta} + \frac{\sin\theta}{\cos\theta}\cdot\frac{\sin\theta}{\sin\theta}$$
$$= \frac{\cos^2\theta + \sin^2\theta}{\sin\theta\cos\theta}$$
$$= \frac{1}{\sin\theta\cos\theta}$$
$$= \frac{1}{\sin\theta}\cdot\frac{1}{\cos\theta}$$
$$= \csc\theta\sec\theta$$

37. Working from the left side: $(1 - \sin\theta)(1 + \sin\theta) = 1 - \sin\theta + \sin\theta - \sin^2\theta = 1 - \sin^2\theta = \cos^2\theta$

38. Working from the left side: $\sin\theta(\csc\theta + \cot\theta) = \sin\theta\left(\dfrac{1}{\sin\theta} + \dfrac{\cos\theta}{\sin\theta}\right) = \sin\theta\cdot\dfrac{1 + \cos\theta}{\sin\theta} = 1 + \cos\theta$

Chapter 2
Right Triangle Trigonometry

2.1 Definition II: Right Triangle Trigonometry

2. Using the Pythagorean Theorem, first find a:

$$a^2 + 5^2 = 13^2$$
$$a^2 + 25 = 169$$
$$a^2 = 144$$
$$a = 12$$

Using $a = 12$, $b = 5$, and $c = 13$, write the six trigonometric functions of A:

$$\sin A = \frac{a}{c} = \tfrac{12}{13} \qquad \cos A = \frac{b}{c} = \tfrac{5}{13} \qquad \tan A = \frac{a}{b} = \tfrac{12}{5}$$

$$\csc A = \frac{c}{a} = \tfrac{13}{12} \qquad \sec A = \frac{c}{b} = \tfrac{13}{5} \qquad \cot A = \frac{b}{a} = \tfrac{5}{12}$$

4. Using the Pythagorean Theorem, first find c:

$$3^2 + 2^2 = c^2$$
$$9 + 4 = c^2$$
$$c^2 = 13$$
$$c = \sqrt{13}$$

Using $a = 3$, $b = 2$, and $c = \sqrt{13}$, write the six trigonometric functions of A:

$$\sin A = \frac{a}{c} = \frac{3}{\sqrt{13}} \qquad \cos A = \frac{b}{c} = \frac{2}{\sqrt{13}} \qquad \tan A = \frac{a}{b} = \tfrac{3}{2}$$

$$\csc A = \frac{c}{a} = \frac{\sqrt{13}}{3} \qquad \sec A = \frac{c}{b} = \frac{\sqrt{13}}{2} \qquad \cot A = \frac{b}{a} = \tfrac{2}{3}$$

6. Using the Pythagorean Theorem, first find c:

$$3^2 + \left(\sqrt{7}\right)^2 = c^2$$
$$9 + 7 = c^2$$
$$c^2 = 16$$
$$c = 4$$

Using $a = 3$, $b = \sqrt{7}$, and $c = 4$, write the six trigonometric functions of A:

$$\sin A = \frac{a}{c} = \tfrac{3}{4} \qquad \cos A = \frac{b}{c} = \frac{\sqrt{7}}{4} \qquad \tan A = \frac{a}{b} = \frac{3}{\sqrt{7}}$$

$$\csc A = \frac{c}{a} = \tfrac{4}{3} \qquad \sec A = \frac{c}{b} = \frac{4}{\sqrt{7}} \qquad \cot A = \frac{b}{a} = \frac{\sqrt{7}}{3}$$

8.   Using the Pythagorean Theorem, first find a:

$$a^2 + 3^2 = 4^2$$
$$a^2 + 9 = 16$$
$$a^2 = 7$$
$$a = \sqrt{7}$$

Using $a = \sqrt{7}$, $b = 3$, and $c = 4$, find the three trigonometric functions of A:

$$\sin A = \frac{a}{c} = \frac{\sqrt{7}}{4} \qquad\qquad \cos A = \frac{b}{c} = \frac{3}{4} \qquad\qquad \tan A = \frac{a}{b} = \frac{\sqrt{7}}{3}$$

Now use the Cofunction Theorem to find the three trigonometric functions of B:

$$\sin B = \cos A = \frac{3}{4} \qquad\qquad \cos B = \sin A = \frac{\sqrt{7}}{4} \qquad\qquad \tan B = \cot A = \frac{b}{a} = \frac{3}{\sqrt{7}}$$

10.   Using the Pythagorean Theorem, first find c:

$$2^2 + 1^2 = c^2$$
$$4 + 1 = c^2$$
$$c^2 = 5$$
$$c = \sqrt{5}$$

Using $a = 2$, $b = 1$, and $c = \sqrt{5}$, find the three trigonometric functions of A:

$$\sin A = \frac{a}{c} = \frac{2}{\sqrt{5}} \qquad\qquad \cos A = \frac{b}{c} = \frac{1}{\sqrt{5}} \qquad\qquad \tan A = \frac{a}{b} = \frac{2}{1} = 2$$

Now use the Cofunction Theorem to find the three trigonometric functions of B:

$$\sin B = \cos A = \frac{1}{\sqrt{5}} \qquad\qquad \cos B = \sin A = \frac{2}{\sqrt{5}} \qquad\qquad \tan B = \cot A = \frac{b}{a} = \frac{1}{2}$$

12.   Using the Pythagorean Theorem, first find c:

$$1^2 + \left(\sqrt{5}\right)^2 = c^2$$
$$1 + 5 = c^2$$
$$c^2 = 6$$
$$c = \sqrt{6}$$

Using $a = 1$, $b = \sqrt{5}$, and $c = \sqrt{6}$, find the three trigonometric functions of A:

$$\sin A = \frac{a}{c} = \frac{1}{\sqrt{6}} \qquad\qquad \cos A = \frac{b}{c} = \frac{\sqrt{5}}{\sqrt{6}} \qquad\qquad \tan A = \frac{a}{b} = \frac{1}{\sqrt{5}}$$

Now use the Cofunction Theorem to find the three trigonometric functions of B:

$$\sin B = \cos A = \frac{\sqrt{5}}{\sqrt{6}} \qquad\qquad \cos B = \sin A = \frac{1}{\sqrt{6}} \qquad\qquad \tan B = \cot A = \frac{b}{a} = \sqrt{5}$$

14.   Using the Pythagorean Theorem, first find c:

$$x^2 + x^2 = c^2$$
$$c^2 = 2x^2$$
$$c = \sqrt{2}\,x$$

Using $a = x$, $b = x$, and $c = \sqrt{2}\,x$, find the three trigonometric functions of A:

$$\sin A = \frac{a}{c} = \frac{x}{\sqrt{2}\,x} = \frac{1}{\sqrt{2}} \qquad\qquad \cos A = \frac{b}{c} = \frac{x}{\sqrt{2}\,x} = \frac{1}{\sqrt{2}} \qquad\qquad \tan A = \frac{a}{b} = \frac{x}{x} = 1$$

Now use the Cofunction Theorem to find the three trigonometric functions of B:

$$\sin B = \cos A = \frac{1}{\sqrt{2}} \qquad\qquad \cos B = \sin A = \frac{1}{\sqrt{2}} \qquad\qquad \tan B = \cot A = \frac{b}{a} = \frac{x}{x} = 1$$

16.   The coordinates of point B are $B(8,6)$. Using the Pythagorean Theorem, first find c:

$$6^2 + 8^2 = c^2$$
$$36 + 64 = c^2$$
$$c^2 = 100$$
$$c = 10$$

Using $a = 6$, $b = 8$, and $c = 10$, find the three trigonometric functions of A:

$$\sin A = \frac{a}{c} = \frac{6}{10} = \frac{3}{5} \qquad \cos A = \frac{b}{c} = \frac{8}{10} = \frac{4}{5} \qquad \tan A = \frac{a}{b} = \frac{6}{8} = \frac{3}{4}$$

18.   Using the Cofunction Theorem, $\cos 40° = \sin 50°$.   **20.**   Using the Cofunction Theorem, $\cot 12° = \tan 78°$.

22.   Using the Cofunction Theorem, $\sin y = \cos(90° - y)$.   **24.**   Using the Cofunction Theorem, $\tan(90° - y) = \cot y$.

26.   Complete the table, using the ratio identity $\sec x = \dfrac{1}{\cos x}$:

x	$\cos x$	$\sec x$
$0°$	1	1
$30°$	$\frac{\sqrt{3}}{2}$	$\frac{2}{\sqrt{3}}$
$45°$	$\frac{1}{\sqrt{2}}$	$\sqrt{2}$
$60°$	$\frac{1}{2}$	2
$90°$	0	undefined

28.   Simplifying the expression: $5\sin^2 30° = 5\left(\frac{1}{2}\right)^2 = 5 \cdot \frac{1}{4} = \frac{5}{4}$

30.   Simplifying the expression: $\sin^3 30° = \left(\frac{1}{2}\right)^3 = \frac{1}{8}$

32.   Simplifying the expression: $\sin^2 60° + \cos^2 60° = \left(\frac{\sqrt{3}}{2}\right)^2 + \left(\frac{1}{2}\right)^2 = \frac{3}{4} + \frac{1}{4} = 1$

Note that, since $\sin^2 \theta + \cos^2 \theta = 1$ is an identity, this value is 1 regardless of the value of θ.

34.   Simplifying the expression: $(\sin 45° - \cos 45°)^2 = \left(\frac{1}{\sqrt{2}} - \frac{1}{\sqrt{2}}\right)^2 = 0^2 = 0$

36.   Simplifying the expression: $\tan^2 45° + \tan^2 60° = 1^2 + \left(\sqrt{3}\right)^2 = 1 + 3 = 4$

38.   Simplifying the expression: $4\cos y = 4\cos 45° = 4 \cdot \frac{\sqrt{2}}{2} = 2\sqrt{2}$

40.   Simplifying the expression: $-2\sin(y + 45°) = -2\sin(45° + 45°) = -2\sin 90° = -2 \cdot 1 = -2$

42.   Simplifying the expression: $3\sin 2y = 3\sin(2 \cdot 45°) = 3\sin 90° = 3 \cdot 1 = 3$

44.   Simplifying the expression: $2\sin(90° - z) = 2\sin(90° - 60°) = 2\sin 30° = 2 \cdot \frac{1}{2} = 1$

46.   Finding the exact value: $\csc 30° = \dfrac{1}{\sin 30°} = \dfrac{1}{\frac{1}{2}} = 2$

48.   Finding the exact value: $\sec 60° = \dfrac{1}{\cos 60°} = \dfrac{1}{\frac{1}{2}} = 2$

50.   Finding the exact value: $\cot 30° = \dfrac{\cos 30°}{\sin 30°} = \dfrac{\frac{\sqrt{3}}{2}}{\frac{1}{2}} = \sqrt{3}$

52.   Finding the exact value: $\csc 45° = \dfrac{1}{\sin 45°} = \dfrac{1}{\frac{1}{\sqrt{2}}} = \sqrt{2}$

54. First find a using the Pythagorean Theorem:
$$a^2 + 8.88^2 = 9.62^2$$
$$a^2 = 9.62^2 - 8.88^2$$
$$a^2 = 13.69$$
$$a = 3.7$$
Now find $\sin A$ and $\cos A$:
$$\sin A = \frac{a}{c} = \frac{3.7}{9.62} \approx 0.38 \qquad \cos A = \frac{b}{c} = \frac{8.88}{9.62} \approx 0.92$$
Using the Cofunction Theorem:
$$\sin B = \cos A \approx 0.92 \qquad \cos B = \sin A \approx 0.38$$

56. First find c using the Pythagorean Theorem:
$$11.28^2 + 8.46^2 = c^2$$
$$c^2 = 198.81$$
$$c = 14.1$$
Now find $\sin A$ and $\cos A$:
$$\sin A = \frac{a}{c} = \frac{11.28}{14.1} = 0.8 \qquad \cos A = \frac{b}{c} = \frac{8.46}{14.1} = 0.6$$
Using the Cofunction Theorem:
$$\sin B = \cos A = 0.6 \qquad \cos B = \sin A = 0.8$$

58. Since $CG = CD = 3$, using the Pythagorean Theorem:
$$(CG)^2 + (CD)^2 = (DG)^2$$
$$3^2 + 3^2 = (DG)^2$$
$$9 + 9 = (DG)^2$$
$$(DG)^2 = 18$$
$$DG = \sqrt{18} = 3\sqrt{2}$$
Now use the Pythagorean Theorem with ΔDGE:
$$(DG)^2 + (GE)^2 = (DE)^2$$
$$\left(3\sqrt{2}\right)^2 + 3^2 = (DE)^2$$
$$18 + 9 = (DE)^2$$
$$(DE)^2 = 27$$
$$DE = \sqrt{27} = 3\sqrt{3}$$
Now, let θ represent the angle formed by diagonals DE and DG. Therefore:
$$\sin \theta = \frac{GE}{DE} = \frac{3}{3\sqrt{3}} = \frac{1}{\sqrt{3}} \qquad \cos \theta = \frac{DG}{DE} = \frac{3\sqrt{2}}{3\sqrt{3}} = \frac{\sqrt{2}}{\sqrt{3}}$$

60. Let $CG = CD = x$, using the Pythagorean Theorem:
$$(CG)^2 + (CD)^2 = (DG)^2$$
$$x^2 + x^2 = (DG)^2$$
$$(DG)^2 = 2x^2$$
$$DG = \sqrt{2x^2} = \sqrt{2}\,x$$

Now use the Pythagorean Theorem with ΔDGE:
$$(DG)^2 + (GE)^2 = (DE)^2$$
$$\left(\sqrt{2}\,x\right)^2 + x^2 = (DE)^2$$
$$2x^2 + x^2 = (DE)^2$$
$$(DE)^2 = 3x^2$$
$$DE = \sqrt{3x^2} = \sqrt{3}\,x$$

Now, let θ represent the angle formed by diagonals DE and DG. Therefore:
$$\sin\theta = \frac{GE}{DE} = \frac{x}{\sqrt{3}\,x} = \frac{1}{\sqrt{3}} \qquad \cos\theta = \frac{DG}{DE} = \frac{\sqrt{2}\,x}{\sqrt{3}\,x} = \frac{\sqrt{2}}{\sqrt{3}}$$

62. Using the distance formula: $r = \sqrt{(-1-3)^2 + (-4+2)^2} = \sqrt{(-4)^2 + (-2)^2} = \sqrt{16+4} = \sqrt{20} = 2\sqrt{5}$

64. To find the x-intercept, let $y = 0$:
$$2x - 3(0) = 6$$
$$2x = 6$$
$$x = 3$$
The x-intercept is $(3,0)$. To find the y-intercept, let $x = 0$:
$$2(0) - 3y = 6$$
$$-3y = 6$$
$$y = -2$$
The y-intercept is $(0,-2)$. Now graphing the line:

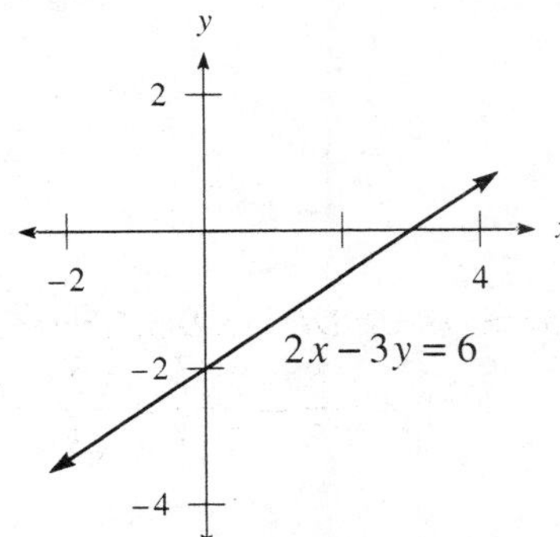

66.   A point on the terminal side is $(-1,1)$. Drawing the angle in standard position:

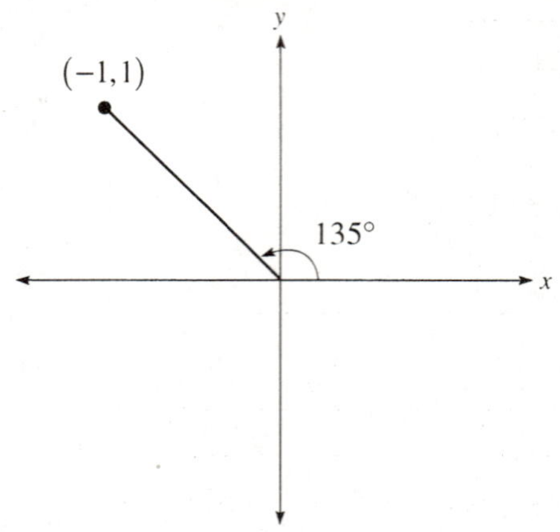

68.   A coterminal angle to $-90°$ is $270°$.
70.   A coterminal angle to $-210°$ is $150°$.

2.2 Calculators and Trigonometric Functions of an Acute Angle

2.   Adding the angles: $41°20' + 32°16' = 73°36'$ **4.**   Adding the angles: $63°38' + 24°52' = 87°90' = 88°30'$
6.   Adding the angles: $77°21' + 23°16' = 100°37'$
8.   Subtracting the angles: $90° - 62°25' = 89°60' - 62°25' = 27°35'$
10.   Subtracting the angles: $180° - 112°19' = 179°60' - 112°19' = 67°41'$
12.   Subtracting the angles: $89°38' - 28°58' = 88°98' - 28°58' = 60°40'$
14.   Subtracting the angles: $80°50' - 50°56' = 79°110' - 50°56' = 29°54'$
16.   Converting to degrees and minutes: $63.2° = 63° + 0.2° = 63° + 0.2(60') = 63°12'$
18.   Converting to degrees and minutes: $18.75° = 18° + 0.75° = 18° + 0.75(60') = 18°45'$
20.   Converting to degrees and minutes: $34.45° = 34° + 0.45° = 34° + 0.45(60') = 34°27'$
22.   Converting to degrees and minutes: $18.8° = 18° + 0.8° = 18° + 0.8(60') = 18°48'$

24.   Converting to decimal degrees: $74°18' = 74° + 18' = 74° + \left(\frac{18}{60}\right)^° = 74.3°$

26.   Converting to decimal degrees: $21°48' = 21° + 48' = 21° + \left(\frac{48}{60}\right)^° = 21.8°$

28.   Converting to decimal degrees: $29°40' = 29° + 40' = 29° + \left(\frac{40}{60}\right)^° \approx 29.67°$

30.   Converting to decimal degrees: $78°21' = 78° + 21' = 78° + \left(\frac{21}{60}\right)^° = 78.35°$

32.   Calculating the value: $\cos 82.9° \approx 0.1236$ **34.**   Calculating the value: $\sin 42° \approx 0.6691$

36.   Calculating the value: $\tan 81.43° \approx 6.6357$ **38.**   Calculating the value: $\cot 24° = \dfrac{1}{\tan 24°} \approx 2.2460$

40.   Calculating the value: $\sec 71.8° = \dfrac{1}{\cos 71.8°} \approx 3.2017$ **42.**   Calculating the value: $\csc 12.21° = \dfrac{1}{\sin 12.21°} \approx 4.7282$

44.   Calculating the value: $\sin 35°10' = \sin\left(35\frac{1}{6}\right)^° \approx 0.5760$

46.   Calculating the value: $\tan 19°45' = \tan\left(19\frac{45}{60}\right)^° = \tan 19.75° \approx 0.3590$

48.   Calculating the value: $\cos 66°40' = \cos\left(66\frac{2}{3}\right)^° \approx 0.3961$

50.   Calculating the value: $\sec 84°48' = \sec\left(84\frac{48}{60}\right)^° = \sec 84.8° = \dfrac{1}{\cos 84.8°} \approx 11.0336$

52. Completing the table:

X	$\cos X$	$\sec X$
$0°$	1	1
$30°$	0.8660	1.1547
$45°$	0.7071	1.4142
$60°$	0.5	2
$90°$	0	error (undefined)

54. Completing the table:

X	$\csc X$	$\sec X$	$\cot X$
$0°$	error (undefined)	1	error (undefined)
$15°$	3.8637	1.0353	3.7321
$30°$	2	1.1547	1.7321
$45°$	1.4142	1.4142	1
$60°$	1.1547	2	0.5774
$75°$	1.0353	3.8637	0.2679
$90°$	1	error (undefined)	0

Note: If you compute $\cot 90° = \dfrac{1}{\tan 90°}$, you will get a domain error message on your calculator.

56. Finding the angle θ: $\theta = \sin^{-1}(0.3971) \approx 23.4°$

58. Finding the angle θ: $\theta = \cos^{-1}(0.5490) \approx 56.7°$

60. Finding the angle θ: $\theta = \tan^{-1}(0.6273) \approx 32.1°$

62. Since $\sec \theta = 1.0801$, $\cos \theta = \dfrac{1}{1.0801}$, so $\theta = \cos^{-1}\left(\dfrac{1}{1.0801}\right) \approx 22.2°$.

64. Since $\csc \theta = 1.4293$, $\sin \theta = \dfrac{1}{1.4293}$, so $\theta = \sin^{-1}\left(\dfrac{1}{1.4293}\right) \approx 44.4°$.

66. Since $\cot \theta = 0.4327$, $\tan \theta = \dfrac{1}{0.4327}$, so $\theta = \tan^{-1}\left(\dfrac{1}{0.4327}\right) \approx 66.6°$.

68. Finding the angle θ: $\theta = \cos^{-1}(0.9153) \approx 23.7516° = 23° + 0.7516(60') = 23°45'$

70. Finding the angle θ: $\theta = \sin^{-1}(0.9954) \approx 84.5023° = 84° + 0.5023(60') = 84°30'$

72. Since $\cot \theta = 4.6252$, $\tan \theta = \dfrac{1}{4.6252}$. Finding the angle θ: $\theta = \tan^{-1}\left(\dfrac{1}{4.6252}\right) \approx 12.2° = 12° + 0.2(60') = 12°12'$

74. Since $\csc \theta = 7.0683$, $\sin \theta = \dfrac{1}{7.0683}$. Finding the angle θ: $\theta = \sin^{-1}\left(\dfrac{1}{7.0683}\right) \approx 8.1333° = 8° + 0.1333(60') = 8°8'$

76. Calculating the values: $\sin 33° \approx 0.5446$ and $\cos 57° \approx 0.5446$

78. Calculating the values: $\sec 56.7° \approx 1.8214$ and $\csc 33.3° \approx 1.8214$

80. Calculating the values: $\tan 10°30' = \tan 10.5° \approx 0.1853$ and $\cot 79°30' = \cot 79.5° \approx 0.1853$

82. Calculating the value: $\cos^2 85° + \sin^2 85° = 1$

84. Calculating the value: $\sin^2 8° + \cos^2 8° = 1$

86. To calculate B, $B = \sin^{-1}(4.321)$, which results in an error message. Since, for any angle B, $\sin B \leq 1$, it is impossible to find an angle B such that $\sin B = 4.321$.

88. To calculate $\cot 0°$, we would find $\tan 0° = 0$ then find the reciprocal. This results in an error message. Since $\dfrac{1}{0}$ is an undefined value, $\cot 0°$ is undefined.

90.   a.   Completing the table:

X	3°	2.5°	2°	1.5°	1°	0.5°	0°
cot X	19.1	22.9	28.6	38.2	57.3	114.6	undefined

b.   Completing the table:

X	0.6°	0.5°	0.4°	0.3°	0.2°	0.1°	0°
cot X	95.5	114.6	143.2	191.0	286.5	573.0	undefined

92.   First find the value of r: $r = \sqrt{\left(-\sqrt{3}\right)^2 + 1^2} = \sqrt{3+1} = \sqrt{4} = 2$

Finding the three trigonometric functions using $x = -\sqrt{3}$, $y = 1$, and $r = 2$:

$$\sin\theta = \frac{y}{r} = \frac{1}{2} \qquad \cos\theta = \frac{x}{r} = -\frac{\sqrt{3}}{2} \qquad \tan\theta = \frac{y}{x} = -\frac{1}{\sqrt{3}}$$

94.   Let $(-1,1)$ be a point on the terminal side of 135°. First find the value of r: $r = \sqrt{(-1)^2 + 1^2} = \sqrt{1+1} = \sqrt{2}$

Finding the three trigonometric functions using $x = -1$, $y = 1$, and $r = \sqrt{2}$:

$$\sin\theta = \frac{y}{r} = \frac{1}{\sqrt{2}} \qquad \cos\theta = \frac{x}{r} = -\frac{1}{\sqrt{2}} \qquad \tan\theta = \frac{y}{x} = \frac{1}{-1} = -1$$

96.   Since $\tan\theta = -\frac{3}{4}$ and θ terminates in quadrant II (where $x < 0$ and $y > 0$), choose $x = -4$ and $y = 3$. Finding r:

$$r = \sqrt{(-4)^2 + 3^2} = \sqrt{16+9} = \sqrt{25} = 5$$

Finding the remaining trigonometric functions using $x = -4$, $y = 3$, and $r = 5$:

$$\sin\theta = \frac{y}{r} = \frac{3}{5} \qquad \cos\theta = \frac{x}{r} = -\frac{4}{5} \qquad \cot\theta = \frac{x}{y} = -\frac{4}{3}$$

$$\csc\theta = \frac{r}{y} = \frac{5}{3} \qquad \sec\theta = \frac{r}{x} = -\frac{5}{4}$$

98.   Since $\sec\theta < 0$, $x < 0$. Thus for $\tan\theta > 0$, we must have $y < 0$. Thus the terminal side of θ lies in quadrant III.

2.3 Solving Right Triangles

2.   Begin by drawing $\triangle ABC$:

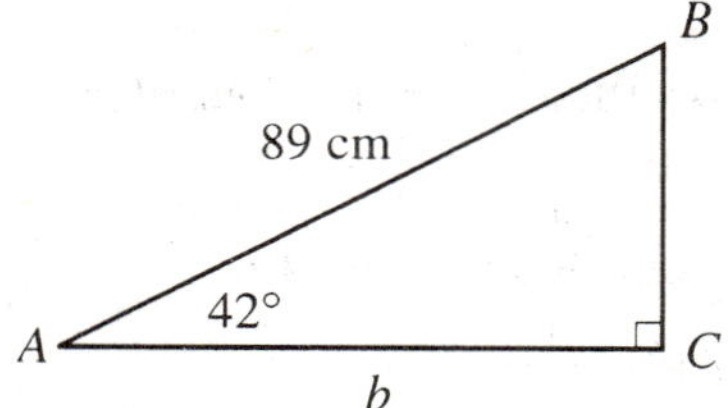

Therefore:

$$\cos 42° = \frac{b}{89}$$
$$b = 89\cos 42° \approx 66 \text{ cm}$$

4.   Begin by drawing $\triangle ABC$:

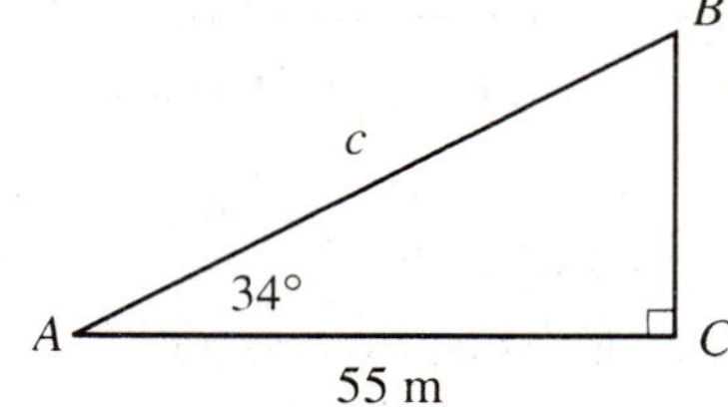

Therefore:

$$\cos 34° = \frac{55}{c}$$
$$c\cos 34° = 55$$
$$c = \frac{55}{\cos 34°} \approx 66 \text{ m}$$

6.   Begin by drawing $\triangle ABC$:

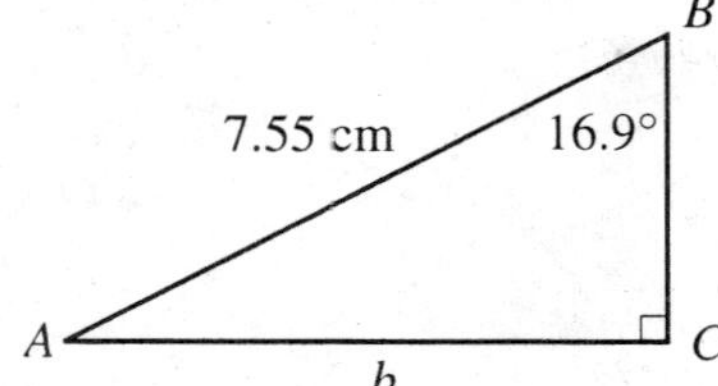

Therefore:

$$\sin 16.9° = \frac{b}{7.55}$$
$$b = 7.55 \sin 16.9° \approx 2.19 \text{ cm}$$

10.   Begin by drawing $\triangle ABC$:

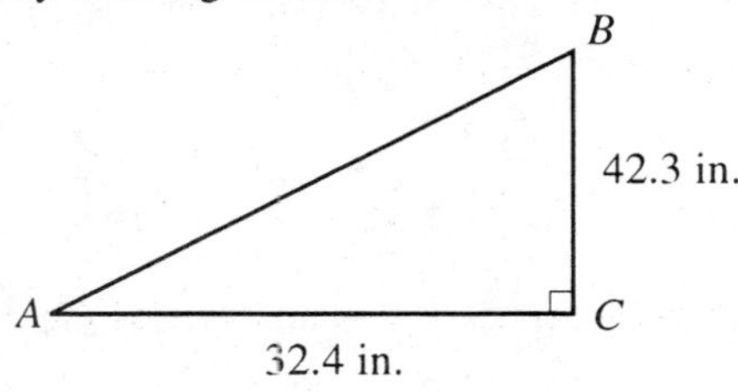

Therefore:

$$\tan B = \frac{32.4}{42.3} \approx 0.7660$$
$$B = \tan^{-1}(0.7660) \approx 37.5°$$

14.   Begin by drawing $\triangle ABC$:

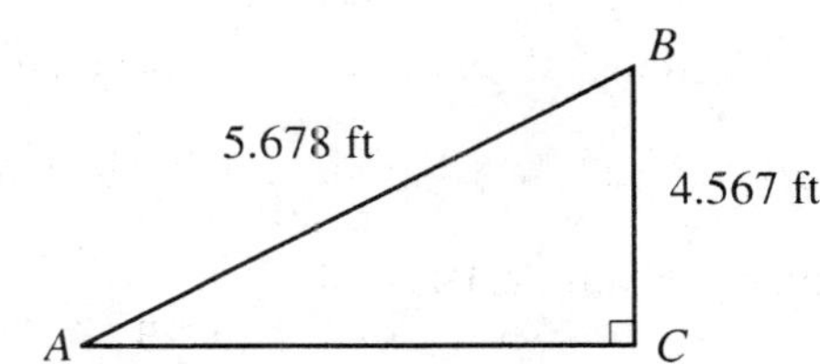

Therefore:

$$\sin A = \frac{4.567}{5.678} \approx 0.8043$$
$$A = \sin^{-1}(0.8043) \approx 53.55°$$

8.   Begin by drawing $\triangle ABC$:

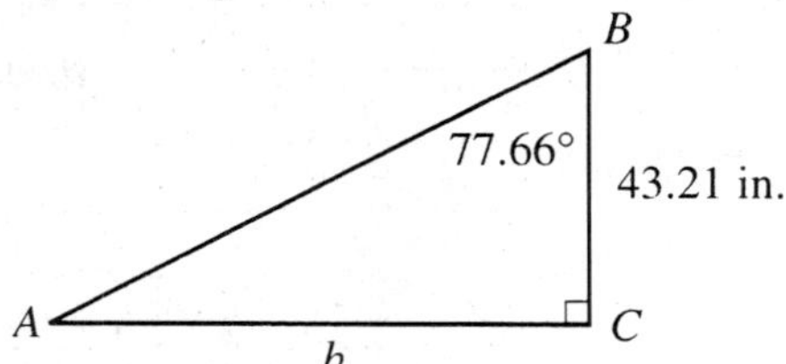

Therefore:

$$\tan 77.66° = \frac{b}{43.21}$$
$$b = 43.21 \tan 77.66° \approx 197.5 \text{ in.}$$

12.   Begin by drawing $\triangle ABC$:

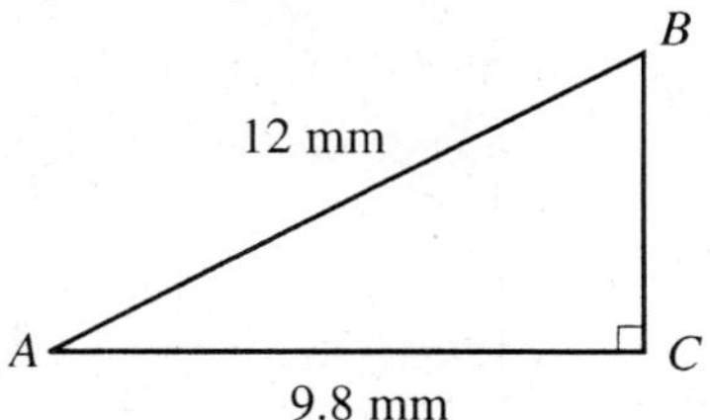

Therefore:

$$\sin B = \frac{9.8}{12} \approx 0.8167$$
$$B = \sin^{-1}(0.8167) \approx 55°$$

16.   Begin by drawing $\triangle ABC$ and label missing information:

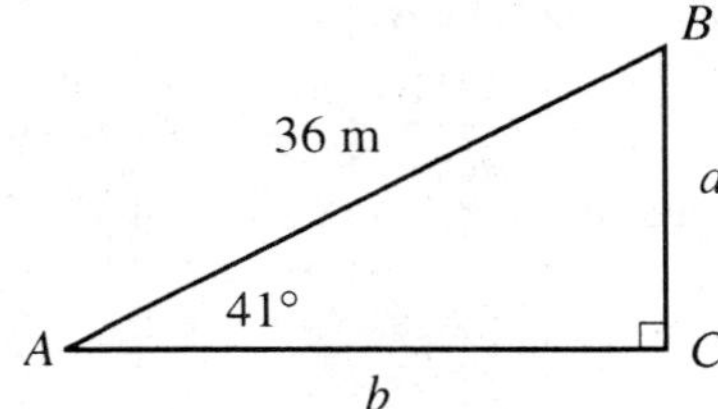

Note that $B = 90° - 41° = 49°$. Therefore:

$$\sin 41° = \frac{a}{36}$$
$$a = 36 \sin 41° \approx 24 \text{ m}$$
$$\cos 41° = \frac{b}{36}$$
$$b = 36 \cos 41° \approx 27 \text{ m}$$

18.   Begin by drawing $\triangle ABC$ and label missing information:

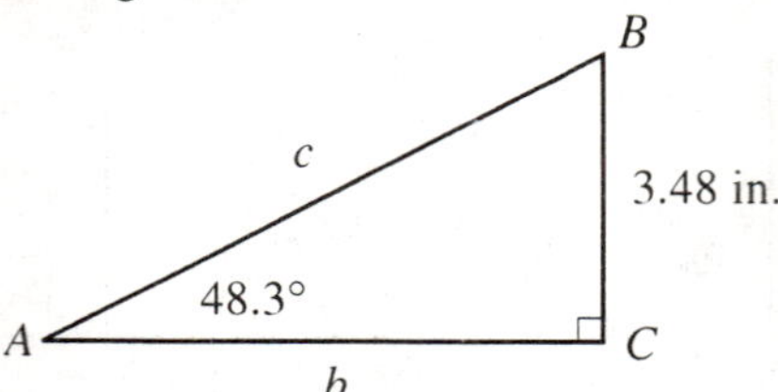

Note that $B = 90° - 48.3° = 41.7°$. Therefore:

$$\sin 48.3° = \frac{3.48}{c}$$
$$c \sin 48.3° = 3.48$$
$$c = \frac{3.48}{\sin 48.3°} \approx 4.66 \text{ in.}$$

$$\cos 48.3° = \frac{b}{4.66}$$
$$b = 4.66 \cos 48.3° \approx 3.10 \text{ in.}$$

20.   Begin by drawing $\triangle ABC$ and label missing information:

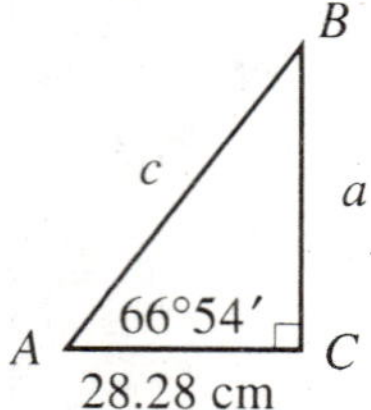

Note that $B = 90° - 66°54' = 89°60' - 66°54' = 23°6'$. So:

$$\cos 66°54' = \frac{28.28}{c}$$
$$c \cos 66°54' = 28.28$$
$$c = \frac{28.28}{\cos 66°54'} \approx 72.08 \text{ cm}$$

$$\sin 66°54' = \frac{a}{72.08}$$
$$a = 72.08 \sin 66°54' \approx 66.30 \text{ cm}$$

22.   Begin by drawing $\triangle ABC$ and label missing information:

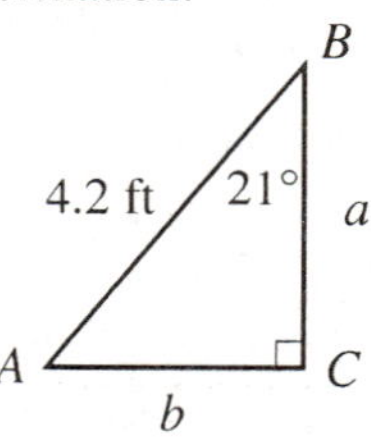

Note that $A = 90° - 21° = 69°$. Therefore:

$$\cos 69° = \frac{b}{4.2}$$
$$b = 4.2 \cos 69° \approx 1.5 \text{ ft}$$

$$\sin 69° = \frac{a}{4.2}$$
$$a = 4.2 \sin 69° \approx 3.9 \text{ ft}$$

24.   Begin by drawing $\triangle ABC$ and label missing information:

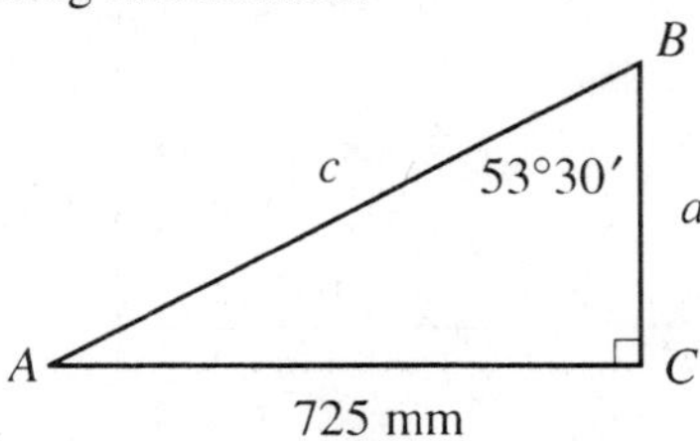

Note that $A = 90° - 53°30' = 89°60' - 53°30' = 36°30'$. So:

$$\cos 36°30' = \frac{725}{c}$$
$$c \cos 36°30' = 725$$
$$c = \frac{725}{\cos 36°30'} \approx 902 \text{ mm}$$

$$\tan 36°30' = \frac{a}{725}$$
$$a = 725 \tan 36°30' \approx 536 \text{ mm}$$

26.   Begin by drawing $\triangle ABC$ and label missing information:

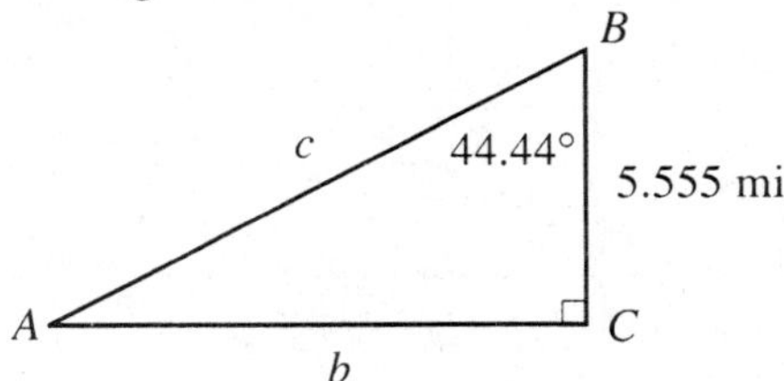

Note that $A = 90° - 44.44° = 45.56°$. Therefore:

$$\sin 45.56° = \frac{5.555}{c}$$
$$c \sin 45.56° = 5.555$$
$$c = \frac{5.555}{\sin 45.56°} \approx 7.780 \text{ mi}$$

$$\tan 45.56° = \frac{5.555}{b}$$
$$b \tan 45.56° = 5.555$$
$$b = \frac{5.555}{\tan 45.56°} \approx 5.447 \text{ mi}$$

28.   Begin by drawing $\triangle ABC$ and label missing information:

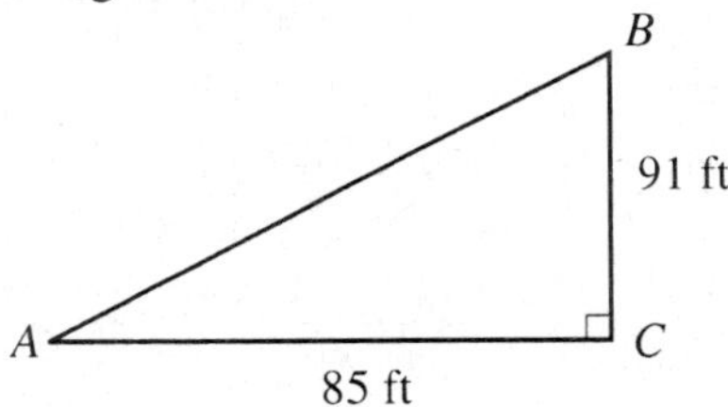

Therefore:

$$c = \sqrt{85^2 + 91^2} \approx 125 \text{ ft}$$
$$\tan A = \frac{91}{85} \approx 1.0706$$
$$A = \tan^{-1}(1.0706) \approx 47°$$
$$B = 90° - 47° = 43°$$

30.   Begin by drawing $\triangle ABC$ and label missing information:

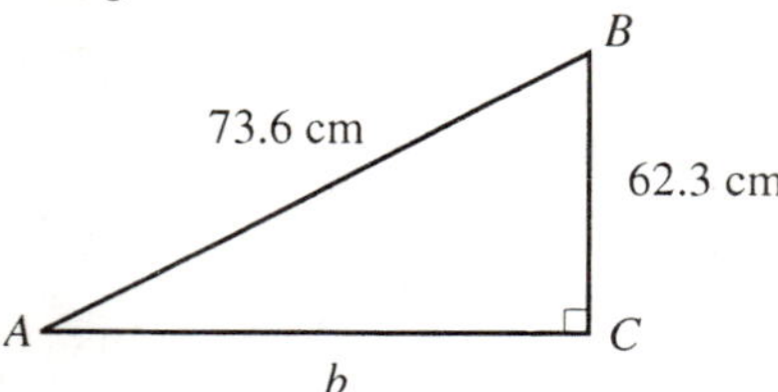

Therefore:
$$b = \sqrt{73.6^2 - 62.3^2} \approx 39.2 \text{ cm}$$
$$\sin A = \frac{62.3}{73.6} \approx 0.8465$$
$$A = \sin^{-1}(0.8465) \approx 57.8°$$
$$B = 90° - 57.8° = 32.2°$$

32.   Begin by drawing $\triangle ABC$ and label missing information:

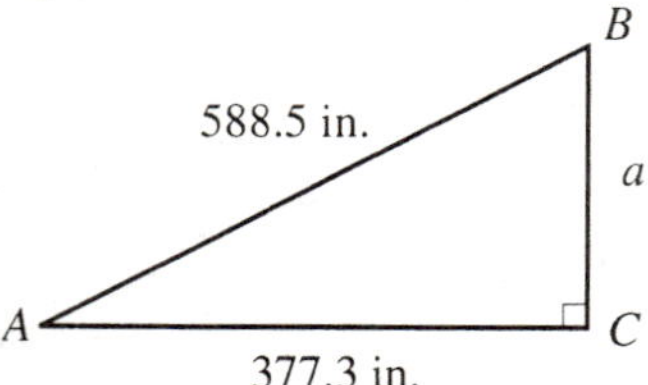

Therefore:
$$a = \sqrt{588.5^2 - 377.3^2} \approx 451.6 \text{ in.}$$
$$\cos A = \frac{377.3}{588.5} \approx 0.6411$$
$$A = \cos^{-1}(0.6411) \approx 50.12°$$
$$B = 90° - 50.12° = 39.88°$$

34.   Since the right triangle is a 45°-45°-90° triangle, its height is 2. Therefore:
$$\tan A = \tfrac{2}{3} \approx 0.6667$$
$$A = \tan^{-1}(0.6667) \approx 34°$$

36.   Re-drawing the figure:

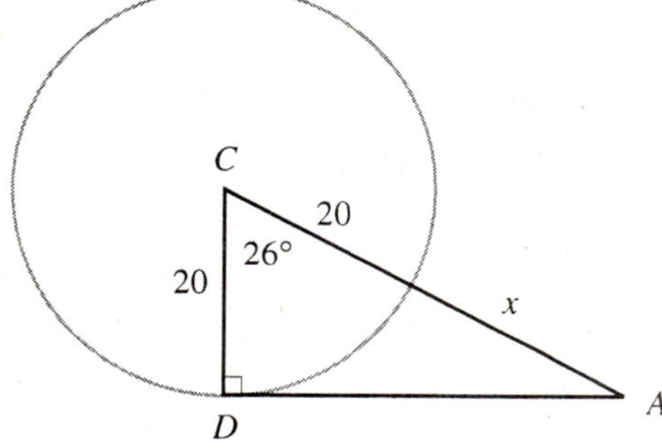

Note that $A = 90° - 26° = 64°$. Therefore:
$$\sin 64° = \frac{20}{20 + x}$$
$$(20 + x)\sin 64° = 20$$
$$20 + x = \frac{20}{\sin 64°}$$
$$x = \frac{20}{\sin 64°} - 20 \approx 2.3$$

38.   Re-drawing the figure:

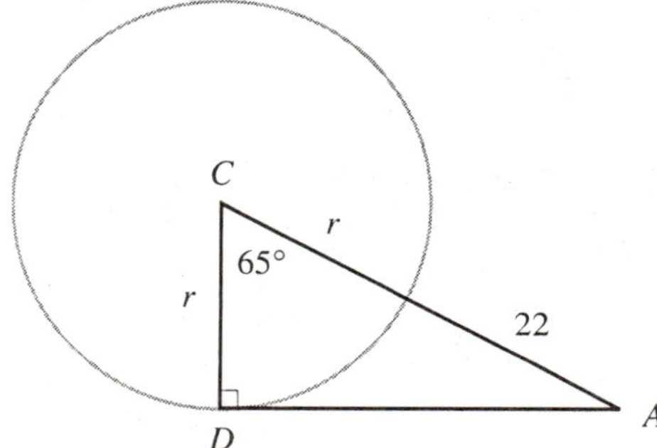

Note that $A = 90° - 65° = 25°$. Therefore:
$$\sin 25° = \frac{r}{r + 22}$$
$$(r + 22)\sin 25° = r$$
$$r\sin 25° + 22\sin 25° = r$$
$$22\sin 25° = r - r\sin 25°$$
$$22\sin 25° = r(1 - \sin 25°)$$
$$r = \frac{22\sin 25°}{1 - \sin 25°} \approx 16$$

40. Re-drawing the figure:

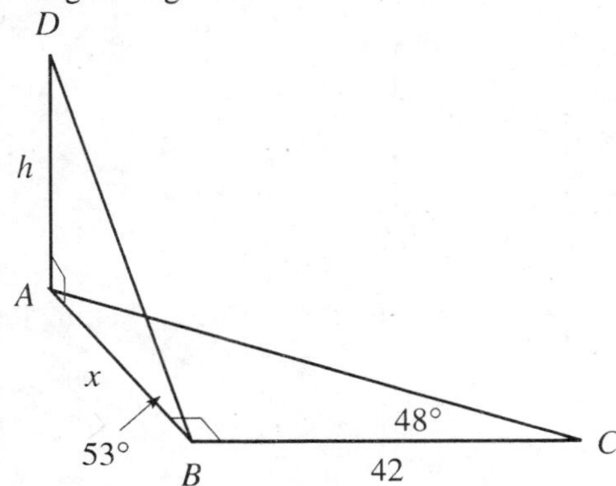

First find x:

$$\tan 48° = \frac{x}{42}$$
$$x = 42\tan 48° \approx 47$$

Now find h:

$$\tan 53° = \frac{h}{47}$$
$$h = 47\tan 53° \approx 62$$

44. Re-drawing the figure:

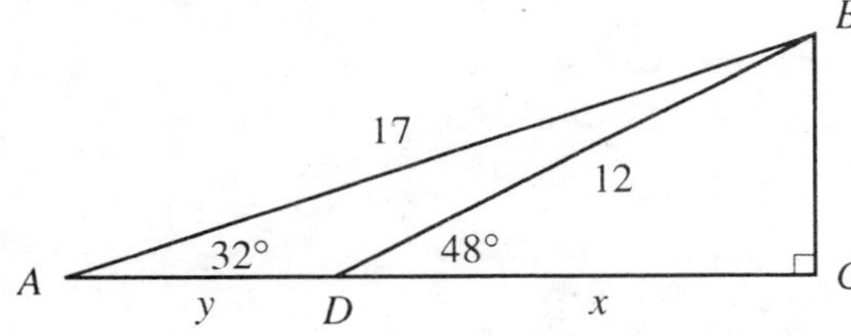

First find x:

$$\cos 48° = \frac{x}{12}$$
$$x = 12\cos 48° \approx 8.0$$

Now find y:

$$\cos 32° = \frac{8+y}{17}$$
$$8 + y = 17\cos 32°$$
$$y = 17\cos 32° - 8 \approx 6.4$$

42. Re-drawing the figure:

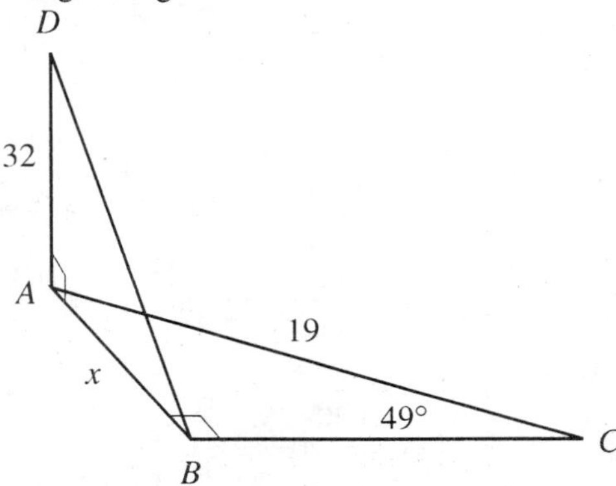

First find x:

$$\sin 49° = \frac{x}{19}$$
$$x = 19\sin 49° \approx 14$$

Now find $\angle ABD$:

$$\tan \angle ABD = \tfrac{32}{14} \approx 2.286$$
$$\angle ABD = \tan^{-1}(2.286) \approx 66°$$

46. Re-drawing the figure:

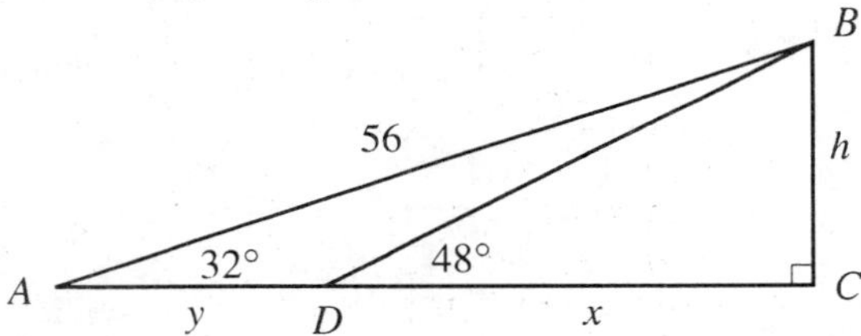

First find h:

$$\sin 32° = \frac{h}{56}$$
$$h = 56\sin 32° \approx 30$$

Now find x:

$$\tan 48° = \frac{30}{x}$$
$$x\tan 48° = 30$$
$$x = \frac{30}{\tan 48°} \approx 27$$

48. Re-drawing the figure:

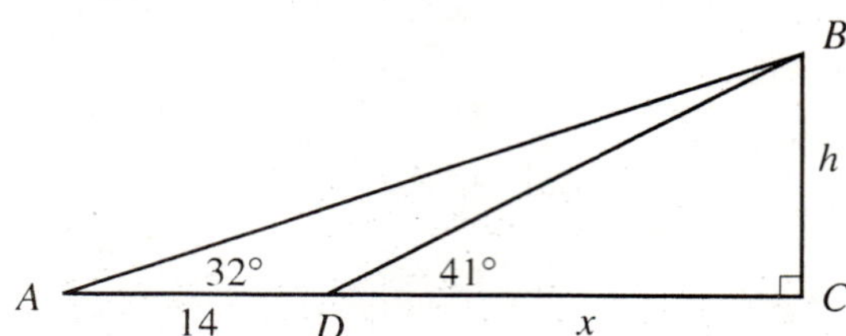

Note that $\tan 41° = \dfrac{h}{x}$, so $h = x\tan 41°$. Also note that $\tan 32° = \dfrac{h}{14 + x}$, so $h = (14 + x)\tan 32°$. Setting these two expressions equal:

$$x\tan 41° = (14 + x)\tan 32°$$
$$x\tan 41° = 14\tan 32° + x\tan 32°$$
$$x\tan 41° - x\tan 32° = 14\tan 32°$$
$$x(\tan 41° - \tan 32°) = 14\tan 32°$$
$$x = \frac{14\tan 32°}{\tan 41° - \tan 32°} \approx 36$$

50. Since $GC = CD = 3$, using the Pythagorean Theorem: $GD = \sqrt{3^2 + 3^2} = \sqrt{18} = 3\sqrt{2}$. Therefore:
$$\tan \angle GDE = \frac{GE}{GD} = \frac{3}{3\sqrt{2}} = \frac{1}{\sqrt{2}}$$
$$\tan \angle GDE \approx 0.7071$$
$$\angle GDE = \tan^{-1}(0.7071) \approx 35.3°$$

52. Let $GC = CD = x$. Using the Pythagorean Theorem: $GD = \sqrt{x^2 + x^2} = \sqrt{2x^2} = x\sqrt{2}$. Therefore:
$$\tan \angle GDE = \frac{GE}{GD} = \frac{x}{x\sqrt{2}} = \frac{1}{\sqrt{2}}$$
$$\tan \angle GDE \approx 0.7071$$
$$\angle GDE = \tan^{-1}(0.7071) \approx 35.3°$$

54. From Example 5, we have:

$$\cos 60° = \frac{139 - h}{125}$$
$$\frac{1}{2} = \frac{139 - h}{125}$$
$$62.5 = 139 - h$$
$$h = 76.5$$

The rider is 76.5 ft above the ground.

56. From Example 5, we have:

$$\cos 135° = \frac{139 - h}{125}$$
$$-\frac{1}{\sqrt{2}} = \frac{139 - h}{125}$$
$$-\frac{125}{\sqrt{2}} = 139 - h$$
$$h = 139 + \frac{125}{\sqrt{2}} \approx 227.4$$

The rider is approximately 227.4 ft above the ground.

58. First note that the distance from the ground to the low point of Colossus is $174 - 165 = 9$ ft. The radius is 82.5 ft. Since $x + h = 82.5 + 9 = 91.5$, $x = 91.5 - h$. Therefore:
$$\cos \theta = \frac{x}{82.5} = \frac{91.5 - h}{82.5}$$
$$91.5 - h = 82.5\cos \theta$$
$$h = 91.5 - 82.5\cos \theta$$

a. Substituting $\theta = 150°$: $h = 91.5 - 82.5\cos 150° \approx 162.9$ ft
b. Substituting $\theta = 240°$: $h = 91.5 - 82.5\cos 240° \approx 132.8$ ft
c. Substituting $\theta = 315°$: $h = 91.5 - 82.5\cos 315° \approx 33.2$ ft

60. Since $\csc B = 3$, $\sin B = \frac{1}{3}$, so $\sin^2 B = \left(\frac{1}{3}\right)^2 = \frac{1}{9}$.

62. Finding $\sin^2\theta$: $\sin^2\theta = 1 - \cos^2\theta = 1 - \left(-\frac{2}{3}\right)^2 = 1 - \frac{4}{9} = \frac{5}{9}$

So $\sin\theta = \pm\dfrac{\sqrt{5}}{3}$. Since θ terminates in quadrant III, where $y < 0$, $\sin\theta < 0$. Thus $\sin\theta = -\dfrac{\sqrt{5}}{3}$.

64. Finding $\cos^2 A$: $\cos^2 A = 1 - \sin^2 A = 1 - \left(\frac{1}{4}\right)^2 = 1 - \frac{1}{16} = \frac{15}{16}$

So $\cos A = \pm\dfrac{\sqrt{15}}{4}$. Since A terminates in quadrant II, where $x < 0$, $\cos A < 0$. Thus $\cos A = -\dfrac{\sqrt{15}}{4}$.

66. First find $\sin\theta$ (note $\sin\theta < 0$ since θ terminates in quadrant IV):
$$\sin\theta = -\sqrt{1 - \cos^2\theta} = -\sqrt{1 - \frac{1}{5}} = -\sqrt{\frac{4}{5}} = -\frac{2}{\sqrt{5}}$$

Now find the other four trigonometric ratios using $x = 1$, $y = -2$, and $r = \sqrt{5}$:

$$\sec\theta = \frac{r}{x} = \sqrt{5} \qquad \csc\theta = \frac{r}{y} = -\frac{\sqrt{5}}{2} \qquad \tan\theta = \frac{y}{x} = -2 \qquad \cot\theta = \frac{x}{y} = -\frac{1}{2}$$

68. Since $\csc\theta = -2$, $\sin\theta = -\frac{1}{2}$. Now find $\cos\theta$ (note that $\cos\theta < 0$ since θ terminates in quadrant III):
$$\cos\theta = -\sqrt{1 - \sin^2\theta} = -\sqrt{1 - \left(-\frac{1}{3}\right)^2} = -\sqrt{1 - \frac{1}{4}} = -\sqrt{\frac{3}{4}} = -\frac{\sqrt{3}}{2}$$

Now find the other three trigonometric ratios using $x = -\sqrt{3}$, $y = -1$, and $r = 2$:

$$\sec\theta = \frac{r}{x} = -\frac{2}{\sqrt{3}} \qquad \tan\theta = \frac{y}{x} = \frac{-1}{-\sqrt{3}} = \frac{1}{\sqrt{3}} \qquad \cot\theta = \frac{x}{y} = \frac{-\sqrt{3}}{-1} = \sqrt{3}$$

70. Entering the functions $Y_1 = -\frac{7}{640}(X - 80)^2 + 70$ and $Y_2 = \tan^{-1}\left(\dfrac{Y_1}{X}\right)$, complete the table:

X	10	5	1	0.5	0.1	0.01
Y_1	16.4063	8.4766	1.7391	0.8723	0.1749	0.0175
Y_2	58.6°	59.5°	60.1°	60.2°	60.2°	60.3°

Based on these results, it appears the angle between the cannon and the horizontal is approximately 60.3°.

2.4 Applications

2. Call x the length of each side. Note the figure:

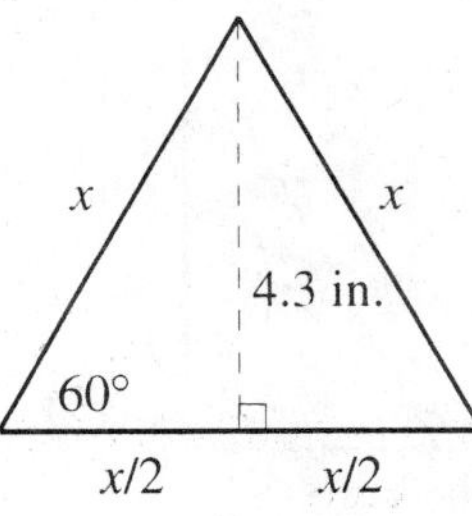

Therefore:
$$\sin 60° = \frac{4.3}{x}$$
$$x \sin 60° = 4.3$$
$$x = \frac{4.3}{\sin 60°} \approx 5.0$$
The length of each side is 5.0 in.

4.   Call w the length of the shorter side (width), and α, β the required angles. Note the figure:

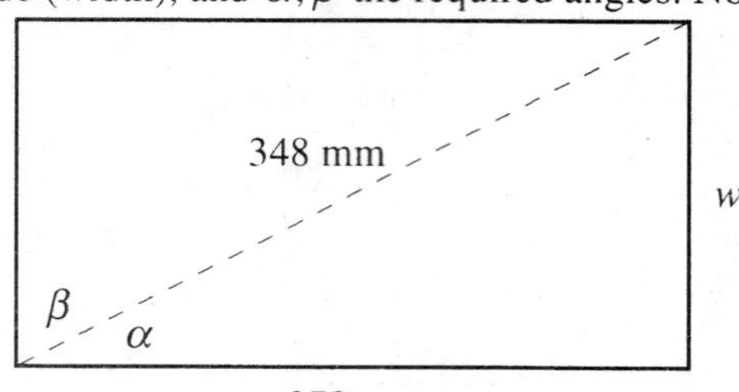

Find w using the Pythagorean Theorem:
$$278^2 + w^2 = 348^2$$
$$w^2 = 348^2 - 278^2 = 43820$$
$$w \approx 209$$
Now find angles α and β:
$$\cos\alpha = \frac{278}{348}$$
$$\cos\alpha = 0.7989$$
$$\alpha = \cos^{-1}(0.7989) \approx 37.0°$$
$$\beta = 90° - 37.0° = 53.0°$$
The shorter side is 209 mm, and the two angles are 37.0° and 53.0°.

6.   Let h represent the height of the hill. Draw the figure:

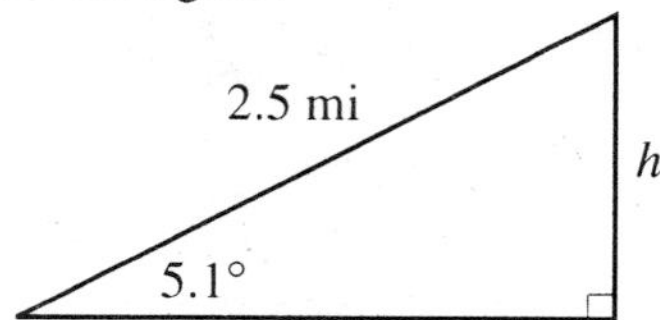

Therefore:
$$\sin 5.1° = \frac{h}{2.5}$$
$$h = 2.5\sin 5.1° \approx 0.22$$
The hill is approximately 0.22 mi high, which is approximately 1,170 feet.

8.   Let θ represent the angle between the ladder and the wall. Draw the figure:

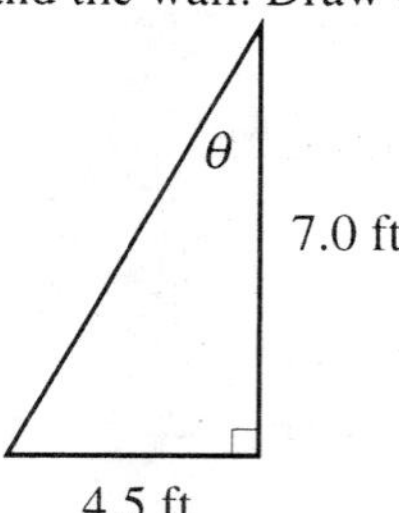

Therefore:
$$\tan\theta = \frac{4.5}{7.0} \approx 0.6429$$
$$\theta = \tan^{-1}(0.6429) \approx 33°$$
The angle between the ladder and the wall is approximately 33°.

10.   Let h represent the height of the building. Draw the figure:

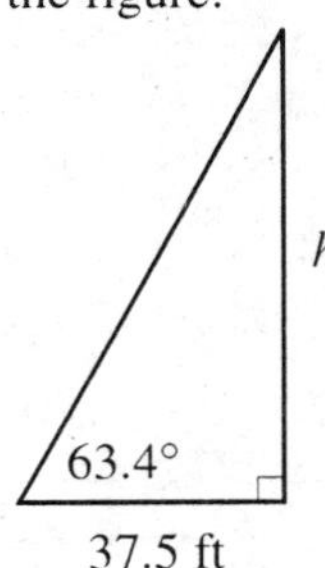

Therefore:
$$\tan 63.4° = \frac{h}{37.5}$$
$$h = 37.5 \tan 63.4° \approx 74.9$$
The height of the building is approximately 74.9 feet.

12.   Draw the figure with the associated labels:

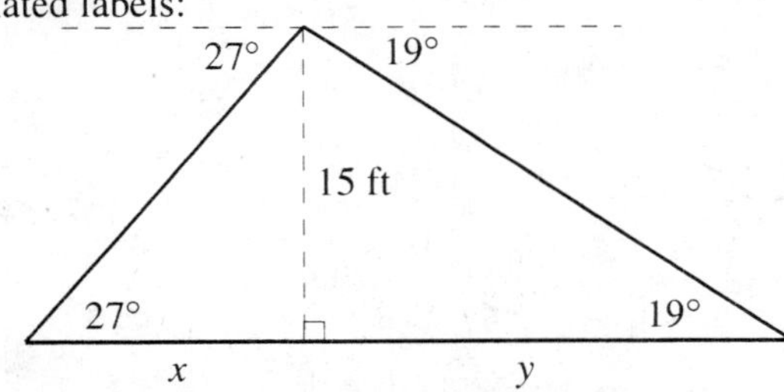

The sum $x + y$ represents the width of the sand pile. Using the two triangles:

$$\tan 27° = \frac{15}{x} \qquad\qquad \tan 19° = \frac{15}{y}$$
$$x \tan 27° = 15 \qquad\qquad y \tan 19° = 15$$
$$x = \frac{15}{\tan 27°} \approx 29.4 \qquad\qquad y = \frac{15}{\tan 19°} \approx 43.6$$

The width of the sand pile is therefore $29.4 + 43.6 = 73.0$ feet.

14.   **a.**   First note that $\frac{5}{8}$ in. $= \frac{5}{8} \cdot 1600 = 1000$ ft, which is the horizontal distance between Stacey and Travis.

b.   There are 8 contour intervals between Stacey and Travis, which corresponds to a vertical distance of $8 \cdot 40 = 320$ ft.

c.   Let θ represent the elevation angle. Construct the triangle:

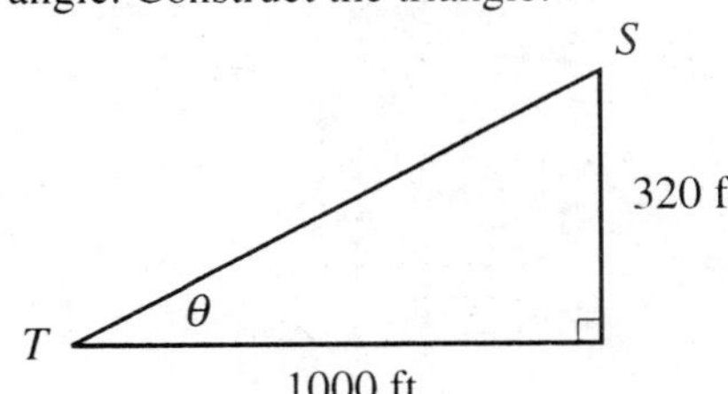

Therefore:
$$\tan \theta = \frac{320}{1000} = 0.32$$
$$\theta = \tan^{-1}(0.32) \approx 17.7°$$
The elevation angle from Travis to Stacey is approximately $17.7°$.

16.   Construct the figure:

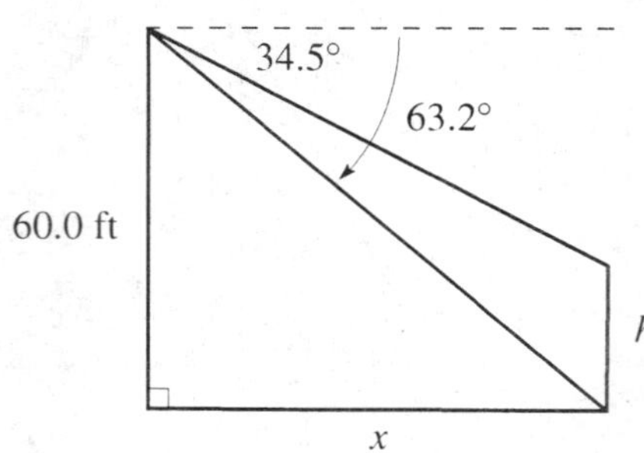

First consider the triangle:

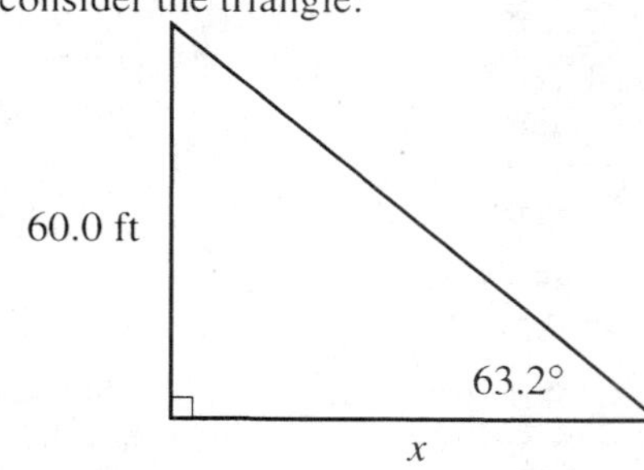

Therefore:
$$\tan 63.2° = \frac{60.0}{x}$$
$$x \tan 63.2° = 60.0$$
$$x = \frac{60.0}{\tan 63.2°} \approx 30.3 \text{ ft}$$

Now consider the triangle:

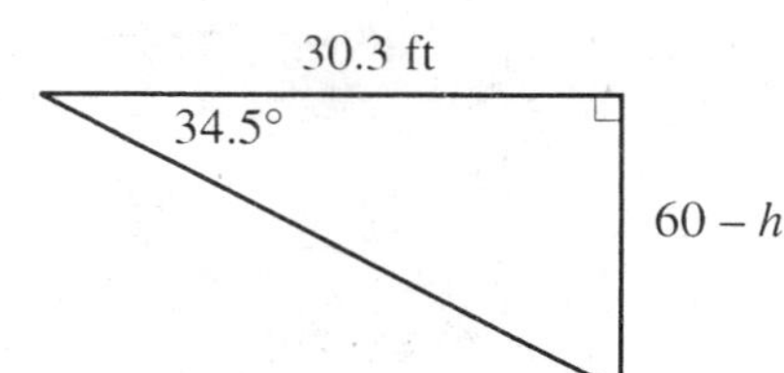

Therefore:
$$\tan 34.5° = \frac{60 - h}{30.3}$$
$$60 - h = 30.3 \tan 34.5°$$
$$h = 60 - 30.3 \tan 34.5° \approx 39.2 \text{ ft}$$

The building next door is approximately 39.2 feet tall.

18.   Construct the figure:

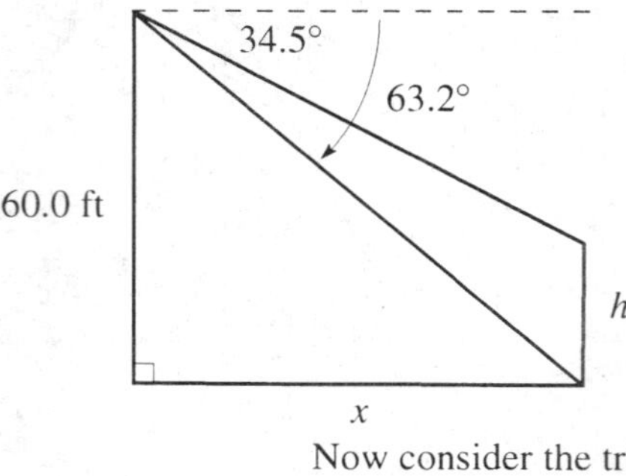

To find his distance from the starting point, use the Pythagorean Theorem:
$$d^2 = 3.5^2 + 2.3^2 = 17.54$$
$$d = \sqrt{17.54} \approx 4.2 \text{ mi}$$
Now find angle θ:
$$\tan \theta = \frac{3.5}{2.3} \approx 1.5217$$
$$\theta = \tan^{-1}(1.5217) \approx 57°$$
His bearing is S 88° W.

20.   Construct the figure:

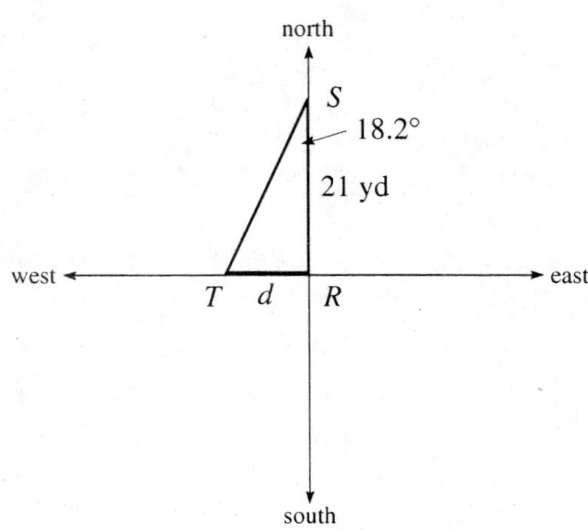

Therefore:

$$\tan 18.2° = \frac{d}{21}$$
$$d = 21 \tan 18.2° \approx 6.9$$

The distance from the tree to the rock is 6.9 yards.

22.   Construct the figure:

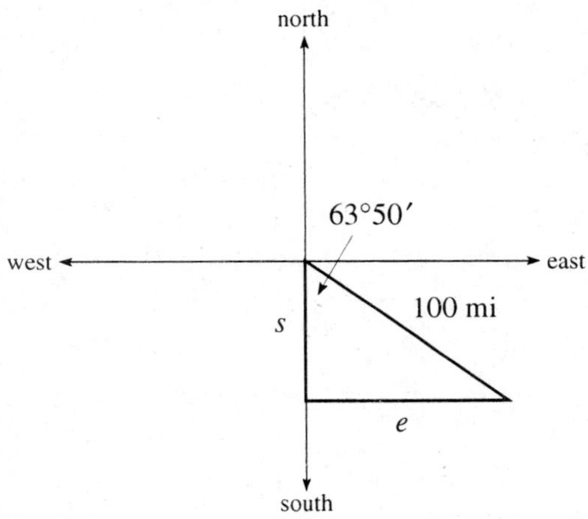

Therefore:

$$\sin 63°50′ = \frac{e}{100} \qquad\qquad \cos 63°50′ = \frac{s}{100}$$
$$e = 100 \sin 63°50′ \approx 89.8 \text{ mi} \qquad s = 100 \cos 63°50′ \approx 44.1 \text{ mi}$$

The boat travels 44.1 mi south and 89.8 mi east.

24. Draw the figure:

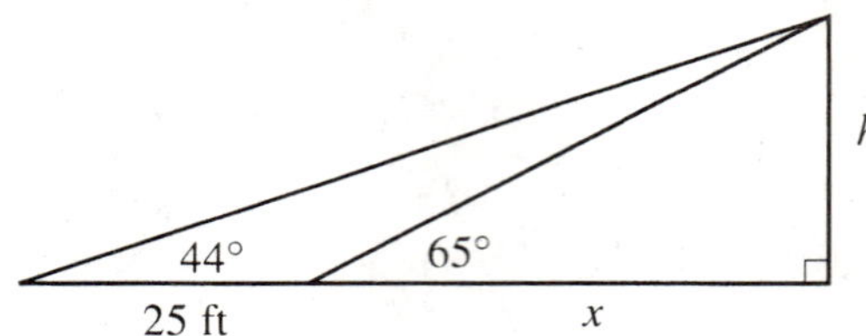

From the smaller triangle:

$$\tan 65° = \frac{h}{x}$$
$$x \tan 65° = h$$
$$x = \frac{h}{\tan 65°}$$

From the larger triangle:

$$\tan 44° = \frac{h}{25 + x}$$
$$(25 + x)\tan 44° = h$$
$$25 + x = \frac{h}{\tan 44°}$$
$$x = \frac{h}{\tan 44°} - 25$$

Setting these two expressions equal:

$$\frac{h}{\tan 65°} = \frac{h}{\tan 44°} - 25$$
$$h\cot 65° = h\cot 44° - 25$$
$$h\cot 65° - h\cot 44° = -25$$
$$h(\cot 65° - \cot 44°) = -25$$
$$h = \frac{25}{\cot 44° - \cot 65°} \approx 44$$

The height of the tree is approximately 44 feet.

26. First find the length CB:

$$\tan 12.3° = \frac{426}{CB}$$
$$CB\tan 12.3° = 426$$
$$CB = \frac{426}{\tan 12.3°} \approx 1954 \text{ ft}$$

Therefore:

$$\sin \angle BCA = \frac{AB}{CB}$$
$$\sin 57.5° = \frac{AB}{1954}$$
$$AB = 1954\sin 57.5° \approx 1648 \text{ ft}$$

A rescue boat at A will have to travel approximately 1648 feet to reach any survivors at point B.

28.   Construct a figure:

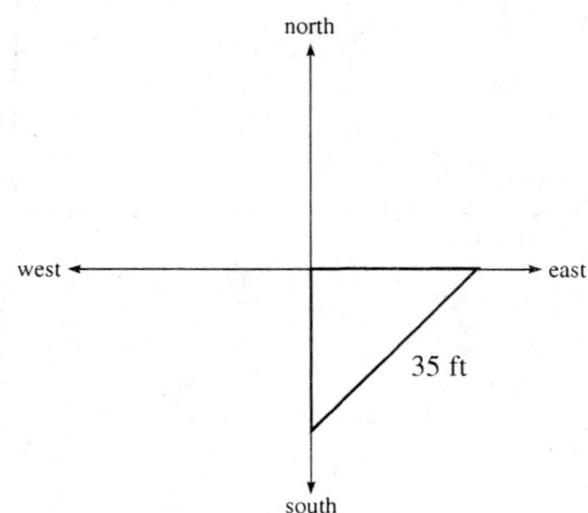

First note that each person is $\dfrac{35}{\sqrt{2}}$ ft ≈ 24.7 ft from the base of the tree. Let h represent the height of the tree. Now construct the triangle:

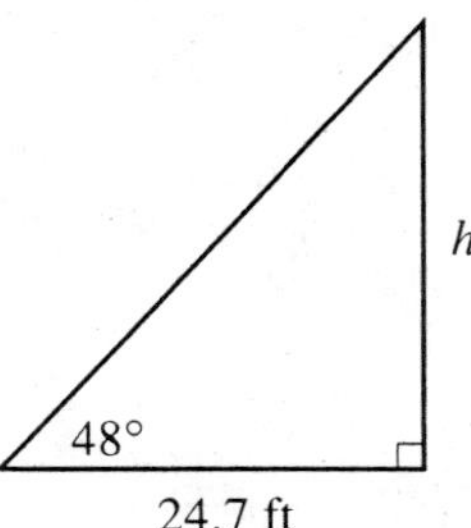

Therefore:
$$\tan 48° = \frac{h}{24.7}$$
$$h = 24.7 \tan 48° \approx 27 \text{ ft}$$
The tree is approximately 27 feet tall.

30.   Construct the figure (not drawn to scale):

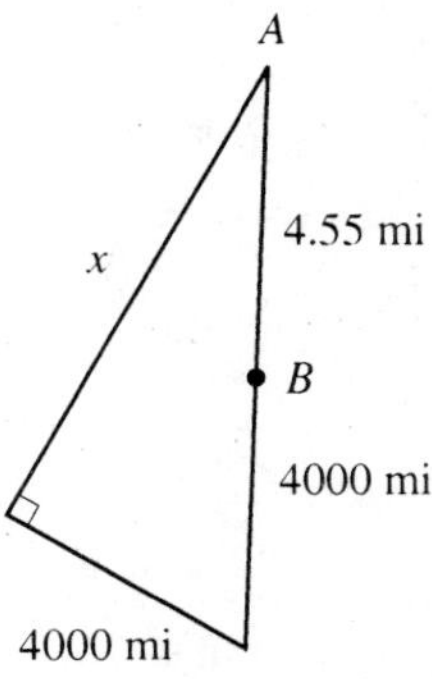

Using the Pythagorean Theorem:
$$x^2 + 4000^2 = 4004.55^2$$
$$x^2 + 16000000 = 16036420.7$$
$$x^2 = 36420.7$$
$$x \approx 191$$
The plane is 191 miles from the horizon. Now find angle A:
$$\sin A = \frac{4000}{4004.55} \approx 0.9989$$
$$A = \sin^{-1}(0.9989) \approx 87.3°$$

32.   Construct the figure:

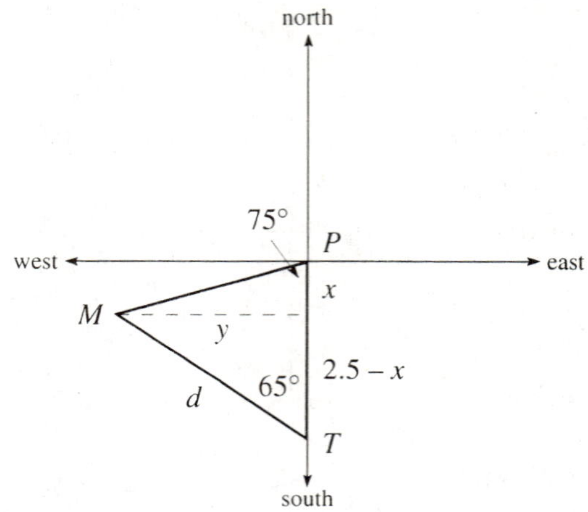

Consider the two triangles:

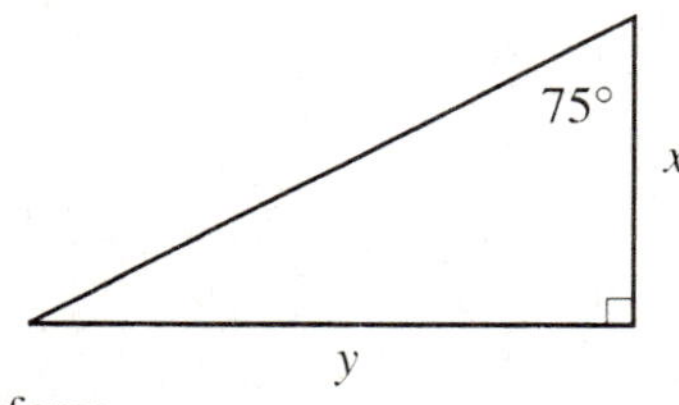

Therefore:

$$\tan 75° = \frac{y}{x}$$
$$y = x\tan 75°$$

$$\tan 65° = \frac{y}{2.5 - x}$$
$$y = (2.5 - x)\tan 65°$$

Setting these two expressions equal:

$$(2.5 - x)\tan 65° = x\tan 75°$$
$$2.5\tan 65° - x\tan 65° = x\tan 75°$$
$$2.5\tan 65° = x\tan 75° + x\tan 65°$$
$$2.5\tan 65° = x(\tan 75° + \tan 65°)$$
$$x = \frac{2.5\tan 65°}{\tan 75° + \tan 65°} \approx 0.91$$

Therefore:

$$\cos 65° = \frac{2.5 - x}{d} = \frac{2.5 - 0.91}{d} = \frac{1.59}{d}$$
$$d\cos 65° = 1.59$$
$$d = \frac{1.59}{\cos 65°} \approx 3.8 \text{ mi}$$

Tim is approximately 3.8 miles from the missile when it is launched.

34.   Note that $\sin\theta_1 = \dfrac{1}{\sqrt{2}}, \sin\theta_2 = \dfrac{1}{\sqrt{3}}, \sin\theta_3 = \dfrac{1}{\sqrt{4}}, \ldots$, thus $\sin\theta_n = \dfrac{1}{\sqrt{n+1}}$.

36.   Simplifying: $\dfrac{1}{\cos\theta} - \cos\theta = \dfrac{1}{\cos\theta} - \cos\theta \cdot \dfrac{\cos\theta}{\cos\theta} = \dfrac{1 - \cos^2\theta}{\cos\theta} = \dfrac{\sin^2\theta}{\cos\theta}$

38.   Working from the left side: $\cos\theta\csc\theta\tan\theta = \cos\theta \cdot \dfrac{1}{\sin\theta} \cdot \dfrac{\sin\theta}{\cos\theta} = \dfrac{\sin\theta\cos\theta}{\sin\theta\cos\theta} = 1$

40.   Working from the left side: $(1 - \cos\theta)(1 + \cos\theta) = 1 - \cos\theta + \cos\theta - \cos^2\theta = 1 - \cos^2\theta = \sin^2\theta$

42.   Working from the left side: $1 - \dfrac{\cos\theta}{\sec\theta} = 1 - \dfrac{\cos\theta}{1/\cos\theta} = 1 - \cos^2\theta = \sin^2\theta$

44.   **a.**   Let x represent the height that is illuminated on the floor. Then:

$$\tan 83.5° = \frac{4}{x}$$

$$x = \frac{4}{\tan 83.5°} \approx 0.46$$

The illuminated area is then: $(0.46)(6.5) \approx 2.96 \text{ ft}^2$.

   b.   Following the procedure from part **a**:

$$\tan 36.5° = \frac{4}{x}$$

$$x = \frac{4}{\tan 36.5°} \approx 5.41$$

The illuminated area is then: $(5.41)(6.5) \approx 35.14 \text{ ft}^2$. The area is much larger on the winter day.

2.5 Vectors: A Geometric Approach

2.   Sketching the vector:

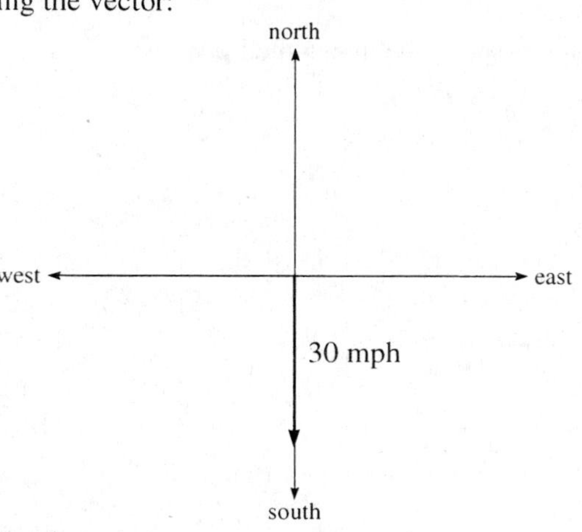

4.   Sketching the vector:

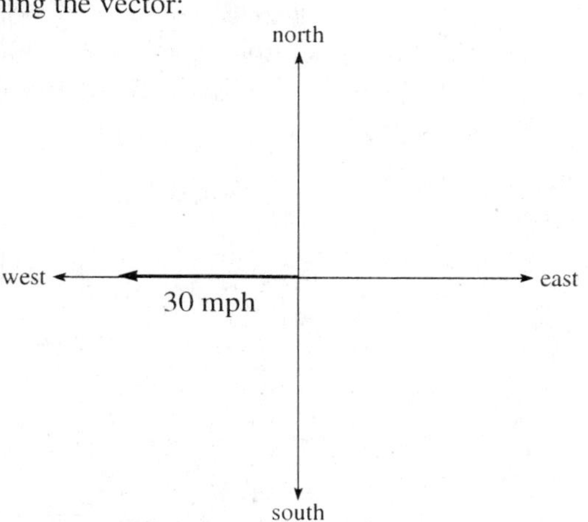

6.   Sketching the vector:

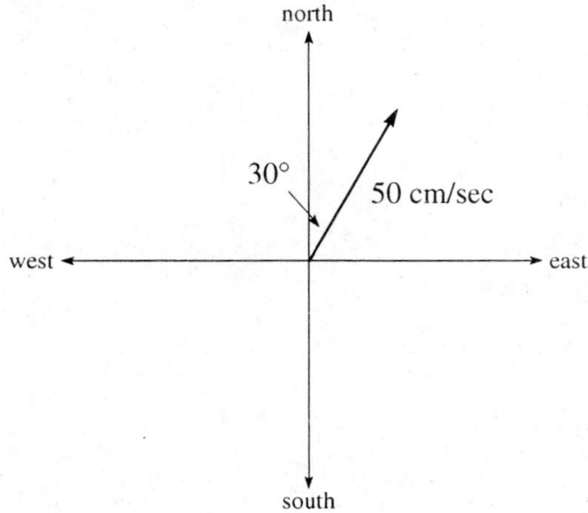

8.   Sketching the vector:

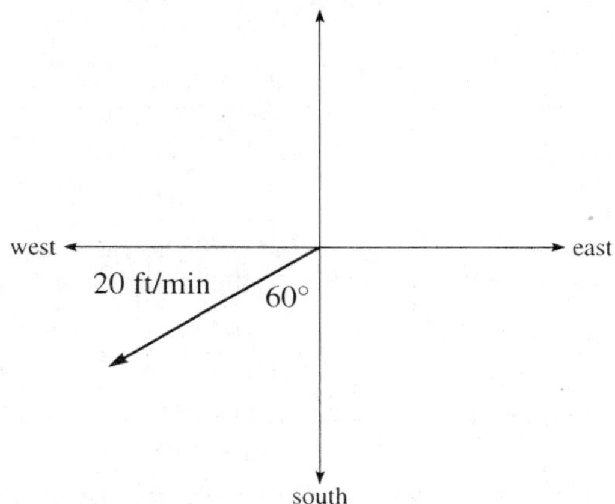

10. Construct the figure for their position after 2 hours (multiply their rates by 2):

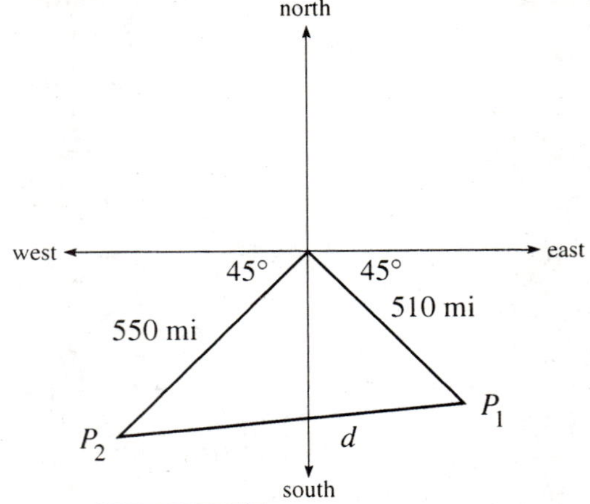

Using the Pythagorean Theorem: $d = \sqrt{550^2 + 510^2} \approx 750$ miles

To find the bearing from P_1 to P_2, first find the vertical (north-south) change in their positions. This is given by:

$$550 \sin 45° - 510 \sin 45° \approx 28.28 \text{ miles}$$

Construct the triangle:

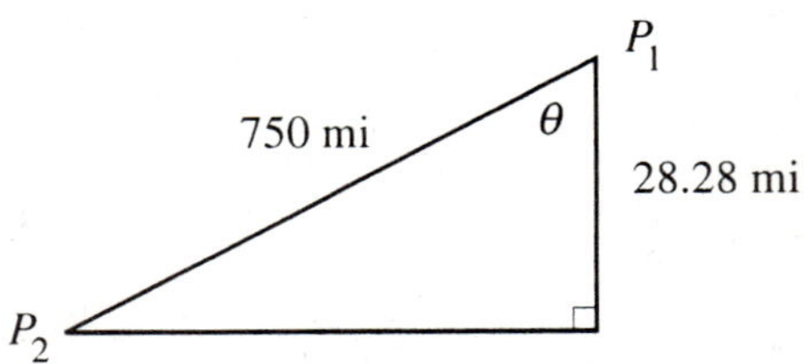

Therefore:

$$\cos \theta = \frac{28.28}{750} \approx 0.0377$$

$$\theta = \cos^{-1}(0.0377) \approx 87.8°$$

The bearing from P_1 to P_2 is S 88° W.

12. Construct the triangle after $t = 2$ hours:

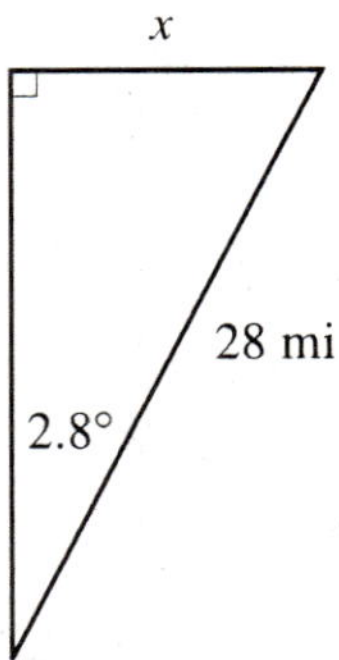

Therefore:

$$\sin 2.8° = \frac{x}{28}$$

$$x = 28 \sin 2.8° \approx 1.4 \text{ miles}$$

The ship will be approximately 1.4 miles off course after 2 hours.

14. Computing the magnitudes of $\mathbf{V}_x$ and $\mathbf{V}_y$:

$$\left|\mathbf{V}_x\right| = 17.6\cos 67.2° \approx 6.8$$

$$\left|\mathbf{V}_y\right| = 17.6\sin 67.2° \approx 16.2$$

16. Computing the magnitudes of $\mathbf{V}_x$ and $\mathbf{V}_y$:

$$\left|\mathbf{V}_x\right| = 380\cos 16°40' \approx 364$$

$$\left|\mathbf{V}_y\right| = 380\sin 16°40' \approx 109$$

18. Computing the magnitudes of $\mathbf{V}_x$ and $\mathbf{V}_y$:

$$\left|\mathbf{V}_x\right| = 48\cos 90° = 0 \qquad \left|\mathbf{V}_y\right| = 48\sin 90° = 48$$

20. Using the Pythagorean Theorem: $\left|\mathbf{V}\right| = \sqrt{45.0^2 + 15.0^2} \approx 47.4$

22. Using the Pythagorean Theorem: $\left|\mathbf{V}\right| = \sqrt{2.2^2 + 5.8^2} \approx 6.2$

24. Computing the magnitudes of $\mathbf{V}_x$ and $\mathbf{V}_y$:

$$\left|\mathbf{V}_x\right| = 1800\cos 60° = 900\tfrac{\text{ft}}{\text{sec}} \qquad \left|\mathbf{V}_y\right| = 1800\sin 60° \approx 1600\tfrac{\text{ft}}{\text{sec}}$$

The horizontal vector is $\langle 900,0\rangle$ and the vertical vector is $\langle 0,1559\rangle$.

26. The horizontal distance traveled is $2 \cdot 900 = 1800$ feet.

28. Using the Pythagorean Theorem: $\left|\mathbf{V}\right| = \sqrt{15.0^2 + 25.0^2} \approx 29.2$ ft/sec

30. Draw the figure corresponding to $t = 3$ hours:

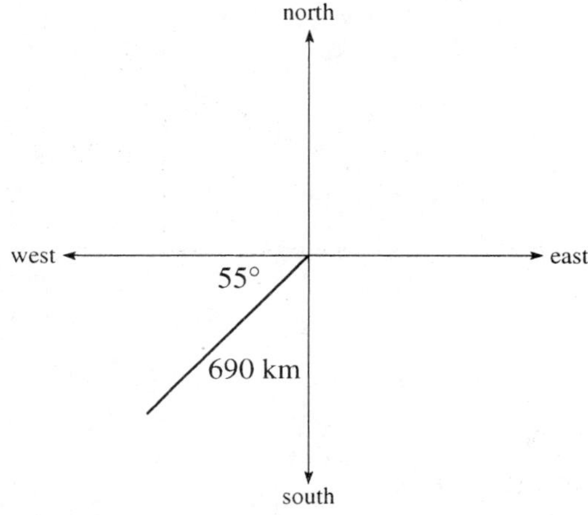

The west and south distances are given by:

west: $690\cos 55° \approx 400$ km   south: $690\sin 55° \approx 570$ km

32. Construct the figure:

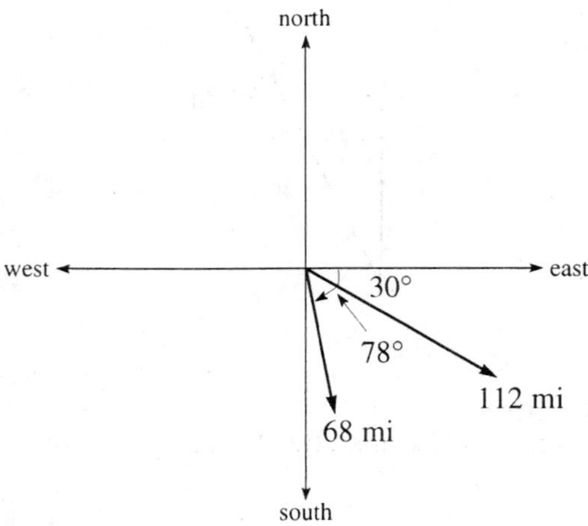

The total distance south and east is given by:

east: $68\cos 78° + 112\cos 30° \approx 110$ miles   south: $68\sin 78° + 112\sin 30° \approx 120$ miles

34.   The corresponding force diagram would be:

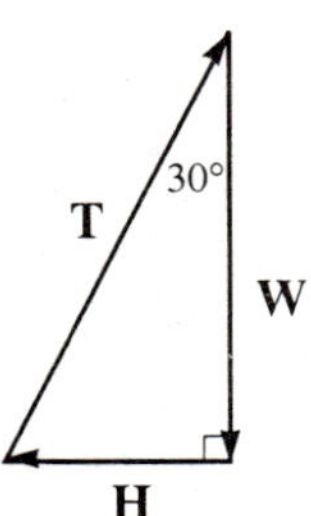

36.   The corresponding force diagram would be:

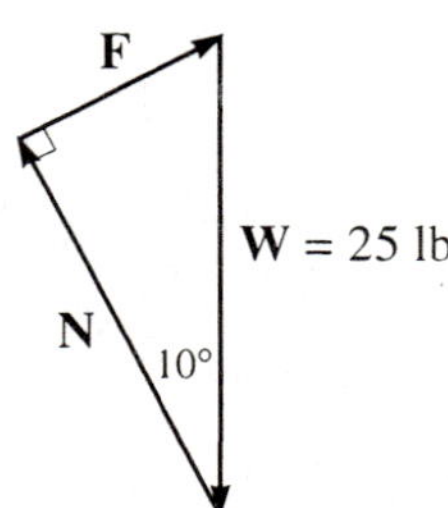

Therefore:

$$\sin 10° = \frac{|\mathbf{F}|}{25.0} \qquad\qquad \cos 10° = \frac{|\mathbf{N}|}{25.0}$$

$$|\mathbf{F}| = 25.0\sin 10° \approx 4.34 \text{ lb} \qquad\qquad |\mathbf{N}| = 25.0\cos 10° \approx 24.6 \text{ lb}$$

38.   The corresponding force diagram would be:

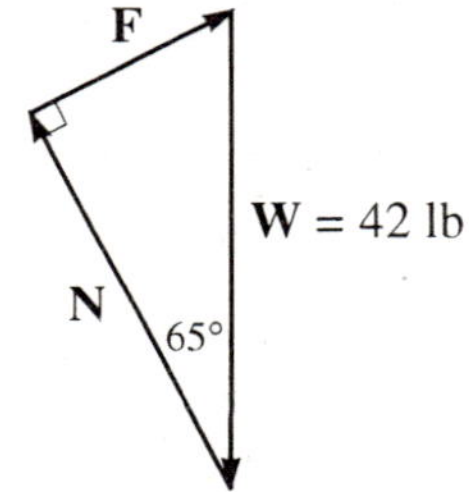

Therefore:

$$\sin 65° = \frac{|\mathbf{F}|}{42}$$

$$|\mathbf{F}| = 42\sin 65° \approx 38.1 \text{ lb}$$

She must pull with a magnitude of 38.1 lb.

40.   The horizontal portion of the force is given by: $|\mathbf{F}_x| = |\mathbf{F}|\cos 25° = 15\cos 25°$ lb

The work is then given by: Work $= (15\cos 25°)(50) \approx 680$ ft-lb

42.   The horizontal portion of the force is given by: $|\mathbf{F}_x| = |\mathbf{F}|\cos 30° = 25\cos 30°$ lb

The work is then given by: Work $= (25\cos 30°)(300) \approx 6500$ ft-lb

44.   Drawing the angle in standard position:

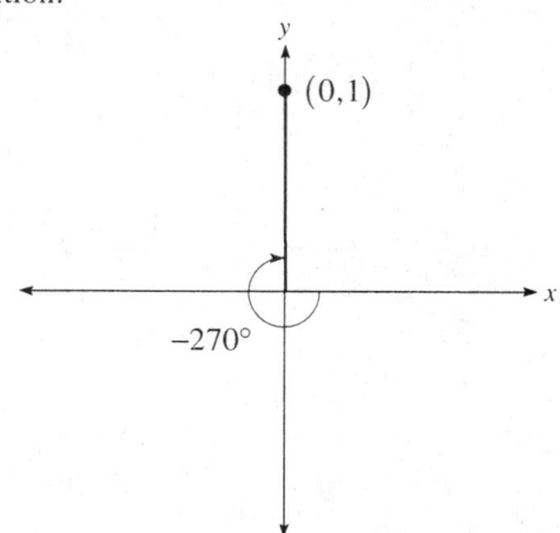

Since $r = 1$, $\sin(-270°) = 1$, $\cos(-270°) = 0$, and $\tan(-270°)$ is undefined.

46.   Choose $(-1,1)$ as a point on the terminal side of θ. Then $r = \sqrt{(-1)^2 + 1^2} = \sqrt{2}$. Therefore:

$$\sin\theta = \frac{y}{r} = \frac{1}{\sqrt{2}} \qquad\qquad \cos\theta = \frac{x}{r} = -\frac{1}{\sqrt{2}}$$

48.   Since $\cos\theta = \dfrac{x}{r} = -\dfrac{3}{5} = -\dfrac{6}{10}$, choose $x = -6$ and $r = 10$. Now find y:

$$(-6)^2 + y^2 = 10^2$$
$$36 + y^2 = 100$$
$$y^2 = 64$$
$$y = \pm 8$$

Chapter 2 Test

1.   First draw the triangle:

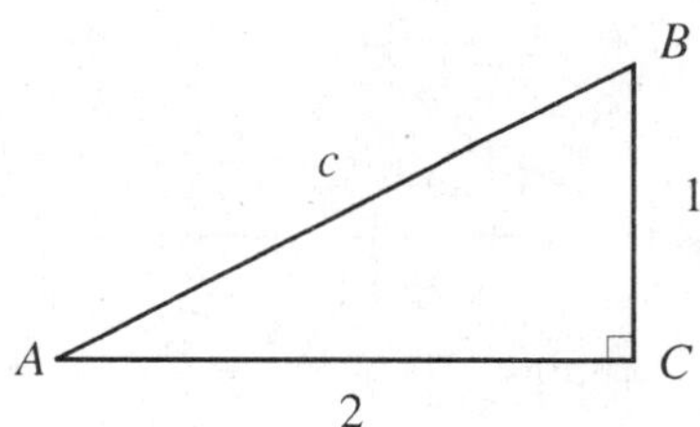

Note that $c = \sqrt{2^2 + 1^2} = \sqrt{5}$. Therefore:

$$\sin A = \frac{1}{\sqrt{5}} \qquad\qquad \cos A = \frac{2}{\sqrt{5}} \qquad\qquad \tan A = \frac{1}{2}$$

$$\sin B = \cos A = \frac{2}{\sqrt{5}} \qquad\qquad \cos B = \sin A = \frac{1}{\sqrt{5}} \qquad\qquad \tan B = \cot A = \frac{2}{1} = 2$$

2.   First draw the triangle:

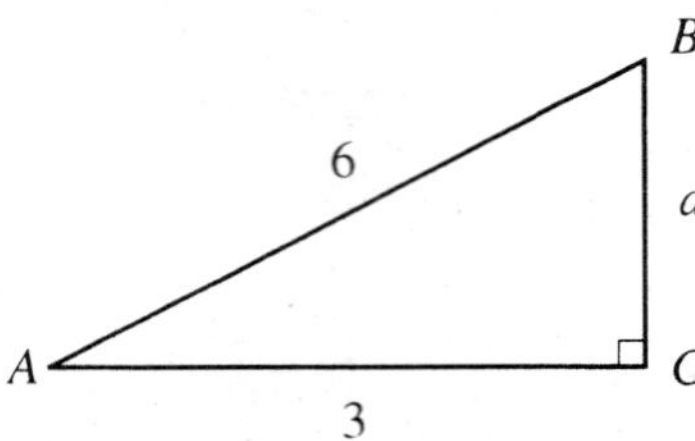

Note that $a = \sqrt{6^2 - 3^2} = \sqrt{27} = 3\sqrt{3}$. Therefore:

$$\sin A = \frac{3\sqrt{3}}{6} = \frac{\sqrt{3}}{2} \qquad\qquad \cos A = \frac{3}{6} = \frac{1}{2} \qquad\qquad \tan A = \frac{3\sqrt{3}}{3} = \sqrt{3}$$

$$\sin B = \cos A = \frac{1}{2} \qquad\qquad \cos B = \sin A = \frac{\sqrt{3}}{2} \qquad\qquad \tan B = \cot A = \frac{3}{3\sqrt{3}} = \frac{1}{\sqrt{3}}$$

3.   First draw the triangle:

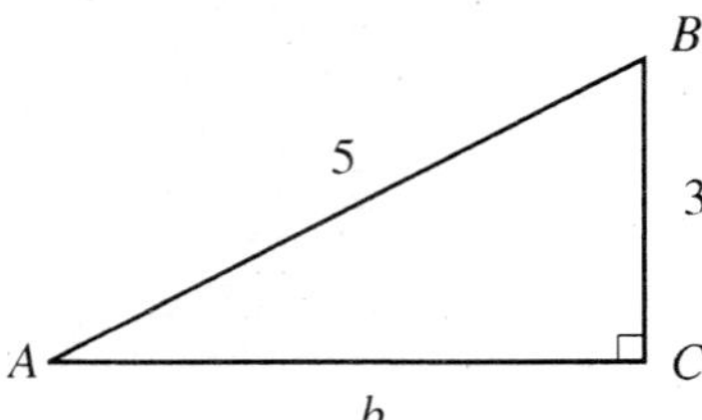

Note that $b = \sqrt{5^2 - 3^2} = \sqrt{16} = 4$. Therefore:

$$\sin A = \frac{3}{5} \qquad\qquad \cos A = \frac{4}{5} \qquad\qquad \tan A = \frac{3}{4}$$

$$\sin B = \cos A = \frac{4}{5} \qquad\qquad \cos B = \sin A = \frac{3}{5} \qquad\qquad \tan B = \cot A = \frac{4}{3}$$

4.   First draw the triangle:

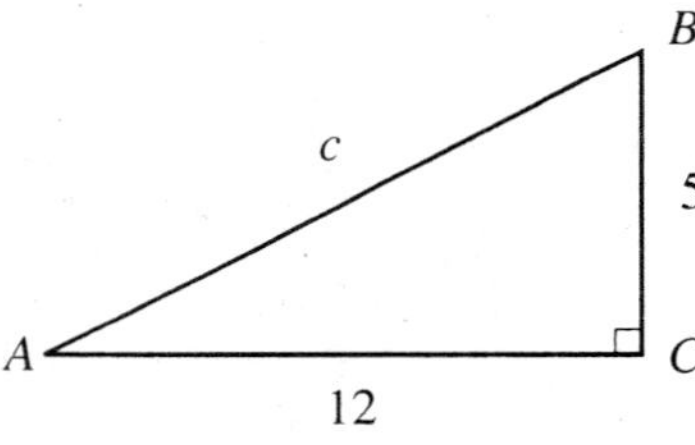

Note that $c = \sqrt{12^2 + 5^2} = \sqrt{169} = 13$. Therefore:

$$\sin A = \frac{5}{13} \qquad\qquad \cos A = \frac{12}{13} \qquad\qquad \tan A = \frac{5}{12}$$

$$\sin B = \cos A = \frac{12}{13} \qquad\qquad \cos B = \sin A = \frac{5}{13} \qquad\qquad \tan B = \cot A = \frac{12}{5}$$

5. $\sin 14° = \cos 76°$ **6.** $\sec 17° = \csc 73°$

7.   Simplifying: $\sin^2 45° + \cos^2 30° = \left(\dfrac{1}{\sqrt{2}}\right)^2 + \left(\dfrac{\sqrt{3}}{2}\right)^2 = \frac{1}{2} + \frac{3}{4} = \frac{5}{4}$

8.   Simplifying: $\tan 45° + \cot 45° = 1 + 1 = 2$

9.   Simplifying: $\sin^2 60° - \cos^2 30° = \left(\dfrac{\sqrt{3}}{2}\right)^2 - \left(\dfrac{\sqrt{3}}{2}\right)^2 = \frac{3}{4} - \frac{3}{4} = 0$

10.   Simplifying: $\dfrac{1}{\sec 30°} = \cos 30° = \dfrac{\sqrt{3}}{2}$ **11.**   Adding: $48°18' + 24°52' = 72°70' = 73°10'$

12.   Subtracting: $25°15' - 15°32' = 24°75' - 15°32' = 9°43'$

13.   Converting to degrees and minutes: $73.2° = 73° + 0.2° = 73° + 0.2(60') = 73°12'$

14.   Converting to degrees and minutes: $16.45° = 16° + 0.45° = 16° + 0.45(60') = 16°27'$

15. Converting to decimal degrees: $2°48' = 2° + 48' = 2° + \left(\frac{48}{60}\right)° = 2.8°$

16. Converting to decimal degrees: $79°30' = 79° + 30' = 79° + \left(\frac{30}{60}\right)° = 79.5°$

17. Calculating the value: $\sin 24°20' = \sin\left(24\frac{1}{3}\right)° \approx 0.4120$

18. Calculating the value: $\cos 37.8° \approx 0.7902$ **19.** Calculating the value: $\tan 63°50' = \tan\left(63\frac{5}{6}\right)° \approx 2.0353$

20. Calculating the value: $\cot 71°20' = \cot\left(71\frac{1}{3}\right)° = \dfrac{1}{\tan\left(71\frac{1}{3}\right)°} \approx 0.3378$

21. Since $\tan\theta = 0.0816$, $\theta = \tan^{-1}(0.0816) \approx 4.7°$.

22. Since $\sec\theta = 1.923$, $\cos\theta = \dfrac{1}{1.923}$, so $\theta = \cos^{-1}\left(\dfrac{1}{1.923}\right) \approx 58.7°$.

23. Since $\sin\theta = 0.9465$, $\theta = \sin^{-1}(0.9465) \approx 71.2°$.   **24.** Since $\cos\theta = 0.9730$, $\theta = \cos^{-1}(0.9730) \approx 13.3°$.

25. 49.35 has four significant digits.   **26.** 0.0028 has two significant digits.

27. First sketch the triangle:

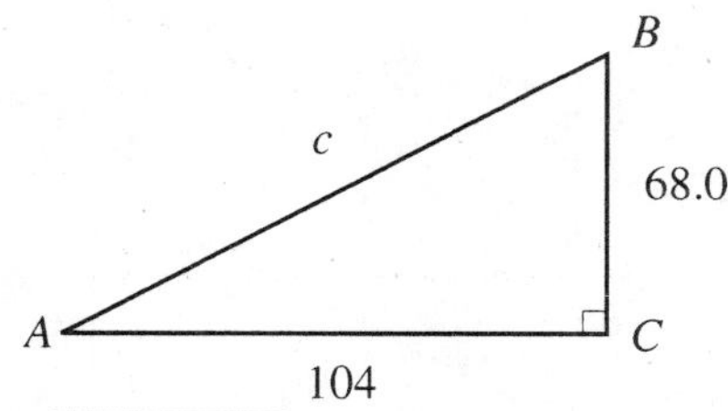

Using the Pythagorean Theorem: $c = \sqrt{104^2 + 68^2} \approx 124$. Therefore:

$\tan A = \frac{68}{104} \approx 0.6538$

$A = \tan^{-1}(0.6538) \approx 33.2°$
$B = 90° - 33.2° = 56.8°$

28. First sketch the triangle:

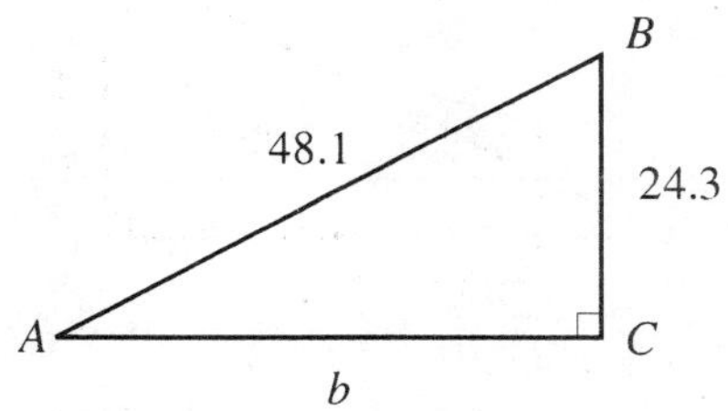

Using the Pythagorean Theorem: $b = \sqrt{48.1^2 - 24.3^2} \approx 41.5$. Therefore:

$\sin A = \dfrac{24.3}{48.1} \approx 0.5052$

$A = \sin^{-1}(0.5052) \approx 30.3°$
$B = 90° - 30.3° = 59.7°$

29.   First sketch the triangle:

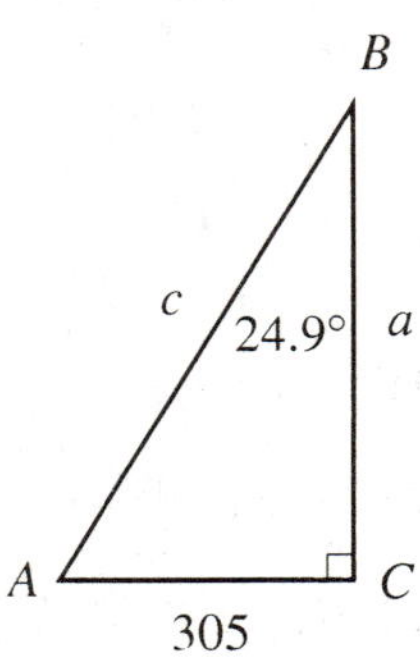

Note that $A = 90° - 24.9° = 65.1°$. Therefore:

$$\tan 65.1° = \frac{a}{305}$$
$$a = 305 \tan 65.1° \approx 657$$

$$\cos 65.1° = \frac{305}{c}$$
$$c \cos 65.1° = 305$$
$$c = \frac{305}{\cos 65.1°} \approx 724$$

30.   First sketch the triangle:

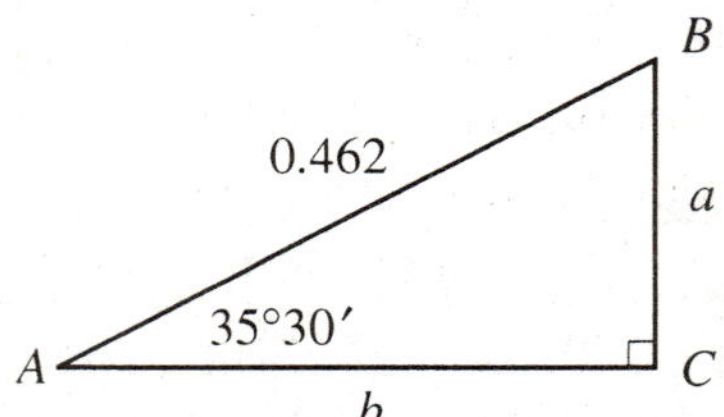

Note that $B = 90° - 35°30' = 89°60' - 35°30' = 54°30'$. Also:

$$\sin 35.5° = \frac{a}{0.462}$$
$$a = 0.462 \sin 35.5° \approx 0.268$$

$$\cos 35.5° = \frac{b}{0.462}$$
$$b = 0.462 \cos 35.5° \approx 0.376$$

31.   First sketch the triangle:

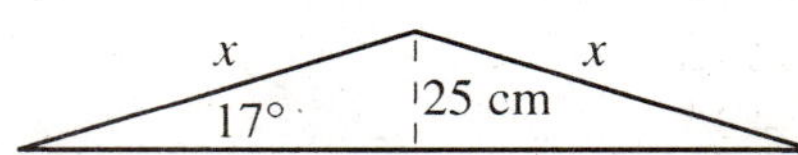

Therefore:

$$\sin 17° = \frac{25}{x}$$
$$x \sin 17° = 25$$
$$x = \frac{25}{\sin 17°} \approx 86 \text{ cm}$$

32.   Sketch the figure:

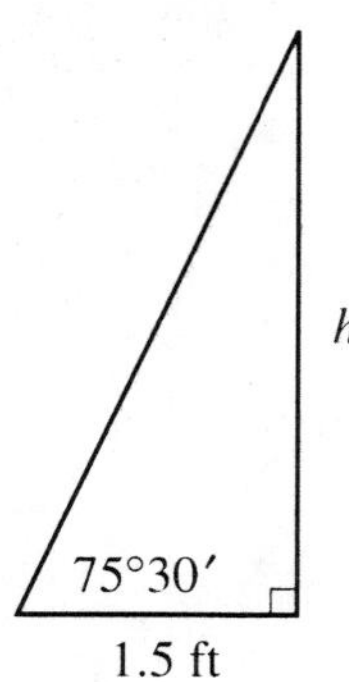

Therefore:

$$\tan 75.5° = \frac{h}{1.5}$$
$$h = 1.5 \tan 75.5° \approx 5.8$$

The post is approximately 5.8 feet tall.

33.   Draw the figure:

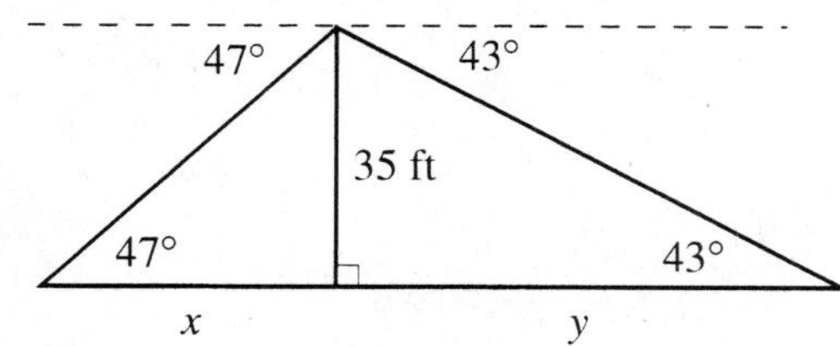

Therefore:

$$\tan 47° = \frac{35}{x}$$
$$x \tan 47° = 35$$
$$x = \frac{35}{\tan 47°} \approx 32.64 \text{ feet}$$

$$\tan 43° = \frac{35}{y}$$
$$y \tan 43° = 35$$
$$y = \frac{35}{\tan 43°} \approx 37.53 \text{ feet}$$

The stakes are $32.6 + 37.5 \approx 70$ feet apart.

34.   The magnitudes are given by:

$$\left| \mathbf{V}_x \right| = 5.0 \cos 30° \approx 4.3$$

$$\left| \mathbf{V}_y \right| = 5.0 \sin 30° = 2.5$$

35.   Let θ represent the required angle. Then:

$$\tan \theta = \tfrac{31}{11} \approx 2.8182$$
$$\theta = \tan^{-1}(2.8182) \approx 70°$$

36.   The magnitudes are given by:

$$\left| \mathbf{V}_x \right| = 800 \cos 62° \approx 380 \text{ ft / sec}$$

$$\left| \mathbf{V}_y \right| = 800 \sin 62° \approx 710 \text{ ft / sec}$$

37.   Draw the figure:

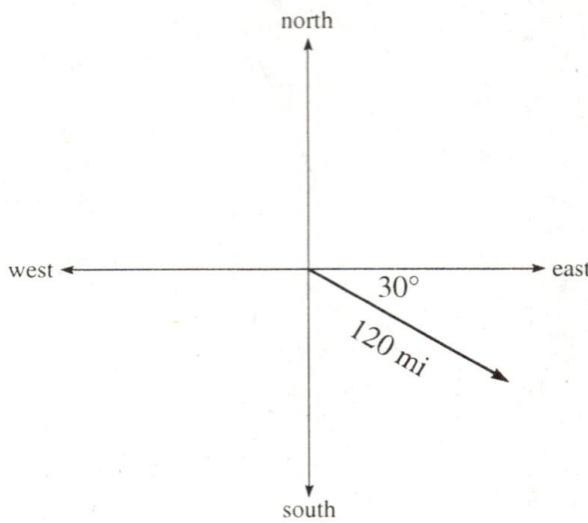

Therefore:

east: $120\cos 30° \approx 100$ miles   south: $120\sin 30° = 60$ miles

38.   Drawing the figure:

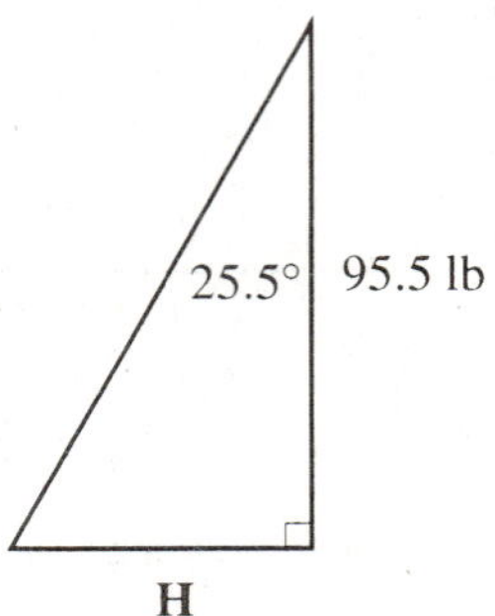

Now find the magnitude of **H**:

$$\tan 25.5° = \frac{|\mathbf{H}|}{95.5}$$

$$|\mathbf{H}| = 95.5\tan 25.5° \approx 45.6$$

Kelly must push horizontally with a force of 45.6 lb.

39.   The corresponding force diagram would be:

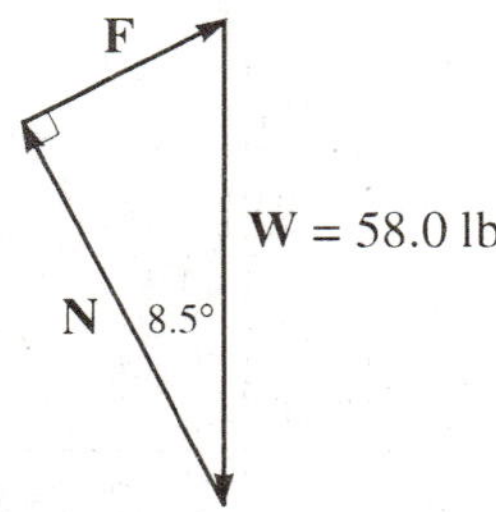

Now find the magnitude of **F**:

$$\sin 8.5° = \frac{|\mathbf{F}|}{58.0}$$

$$|\mathbf{F}| = 58.0\sin 8.5° \approx 8.57 \text{ lb}$$

Tyler must push with a force of 8.57 lb.

40.   The horizontal portion of the force is given by: $\left|\mathbf{F}_x\right| = |\mathbf{F}|\cos 40° = 35\cos 40°$ lb

The work is then given by: Work $= (35\cos 40°)(80) \approx 2100$ ft-lb

Chapter 3
Radian Measure

3.1 Reference Angle

2. The reference angle for $150°$ is $180° - 150° = 30°$:

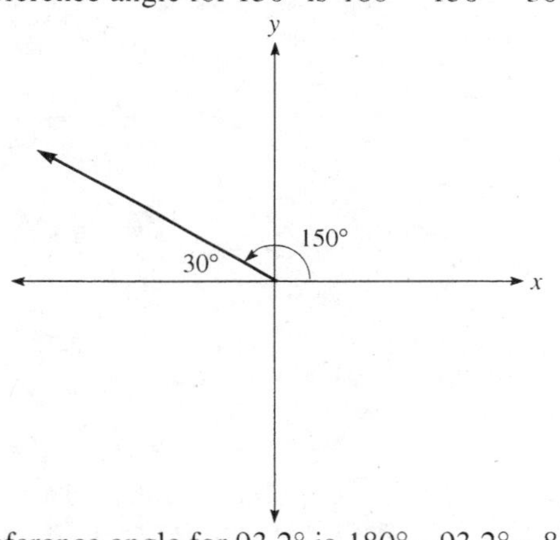

4. The reference angle for $253.8°$ is $253.8° - 180° = 73.8°$:

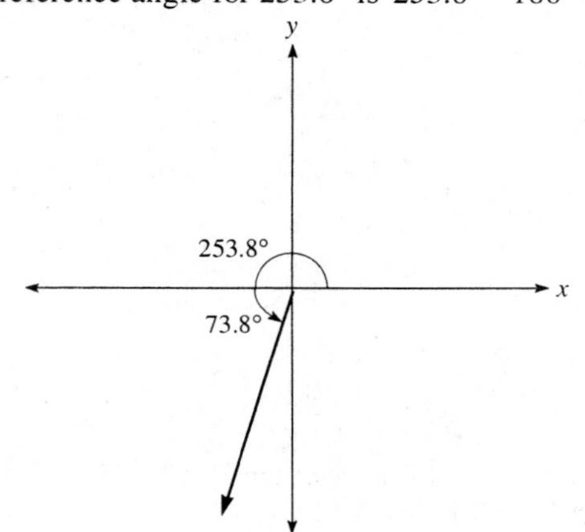

6. The reference angle for $93.2°$ is $180° - 93.2° = 86.8°$:

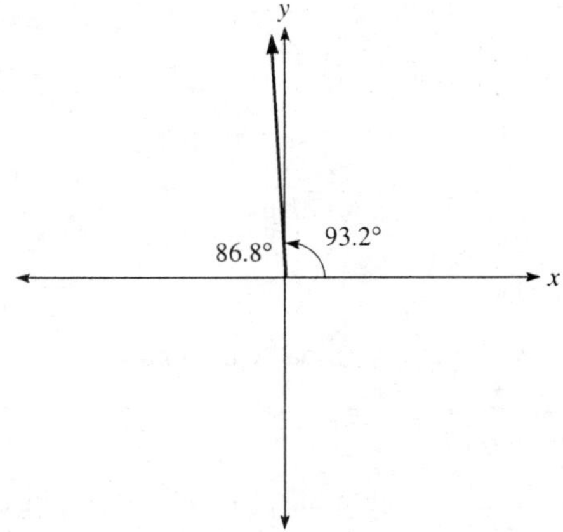

8.   The reference angle for $171°40'$ is $180° - 171°40' = 8°20'$:

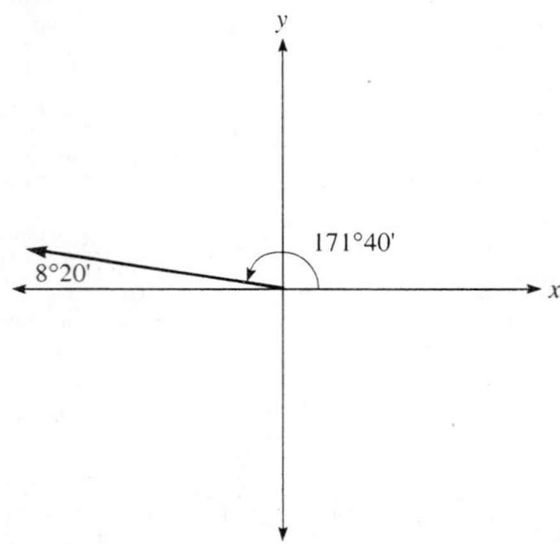

10.   The reference angle for $-330°$ is $-330° + 360° = 30°$:   **12.**   The reference angle for $-150°$ is $-150° + 180° = 30°$:

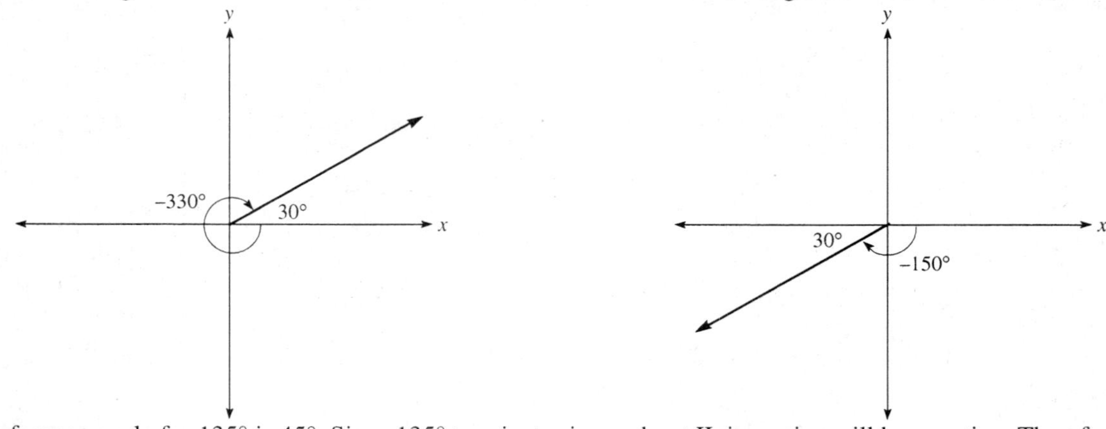

14.   The reference angle for $135°$ is $45°$. Since $135°$ terminates in quadrant II, its cosine will be negative. Therefore:
$$\cos 135° = -\cos 45° = -\frac{1}{\sqrt{2}}$$

16.   The reference angle for $210°$ is $30°$. Since $210°$ terminates in quadrant III, its sine will be negative. Therefore:
$$\sin 210° = -\sin 30° = -\frac{1}{2}$$

18.   The reference angle for $315°$ is $45°$. Since $315°$ terminates in quadrant IV, its tangent will be negative. Therefore:
$$\tan 315° = -\tan 45° = -1$$

20.   The reference angle for $150°$ is $30°$. Since $150°$ terminates in quadrant II, its cosine will be negative. Therefore:
$$\cos 150° = -\cos 30° = -\frac{\sqrt{3}}{2}$$

22.   The reference angle for $330°$ is $30°$. Since $330°$ terminates in quadrant IV, its secant is positive. Therefore:
$$\sec 330° = \sec 30° = \frac{1}{\cos 30°} = \frac{1}{\sqrt{3}/2} = \frac{2}{\sqrt{3}}$$

24.   The reference angle for $300°$ is $60°$. Since $300°$ terminates in quadrant IV, its cosecant is negative. Therefore:
$$\csc 300° = -\csc 60° = -\frac{1}{\sin 60°} = -\frac{1}{\sqrt{3}/2} = -\frac{2}{\sqrt{3}}$$

26.   The reference angle for $420°$ is $60°$. Since $420°$ terminates in quadrant I, its cosine is positive. Therefore:
$$\cos 420° = \cos 60° = \frac{1}{2}$$

28. The reference angle for $510°$ is $30°$. Since $510°$ terminates in quadrant II, its cotangent is negative. Therefore:

$$\cot 510° = -\cot 30° = -\frac{\cos 30°}{\sin 30°} = -\frac{\sqrt{3}/2}{1/2} = -\sqrt{3}$$

30. Calculating the value: $\cos 238° \approx -0.5299$

32. Calculating the value: $\csc 166.7° = \dfrac{1}{\sin 166.7°} \approx 4.3469$

34. Calculating the value: $\tan 253.8° \approx 3.4420$

36. Calculating the value: $\csc 93.2° = \dfrac{1}{\sin 93.2°} \approx 1.0016$

38. Calculating the value: $\cot 420° = \dfrac{1}{\tan 420°} \approx 0.5774$

40. Calculating the value: $\sec 590.9° = \dfrac{1}{\cos 590.9°} \approx -1.5856$

42. Calculating the value: $\sin(-225°) \approx 0.7071$

44. Calculating the value: $\tan 171°40' = \tan\left(171\frac{2}{3}\right)° \approx -0.1465$

46. Calculating the value: $\csc 670°20' = \csc\left(670\frac{1}{3}\right)° = \dfrac{1}{\sin\left(670\frac{1}{3}\right)°} \approx -1.3118$

48. Calculating the value: $\cos(-150°) \approx -0.8660$

50. The reference angle is given by $\hat{\theta} = \sin^{-1}(0.3090) \approx 18.0°$. Since θ terminates in quadrant IV, $\theta = 360° - 18.0° = 342.0°$.

52. The reference angle is given by $\hat{\theta} = \cos^{-1}(0.7660) \approx 40.0°$. Since θ terminates in quadrant III, $\theta = 180° + 40.0° = 220.0°$.

54. The reference angle is given by $\hat{\theta} = \tan^{-1}(0.5890) \approx 30.5°$. Since θ terminates in quadrant I, $\theta = 30.5°$.

56. The reference angle is given by $\hat{\theta} = \cos^{-1}(0.2644) \approx 74.7°$. Since θ terminates in quadrant IV, $\theta = 360° - 74.7° = 285.3°$.

58. The reference angle is given by $\hat{\theta} = \sin^{-1}(0.9652) \approx 74.8°$. Since θ terminates in quadrant I, $\theta = 74.8°$.

60. Since $\csc\theta = 1.4325$, $\sin\theta = \dfrac{1}{1.4325}$. The reference angle is given by $\hat{\theta} = \sin^{-1}\left(\dfrac{1}{1.4325}\right) \approx 44.3°$. Since θ terminates in quadrant II, $\theta = 180° - 44.3° = 135.7°$.

62. Since $\sec\theta = -3.4159$, $\cos\theta = -\dfrac{1}{3.4159}$. The reference angle is given by $\hat{\theta} = \cos^{-1}\left(\dfrac{1}{3.4159}\right) \approx 73.0°$. Since θ terminates in quadrant II, $\theta = 180° - 73.0° = 107.0°$.

64. Since $\cot\theta = -0.1234$, $\tan\theta = -\dfrac{1}{0.1234}$. The reference angle is given by $\hat{\theta} = \tan^{-1}\left(\dfrac{1}{0.1234}\right) \approx 83.0°$. Since θ terminates in quadrant IV, $\theta = 360° - 83.0° = 277.0°$.

66. Since $\csc\theta = -1.7876$, $\sin\theta = -\dfrac{1}{1.7876}$. The reference angle is given by $\hat{\theta} = \sin^{-1}\left(\dfrac{1}{1.7876}\right) \approx 34.0°$. Since θ terminates in quadrant III, $\theta = 180° + 34.0° = 214.0°$.

68. The reference angle is $45°$. Since θ terminates in quadrant III, $\theta = 180° + 45° = 225°$.

70. The reference angle is $30°$. Since θ terminates in quadrant III, $\theta = 180° + 30° = 210°$.

72. The reference angle is $45°$. Since θ terminates in quadrant II, $\theta = 180° - 45° = 135°$.

74. The reference angle is $30°$. Since θ terminates in quadrant III, $\theta = 180° + 30° = 210°$.

76. Since $\csc\theta = 2$, $\sin\theta = \frac{1}{2}$, so the reference angle is $30°$. Since θ terminates in quadrant II, $\theta = 180° - 30° = 150°$.

78. Since $\sec\theta = \sqrt{2}$, $\cos\theta = \dfrac{1}{\sqrt{2}}$, so the reference angle is $45°$. Since θ terminates in quadrant IV, $\theta = 360° - 45° = 315°$.

80. Since $\cot\theta = \sqrt{3}$, $\tan\theta = \dfrac{1}{\sqrt{3}}$, so the reference angle is $30°$. Since θ terminates in quadrant III, $\theta = 180° + 30° = 210°$.

82. The complement of $120°$ is $90° - 120° = -30°$, and the supplement is $180° - 120° = 60°$.

84. The complement of $90° - y$ is $90° - (90° - y) = y$, and the supplement is $180° - (90° - y) = 90° + y$.

86. The hypotenuse is $\frac{3}{4}\sqrt{2}$.

88. Simplifying the expression: $4\sin 60° - 2\cos 30° = 4 \cdot \dfrac{\sqrt{3}}{2} - 2 \cdot \dfrac{\sqrt{3}}{2} = 2\sqrt{3} - \sqrt{3} = \sqrt{3}$

90. Simplifying the expression: $(\sin 45° + \cos 45°)^2 = \left(\dfrac{1}{\sqrt{2}} + \dfrac{1}{\sqrt{2}}\right)^2 = \left(\dfrac{2}{\sqrt{2}}\right)^2 = \left(\sqrt{2}\right)^2 = 2$

3.2 Radians and Degrees

2. Computing the radian measure: $\theta = \dfrac{s}{r} = \dfrac{3}{6} = \dfrac{1}{2}$ radian

4. Computing the radian measure: $\theta = \dfrac{s}{r} = \dfrac{10}{5} = 2$ radians

6. Computing the radian measure: $\theta = \dfrac{s}{r} = \dfrac{12}{3} = 4$ radians

8. Computing the radian measure: $\theta = \dfrac{s}{r} = \dfrac{1/8}{1/4} = \dfrac{1}{8} \cdot \dfrac{4}{1} = \dfrac{1}{2}$ radian

10. Computing the radian measure: $\theta = \dfrac{s}{r} = \dfrac{2500}{4000} = \dfrac{5}{8}$ radian

12. **a.** Drawing the angle in standard position:

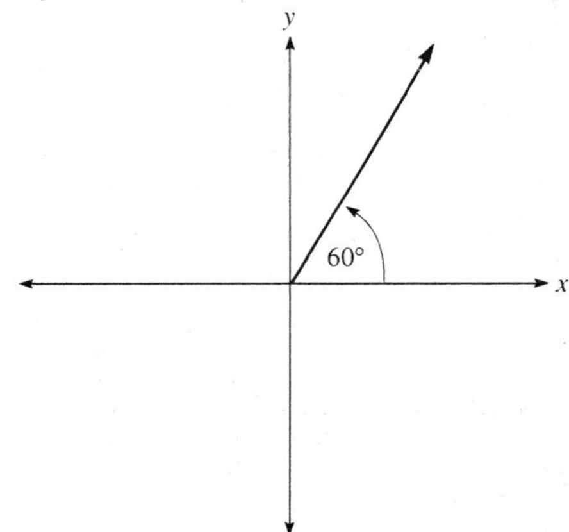

   b. Converting to radian measure: $60° = 60 \cdot \dfrac{\pi}{180} = \dfrac{\pi}{3}$ radians

   c. The reference angle is $60° = \dfrac{\pi}{3}$ radians.

14. **a.** Drawing the angle in standard position:

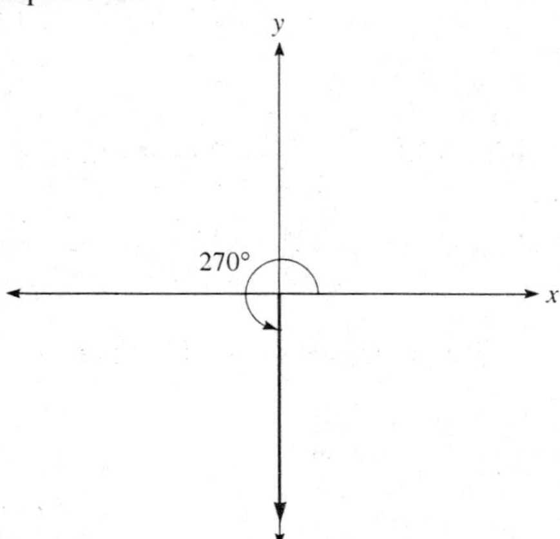

b. Converting to radian measure: $270° = 270 \cdot \dfrac{\pi}{180} = \dfrac{3\pi}{2}$ radians

c. The reference angle is undefined.

16. **a.** Drawing the angle in standard position:

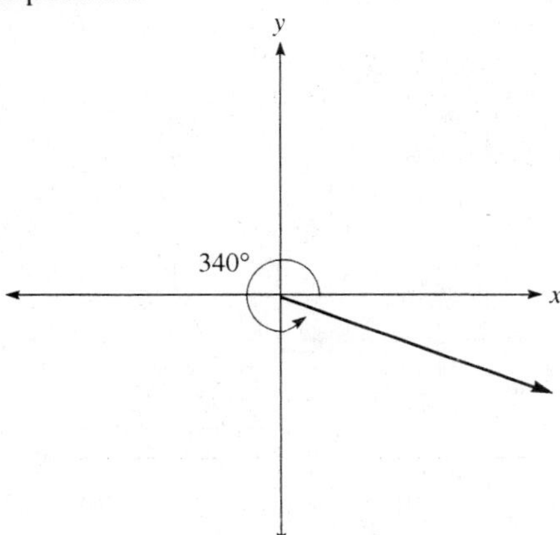

b. Converting to radian measure: $340° = 340 \cdot \dfrac{\pi}{180} = \dfrac{17\pi}{9}$ radians

c. The reference angle is $20° = \dfrac{\pi}{9}$ radians.

18.   a.   Drawing the angle in standard position:

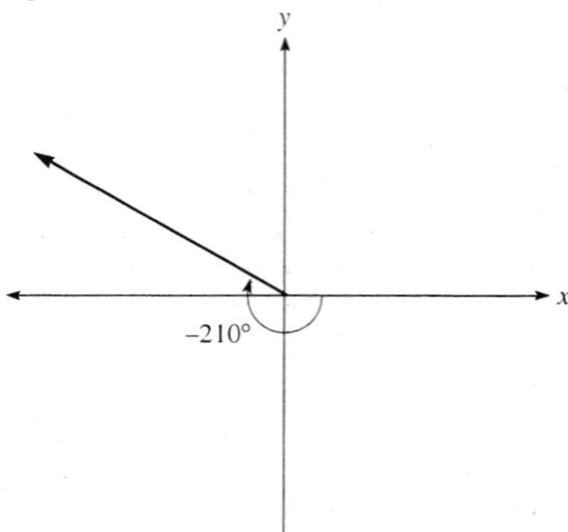

b.   Converting to radian measure: $-210° = -210 \cdot \dfrac{\pi}{180} = -\dfrac{7\pi}{6}$ radians

c.   The reference angle is $30° = \dfrac{\pi}{6}$ radians.

20.   a.   Drawing the angle in standard position:

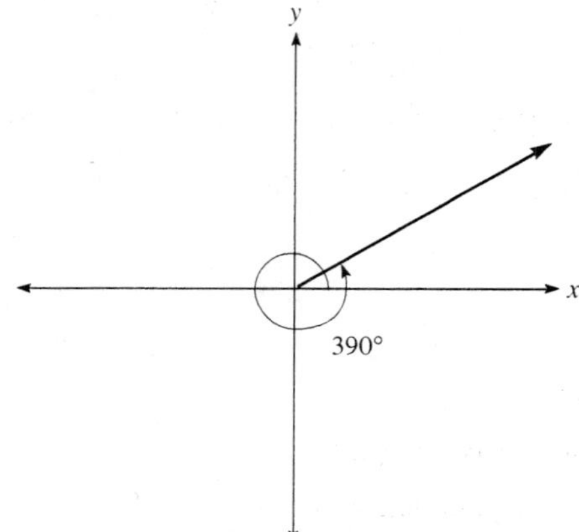

b.   Converting to radian measure: $390° = 390 \cdot \dfrac{\pi}{180} = \dfrac{13\pi}{6}$ radians

c.   The reference angle is $30° = \dfrac{\pi}{6}$ radians.

22. **a.** Drawing the angle in standard position:

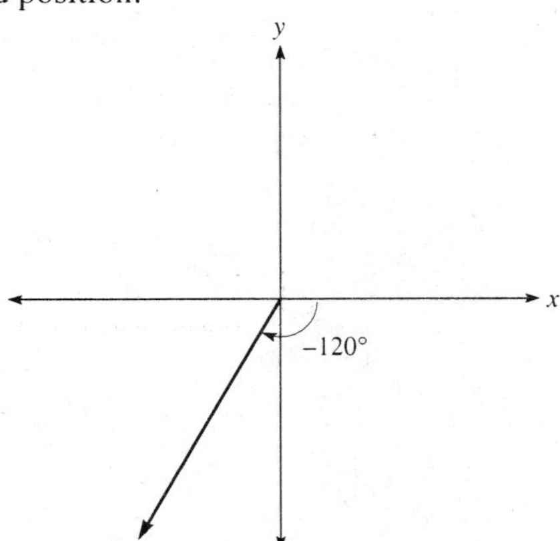

b. Converting to radian measure: $-120° = -120 \cdot \dfrac{\pi}{180} = -\dfrac{2\pi}{3}$ radians

c. The reference angle is $60° = \dfrac{\pi}{3}$ radians.

24. Converting to radian measure: $256°20' = 256\tfrac{1}{3}° = 256\tfrac{1}{3} \cdot \dfrac{\pi}{180} \approx 4.47$ radians

26. Converting to radian measure: $1° = 1 \cdot \dfrac{\pi}{180} \approx 0.0175$ radians

28. Let S represent the distance (in statute miles) apart. Since $\theta = 20' = \tfrac{1}{3}°$, convert to radians:

$\theta = \tfrac{1}{3}° = \tfrac{1}{3} \cdot \dfrac{\pi}{180} = \dfrac{\pi}{540}$ radians. Therefore:

$$\dfrac{\pi}{540} = \dfrac{S}{4000}$$
$$540S = 4000\pi$$
$$S = \dfrac{4000\pi}{540} \approx 23.27$$

The ships are approximately 23.27 statute miles apart.

30. Since a 25 minute period is $\dfrac{25}{60} = \dfrac{5}{12}$ of a complete revolution, the number of radians is: $\dfrac{5}{12} \cdot 2\pi = \dfrac{5\pi}{6}$

32. **a.** Converting to degree measure: $\dfrac{\pi}{4}$ radians $= \dfrac{\pi}{4} \cdot \dfrac{180°}{\pi} = 45°$

b. Drawing the angle in standard position:

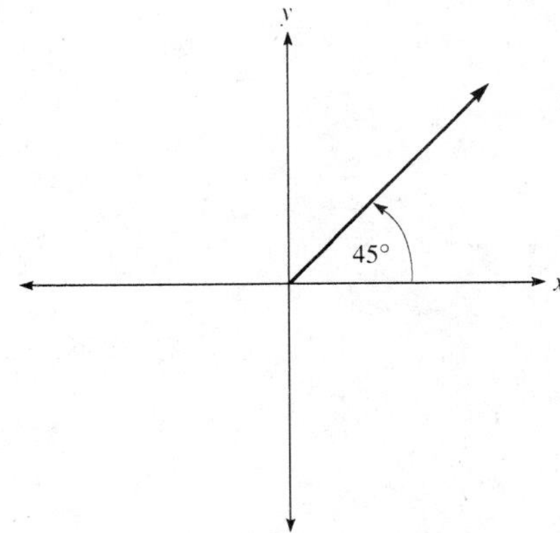

c. The reference angle is $45° = \dfrac{\pi}{4}$ radians.

34.   **a.**   Converting to degree measure: $\dfrac{3\pi}{4}$ radians $= \dfrac{3\pi}{4} \cdot \dfrac{180°}{\pi} = 135°$

     b.   Drawing the angle in standard position:

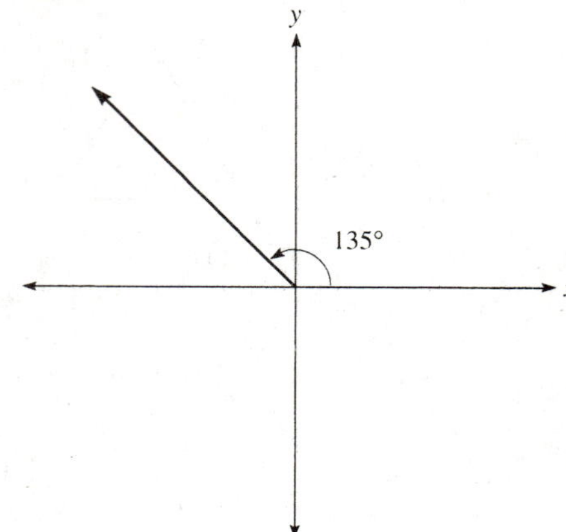

     c.   The reference angle is $45° = \dfrac{\pi}{4}$ radians.

36.   **a.**   Converting to degree measure: $-\dfrac{5\pi}{6}$ radians $= -\dfrac{5\pi}{6} \cdot \dfrac{180°}{\pi} = -150°$

     b.   Drawing the angle in standard position:

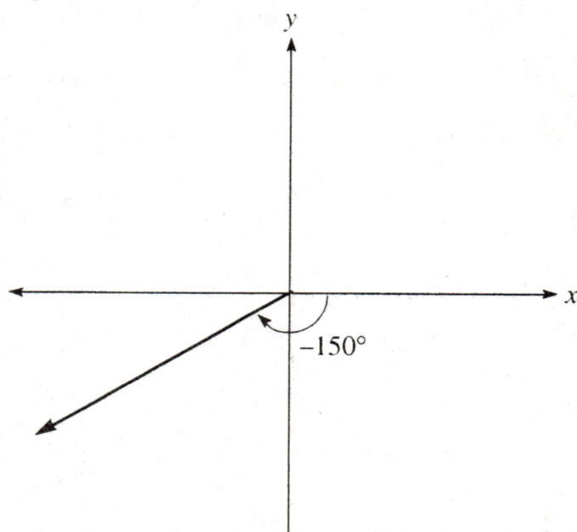

     c.   The reference angle is $30° = \dfrac{\pi}{6}$ radians.

38. **a.** Converting to degree measure: $\dfrac{7\pi}{3}$ radians $= \dfrac{7\pi}{3} \cdot \dfrac{180°}{\pi} = 420°$

    b. Drawing the angle in standard position:

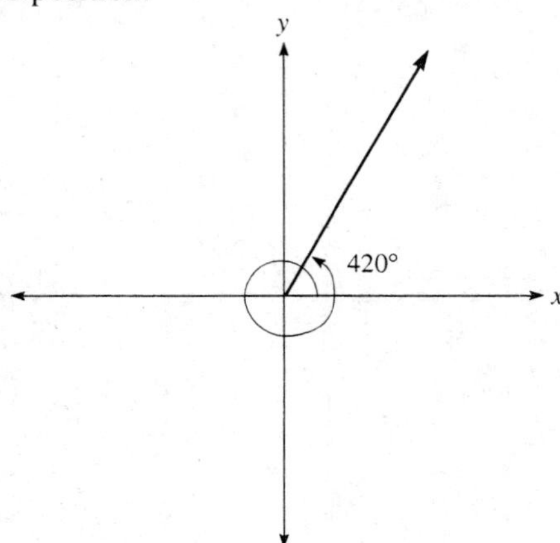

    c. The reference angle is $60° = \dfrac{\pi}{3}$ radians.

40. **a.** Converting to degree measure: 3π radians $= 3\pi \cdot \dfrac{180°}{\pi} = 540°$

    b. Drawing the angle in standard position:

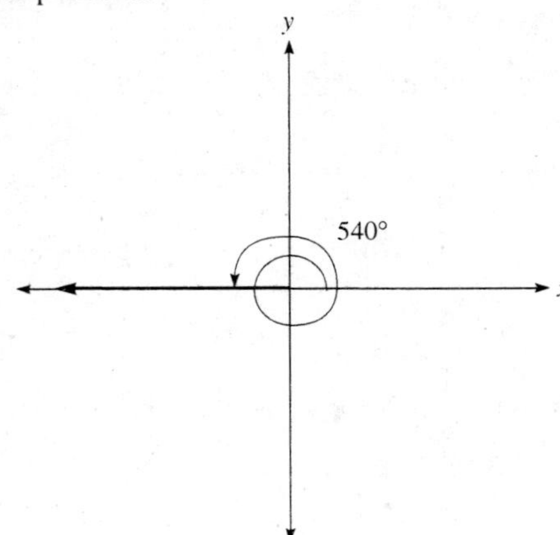

    c. The reference angle is undefined.

42. **a.** Converting to degree measure: $\dfrac{5\pi}{12}$ radians $= \dfrac{5\pi}{12} \cdot \dfrac{180°}{\pi} = 75°$

b. Drawing the angle in standard position:

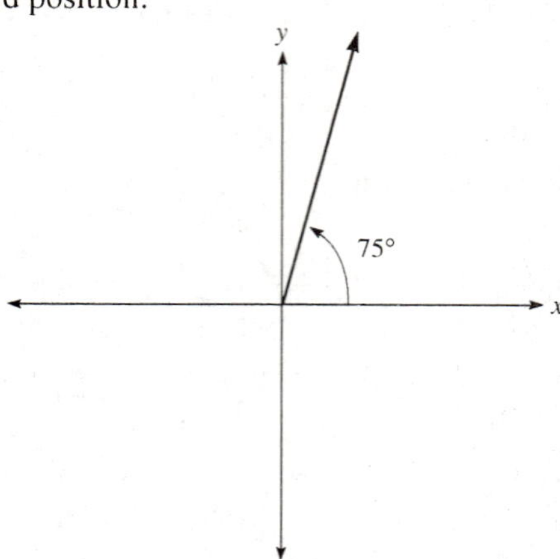

c. The reference angle is $75° = \dfrac{5\pi}{12}$ radians.

44. Converting to degree measure: 2 radians $= 2 \cdot \dfrac{180°}{\pi} = \dfrac{360°}{\pi} \approx 114.6°$

46. Converting to degree measure: 2.4 radians $= 2.4 \cdot \dfrac{180°}{\pi} = \dfrac{432°}{\pi} \approx 137.5°$

48. Converting to degree measure: 0.25 radians $= 0.25 \cdot \dfrac{180°}{\pi} = \dfrac{45°}{\pi} \approx 14.3°$

50. Converting to degree measure: 6 radians $= 6 \cdot \dfrac{180°}{\pi} = \dfrac{1080°}{\pi} \approx 343.8°$

52. Since $\dfrac{4\pi}{3}$ terminates in quadrant III, its cosine will be negative. The reference angle is $\dfrac{4\pi}{3} - \pi = \dfrac{\pi}{3}$, therefore:

$$\cos\dfrac{4\pi}{3} = -\cos\dfrac{\pi}{3} = -\dfrac{1}{2}$$

54. Since $\dfrac{\pi}{3}$ terminates in quadrant I, its cotangent will be positive. The reference angle is $\dfrac{\pi}{3}$, therefore:

$$\cot\dfrac{\pi}{3} = \dfrac{\cos\dfrac{\pi}{3}}{\sin\dfrac{\pi}{3}} = \dfrac{\frac{1}{2}}{\frac{\sqrt{3}}{2}} = \dfrac{1}{\sqrt{3}}$$

56. Using the point $(0,-1)$ as a point on the terminal side of $\dfrac{3\pi}{2}$: $\csc\dfrac{3\pi}{2} = \dfrac{1}{\sin\dfrac{3\pi}{2}} = \dfrac{1}{-1} = -1$

58. Since $\dfrac{5\pi}{6}$ terminates in quadrant II, its secant will be negative. The reference angle is $\pi - \dfrac{5\pi}{6} = \dfrac{\pi}{6}$, therefore:

$$\sec\dfrac{5\pi}{6} = -\sec\dfrac{\pi}{6} = -\dfrac{1}{\cos\dfrac{\pi}{6}} = -\dfrac{1}{\frac{\sqrt{3}}{2}} = -\dfrac{2}{\sqrt{3}}$$

60. Since $-\dfrac{\pi}{4}$ terminates in quadrant IV, its cosine will be positive. The reference angle is $\dfrac{\pi}{4}$, therefore:

$$4\cos\left(-\dfrac{\pi}{4}\right) = 4\cos\dfrac{\pi}{4} = 4 \cdot \dfrac{\sqrt{2}}{2} = 2\sqrt{2}$$

62. Computing the value: $-\cos\dfrac{\pi}{4} = -\dfrac{1}{\sqrt{2}}$

64. Computing the value: $2\sin\dfrac{\pi}{6} = 2\bullet\dfrac{1}{2} = 1$

66. Evaluating the expression: $\sin\left(3\bullet\dfrac{\pi}{6}\right) = \sin\dfrac{\pi}{2} = 1$

68. Evaluating the expression: $6\cos\left(2\bullet\dfrac{\pi}{6}\right) = 6\cos\dfrac{\pi}{3} = 6\bullet\dfrac{1}{2} = 3$

70. Evaluating the expression: $\sin\left(\dfrac{\pi}{6} - \dfrac{\pi}{2}\right) = \sin\left(-\dfrac{\pi}{3}\right) = -\sin\dfrac{\pi}{3} = -\dfrac{\sqrt{3}}{2}$

72. Evaluating the expression: $4\cos\left(3\bullet\dfrac{\pi}{6} + \dfrac{\pi}{6}\right) = 4\cos\dfrac{2\pi}{3} = -4\cos\dfrac{\pi}{3} = -4\bullet\dfrac{1}{2} = -2$

74. Computing the y-coordinates: $\cos 0 = 1$, $\cos\dfrac{\pi}{4} = \dfrac{1}{\sqrt{2}}$, $\cos\dfrac{\pi}{2} = 0$, $\cos\dfrac{3\pi}{4} = -\dfrac{1}{\sqrt{2}}$, $\cos\pi = -1$

The ordered pairs are $(0,1)$, $\left(\dfrac{\pi}{4}, \dfrac{1}{\sqrt{2}}\right)$, $\left(\dfrac{\pi}{2}, 0\right)$, $\left(\dfrac{3\pi}{4}, -\dfrac{1}{\sqrt{2}}\right)$, and $(\pi, -1)$.

76. Computing the y-coordinates: $\dfrac{1}{2}\cos 0 = \dfrac{1}{2}$, $\dfrac{1}{2}\cos\dfrac{\pi}{2} = 0$, $\dfrac{1}{2}\cos\pi = -\dfrac{1}{2}$, $\dfrac{1}{2}\cos\dfrac{3\pi}{2} = 0$, $\dfrac{1}{2}\cos 2\pi = \dfrac{1}{2}$

The ordered pairs are $\left(0, \dfrac{1}{2}\right)$, $\left(\dfrac{\pi}{2}, 0\right)$, $\left(\pi, -\dfrac{1}{2}\right)$, $\left(\dfrac{3\pi}{2}, 0\right)$, and $\left(2\pi, \dfrac{1}{2}\right)$.

78. Computing the y-coordinates:
$$\cos(3\bullet 0) = \cos 0 = 1, \quad \cos\left(3\bullet\dfrac{\pi}{6}\right) = \cos\dfrac{\pi}{2} = 0, \quad \cos\left(3\bullet\dfrac{\pi}{3}\right) = \cos\pi = -1,$$
$$\cos\left(3\bullet\dfrac{\pi}{2}\right) = \cos\dfrac{3\pi}{2} = 0, \quad \cos\left(3\bullet\dfrac{2\pi}{3}\right) = \cos 2\pi = 1$$

The ordered pairs are $(0,1)$, $\left(\dfrac{\pi}{6}, 0\right)$, $\left(\dfrac{\pi}{3}, -1\right)$, $\left(\dfrac{\pi}{2}, 0\right)$, and $\left(\dfrac{2\pi}{3}, 1\right)$.

80. Computing the y-coordinates:
$$\cos\left(\dfrac{\pi}{6} - \dfrac{\pi}{6}\right) = \cos 0 = 1, \quad \cos\left(\dfrac{\pi}{3} - \dfrac{\pi}{6}\right) = \cos\dfrac{\pi}{6} = \dfrac{\sqrt{3}}{2}, \quad \cos\left(\dfrac{2\pi}{3} - \dfrac{\pi}{6}\right) = \cos\dfrac{\pi}{2} = 0,$$
$$\cos\left(\pi - \dfrac{\pi}{6}\right) = \cos\dfrac{5\pi}{6} = -\dfrac{\sqrt{3}}{2}, \quad \cos\left(\dfrac{7\pi}{6} - \dfrac{\pi}{6}\right) = \cos\pi = -1$$

The ordered pairs are $\left(\dfrac{\pi}{6}, 1\right)$, $\left(\dfrac{\pi}{3}, \dfrac{\sqrt{3}}{2}\right)$, $\left(\dfrac{2\pi}{3}, 0\right)$, $\left(\pi, -\dfrac{\sqrt{3}}{2}\right)$, and $\left(\dfrac{7\pi}{6}, -1\right)$.

82. Computing the y-coordinates:
$$5\cos\left(2\bullet\dfrac{\pi}{6} - \dfrac{\pi}{3}\right) = 5\cos 0 = 5, \quad 5\cos\left(2\bullet\dfrac{\pi}{3} - \dfrac{\pi}{3}\right) = 5\cos\dfrac{\pi}{3} = 2.5, \quad 5\cos\left(2\bullet\dfrac{2\pi}{3} - \dfrac{\pi}{3}\right) = 5\cos\pi = -5,$$
$$5\cos\left(2\bullet\pi - \dfrac{\pi}{3}\right) = 5\cos\dfrac{5\pi}{3} = 2.5, \quad 5\cos\left(2\bullet\dfrac{7\pi}{6} - \dfrac{\pi}{3}\right) = 5\cos 2\pi = 5$$

The ordered pairs are $\left(\dfrac{\pi}{6}, 5\right)$, $\left(\dfrac{\pi}{3}, 2.5\right)$, $\left(\dfrac{2\pi}{3}, -5\right)$, $(\pi, 2.5)$, and $\left(\dfrac{7\pi}{6}, 5\right)$.

84. The central angle measurement is: $\dfrac{2\pi}{18} = \dfrac{\pi}{9}$

86. First find $r = \sqrt{(-1)^2 + 3^2} = \sqrt{1+9} = \sqrt{10}$. Using $x = -1$, $y = 3$, and $r = \sqrt{10}$ in the definitions:

$$\sin\theta = \dfrac{y}{r} = \dfrac{3}{\sqrt{10}} \qquad\qquad \cos\theta = \dfrac{x}{r} = -\dfrac{1}{\sqrt{10}} \qquad\qquad \tan\theta = \dfrac{y}{x} = \dfrac{3}{-1} = -3$$

$$\csc\theta = \dfrac{r}{y} = \dfrac{\sqrt{10}}{3} \qquad\qquad \sec\theta = \dfrac{r}{x} = -\sqrt{10} \qquad\qquad \cot\theta = \dfrac{x}{y} = -\dfrac{1}{3}$$

88.   First find $r = \sqrt{a^2 + b^2}$. Using $x = a$, $y = b$, and $r = \sqrt{a^2 + b^2}$ in the definitions:

$$\sin\theta = \frac{y}{r} = \frac{b}{\sqrt{a^2 + b^2}} \qquad \cos\theta = \frac{x}{r} = \frac{a}{\sqrt{a^2 + b^2}} \qquad \tan\theta = \frac{y}{x} = \frac{b}{a}$$

$$\csc\theta = \frac{r}{y} = \frac{\sqrt{a^2 + b^2}}{b} \qquad \sec\theta = \frac{r}{x} = \frac{\sqrt{a^2 + b^2}}{a} \qquad \cot\theta = \frac{x}{y} = \frac{a}{b}$$

90.   Choose $x = -1$ and $r = \sqrt{2}$. Find y using $x^2 + y^2 = r^2$:

$$(-1)^2 + y^2 = \left(\sqrt{2}\right)^2$$
$$1 + y^2 = 2$$
$$y^2 = 1$$
$$y = \pm 1$$

Since θ terminates in quadrant II, $y > 0$ and thus $y = 1$. Using $x = -1$, $y = 1$, and $r = \sqrt{2}$ in the definitions:

$$\sin\theta = \frac{y}{r} = \frac{1}{\sqrt{2}} \qquad \tan\theta = \frac{y}{x} = \frac{1}{-1} = -1 \qquad \sec\theta = \frac{r}{x} = \frac{\sqrt{2}}{-1} = -\sqrt{2}$$

$$\csc\theta = \frac{r}{y} = \sqrt{2} \qquad \cot\theta = \frac{x}{y} = \frac{-1}{1} = -1$$

92.   Since θ terminates in quadrant III, $x < 0$ and $y < 0$. Choose $x = -1$ and $y = -2$. Find
$r = \sqrt{(-1)^2 + (-2)^2} = \sqrt{1 + 4} = \sqrt{5}$. Using $x = -1$, $y = -2$, and $r = \sqrt{5}$ in the definitions:

$$\sin\theta = \frac{y}{r} = -\frac{2}{\sqrt{5}} \qquad \cos\theta = \frac{x}{r} = -\frac{1}{\sqrt{5}} \qquad \tan\theta = \frac{y}{x} = \frac{-2}{-1} = 2$$

$$\csc\theta = \frac{r}{y} = -\frac{\sqrt{5}}{2} \qquad \sec\theta = \frac{r}{x} = -\sqrt{5} \qquad \cot\theta = \frac{x}{y} = \frac{-1}{-2} = \frac{1}{2}$$

3.3 Definition III: Circular Functions

2.   Finding the six trigonometric functions of $135°$:

$$\sin 135° = \frac{1}{\sqrt{2}} \qquad \tan 135° = \frac{\sin 135°}{\cos 135°} = \frac{1/\sqrt{2}}{-1/\sqrt{2}} = -1 \qquad \csc 135° = \frac{1}{\sin 135°} = \frac{1}{1/\sqrt{2}} = \sqrt{2}$$

$$\cos 135° = -\frac{1}{\sqrt{2}} \qquad \cot 135° = \frac{\cos 135°}{\sin 135°} = \frac{-1/\sqrt{2}}{1/\sqrt{2}} = -1 \qquad \sec 135° = \frac{1}{\cos 135°} = \frac{1}{-1/\sqrt{2}} = -\sqrt{2}$$

4.   Finding the six trigonometric functions of $\dfrac{5\pi}{3}$:

$$\sin\frac{5\pi}{3} = -\frac{\sqrt{3}}{2} \qquad \tan\frac{5\pi}{3} = \frac{\sin\frac{5\pi}{3}}{\cos\frac{5\pi}{3}} = \frac{-\sqrt{3}/2}{1/2} = -\sqrt{3} \qquad \csc\frac{5\pi}{3} = \frac{1}{\sin\frac{5\pi}{3}} = \frac{1}{-\sqrt{3}/2} = -\frac{2}{\sqrt{3}}$$

$$\cos\frac{5\pi}{3} = \frac{1}{2} \qquad \cot\frac{5\pi}{3} = \frac{\cos\frac{5\pi}{3}}{\sin\frac{5\pi}{3}} = \frac{1/2}{-\sqrt{3}/2} = -\frac{1}{\sqrt{3}} \qquad \sec\frac{5\pi}{3} = \frac{1}{\cos\frac{5\pi}{3}} = \frac{1}{1/2} = 2$$

6.   Finding the six trigonometric functions of $270°$:

$$\sin 270° = -1 \qquad \tan 270° = \frac{\sin 270°}{\cos 270°} = \frac{-1}{0}, \text{ which is undefined} \qquad \csc 270° = \frac{1}{\sin 270°} = \frac{1}{-1} = -1$$

$$\cos 270° = 0 \qquad \cot 270° = \frac{\cos 270°}{\sin 270°} = \frac{0}{-1} = 0 \qquad \sec 270° = \frac{1}{\cos 270°} = \frac{1}{0}, \text{ which is undefined}$$

8. Finding the six trigonometric functions of $\dfrac{5\pi}{4}$:

$$\sin\frac{5\pi}{4} = -\frac{1}{\sqrt{2}} \qquad \tan\frac{5\pi}{4} = \frac{\sin\frac{5\pi}{4}}{\cos\frac{5\pi}{4}} = \frac{-\frac{1}{\sqrt{2}}}{-\frac{1}{\sqrt{2}}} = 1 \qquad \csc\frac{5\pi}{4} = \frac{1}{\sin\frac{5\pi}{4}} = \frac{1}{-\frac{1}{\sqrt{2}}} = -\sqrt{2}$$

$$\cos\frac{5\pi}{4} = -\frac{1}{\sqrt{2}} \qquad \cot\frac{5\pi}{4} = \frac{\cos\frac{5\pi}{4}}{\sin\frac{5\pi}{4}} = \frac{-\frac{1}{\sqrt{2}}}{-\frac{1}{\sqrt{2}}} = 1 \qquad \sec\frac{5\pi}{4} = \frac{1}{\cos\frac{5\pi}{4}} = \frac{1}{-\frac{1}{\sqrt{2}}} = -\sqrt{2}$$

10. Since cosine is an even function: $\cos(-120°) = \cos 120° = -\frac{1}{2}$

12. Since cosine is an even function: $\cos\left(-\dfrac{4\pi}{3}\right) = \cos\dfrac{4\pi}{3} = -\frac{1}{2}$

14. Since sine is an odd function: $\sin(-90°) = -\sin 90° = -1$

16. Since sine is an odd function: $\sin\left(-\dfrac{7\pi}{4}\right) = -\sin\dfrac{7\pi}{4} = -\left(-\dfrac{1}{\sqrt{2}}\right) = \dfrac{1}{\sqrt{2}}$

18. From the unit circle, $\sin\theta = -\frac{1}{2}$ when $\theta = \dfrac{7\pi}{6}$ and $\theta = \dfrac{11\pi}{6}$.

20. From the unit circle, $\cos\theta = 0$ when $\theta = \dfrac{\pi}{2}$ and $\theta = \dfrac{3\pi}{2}$.

22. From the unit circle, $\cot\theta = \sqrt{3}$ when $\theta = \dfrac{\pi}{6}$ and $\theta = \dfrac{7\pi}{6}$.

24. Finding the values: $\sin\dfrac{4\pi}{3} \approx -0.8660, \cos\dfrac{4\pi}{3} = -0.5$

26. Finding the values: $\sin\dfrac{5\pi}{12} \approx 0.9659, \cos\dfrac{5\pi}{12} \approx 0.2588$

28. Finding the values: $t \approx 2.0944, 4.1888$

30. Finding the values: $t \approx 1.5708$

32. Finding the values: $\sin 225° \approx -0.7071, \cos 225° \approx -0.7071$

34. Finding the values: $\sin 310° \approx -0.7660, \cos 310° \approx 0.6428$

36. Finding the values: $t = 90°, 270°$

38. Finding the values: $t = 135°, 315°$

40. Since $x = \cos\theta$ and $y = \sin\theta$, $\sin\theta = -\dfrac{3}{\sqrt{10}}$, $\cos\theta = -\dfrac{1}{\sqrt{10}}$, and $\tan\theta = \dfrac{-\frac{3}{\sqrt{10}}}{-\frac{1}{\sqrt{10}}} = 3$.

42. Since cosine is an even function: $\cos(-\theta) = \cos\theta = -\frac{1}{3}$

44. Computing the value: $\cos\left(2\pi + \dfrac{\pi}{2}\right) = \cos\dfrac{\pi}{2} = 0$

46. Computing the value: $\cos\left(2\pi + \dfrac{\pi}{3}\right) = \cos\dfrac{\pi}{3} = \frac{1}{2}$

48. Computing the value: $\cos\dfrac{5\pi}{2} = \cos\left(2\pi + \dfrac{\pi}{2}\right) = \cos\dfrac{\pi}{2} = 0$

50. Computing the value: $\cos\dfrac{7\pi}{3} = \cos\left(2\pi + \dfrac{\pi}{3}\right) = \cos\dfrac{\pi}{3} = \frac{1}{2}$

52.   Drawing the figure:

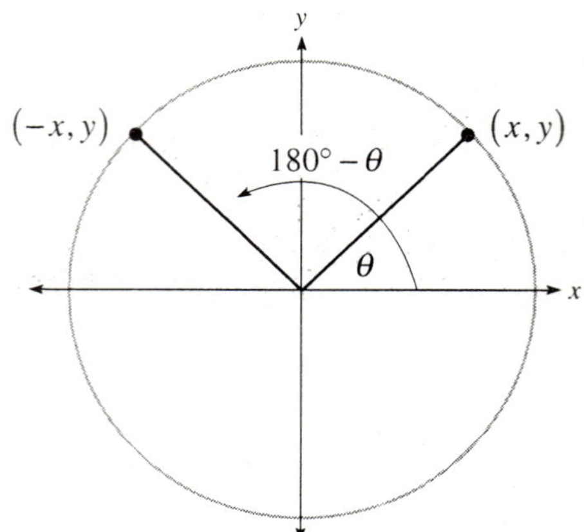

Since $\cos(180° - \theta) = -x$ while $\cos\theta = x$, $\cos(180° - \theta) = -\cos\theta$.

54.   Since sine is an odd function and cosine is an even function:

$$\cot(-\theta) = \frac{\cos(-\theta)}{\sin(-\theta)} = \frac{\cos\theta}{-\sin\theta} = -\frac{\cos\theta}{\sin\theta} = -\cot\theta$$

Thus cotangent is an odd function.

56.   Working from the left side: $\cos(-\theta)\tan\theta = \cos\theta \cdot \dfrac{\sin\theta}{\cos\theta} = \sin\theta$

58.   Working from the left side: $\cos(-\theta)\csc(-\theta)\tan(-\theta) = \cos\theta \cdot \dfrac{1}{\sin(-\theta)} \cdot \dfrac{\sin(-\theta)}{\cos(-\theta)} = \cos\theta \cdot \dfrac{1}{-\sin\theta} \cdot \dfrac{-\sin\theta}{\cos\theta} = 1$

60.   Working from the left side: $\sec\theta - \cos(-\theta) = \dfrac{1}{\cos\theta} - \cos\theta \cdot \dfrac{\cos\theta}{\cos\theta} = \dfrac{1 - \cos^2\theta}{\cos\theta} = \dfrac{\sin^2\theta}{\cos\theta}$

62.   Drawing the figure:

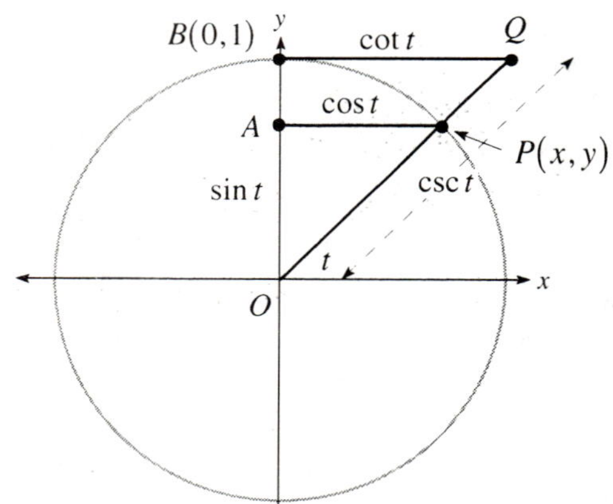

Since ΔOAP is similar to ΔOBQ, $\dfrac{AP}{AO} = \dfrac{BQ}{BO}$. Since $AP = \cos t$, $AO = \sin t$, and $BO = 1$, $\dfrac{\cos t}{\sin t} = \dfrac{BQ}{1}$, so

$BQ = \dfrac{\cos t}{\sin t} = \cot t$. Using the same triangles, $\dfrac{OP}{OQ} = \dfrac{OA}{OB}$. Since $OP = 1$, $OA = \sin t$, and $OB = 1$, $\dfrac{1}{OQ} = \dfrac{\sin t}{1}$, so

$OQ = \dfrac{1}{\sin t} = \csc t$.

64. Sketching the triangle:

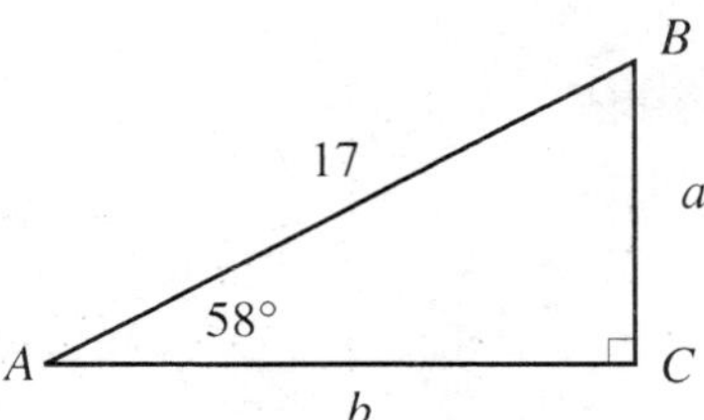

Note that $B = 90° - 58° = 32°$. Finding a:

$$\sin 58° = \frac{a}{17}$$
$$a = 17\sin 58° \approx 14$$

Finding b:

$$\cos 58° = \frac{b}{17}$$
$$b = 17\cos 58° \approx 9.0$$

66. Sketching the triangle:

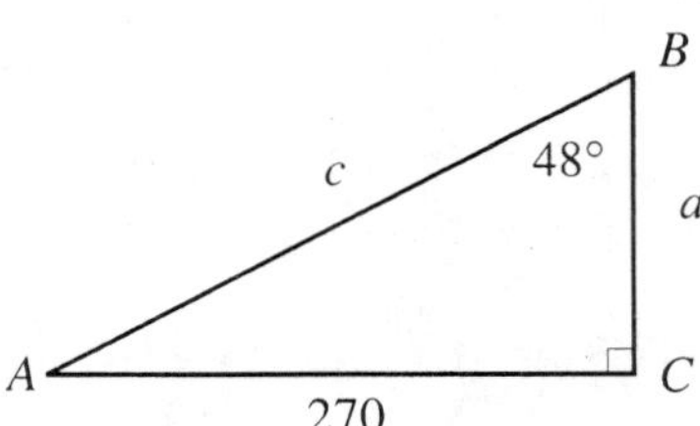

Note that $A = 90° - 48° = 42°$. Finding a:

$$\tan 48° = \frac{270}{a}$$
$$a\tan 48° = 270$$
$$a = \frac{270}{\tan 48°} \approx 240$$

Finding c:

$$\sin 48° = \frac{270}{c}$$
$$c\sin 48° = 270$$
$$c = \frac{270}{\sin 48°} \approx 360$$

68. Sketching the triangle:

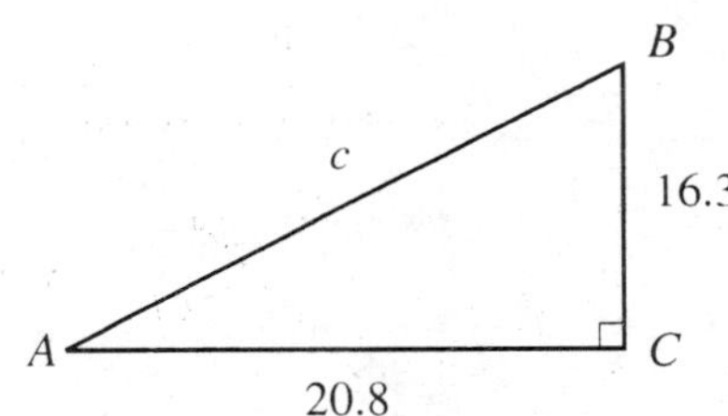

Using the Pythagorean Theorem: $c = \sqrt{(20.8)^2 + (16.3)^2} = \sqrt{432.64 + 265.69} = \sqrt{698.33} \approx 26.4$

Finding angle A:

$$\tan A = \frac{16.3}{20.8} \approx 0.7837$$
$$A = \tan^{-1}(0.7837) \approx 38.1°$$
$$B = 90° - 38.1° = 51.9°$$

70. Sketching the triangle:

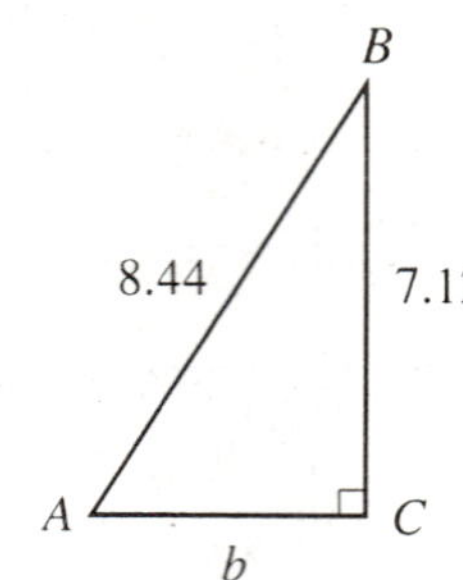

Using the Pythagorean Theorem: $b = \sqrt{(8.44)^2 - (7.12)^2} = \sqrt{71.2336 - 50.6944} = \sqrt{20.5392} \approx 4.53$

Finding angle A:

$$\sin A = \frac{7.12}{8.44} \approx 0.8436$$

$$A = \sin^{-1}(0.8436) \approx 57.5°$$

$$B = 90° - 57.5° = 32.5°$$

3.4 Arc Length and Area of a Sector

2. Finding the arc length: $s = r\theta = 2 \cdot 3 = 6$ inches

4. Finding the arc length: $s = r\theta = 1.8 \cdot 2.4 = 4.32$ feet

6. Finding the arc length: $s = r\theta = 12 \cdot \dfrac{\pi}{3} = 4\pi \approx 12.6$ cm

8. Note that $30° = \dfrac{\pi}{6}$ radians, so: $s = r\theta = 4 \cdot \dfrac{\pi}{6} = \dfrac{2\pi}{3} \approx 2.09$ mm

10. Note that $315° = \dfrac{7\pi}{4}$ radians, so: $s = r\theta = 5 \cdot \dfrac{7\pi}{4} = \dfrac{35\pi}{4} \approx 27.5$ inches

12. Since 40 minutes $= \dfrac{40}{60} \cdot 2\pi = \dfrac{4\pi}{3}$ radians, the arc length is given by $s = r\theta = 1.2 \cdot \dfrac{4\pi}{3} = 1.6\pi \approx 5.03$ cm.

14. Using $r = 4200$ mi, $\theta = \dfrac{s}{r} = \dfrac{8400}{4200} = 2$ radians. Set up the proportion:

$$\frac{2 \text{ radians}}{2\pi \text{ radians}} = \frac{x \text{ hours}}{6 \text{ hours}}$$

$$2\pi x = 12$$

$$x = \frac{12}{2\pi} = \frac{6}{\pi} \approx 1.91 \text{ hours}$$

16. Since $20° = 20 \cdot \dfrac{\pi}{180} = \dfrac{\pi}{9}$ radians, the length the pendulum swings in 1 second is $s = r\theta = 4 \cdot \dfrac{\pi}{9} = \dfrac{4\pi}{9}$ feet. In 60 seconds, the total distance traveled is $60 \cdot \dfrac{4\pi}{9} = \dfrac{80\pi}{3} \approx 83.8$ feet.

18. Since $270° = 270 \cdot \dfrac{\pi}{180} = \dfrac{3\pi}{2}$ radians, the length of the cable riding on one of the drive sheaves is:

$$s = r\theta = 6 \cdot \frac{3\pi}{2} = 9\pi \approx 28.3 \text{ feet}$$

20. Using $s = r\theta$, we have:

$$335 = 160\theta$$

$$\theta = \frac{335}{160} \approx 2.09 \text{ radians}$$

Converting to degrees: $\theta = 2.09 \cdot \dfrac{180°}{\pi} \approx 120°$

22. Since $0.5° = 0.5 \cdot \dfrac{\pi}{180} = \dfrac{\pi}{360}$ radians, the diameter of the sun can be approximated by the arc length:

$$s = r\theta = 93,000,000 \cdot \dfrac{\pi}{360} \approx 810,000 \text{ miles}$$

24. Since $\theta = 60° = \dfrac{\pi}{3}$ radians: $s = r\theta = 125 \cdot \dfrac{\pi}{3} = \dfrac{125\pi}{3} \approx 131$ feet

26. Since $\theta = 315° = \dfrac{7\pi}{4}$ radians: $s = r\theta = 125 \cdot \dfrac{7\pi}{4} = \dfrac{875\pi}{4} \approx 687$ feet

28. **a.**   Since $\theta = 150° = \dfrac{5\pi}{6}$ radians: $s = r\theta = 82.5 \cdot \dfrac{5\pi}{6} \approx 216$ feet

   b.   Since $\theta = 240° = \dfrac{4\pi}{3}$ radians: $s = r\theta = 82.5 \cdot \dfrac{4\pi}{3} \approx 346$ feet

   c.   Since $\theta = 345° = 345 \cdot \dfrac{\pi}{180} = \dfrac{23\pi}{12}$ radians: $s = r\theta = 82.5 \cdot \dfrac{23\pi}{12} \approx 497$ feet

30. Using $s = r\theta$:
   $$2 = r \cdot 1$$
   $$r = 2 \text{ feet}$$

32. Using $s = r\theta$:
   $$10.2 = r \cdot 5.1$$
   $$r = 2 \text{ inches}$$

34. Using $s = r\theta$:
   $$\pi = r \cdot \dfrac{3\pi}{4}$$
   $$4\pi = r \cdot 3\pi$$
   $$r = \tfrac{4}{3} \text{ cm}$$

36. Note that $\theta = 180° = \pi$ radians. Using $s = r\theta$:
   $$\dfrac{\pi}{2} = r \cdot \pi$$
   $$r = \tfrac{1}{2} \text{ m}$$

38. Note that $\theta = 150° = \dfrac{5\pi}{6}$ radians. Using $s = r\theta$:
   $$5 = r \cdot \dfrac{5\pi}{6}$$
   $$30 = r \cdot 5\pi$$
   $$r = \dfrac{6}{\pi} \approx 1.91 \text{ km}$$

40. Finding the sector area: $A = \tfrac{1}{2} r^2 \theta = \tfrac{1}{2}(2)^2 \cdot 3 = 6 \text{ cm}^2$

42. Finding the sector area: $A = \tfrac{1}{2} r^2 \theta = \tfrac{1}{2}(2)^2 \cdot 1.8 = 3.6 \text{ in.}^2$

44. Finding the sector area: $A = \tfrac{1}{2} r^2 \theta = \tfrac{1}{2}(3)^2 \cdot \dfrac{2\pi}{5} = \dfrac{9\pi}{5} \approx 5.65 \text{ m}^2$

46. Note that $\theta = 15° = 15 \cdot \dfrac{\pi}{180} = \dfrac{\pi}{12}$ radians. Finding the sector area: $A = \tfrac{1}{2} r^2 \theta = \tfrac{1}{2}(10)^2 \cdot \dfrac{\pi}{12} = \dfrac{25\pi}{6} \approx 13.1 \text{ m}^2$

48. First find the radius of the circle. Given $s = 3$ ft and $\theta = \dfrac{\pi}{4}$ radians:

   $$3 = r \cdot \dfrac{\pi}{4}$$
   $$12 = \pi r$$
   $$r = \dfrac{12}{\pi} \text{ ft}$$

   Finding the area of the sector: $A = \tfrac{1}{2} r^2 \theta = \tfrac{1}{2}\left(\dfrac{12}{\pi}\right)^2 \cdot \dfrac{\pi}{4} = \dfrac{18}{\pi} \approx 5.73 \text{ ft}^2$

50. First find the radius of the circle. Given $A = \dfrac{\pi}{3}$ cm^2 and $\theta = 30° = \dfrac{\pi}{6}$ radians:

$$\dfrac{\pi}{3} = \tfrac{1}{2}r^2 \cdot \dfrac{\pi}{6}$$
$$\dfrac{\pi}{3} = \dfrac{\pi}{12}r^2$$
$$r^2 = 4$$
$$r = 2 \text{ cm}$$

The arc length is therefore: $s = r\theta = 2 \cdot \dfrac{\pi}{6} = \dfrac{\pi}{3} \approx 1.05$ cm

52. Given $A = 25$ in.^{2} and $\theta = 4$ radians:

$$25 = \tfrac{1}{2}r^2 \cdot 4$$
$$25 = 2r^2$$
$$r^2 = 12.5$$
$$r = \sqrt{12.5} \approx 3.54 \text{ in.}$$

54. First note that $\theta = 60° = \dfrac{\pi}{3}$ radians. Treat the area being cleaned as a large sector area (with radius 10 in.) minus a smaller sector area (with radius 1 in.). Therefore:

$$A = \tfrac{1}{2}(10)^2 \cdot \dfrac{\pi}{3} - \tfrac{1}{2}(1)^2 \cdot \dfrac{\pi}{3} = \dfrac{50\pi}{3} - \dfrac{\pi}{6} = \dfrac{33\pi}{2} \approx 51.8 \text{ in.}^2$$

56. The radian measure formed by two adjacent spokes is: $\dfrac{2\pi}{18} = \dfrac{\pi}{9}$

The arc length is therefore: $s = r\theta = 350\left(\dfrac{\pi}{9}\right) = \dfrac{350\pi}{9} \approx 122$ mm

58. Let h represent the height of the hill. Drawing the triangle:

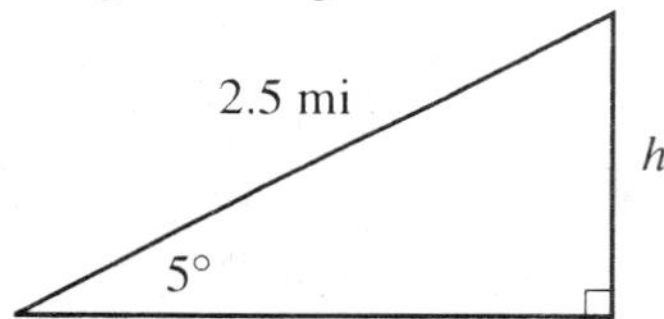

Therefore:

$$\sin 5° = \dfrac{h}{2.5}$$
$$h = 2.5\sin 5° \approx 0.218$$

The hill is approximately 0.218 miles high.

60.   Drawing the figure:

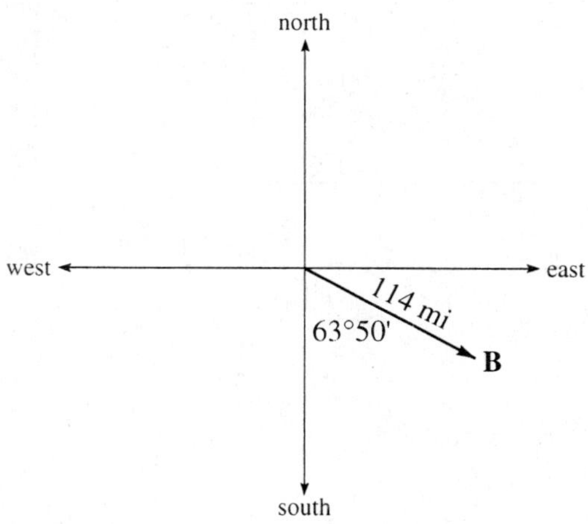

The components south and east are given by:

$$\mathbf{B}_S = 114\cos 63°50' = 114\cos\left(63\tfrac{5}{6}\right)° \approx 50.3 \text{ miles}$$

$$\mathbf{B}_E = 114\sin 63°50' = 114\sin\left(63\tfrac{5}{6}\right)° \approx 102 \text{ miles}$$

62.   Let $\mathbf{A}$ and $\mathbf{B}$ represent the two headings. Drawing the figure:

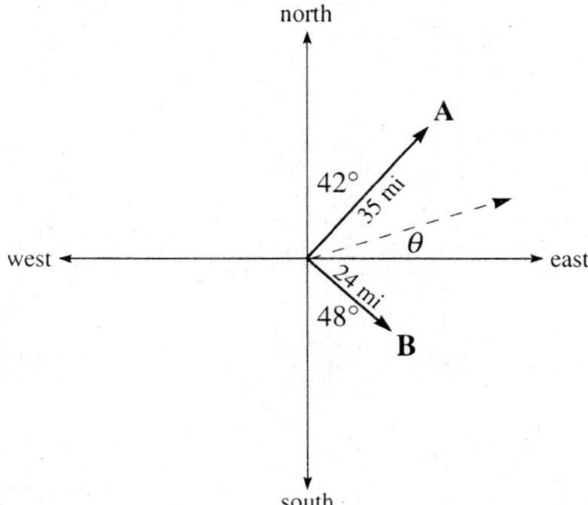

Representing A and B in vector form:

$$\mathbf{A} = \langle 35\sin 42°, 35\cos 42° \rangle \qquad\qquad \mathbf{B} = \langle 24\sin 48°, -24\cos 48° \rangle$$

The resultant vector is therefore given by: $\langle 35\sin 42° + 24\sin 48°, 35\cos 42° - 24\cos 48° \rangle \approx \langle 41.26, 9.95 \rangle$

The distance from the ship from the harbor entrance is given by: $r = \sqrt{(41.26)^2 + (9.95)^2} = \sqrt{1801.39} \approx 42.4$ miles

To find the bearing of the ship:

$$\tan\theta = \frac{9.95}{41.26} \approx 0.2412$$

$$\theta = \tan^{-1}(0.2412) \approx 13.6°$$

The bearing of the ship (from the harbor entrance) is N 76.4° E.

64.   Let h represent the height of the tree. Drawing the figure:

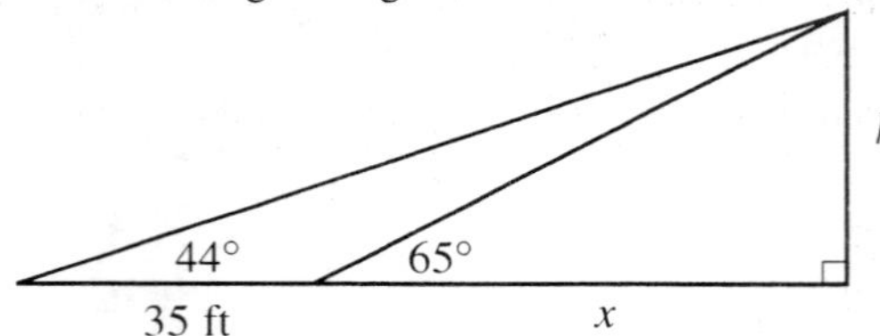

Using the smaller triangle:

$$\tan 65° = \frac{h}{x}$$
$$x \tan 65° = h$$
$$x = \frac{h}{\tan 65°} = h \cot 65°$$

Using the larger triangle:

$$\tan 44° = \frac{h}{35 + x}$$
$$(35 + x)\tan 44° = h$$
$$35 + x = \frac{h}{\tan 44°} = h \cot 44°$$
$$x = h \cot 44° - 35$$

Setting these two expressions equal:
$$h \cot 65° = h \cot 44° - 35$$
$$35 = h \cot 44° - h \cot 65°$$
$$35 = h(\cot 44° - \cot 65°)$$
$$h = \frac{35}{\cot 44° - \cot 65°} \approx 61.5 \text{ ft}$$

The tree is approximately 61.5 feet tall.

66.   Since $\theta = \dfrac{s}{r} = \dfrac{3470}{384000}$ is the central angle, the distance x the friend must stand is given by:

$$\frac{6 \text{ feet}}{x} = \frac{3470}{384000}$$
$$3470x = 2304000$$
$$x \approx 664 \text{ feet}$$

The friend must stand approximately 664 feet away.

68.   The smallest apparent diameter is: $\theta = \dfrac{s}{r} = \dfrac{3470}{406630} \cdot \dfrac{180°}{\pi} \approx 0.489°$

The largest apparent diameter is: $\theta = \dfrac{s}{r} = \dfrac{3470}{356340} \cdot \dfrac{180°}{\pi} \approx 0.558°$

3.5 Velocities

2.   The linear velocity is given by: $v = \dfrac{s}{t} = \dfrac{10 \text{ ft}}{2 \text{ min}} = 5 \text{ ft/min}$

4.   The linear velocity is given by: $v = \dfrac{s}{t} = \dfrac{12 \text{ cm}}{2 \text{ sec}} = 6 \text{ cm/sec}$

6.   The linear velocity is given by: $v = \dfrac{s}{t} = \dfrac{100 \text{ mi}}{4 \text{ hr}} = 25 \text{ mi/hr}$

8.   The distance is given by: $s = vt = 10 \frac{\text{ft}}{\text{sec}} \cdot 4 \text{ sec} = 40 \text{ feet}$

10.   The distance is given by: $s = vt = 55 \frac{\text{mi}}{\text{hr}} \cdot \frac{1}{2} \text{ hr} = 27.5 \text{ miles}$

12.   The distance is given by: $s = vt = 63 \frac{\text{mi}}{\text{hr}} \cdot 10 \text{ sec} \cdot \dfrac{1 \text{ hr}}{3600 \text{ sec}} = 0.175 \text{ miles}$

14.   The angular velocity is given by: $\omega = \dfrac{\theta}{t} = \dfrac{\frac{3\pi}{4} \text{ radians}}{5 \text{ sec}} = \dfrac{3\pi}{20} \approx 0.471 \text{ radians / sec}$

16.   The angular velocity is given by: $\omega = \dfrac{\theta}{t} = \dfrac{24 \text{ radians}}{6 \text{ min}} = 4 \text{ radians/min}$

18.   The angular velocity is given by: $\omega = \dfrac{\theta}{t} = \dfrac{12\pi \text{ radians}}{5\pi \text{ sec}} = 2.4 \text{ radians/sec}$

20. The angular velocity is given by: $\omega = \dfrac{\theta}{t} = \dfrac{24\pi \text{ radians}}{1.8 \text{ hr}} = \dfrac{40\pi}{3} \approx 41.9 \text{ radians / hr}$

22. The angular velocity is given by $\omega = \dfrac{\theta}{t} = \dfrac{2\pi \text{ radians}}{4 \text{ sec}} = \dfrac{\pi}{2}$ radians/sec. In t seconds, the angle θ is given by $\theta = \dfrac{\pi}{2}t$.

Therefore:

$$\cos\left(\frac{\pi}{2}t\right) = \frac{100}{l}$$

$$l\cos\left(\frac{\pi}{2}t\right) = 100$$

$$l = \frac{100}{\cos\left(\dfrac{\pi}{2}t\right)} = 100\sec\left(\frac{\pi}{2}t\right)$$

Substituting values for t:

$$t = 0.5: \quad l = \left|100\sec\frac{\pi}{4}\right| = \left|100\sqrt{2}\right| \approx 141 \text{ ft}$$

$$t = 1.0: \quad l = \left|100\sec\frac{\pi}{2}\right|, \text{ which is undefined}$$

$$t = 1.5: \quad l = \left|100\sec\frac{3\pi}{4}\right| = \left|-100\sqrt{2}\right| \approx 141 \text{ ft}$$

24. First find $v = r\omega = 4 \cdot 2 = 8 \,\frac{\text{inches}}{\text{sec}}$. Thus $s = vt = 8 \,\frac{\text{inches}}{\text{sec}} \cdot 5 \text{ sec} = 40 \text{ inches}$.

26. First find $v = r\omega = 8 \cdot \dfrac{4\pi}{3} = \dfrac{32\pi}{3} \,\frac{\text{m}}{\text{sec}}$. Thus $s = vt = \dfrac{32\pi}{3}\,\frac{\text{m}}{\text{sec}} \cdot 20 \text{ sec} = \dfrac{640\pi}{3} \approx 670 \text{ m}$.

28. First find $v = r\omega = 6 \cdot 10 = 60 \,\frac{\text{ft}}{\text{sec}}$. Thus $s = vt = 60 \,\frac{\text{ft}}{\text{sec}} \cdot 2 \text{ min} \cdot \dfrac{60 \text{ sec}}{1 \text{ min}} = 7200 \text{ ft}$.

30. Finding the angular velocity: $\omega = 20 \,\frac{\text{rev}}{\text{min}} \cdot 2\pi \,\frac{\text{radians}}{\text{rev}} = 40\pi \approx 126 \,\frac{\text{radians}}{\text{min}}$

32. Finding the angular velocity: $\omega = 16\frac{2}{3} \,\frac{\text{rev}}{\text{min}} \cdot 2\pi \,\frac{\text{radians}}{\text{rev}} = \dfrac{100\pi}{3} \approx 105 \,\frac{\text{radians}}{\text{min}}$

34. Finding the angular velocity: $\omega = 7.2 \,\frac{\text{rev}}{\text{min}} \cdot 2\pi \,\frac{\text{radians}}{\text{rev}} = 14.4\pi \approx 45.2 \,\frac{\text{radians}}{\text{min}}$

36. Finding the linear velocity: $v = r\omega = 8 \cdot 4 = 32 \,\frac{\text{inches}}{\text{sec}}$

38. Finding the angular velocity: $\omega = \dfrac{v}{r} = \dfrac{8}{3} \,\frac{\text{radians}}{\text{sec}}$

40. The angular velocity is $\omega = 20 \,\frac{\text{rev}}{\text{min}} \cdot 2\pi \,\frac{\text{radians}}{\text{rev}} = 40\pi \,\frac{\text{radians}}{\text{min}}$. Thus the linear velocity is given by $v = r\omega = 1 \cdot 40\pi = 40\pi \approx 126 \,\frac{\text{feet}}{\text{min}}$.

42. The angular velocity is $\omega = \frac{1}{24} \,\frac{\text{rev}}{\text{hr}} \cdot 2\pi \,\frac{\text{radians}}{\text{rev}} = \dfrac{\pi}{12} \,\frac{\text{radians}}{\text{hr}}$. The linear velocity is given by

$$v = r\omega = 4000 \cdot \frac{\pi}{12} = \frac{1000\pi}{3} \approx 1050 \text{ mph}.$$

44. The linear velocity is the blade is $v = 1100 \,\frac{\text{ft}}{\text{sec}}$. Since $r = 1$ ft, $\omega = \dfrac{v}{r} = \dfrac{1100 \,\frac{\text{ft}}{\text{sec}}}{1 \text{ ft}} = 1100 \,\frac{\text{radians}}{\text{sec}}$.

Now convert to rpm: $\omega = 1100 \,\frac{\text{radians}}{\text{sec}} \cdot \dfrac{1 \text{ rev}}{2\pi \text{ radians}} \cdot \dfrac{60 \text{ sec}}{1 \text{ min}} \approx 10,500 \text{ rpm}$

46. Given 18 rpm, the angular velocity is $\omega = 18 \,\frac{\text{rev}}{\text{min}} \cdot 2\pi \,\frac{\text{radians}}{\text{rev}} = 36\pi \,\frac{\text{radians}}{\text{min}}$. The linear velocity would be

$$v = r\omega = 6.5 \text{ ft} \cdot 36\pi \,\frac{\text{radians}}{\text{min}} = 234\pi \,\frac{\text{ft}}{\text{min}}. \text{ Converting to mph: } 234\pi \,\frac{\text{ft}}{\text{min}} \cdot \frac{1 \text{ mi}}{5280 \text{ ft}} \cdot \frac{60 \text{ min}}{1 \text{ hr}} \approx 8.35 \text{ mph}.$$

48. First convert the linear velocity to feet per minute: $12 \frac{\text{mi}}{\text{hr}} \cdot \frac{5280 \text{ ft}}{1 \text{ mi}} \cdot \frac{1 \text{ hr}}{60 \text{ min}} = 1056 \frac{\text{ft}}{\text{min}}$

Thus the angular velocity is: $\omega = \frac{v}{r} = \frac{1056 \frac{\text{ft}}{\text{min}}}{7 \text{ ft}} \approx 150.86 \frac{\text{radians}}{\text{min}}$

Now converting to rpm: $150.86 \frac{\text{radians}}{\text{min}} \cdot \frac{1 \text{ rev}}{2\pi \text{ radians}} \approx 24.0 \text{ rpm}$

50. The angular velocity is: $\omega = 10 \frac{\text{rev}}{\text{min}} \cdot 2\pi \frac{\text{radians}}{\text{rev}} = 20\pi \frac{\text{radians}}{\text{min}}$.

Converting units: $\omega = 20\pi \frac{\text{radians}}{\text{min}} \cdot \frac{1 \text{ min}}{60 \text{ sec}} = \frac{\pi}{3} \frac{\text{radians}}{\text{sec}}$

Therefore the radius is given by: $r = \frac{v}{\omega} = \frac{2.0 \frac{\text{meters}}{\text{sec}}}{\frac{\pi}{3} \frac{\text{radians}}{\text{sec}}} \approx 1.91 \text{ meters}$.

The diameter of the pulley should be 3.82 meters.

52. Given 1.5 rpm, the angular velocity is $\omega = 1.5 \frac{\text{rev}}{\text{min}} \cdot 2\pi \frac{\text{radians}}{\text{rev}} = 3\pi \frac{\text{radians}}{\text{min}}$. The linear velocity would be

$v = r\omega = 82.5 \text{ ft} \cdot 3\pi \frac{\text{radians}}{\text{min}} = 247.5\pi \frac{\text{ft}}{\text{min}}$. Converting to mph: $247.5\pi \frac{\text{ft}}{\text{min}} \cdot \frac{1 \text{ mi}}{5280 \text{ ft}} \cdot \frac{60 \text{ min}}{1 \text{ hr}} \approx 8.84 \text{ mph}$

The 10 mph figure is too large.

54. In 1 hour the bicycle travels 16,000 m, so the linear velocity is $v = 16000 \frac{\text{m}}{\text{hr}}$. Thus the angular velocity is

$\omega = \frac{v}{r} = \frac{16,000 \frac{\text{m}}{\text{hr}}}{0.3 \text{ m}} \approx 53,333 \frac{\text{radians}}{\text{hr}}$. Converting to rpm:

$\omega = 53,333 \frac{\text{radians}}{\text{hr}} \cdot \frac{1 \text{ rev}}{2\pi \text{ radians}} \approx 8,488 \frac{\text{rev}}{\text{hr}} \cdot \frac{1 \text{ hr}}{60 \text{ min}} \approx 141 \text{ rpm}$

56. First find the angular velocity of the chainring: $\omega = 80 \frac{\text{rev}}{\text{min}} \cdot 2\pi \frac{\text{rad}}{\text{rev}} = 160\pi \frac{\text{rad}}{\text{min}}$

The linear velocity of the chain is given by: $v = r\omega = (105 \text{ mm}) \cdot \left(160\pi \frac{\text{rad}}{\text{min}}\right) = 16800\pi \frac{\text{mm}}{\text{min}}$

The angular velocity of the rear sprocket is: $\omega = \frac{v}{r} = \frac{16800\pi \frac{\text{mm}}{\text{min}}}{20 \text{ mm}} = 840\pi \frac{\text{rad}}{\text{min}}$

The linear velocity of the rear tire is: $v = r\omega = (350 \text{ mm}) \cdot \left(840\pi \frac{\text{rad}}{\text{min}}\right) = 294000\pi \frac{\text{mm}}{\text{min}}$

Converting to kilometers per hour: $v = 294000\pi \frac{\text{mm}}{\text{min}} \cdot \frac{60 \text{ min}}{1 \text{ hr}} \cdot \frac{1 \text{ km}}{1000000 \text{ mm}} \approx 55.4 \frac{\text{km}}{\text{hour}}$

58. First find the angular velocity of the chainring: $\omega = 95 \frac{\text{rev}}{\text{min}} \cdot 2\pi \frac{\text{rad}}{\text{rev}} = 190\pi \frac{\text{rad}}{\text{min}}$

The linear velocity of the chain is given by: $v = r\omega = (75 \text{ mm}) \cdot \left(190\pi \frac{\text{rad}}{\text{min}}\right) = 14250\pi \frac{\text{mm}}{\text{min}}$

The linear velocity of the rear tire is: $24 \frac{\text{km}}{\text{hour}} \cdot \frac{1000000 \text{ mm}}{1 \text{ km}} \cdot \frac{1 \text{ hour}}{60 \text{ min}} = 400,000 \frac{\text{mm}}{\text{min}}$

The angular velocity of the rear tire is: $\omega = \frac{v}{r} = \frac{400000 \frac{\text{mm}}{\text{min}}}{350 \text{ mm}} \approx 1142.86 \frac{\text{rad}}{\text{min}}$

Now let R represent the radius of the sprocket. Since $v = r\omega$:
$$14250\pi = R \cdot 1142.86$$

$$R = \frac{14250\pi \frac{\text{mm}}{\text{min}}}{1142.86 \frac{\text{rad}}{\text{min}}} \approx 39.17 \text{ mm}$$

The diameter of the sprocket would need to be 78.3 mm.

60.   The linear velocity of the rear tire is: $40\,\frac{\text{km}}{\text{hour}} \cdot \frac{1000000\ \text{mm}}{1\ \text{km}} \cdot \frac{1\ \text{hour}}{60\ \text{min}} = 666666.7\,\frac{\text{mm}}{\text{min}}$

Thus the angular velocity of the rear tire is: $\omega = \dfrac{v}{r} = \dfrac{666666.7\,\frac{\text{mm}}{\text{min}}}{350\ \text{mm}} \approx 1904.76\,\frac{\text{rad}}{\text{min}}$

The linear velocity of the chain is: $v = r\omega = (20\ \text{mm}) \cdot \left(1904.76\,\frac{\text{rad}}{\text{min}}\right) \approx 38095.24\,\frac{\text{mm}}{\text{min}}$

The angular velocity of the chainring is: $\omega = \dfrac{v}{r} = \dfrac{38095.24\,\frac{\text{mm}}{\text{min}}}{105\ \text{mm}} \approx 362.81\,\frac{\text{rad}}{\text{min}}$

Converting to rpm: $\omega = 362.81\,\frac{\text{rad}}{\text{min}} \cdot \dfrac{1\ \text{rev}}{2\pi\ \text{rad}} \approx 57.7\ \text{rpm}$

62.   Draw the triangle, where x represents the distance off course:

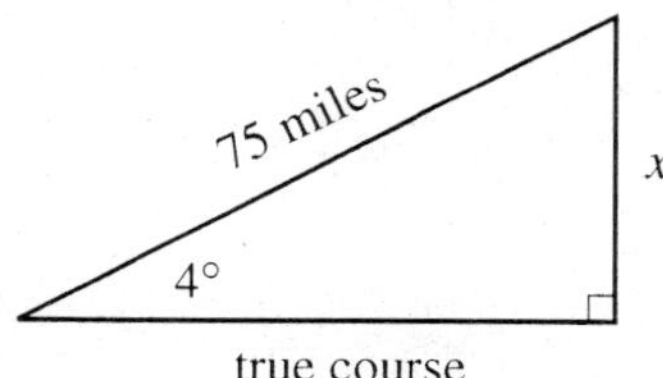

Therefore:

$$\sin 4° = \frac{x}{75}$$
$$x = 75\sin 4° \approx 5.23\ \text{mi}$$

The ship is approximately 5.23 miles off course.

64.   The magnitude of the vector is given by: $\sqrt{75^2 + 45^2} = \sqrt{7650} \approx 87.5$

66.   After 2 hours the plane has flown 571 miles. Drawing the figure:

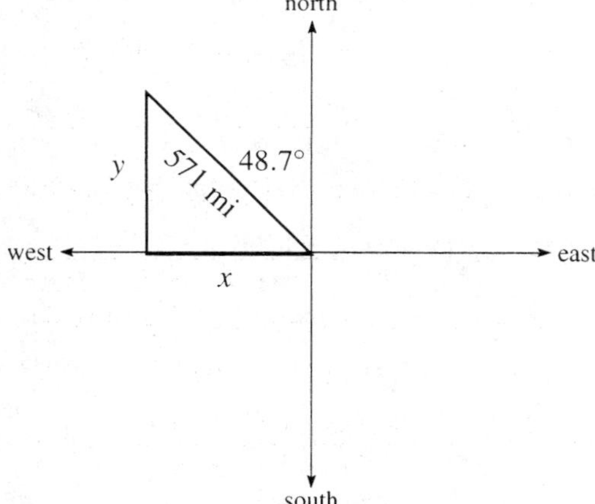

The north and west directions are given by:

$\quad y = 571\cos 48.7° \approx 377\ \text{mi north}$ $\qquad\qquad x = 571\sin 48.7° \approx 429\ \text{mi west}$

Chapter 3 Test

1.   The reference angle for $235°$ is $235° - 180° = 55°$:

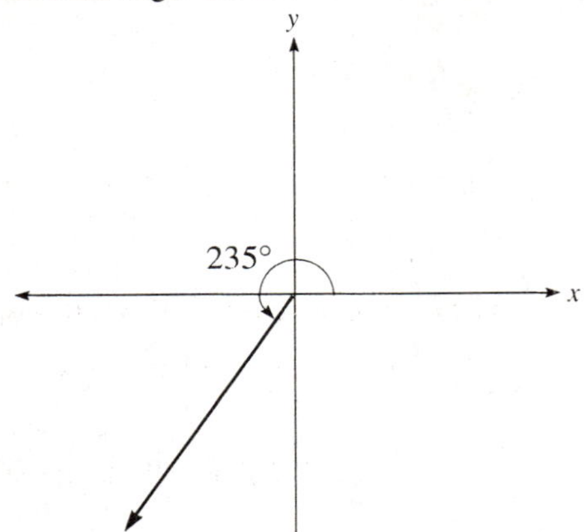

2.   The reference angle for $117.8°$ is $180° - 117.8° = 62.2°$:

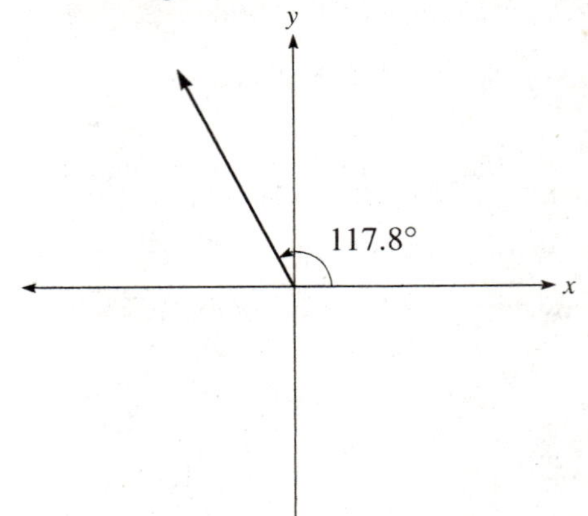

3.   The reference angle for $410°20'$ is $410°20' - 360° = 50°20'$:

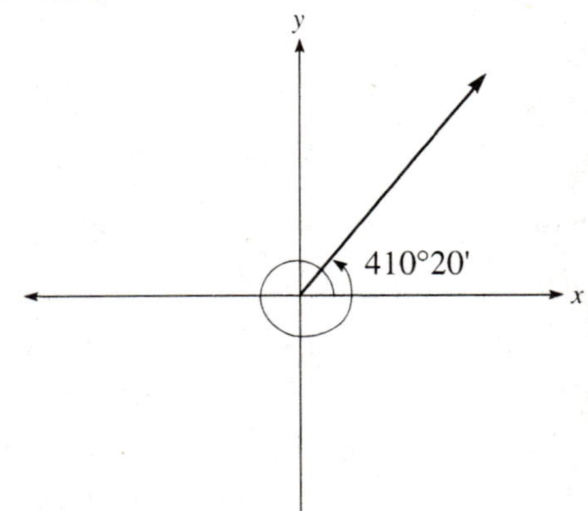

4.   The reference angle for $-225°$ is $-225° + 270° = 45°$:

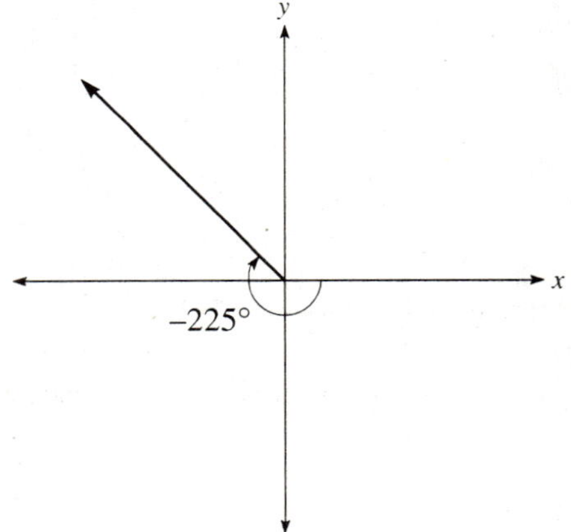

5.   Calculating the value: $\cot 320° = \dfrac{1}{\tan 320°} \approx -1.1918$

6.   Calculating the value: $\cot(-25°) = \dfrac{1}{\tan(-25°)} \approx -2.1445$

7. Calculating the value: $\csc(-236.7°) = \dfrac{1}{\sin(-236.7°)} \approx 1.1964$

8. Calculating the value: $\sec 322.3° = \dfrac{1}{\cos 322.3°} \approx 1.2639$

9. Calculating the value: $\sec 140°20' = \sec\left(140\tfrac{1}{3}\right)° = \dfrac{1}{\cos\left(140\tfrac{1}{3}\right)°} \approx -1.2991$

10. Calculating the value: $\csc 188°50' = \csc\left(188\tfrac{5}{6}\right)° = \dfrac{1}{\sin\left(188\tfrac{5}{6}\right)°} \approx -6.5121$

11. The reference angle is given by $\hat{\theta} = \sin^{-1}(0.1045) \approx 6.0°$. Since θ terminates in quadrant II, $\theta = 180° - 6.0° = 174.0°$.

12. The reference angle is given by $\hat{\theta} = \cos^{-1}(0.4772) \approx 61.5°$. Since θ terminates in quadrant III, $\theta = 180° + 61.5° = 241.5°$.

13. Since $\cot\theta = 0.9659$, $\tan\theta = \dfrac{1}{0.9659}$. The reference angle is given by $\hat{\theta} = \tan^{-1}\left(\dfrac{1}{0.9659}\right) \approx 46.0°$. Since θ terminates in quadrant III, $\theta = 180° + 46.0° = 226.0°$.

14. Since $\sec\theta = 1.545$, $\cos\theta = \dfrac{1}{1.545}$. The reference angle is given by $\hat{\theta} = \cos^{-1}\left(\dfrac{1}{1.545}\right) \approx 49.7°$. Since θ terminates in quadrant IV, $\theta = 360° - 49.7° = 310.3°$.

15. The reference angle for $225°$ is $45°$. Since $225°$ terminates in quadrant III, its sine will be negative. Therefore:
$$\sin 225° = -\sin 45° = -\dfrac{1}{\sqrt{2}}$$

16. The reference angle for $135°$ is $45°$. Since $135°$ terminates in quadrant II, its cosine will be negative. Therefore:
$$\cos 135° = -\cos 45° = -\dfrac{1}{\sqrt{2}}$$

17. The reference angle for $330°$ is $30°$. Since $330°$ terminates in quadrant IV, its tangent will be negative. Therefore:
$$\tan 330° = -\tan 30° = -\dfrac{\sin 30°}{\cos 30°} = -\dfrac{\tfrac{1}{2}}{\tfrac{\sqrt{3}}{2}} = -\dfrac{1}{\sqrt{3}}$$

18. The reference angle for $390°$ is $30°$. Since $390°$ terminates in quadrant I, its secant will be positive. Therefore:
$$\sec 390° = \sec 30° = \dfrac{1}{\cos 30°} = \dfrac{1}{\tfrac{\sqrt{3}}{2}} = \dfrac{2}{\sqrt{3}}$$

19. Converting to radian measure: $250° = 250 \cdot \dfrac{\pi}{180} = \dfrac{25\pi}{18}$ radians

20. Converting to radian measure: $-390° = -390 \cdot \dfrac{\pi}{180} = -\dfrac{13\pi}{6}$ radians

21. Converting to degree measure: $\dfrac{4\pi}{3}$ radians $= \dfrac{4\pi}{3} \cdot \dfrac{180°}{\pi} = 240°$

22. Converting to degree measure: $\dfrac{7\pi}{12}$ radians $= \dfrac{7\pi}{12} \cdot \dfrac{180°}{\pi} = 105°$

23. Since $\dfrac{2\pi}{3}$ terminates in quadrant II, its sine will be positive. The reference angle is $\pi - \dfrac{2\pi}{3} = \dfrac{\pi}{3}$. Therefore:
$$\sin\dfrac{2\pi}{3} = \sin\dfrac{\pi}{3} = \dfrac{\sqrt{3}}{2}$$

24. Since $\dfrac{2\pi}{3}$ terminates in quadrant II, its cosine will be negative. The reference angle is $\pi - \dfrac{2\pi}{3} = \dfrac{\pi}{3}$. Therefore:
$$\cos\dfrac{2\pi}{3} = -\cos\dfrac{\pi}{3} = -\dfrac{1}{2}$$

25. Computing the value: $4\cos\left(-\dfrac{3\pi}{4}\right) = 4\cos\dfrac{3\pi}{4} = -4\cos\dfrac{\pi}{4} = -4 \cdot \dfrac{\sqrt{2}}{2} = -2\sqrt{2}$

26. Computing the value: $2\cos\left(-\dfrac{5\pi}{3}\right) = 2\cos\dfrac{\pi}{3} = 2 \cdot \dfrac{1}{2} = 1$

27. Computing the value: $\sec\dfrac{5\pi}{6} = \dfrac{1}{\cos\dfrac{5\pi}{6}} = -\dfrac{1}{\cos\dfrac{\pi}{6}} = -\dfrac{1}{\sqrt{3}\big/2} = -\dfrac{2}{\sqrt{3}}$

28. Computing the value: $\csc\dfrac{5\pi}{6} = \dfrac{1}{\sin\dfrac{5\pi}{6}} = \dfrac{1}{\sin\dfrac{\pi}{6}} = \dfrac{1}{1/2} = 2$

29. Substituting $x = \dfrac{\pi}{3}$: $2\cos\left(3 \cdot \dfrac{\pi}{3} - \dfrac{\pi}{2}\right) = 2\cos\left(\pi - \dfrac{\pi}{2}\right) = 2\cos\dfrac{\pi}{2} = 2 \cdot 0 = 0$

30. Substituting $x = \dfrac{\pi}{4}$: $4\sin\left(2 \cdot \dfrac{\pi}{4} + \dfrac{\pi}{4}\right) = 4\sin\left(\dfrac{\pi}{2} + \dfrac{\pi}{4}\right) = 4\sin\dfrac{3\pi}{4} = 4 \cdot \dfrac{\sqrt{2}}{2} = 2\sqrt{2}$

31. Since sine is an odd function and cosine is an even function: $\cot(-\theta) = \dfrac{\cos(-\theta)}{\sin(-\theta)} = \dfrac{\cos\theta}{-\sin\theta} = -\dfrac{\cos\theta}{\sin\theta} = -\cot\theta$

32. Working from the left side: $\sin(-\theta)\sec(-\theta)\cot(-\theta) = -\sin\theta \cdot \dfrac{1}{\cos(-\theta)} \cdot \dfrac{\cos(-\theta)}{\sin(-\theta)} = -\sin\theta \cdot \dfrac{1}{\cos\theta} \cdot \dfrac{\cos\theta}{-\sin\theta} = 1$

33. Finding the arc length: $s = r\theta = 12 \cdot \dfrac{\pi}{6} = 2\pi \approx 6.28$ meters

34. Note that $60° = \dfrac{\pi}{3}$ radians, so: $s = r\theta = 6 \cdot \dfrac{\pi}{3} = 2\pi \approx 6.28$ feet

35. Using $s = r\theta$:

$$\pi = r \cdot \dfrac{\pi}{4}$$
$$4\pi = r\pi$$
$$r = 4 \text{ cm}$$

36. Using $s = r\theta$:

$$\dfrac{\pi}{4} = r \cdot \dfrac{2\pi}{3}$$
$$3\pi = r \cdot 8\pi$$
$$r = \tfrac{3}{8} \text{ cm}$$

37. Note that $\theta = 90° = \dfrac{\pi}{2}$ radians. Finding the sector area: $A = \tfrac{1}{2}r^2\theta = \tfrac{1}{2}(4)^2 \cdot \dfrac{\pi}{2} = 4\pi \approx 12.6$ in.^{2}

38. Finding the sector area: $A = \tfrac{1}{2}r^2\theta = \tfrac{1}{2}(3)^2(2.4) = 10.8$ cm^2

39. In 30 minutes the minute hand subtends a central angle of π radians.
Finding the arc length: $s = r\theta = 2 \cdot \pi = 2\pi \approx 6.28$ cm

40. In 2 minutes, the string has twirled $\theta = \left(0.5\tfrac{\text{rev}}{\text{min}}\right) \cdot (2 \text{ min}) = 1 \text{ rev} = 2\pi$ radians.

Thus the plane has traveled: $s = r\theta = (5 \text{ feet}) \cdot (2\pi \text{ radians}) = 10\pi \approx 31.4$ feet

41. Using $s = r\theta$ to find the central angle:
$$8 = r \cdot 4$$
$$r = 2 \text{ in.}$$
Now finding the sector area: $A = \tfrac{1}{2}r^2\theta = \tfrac{1}{2}(2)^2(4) = 8$ in.^{2}

42. The distance is given by: $s = vt = 30 \cdot 3 = 90$ feet

43. Note that 1 min = 60 sec. The distance is given by: $s = vt = 66 \cdot 60 = 3,960$ feet

44. Since $v = r\omega$, $v = 3 \cdot 4 = 12 \tfrac{\text{in.}}{\text{sec}}$. Thus $s = vt = 12 \cdot 6 = 72$ inches.

45. Since $v = r\omega$, $v = 8 \cdot \dfrac{3\pi}{4} = 6\pi \tfrac{\text{ft}}{\text{sec}}$. Thus $s = vt = 6\pi \cdot 20 = 120\pi \approx 377$ feet.

46. The angular velocity is given by: $\omega = 6\tfrac{\text{rev}}{\text{min}} \cdot 2\pi\tfrac{\text{radians}}{\text{rev}} = 12\pi \approx 37.7\tfrac{\text{radians}}{\text{min}}$

47. The angular velocity is given by: $\omega = 2\tfrac{\text{rev}}{\text{min}} \cdot 2\pi\tfrac{\text{radians}}{\text{rev}} = 4\pi \approx 12.6\tfrac{\text{radians}}{\text{min}}$

48. The angular velocity is given by: $\omega = \dfrac{v}{r} = \dfrac{5 \text{ cm/sec}}{10 \text{ cm}} = \tfrac{1}{2}\tfrac{\text{radians}}{\text{sec}}$

49. The angular velocity is given by: $\omega = \dfrac{v}{r} = \dfrac{5 \text{ cm/sec}}{3 \text{ cm}} = \tfrac{5}{3}\tfrac{\text{radians}}{\text{sec}}$

50. First find $\omega = 20 \frac{\text{rev}}{\text{min}} \cdot 2\pi \frac{\text{radians}}{\text{rev}} = 40\pi \frac{\text{radians}}{\text{min}}$. Thus: $v = r\omega = 2 \text{ ft} \cdot 40\pi \frac{\text{radians}}{\text{min}} = 80\pi \approx 251 \frac{\text{feet}}{\text{min}}$

51. First find $\omega = 10 \frac{\text{rev}}{\text{min}} \cdot 2\pi \frac{\text{radians}}{\text{rev}} = 20\pi \frac{\text{radians}}{\text{min}}$. Thus: $v = r\omega = 1 \text{ ft} \cdot 20\pi \frac{\text{radians}}{\text{min}} = 20\pi \approx 62.8 \frac{\text{feet}}{\text{min}}$

52. For the smaller pulley: $\omega = \frac{v}{r} = \frac{24 \text{ cm/sec}}{6 \text{ cm}} = 4 \frac{\text{radians}}{\text{sec}}$

For the larger pulley: $\omega = \frac{v}{r} = \frac{24 \text{ cm/sec}}{8 \text{ cm}} = 3 \frac{\text{radians}}{\text{sec}}$

53. First find $\omega = 900 \frac{\text{rev}}{\text{min}} \cdot 2\pi \frac{\text{radians}}{\text{rev}} = 1800\pi \frac{\text{radians}}{\text{min}}$. Thus the linear velocity is given by:

$$v = r\omega = 1.5 \text{ ft} \cdot 1800\pi \frac{\text{radians}}{\text{min}} = 2700\pi \frac{\text{ft}}{\text{min}} \approx 8,480 \frac{\text{ft}}{\text{min}}$$

54. The angular velocity is $\omega = 300 \frac{\text{rev}}{\text{min}} \cdot 2\pi \frac{\text{radians}}{\text{rev}} = 600\pi \frac{\text{radians}}{\text{min}}$. The linear velocity is given by:

$$v = r\omega = 1.5 \text{ in.} \cdot 600\pi \frac{\text{radians}}{\text{min}} = 900\pi \approx 2,830 \frac{\text{in.}}{\text{min}}.$$

55. The angular velocity is $\omega = 2941 \frac{\text{rev}}{\text{min}} \cdot 2\pi \frac{\text{radians}}{\text{rev}} = 5882\pi \frac{\text{radians}}{\text{min}}$. The linear velocities are given by:

$$v = r\omega = 1 \text{ in.} \cdot 5882\pi \frac{\text{radians}}{\text{min}} = 5882\pi \frac{\text{in.}}{\text{min}} \cdot \frac{60 \text{ min.}}{1 \text{ hr}} \cdot \frac{1 \text{ ft}}{12 \text{ in.}} \cdot \frac{1 \text{ mi}}{5280 \text{ ft}} \approx 17.5 \text{ mph}$$

$$v = r\omega = 1.5 \text{ in.} \cdot 5882\pi \frac{\text{radians}}{\text{min}} = 5882\pi \frac{\text{in.}}{\text{min}} \cdot \frac{60 \text{ min.}}{1 \text{ hr}} \cdot \frac{1 \text{ ft}}{12 \text{ in.}} \cdot \frac{1 \text{ mi}}{5280 \text{ ft}} \approx 26.2 \text{ mph}$$

56. The linear velocity is: $11 \frac{\text{mi}}{\text{hour}} \cdot \frac{5280 \text{ ft}}{1 \text{ mi}} \cdot \frac{1 \text{ hour}}{60 \text{ min}} = 968 \frac{\text{ft}}{\text{min}}$

Finding the angular velocity: $\omega = \frac{v}{r} = \frac{968 \frac{\text{ft}}{\text{min}}}{7 \text{ ft}} \approx 138.29 \frac{\text{radians}}{\text{min}}$

Converting to rpm: $\omega = 138.29 \frac{\text{radians}}{\text{min}} \cdot \frac{1 \text{ rev}}{2\pi \text{ radians}} \approx 22.0 \text{ rpm}$

57. First find the angular velocity of the chainring: $\omega = 75 \frac{\text{rev}}{\text{min}} \cdot 2\pi \frac{\text{rad}}{\text{rev}} = 150\pi \frac{\text{rad}}{\text{min}}$

The linear velocity of the chain is given by: $v = r\omega = (105 \text{ mm}) \cdot \left(150\pi \frac{\text{rad}}{\text{min}}\right) = 15750\pi \frac{\text{mm}}{\text{min}}$

The angular velocity of the rear sprocket is: $\omega = \frac{v}{r} = \frac{15750\pi \frac{\text{mm}}{\text{min}}}{25 \text{ mm}} = 630\pi \frac{\text{rad}}{\text{min}}$

The linear velocity of the rear tire is: $v = r\omega = (350 \text{ mm}) \cdot \left(630\pi \frac{\text{rad}}{\text{min}}\right) = 220500\pi \frac{\text{mm}}{\text{min}}$

Converting to kilometers per hour: $v = 220500\pi \frac{\text{mm}}{\text{min}} \cdot \frac{60 \text{ min}}{1 \text{ hr}} \cdot \frac{1 \text{ km}}{1000000 \text{ mm}} \approx 41.6 \frac{\text{km}}{\text{hour}}$

Chapter 4
Graphing and Inverse Functions

4.1 Basic Graphs

2. Making a table of values:

x	0	$\pi/4$	$\pi/2$	$3\pi/4$	π	$5\pi/4$	$3\pi/2$	$7\pi/4$	2π
$y = \cot x$	undefined	1	0	-1	undefined	1	0	-1	undefined

Sketching the graph:

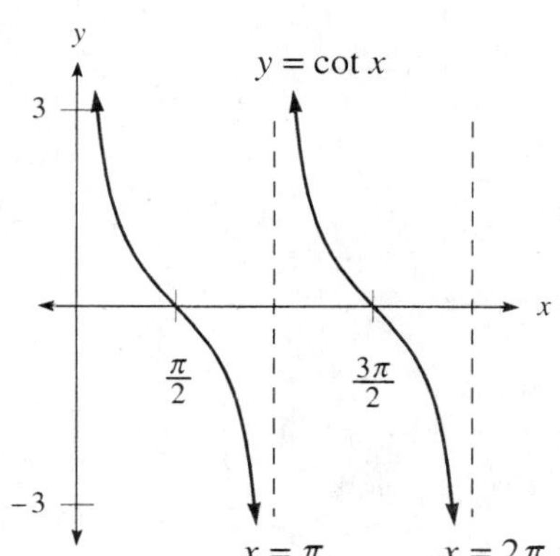

4. Making a table of values:

x	0	$\pi/4$	$\pi/2$	$3\pi/4$	π	$5\pi/4$	$3\pi/2$	$7\pi/4$	2π
$y = \sin x$	0	$\frac{1}{\sqrt{2}} \approx 0.7$	1	$\frac{1}{\sqrt{2}} \approx 0.7$	0	$-\frac{1}{\sqrt{2}} \approx -0.7$	-1	$-\frac{1}{\sqrt{2}} \approx -0.7$	0

Sketching the graph:

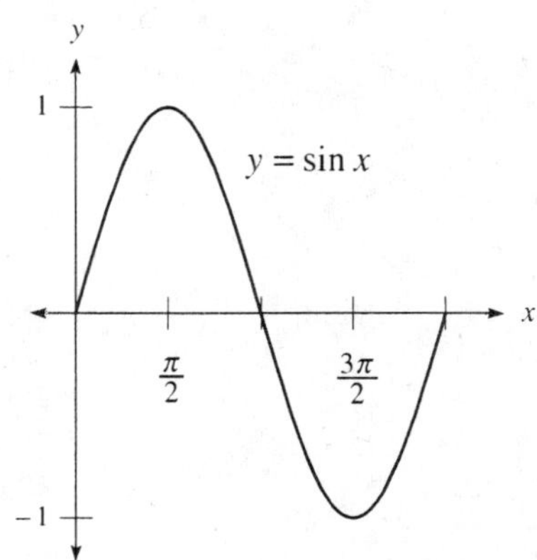

6.   Making a table of values:

x	0	$\pi/4$	$\pi/2$	$3\pi/4$	π	$5\pi/4$	$3\pi/2$	$7\pi/4$	2π
$y = \sec x$	1	$\sqrt{2} \approx 1.4$	undefined	$-\sqrt{2} \approx -1.4$	-1	$-\sqrt{2} \approx -1.4$	undefined	$\sqrt{2} \approx 1.4$	1

Sketching the graph:

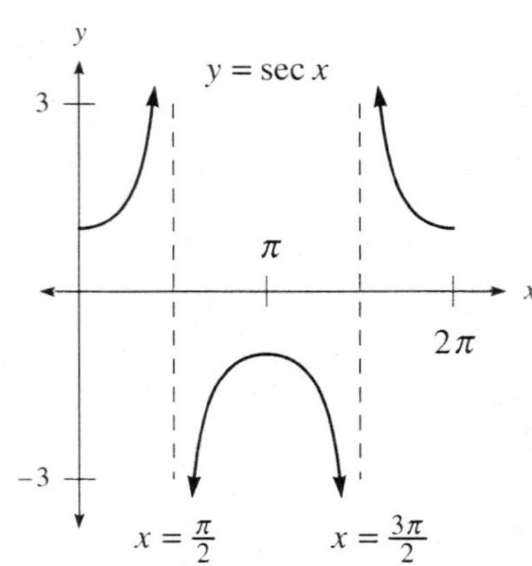

8.   Sketching $y = \cos x$ from $x = -4\pi$ to $x = 4\pi$:

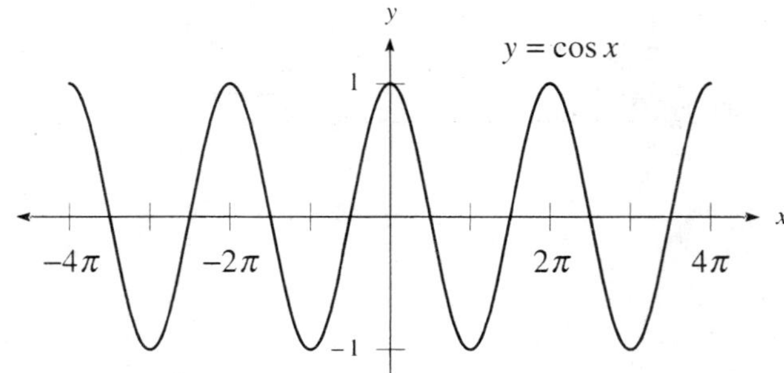

10.   Sketching $y = \csc x$ from $x = -4\pi$ to $x = 4\pi$:

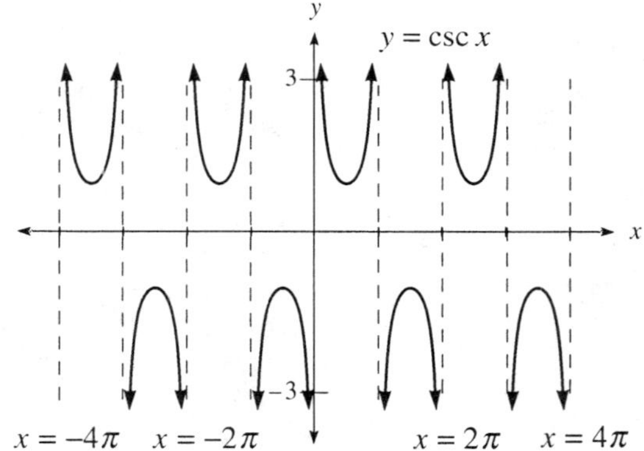

12.   Sketching $y = \tan x$ from $x = -4\pi$ to $x = 4\pi$:

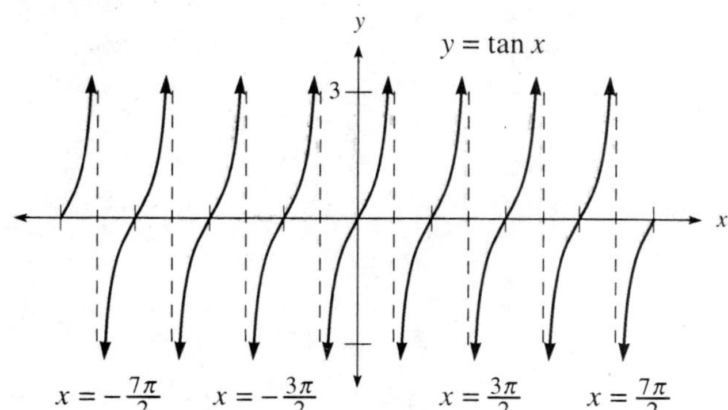

14. $\sin x = 0$ when x is any multiple of π. So $\sin x = 0$ when $x = k\pi$, where k is any integer.

16. $\cos x = 1$ when x is any multiple of 2π. So $\cos x = 1$ when $x = 2k\pi$, where k is any integer.

18. $\cot x = 0$ whenever $\cos x = 0$, which occurs when x is any odd multiple of $\dfrac{\pi}{2}$. So $\cot x = 0$ when $x = k \cdot \dfrac{\pi}{2}$, where

k is any odd integer. This can also be written as $x = \dfrac{\pi}{2} + k\pi$, where k is any integer.

20. $\sec x = 1$ whenever $\cos x = 1$, which occurs when x is any multiple of 2π. So $\cos x = 1$ when $x = 2k\pi$, where k is any integer.

22.   The function $\cot x$ is undefined when $\sin x = 0$, which occurs when $x = 0, \pi, 2\pi, \dots$. We can write this as $x = k\pi$, where k is any integer.

24.   The function $\sec x$ is undefined when $\cos x = 0$, which occurs when $x = \dfrac{\pi}{2}, \dfrac{3\pi}{2}, \dfrac{5\pi}{2}, \dots$. We can write this as

$x = k \cdot \dfrac{\pi}{2}$, where k is any odd integer. This can also be written as $x = \dfrac{\pi}{2} + k\pi$, where k is any integer.

26.   The amplitude is $A = \frac{1}{2}|4 - (-4)| = \frac{1}{2}(8) = 4$, and the period is 4π.

28.   The amplitude is $A = \frac{1}{2}|2 - (-2)| = \frac{1}{2}(4) = 2$, and the period is 2π.

30.   The amplitude is $A = \frac{1}{2}|5 - (-5)| = \frac{1}{2}(10) = 5$, and the period is 3.

32.   Making a table of values:

x	0	$\pi/2$	π	$3\pi/2$	2π
$y = \frac{1}{2}\cos x$	$\frac{1}{2}$	0	$-\frac{1}{2}$	0	$\frac{1}{2}$

Sketching the graph:

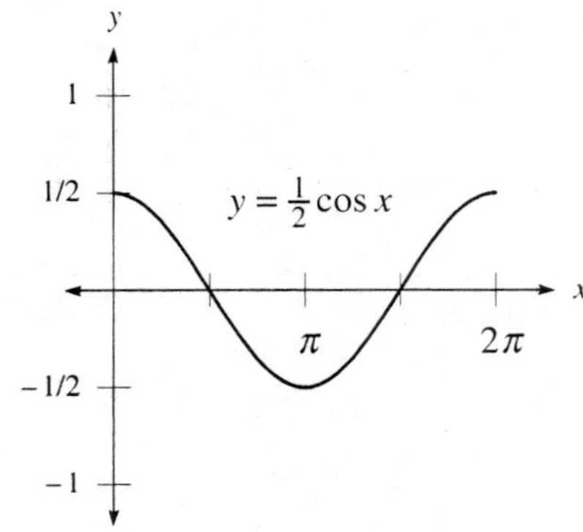

The amplitude is $A = \frac{1}{2}\left|\frac{1}{2} - \left(-\frac{1}{2}\right)\right| = \frac{1}{2}(1) = \frac{1}{2}$.

34.   Making a table of values:

x	0	$\pi/6$	$\pi/3$	$\pi/2$	$2\pi/3$	$5\pi/6$	π	$7\pi/6$	$4\pi/3$	$3\pi/2$	$5\pi/3$	$11\pi/6$	2π
$y = \sin 3x$	0	1	0	-1	0	1	0	-1	0	1	0	-1	0

Sketching the graph:

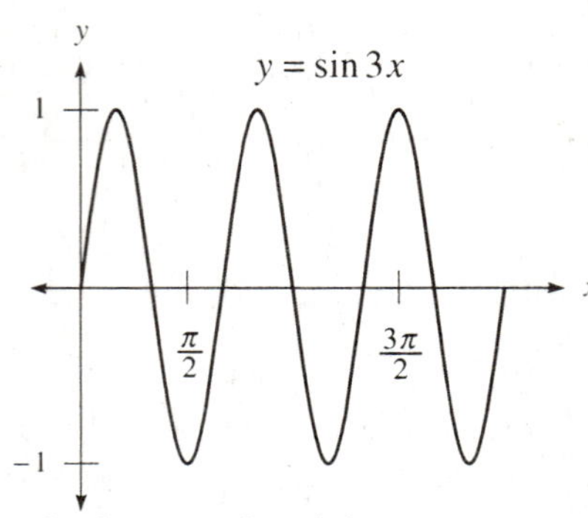

The graph goes through 3 complete cycles between 0 and 2π.

36.   The value of A effects the amplitude of the curve:

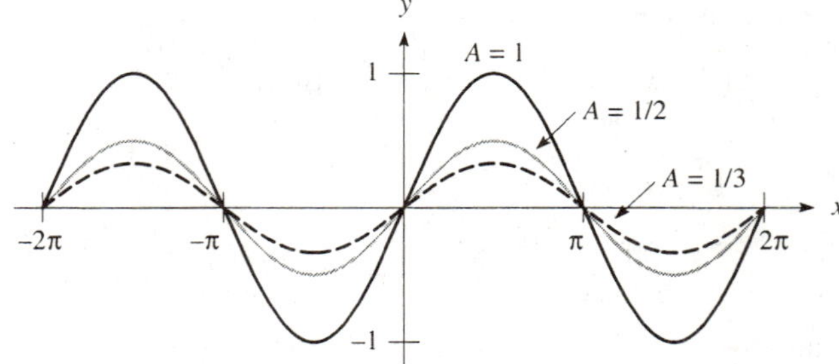

38.   The value of A effects the amplitude of the curve:

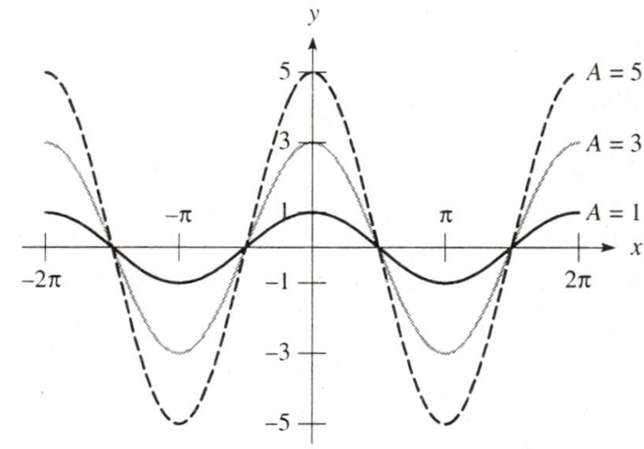

40.   The value of *A* effects the steepness (slope) of the curve:

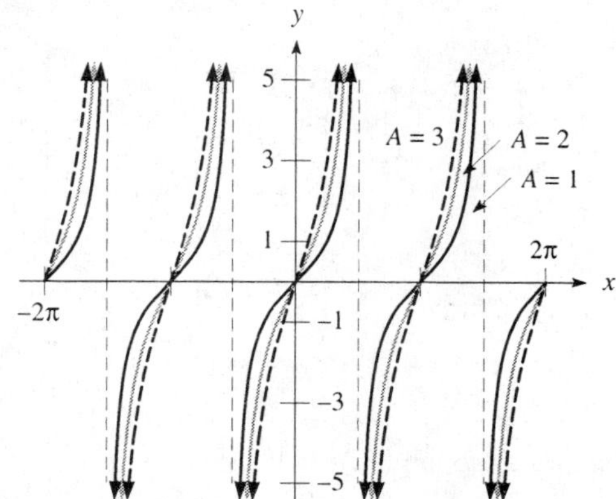

42   The value of *A* effects the highest and lowest values of each branch of the curve:

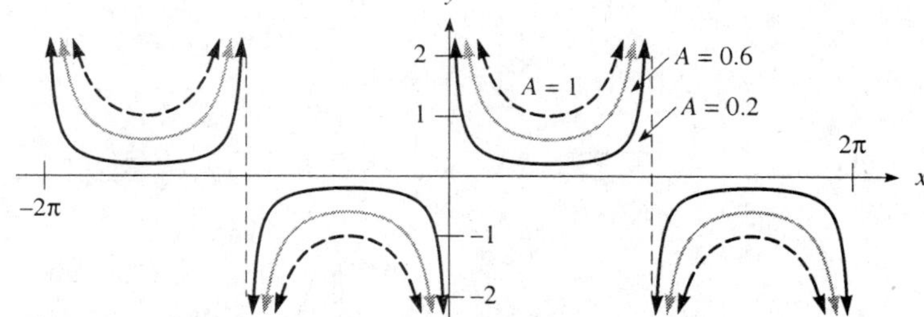

44.   The value of *B* effects how many cycles are completed in the given interval:

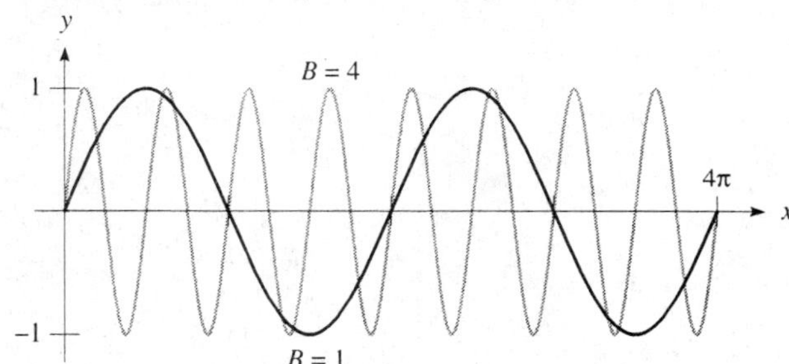

46.   The value of *B* effects how many cycles are completed in the given interval:

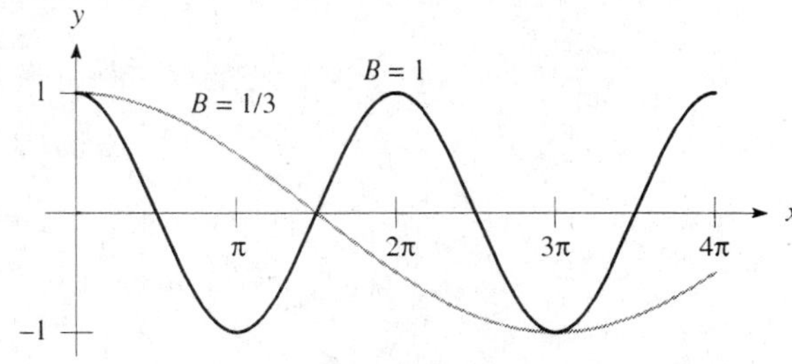

48.   The value of B effects how many cycles are completed in the given interval:

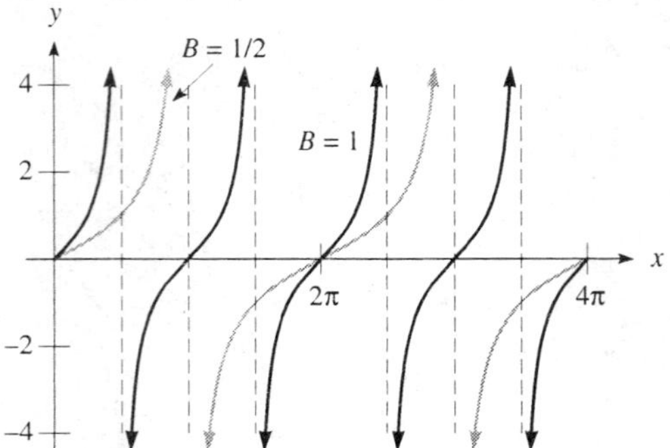

50.   The value of B effects how many cycles are completed in the given interval:

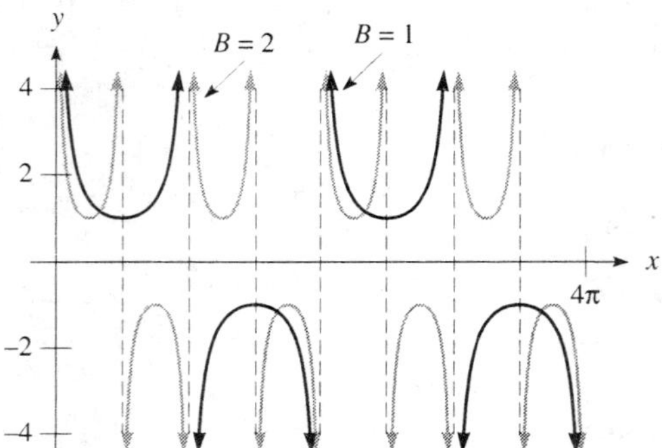

52.   The amplitude is 6:

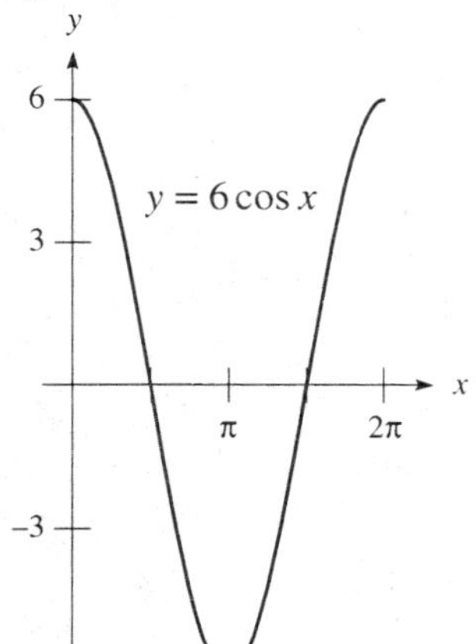

54.   The amplitude is 1/3:

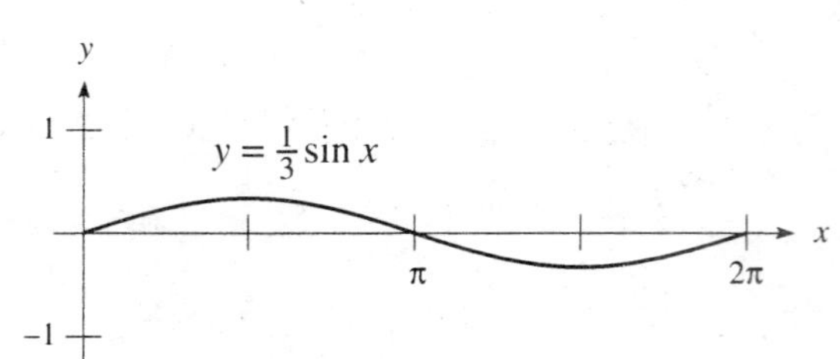

56. Graphing one cycle:

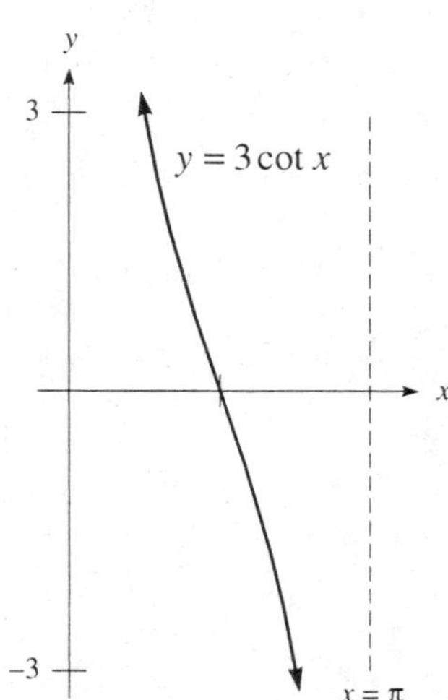

58. Graphing one cycle:

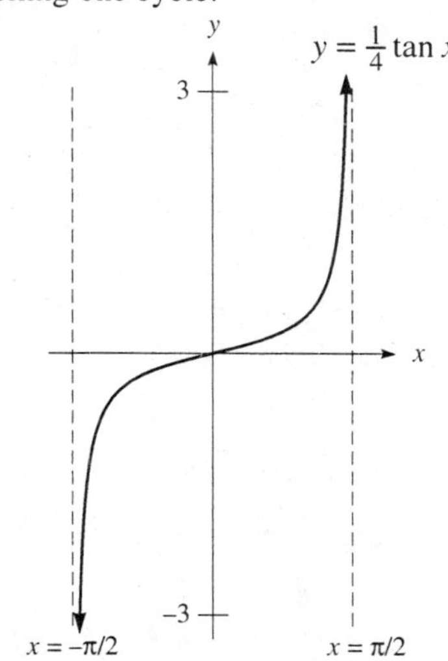

60. Graphing one cycle:

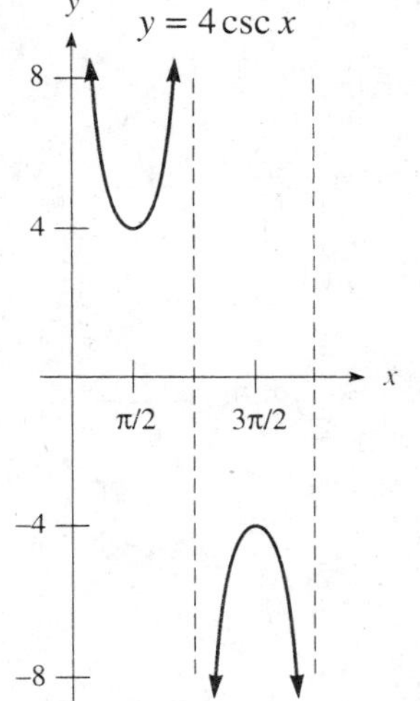

62. Graphing one cycle:

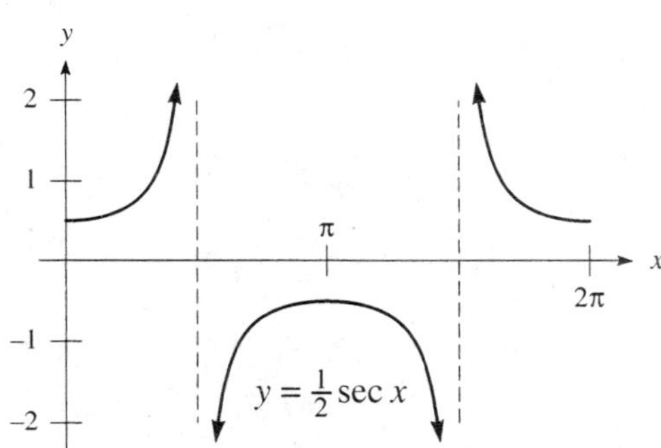

64. The negative sign reflects the curve across the x-axis:

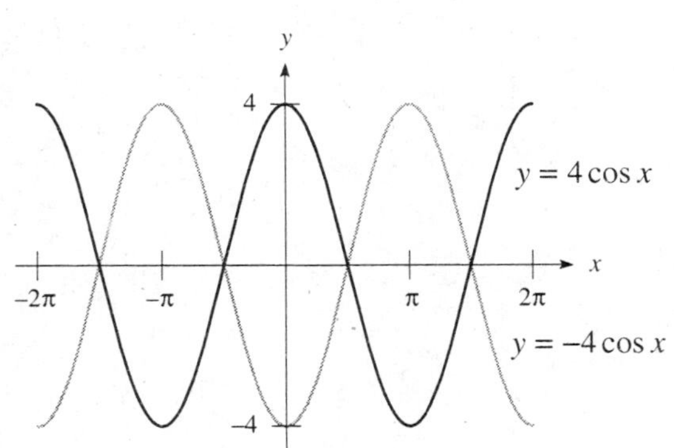

66. The negative sign reflects the curve across the x-axis:

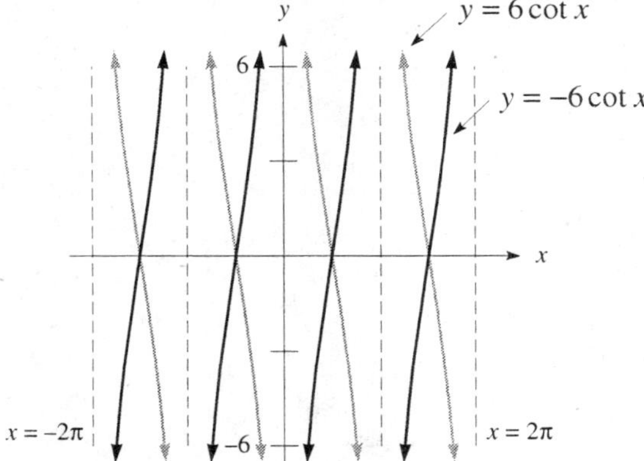

68.   The negative sign reflects the curve across the *x*-axis:

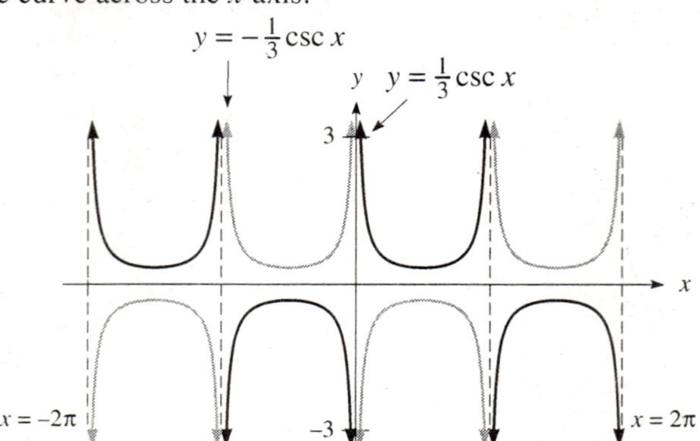

70.   Working from the left side: $\sin\theta\tan\theta + \cos\theta = \sin\theta\cdot\dfrac{\sin\theta}{\cos\theta} + \cos\theta\cdot\dfrac{\cos\theta}{\cos\theta} = \dfrac{\sin^2\theta + \cos^2\theta}{\cos\theta} = \dfrac{1}{\cos\theta} = \sec\theta$

72.   Working from the left side:
$$(\sin\theta + \cos\theta)^2 = (\sin\theta + \cos\theta)(\sin\theta + \cos\theta) = \sin^2\theta + \sin\theta\cos\theta + \sin\theta\cos\theta + \cos^2\theta = 1 + 2\sin\theta\cos\theta$$

74.   Working from the left side: $\sec\theta - \cos(-\theta) = \dfrac{1}{\cos\theta} - \cos\theta\cdot\dfrac{\cos\theta}{\cos\theta} = \dfrac{1 - \cos^2\theta}{\cos\theta} = \dfrac{\sin^2\theta}{\cos\theta}$

76.   Converting to degrees: $\dfrac{\pi}{6}$ radians $= \dfrac{\pi}{6}\cdot\dfrac{180°}{\pi} = 30°$ **78.**   Converting to degrees: $\dfrac{\pi}{2}$ radians $= \dfrac{\pi}{2}\cdot\dfrac{180°}{\pi} = 90°$

80.   Converting to degrees: $\dfrac{5\pi}{3}$ radians $= \dfrac{5\pi}{3}\cdot\dfrac{180°}{\pi} = 300°$ **82.**   Converting to degrees: $\dfrac{7\pi}{6}$ radians $= \dfrac{7\pi}{6}\cdot\dfrac{180°}{\pi} = 210°$

4.2 Period, Reflection, and Vertical Translation

2.   The period is $\dfrac{2\pi}{\frac{1}{2}} = 4\pi$. Sketching the graph: **4.**   The period is $\dfrac{2\pi}{3}$. Sketching the graph:

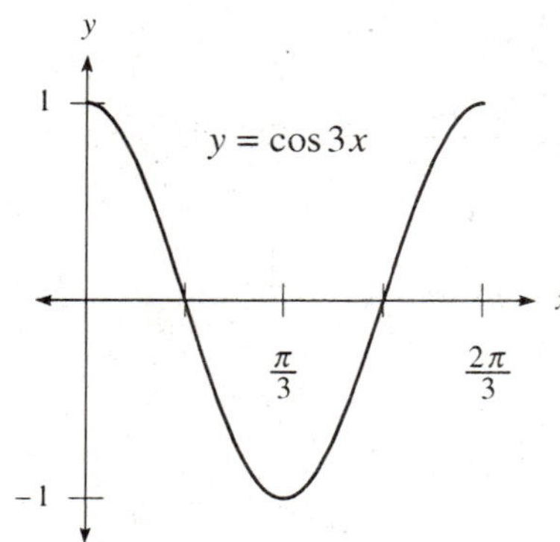

6.   The period is $\dfrac{2\pi}{\pi}=2$. Sketching the graph:

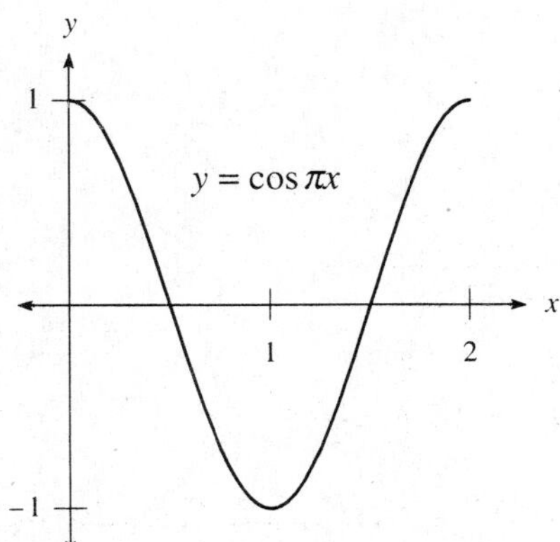

8.   The period is $\dfrac{2\pi}{\pi/2}=4$. Sketching the graph:

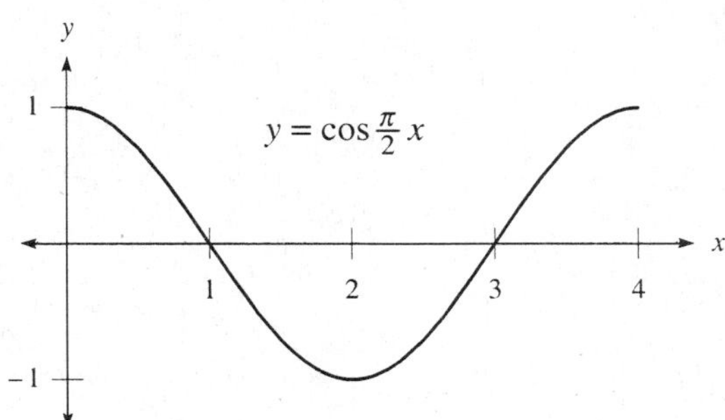

10.   The period is $\dfrac{2\pi}{4}=\dfrac{\pi}{2}$. Sketching the graph:

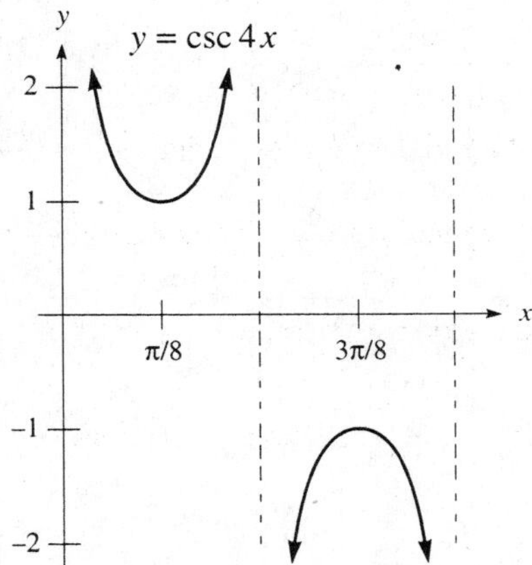

12.   The period is $\dfrac{2\pi}{1/4}=8\pi$. Sketching the graph:

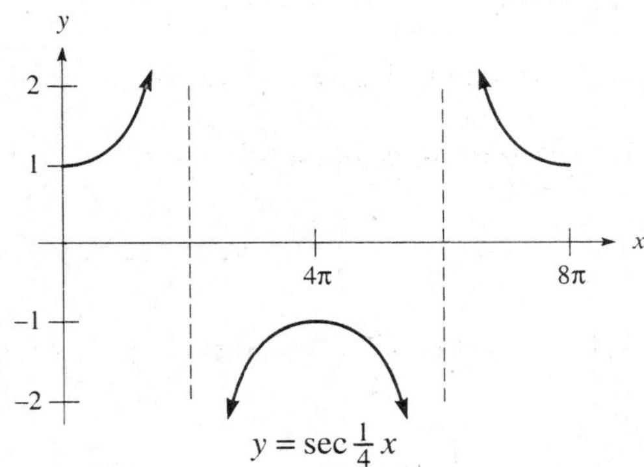

14.   The period is $\dfrac{\pi}{1/3}=3\pi$. Sketching the graph:

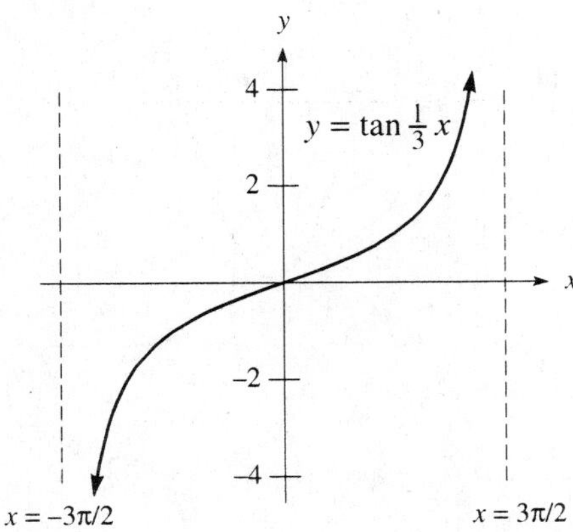

16.   The period is $\dfrac{\pi}{\pi}=1$. Sketching the graph:

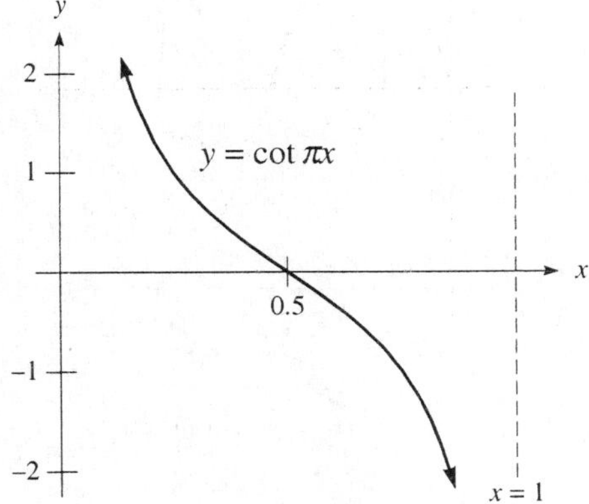

18.   The amplitude is 2 and the period is $\dfrac{2\pi}{4} = \dfrac{\pi}{2}$:

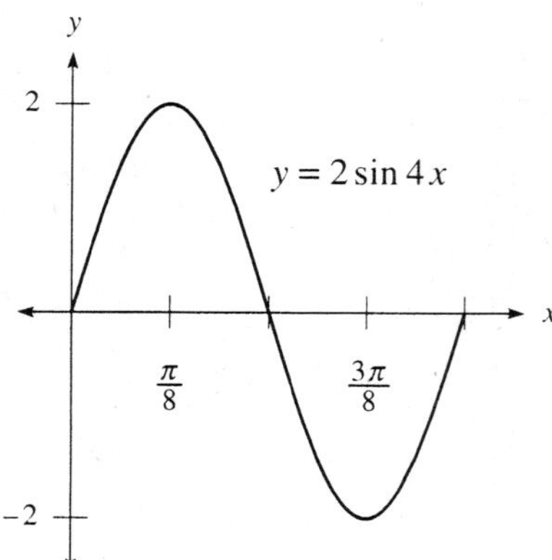

20.   The amplitude is 2 and the period is $\dfrac{2\pi}{\frac{1}{3}} = 6\pi$:

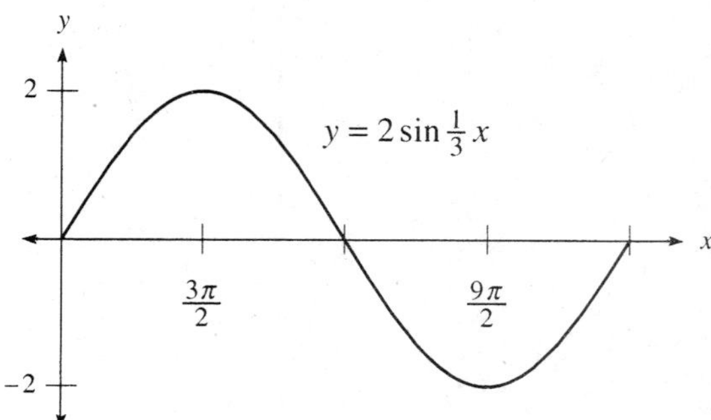

22.   The amplitude is $\frac{1}{2}$ and the period is $\dfrac{2\pi}{3}$:

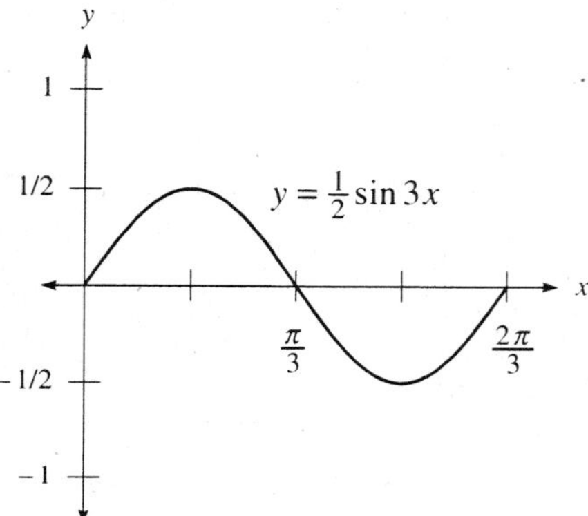

24.   The amplitude is 2 and the period is $\dfrac{2\pi}{\pi/2} = 4$:

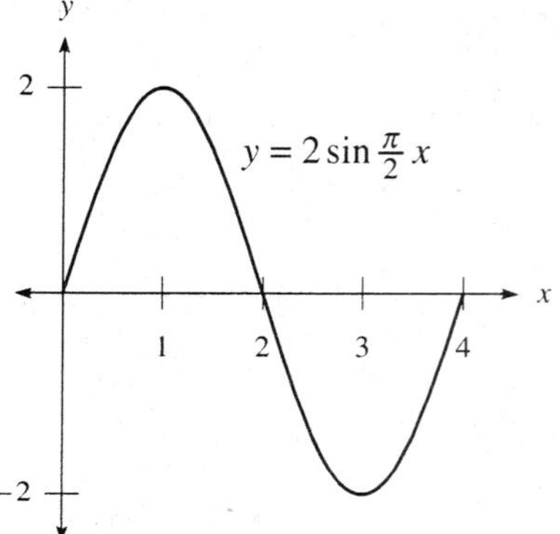

26.   The period is $\dfrac{2\pi}{3}$:

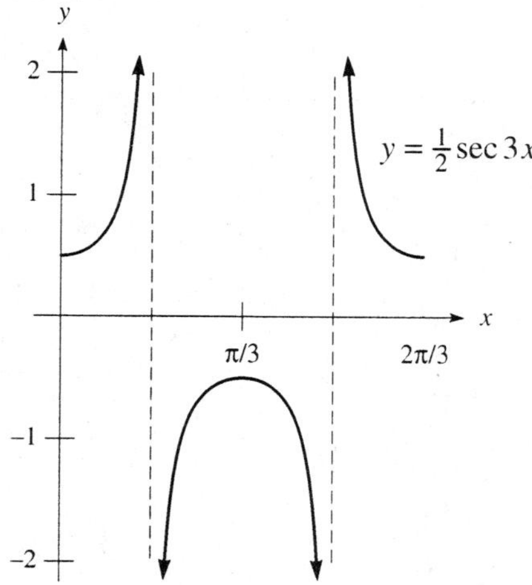

28.   The period is $\dfrac{2\pi}{1/2} = 4\pi$:

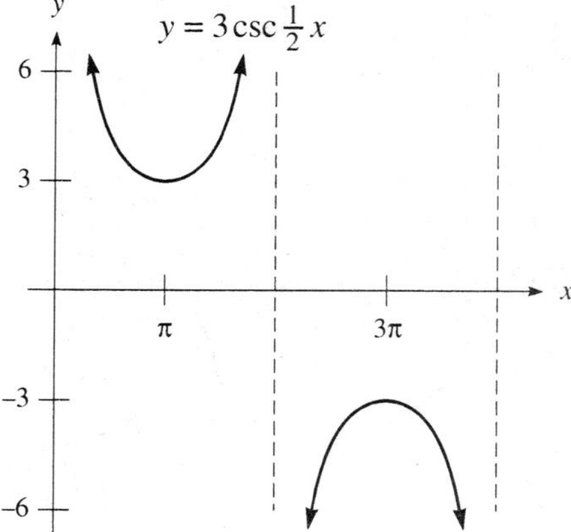

30. The period is $\dfrac{\pi}{2}$:

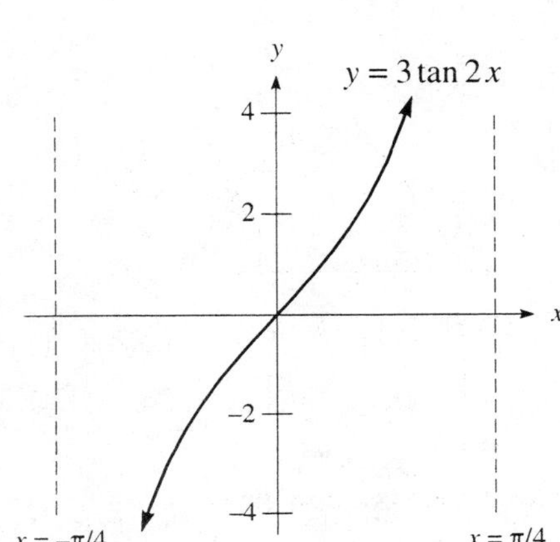

32. The period is $\dfrac{\pi}{1/2} = 2\pi$:

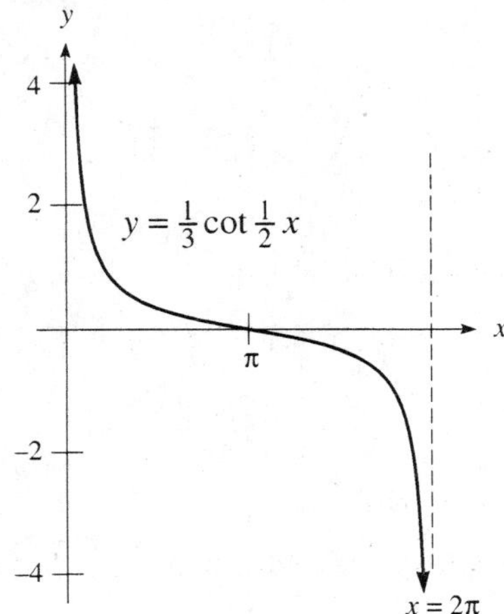

34. Displacing the graph from Problem 18 by −2 units:

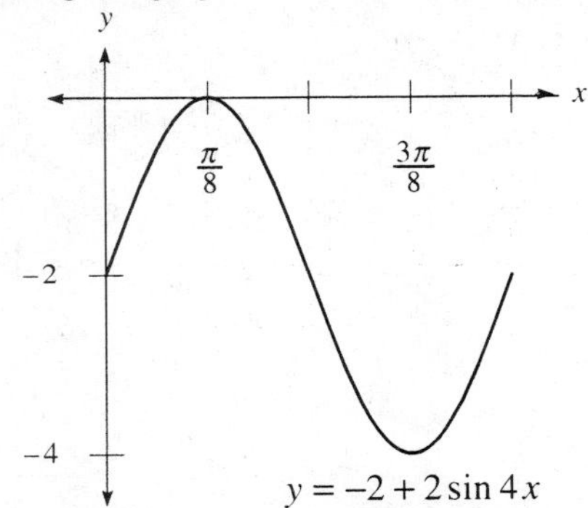

36. Displacing the graph from Problem 22 by 1 unit:

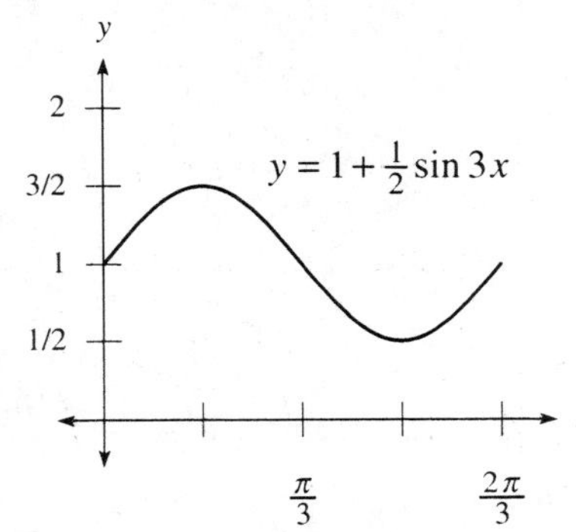

38. Displacing the graph from Problem 28 by 3 units:

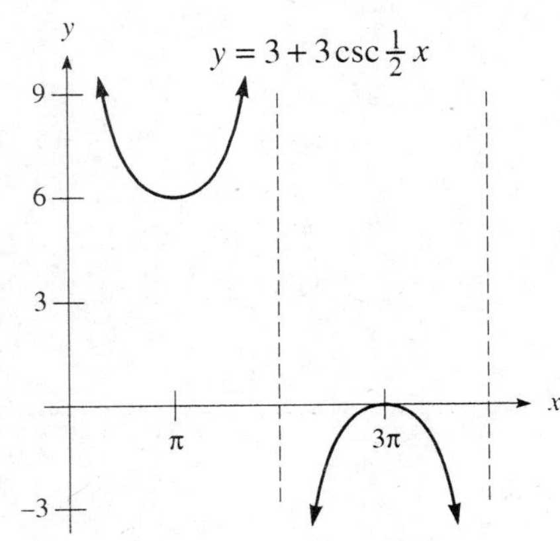

40. Displacing the graph from Problem 30 by −4 units:

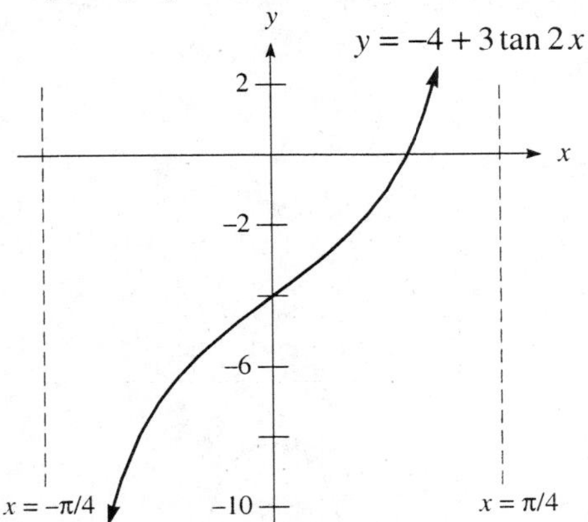

42. The amplitude is 3 and the period is $\dfrac{2\pi}{\pi} = 2$:

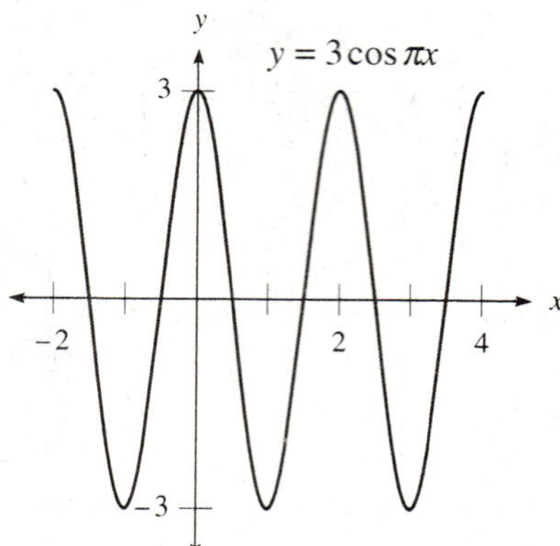

44. The amplitude is 3 and the period is $\dfrac{2\pi}{2} = \pi$. Note the reflection across the x-axis:

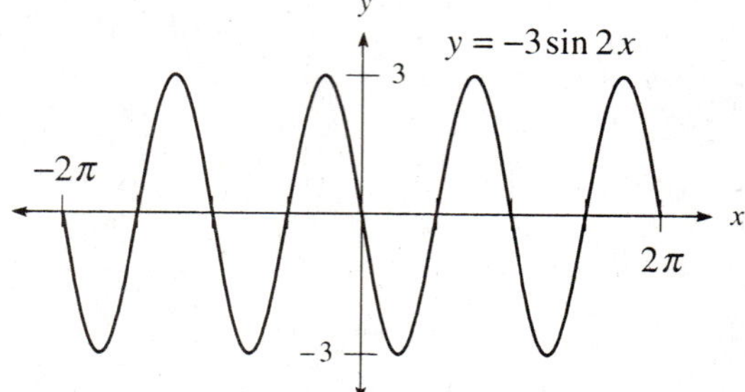

46. The amplitude is 3 and the period is $\dfrac{2\pi}{\frac{1}{2}} = 4\pi$. Sketching the graph:

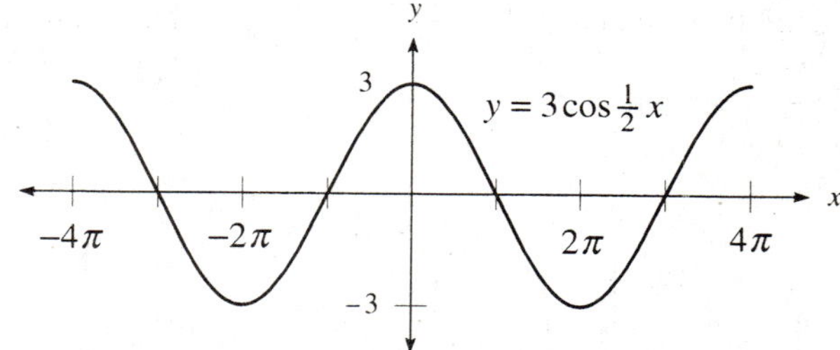

48. Note this graph is the same as $y = -2\cos 3x$, since cosine is an even function. The amplitude is 2 and the period is $\dfrac{2\pi}{3}$. Also note the reflection across the x-axis:

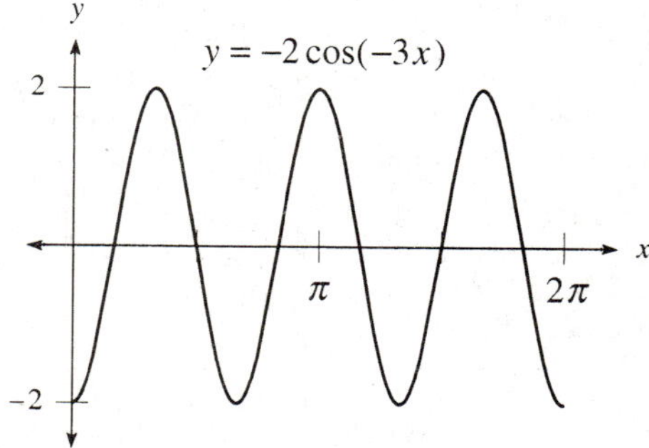

50.   The period is $\dfrac{2\pi}{3}$. Sketching the graph for $0 \le x \le 2\pi$:

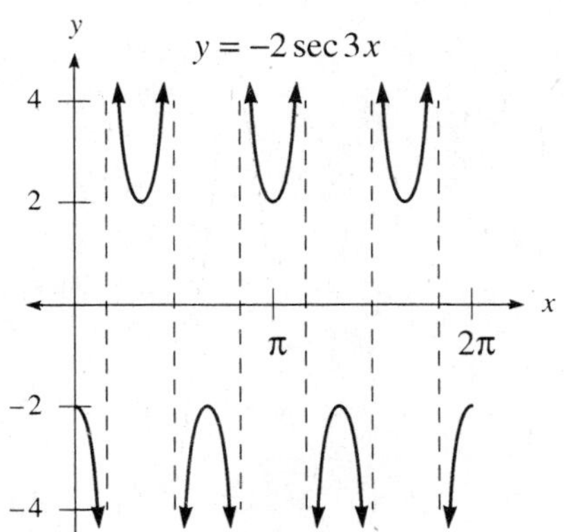

52.   The period is $\dfrac{\pi}{4}$. Note this is a reflection of $y = \tan 4x$ across the x-axis. Sketching the graph:

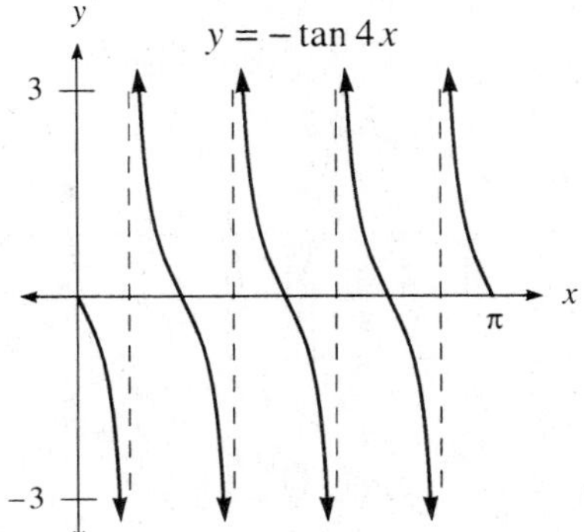

54.   The maximum velocity is the amplitude, which is $3.5\,\dfrac{\text{meters}}{\text{sec}}$. The time it takes to move from lowest to highest position is $\tfrac{1}{2}$ of the period, which is $\dfrac{1}{2}\left(\dfrac{2\pi}{2\pi}\right) = \tfrac{1}{2}$ second.

56.   Sketching the graph:

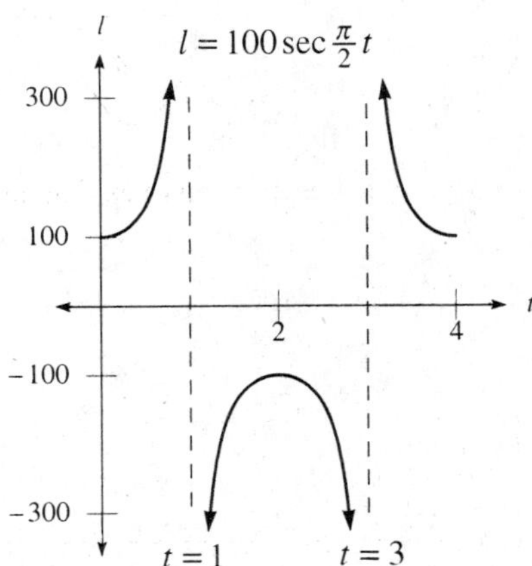

58.   **a.**   Sketching the graphs:

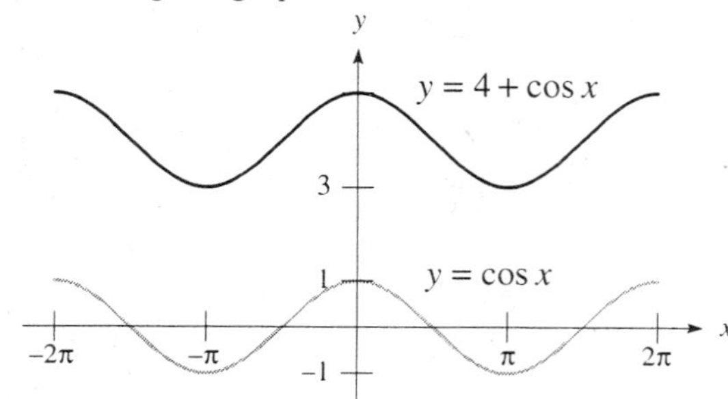

b.   Sketching the graphs:

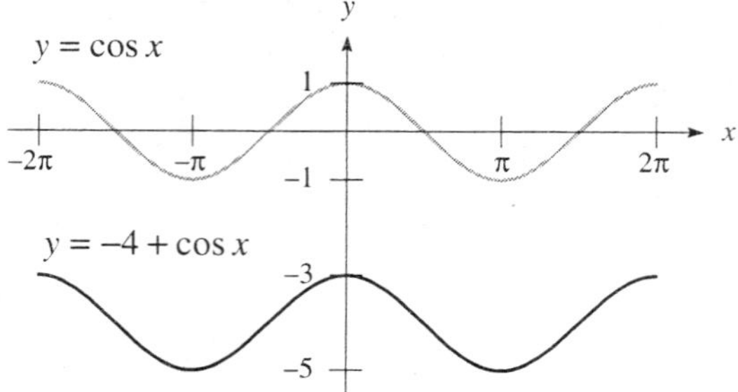

c.   Sketching the graphs:

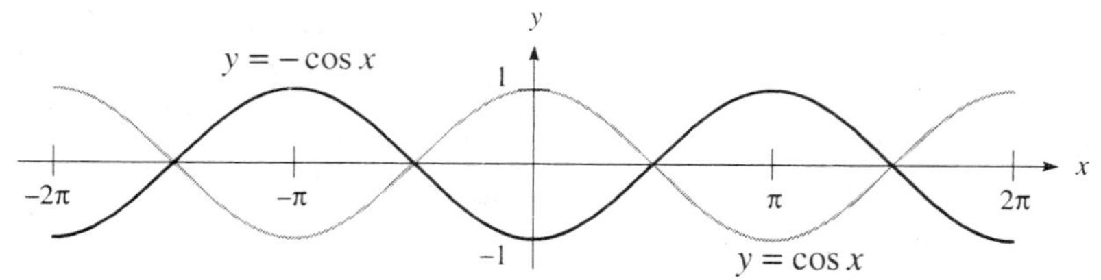

60.   **a.**   Sketching the graphs:

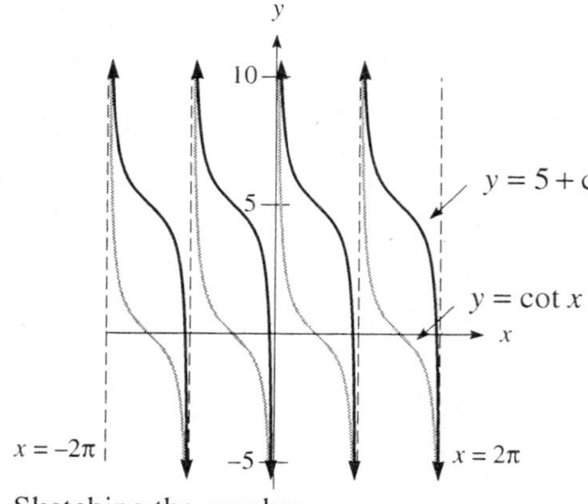

b.   Sketching the graphs:

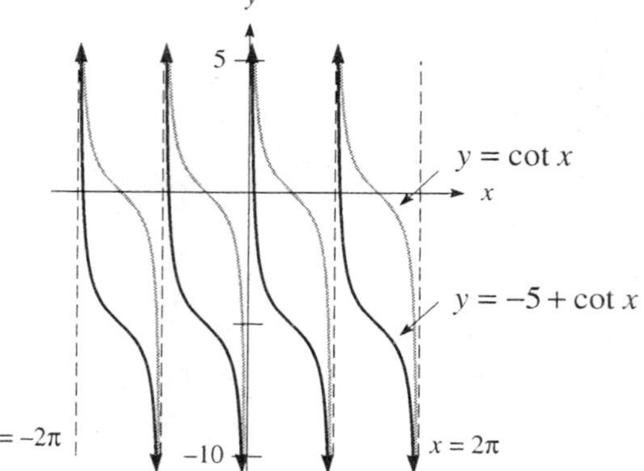

c.   Sketching the graphs:

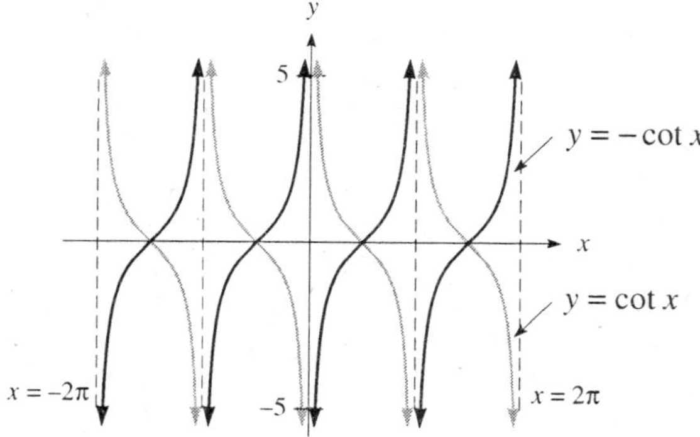

62.   **a.**   Sketching the graphs:

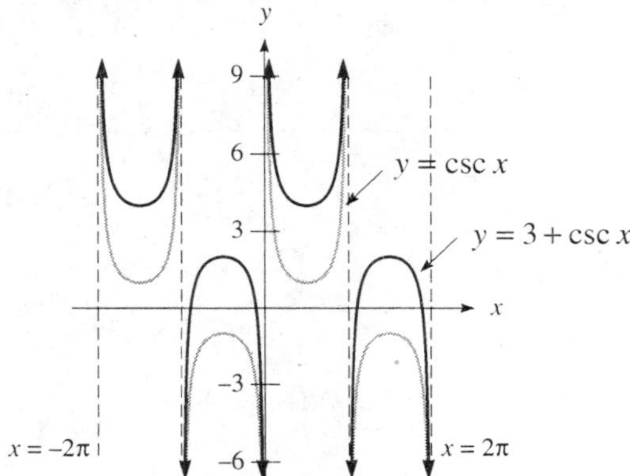

b.   Sketching the graphs:

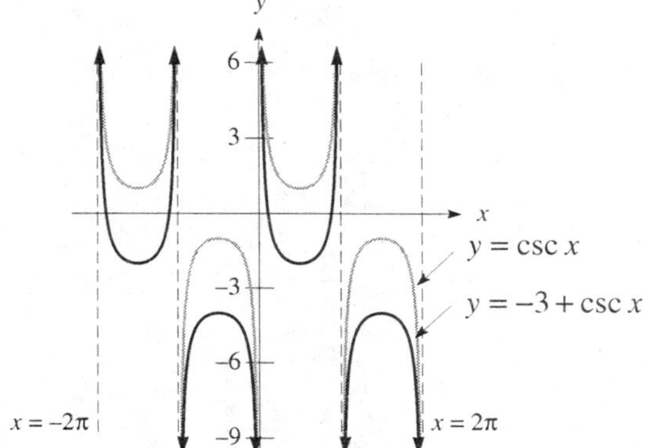

c.   Sketching the graphs:

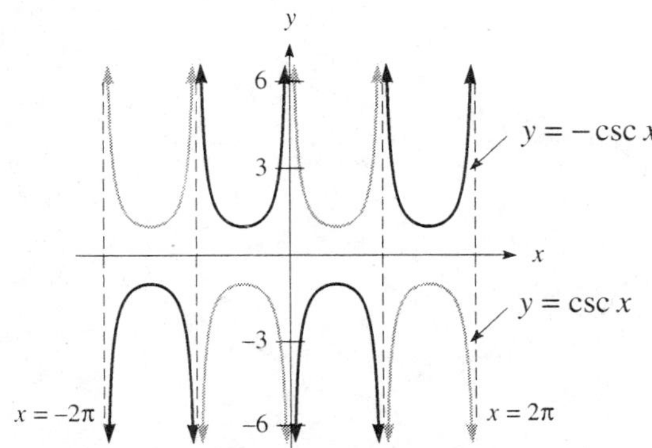

64.   The value of C affects the graph by shifting it $-C$ units along the x-axis:

a.   Sketching the graphs:

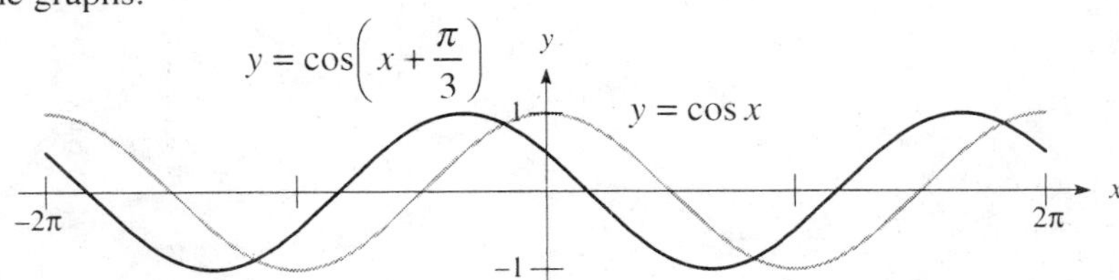

b.   Sketching the graphs:

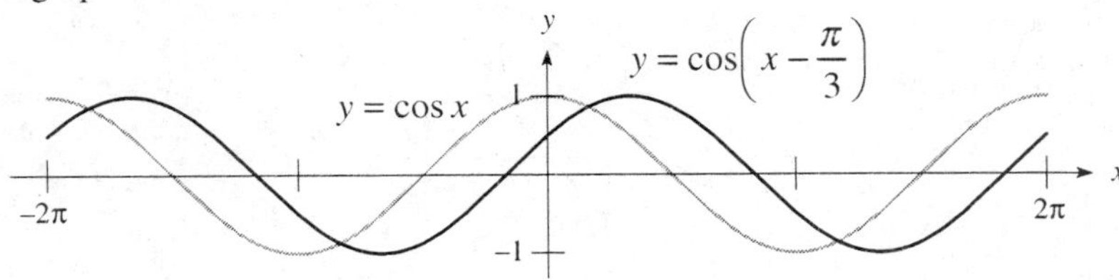

66.   The value of C affects the graph by shifting it $-C$ units along the x-axis:
    a.   Sketching the graphs: **b.**   Sketching the graphs:

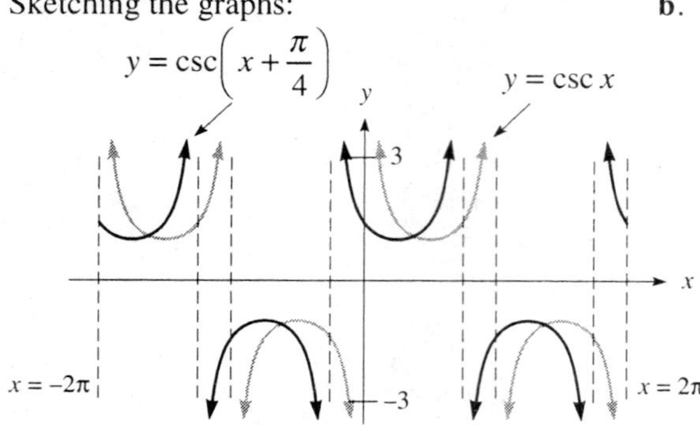

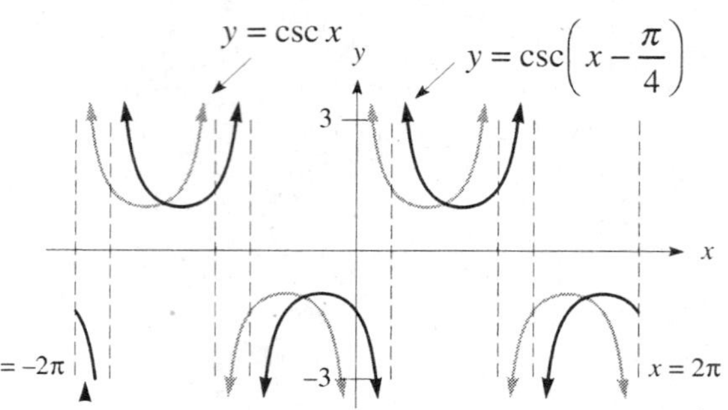

68.   Evaluating when $x = \dfrac{\pi}{2}$: $\sin\left(\dfrac{\pi}{2} - \dfrac{\pi}{2} \right) = \sin 0 = 0$ **70.**   Evaluating when $y = \dfrac{\pi}{6}$: $\cos\left(\dfrac{\pi}{6} + \dfrac{\pi}{6} \right) = \cos\dfrac{\pi}{3} = \dfrac{1}{2}$

72.   Evaluating when $x = \dfrac{\pi}{2}$ and $y = \dfrac{\pi}{6}$: $\cos\left(\dfrac{\pi}{2} + \dfrac{\pi}{6} \right) = \cos\dfrac{2\pi}{3} = -\dfrac{1}{2}$

74.   Evaluating when $x = \dfrac{\pi}{2}$ and $y = \dfrac{\pi}{6}$: $\cos\dfrac{\pi}{2} + \cos\dfrac{\pi}{6} = 0 + \dfrac{\sqrt{3}}{2} = \dfrac{\sqrt{3}}{2}$

76.   Converting to radians: $30° = 30° \cdot \dfrac{\pi}{180°} = \dfrac{\pi}{6}$ radians   **78.**   Converting to radians: $90° = 90° \cdot \dfrac{\pi}{180°} = \dfrac{\pi}{2}$ radians

80.   Converting to radians: $300° = 300° \cdot \dfrac{\pi}{180°} = \dfrac{5\pi}{3}$ radians

82.   Converting to radians: $120° = 120° \cdot \dfrac{\pi}{180°} = \dfrac{2\pi}{3}$ radians

4.3 Phase Shift

2.   The phase shift is $-\dfrac{\pi}{6}$:   **4.**   The phase shift is $\dfrac{\pi}{6}$:

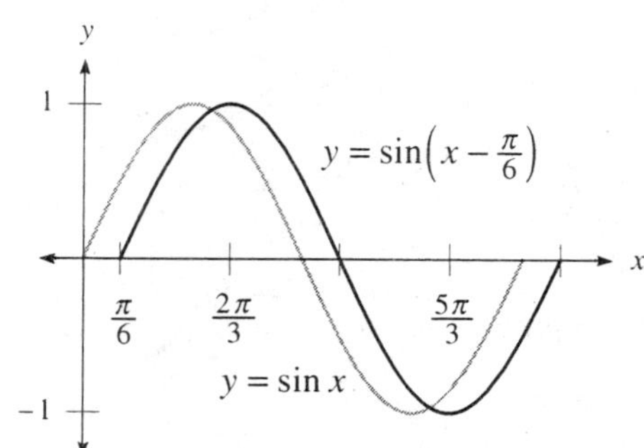

6.   The phase shift is $\dfrac{\pi}{3}$:

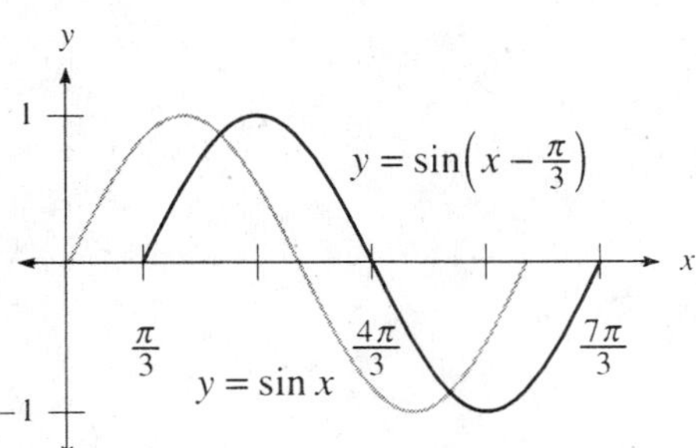

8.   The phase shift is $-\dfrac{\pi}{2}$:

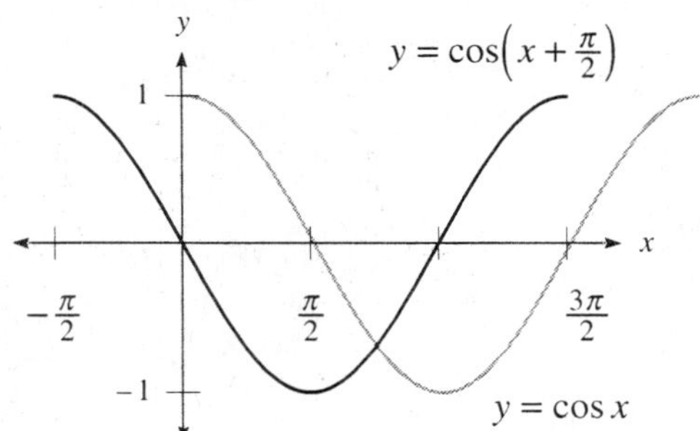

10.   The phase shift is $\dfrac{\pi}{4}$:

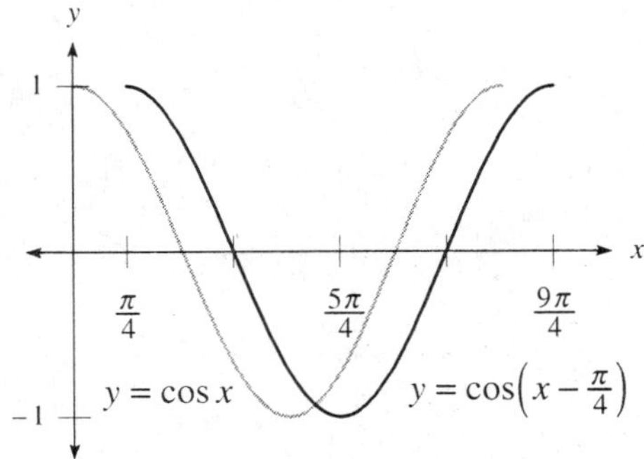

12.   The amplitude is 1, the period is $\dfrac{2\pi}{2} = \pi$, and the phase shift is $-\dfrac{\pi}{2}$:

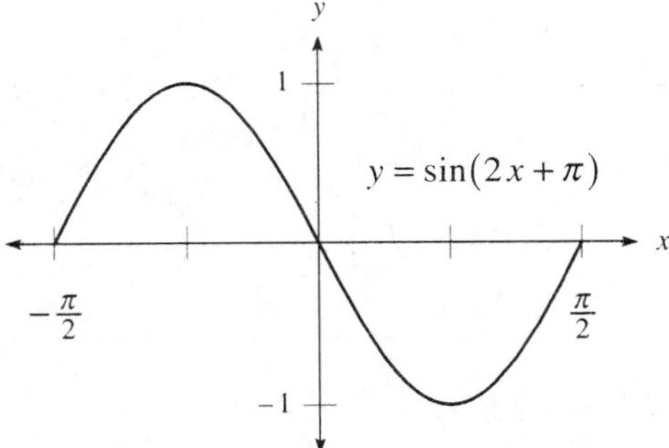

14.   The amplitude is 1, the period is $\dfrac{2\pi}{\pi} = 2$, and the phase shift is $\dfrac{\pi/2}{\pi} = \frac{1}{2}$:

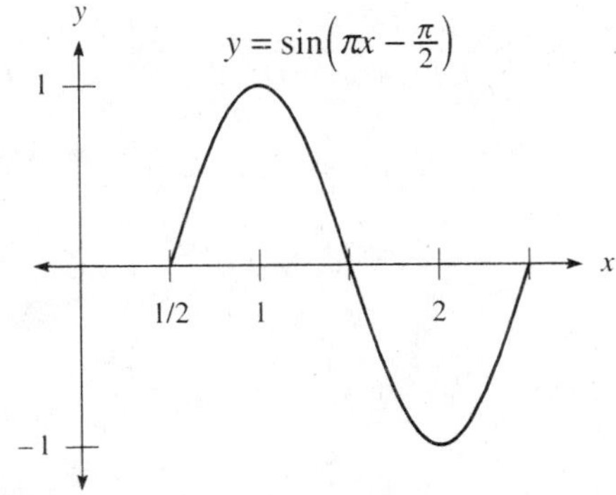

16.   The amplitude is 1, the period is $\dfrac{2\pi}{2} = \pi$, and the phase shift is $\dfrac{\pi/2}{2} = \dfrac{\pi}{4}$. Note this is a reflection of $y = \cos\left(2x - \dfrac{\pi}{2}\right)$ across the x-axis:

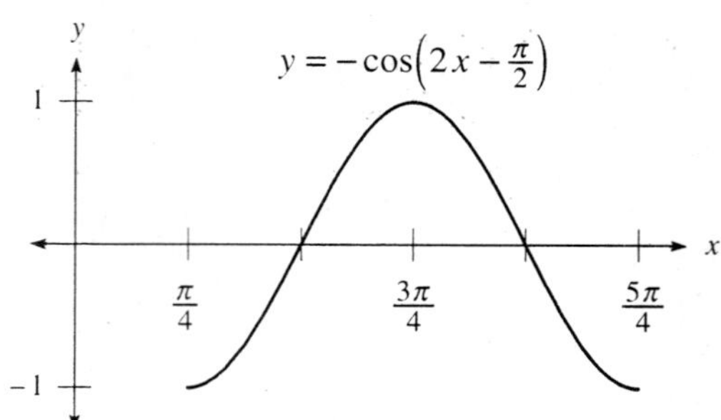

18.   The amplitude is 3, the period is $\dfrac{2\pi}{1/2} = 4\pi$, and the phase shift is $\dfrac{-\pi/3}{1/2} = -\dfrac{2\pi}{3}$:

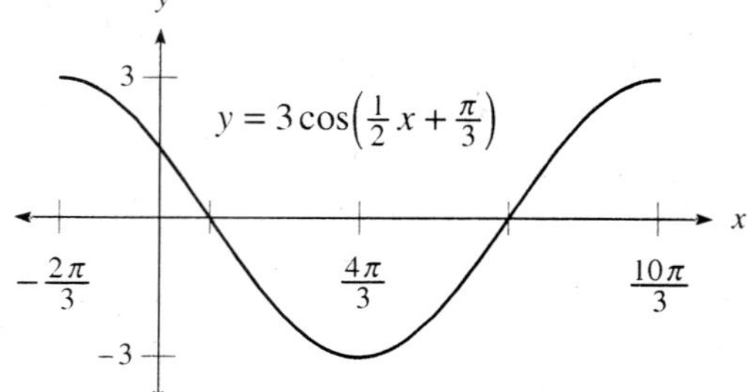

20.   The amplitude is $\frac{4}{3}$, the period is $\dfrac{2\pi}{3}$, and the phase shift is $\dfrac{-\pi/2}{3} = -\dfrac{\pi}{6}$:

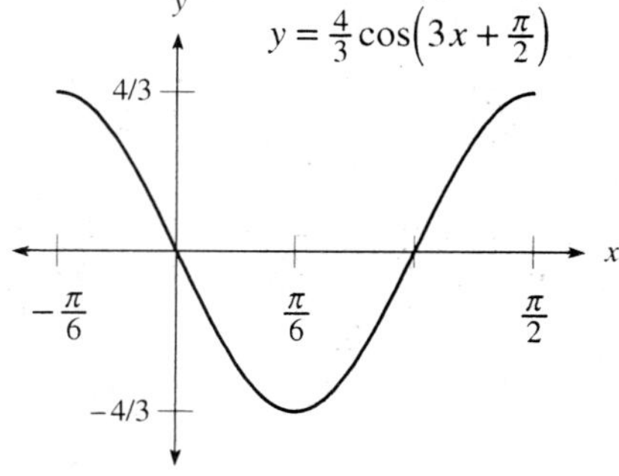

22.   The amplitude is 3, the period is $\dfrac{2\pi}{\pi/3} = 6$, and the phase shift is $\dfrac{\pi/3}{\pi/3} = 1$:

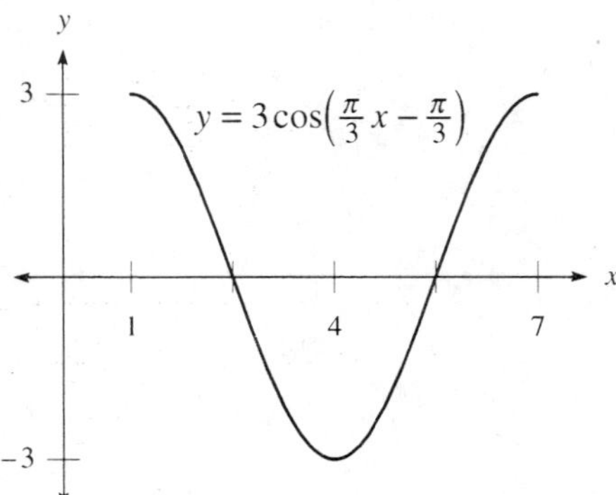

24.   Displacing the graph from Problem 12 by −1 units:

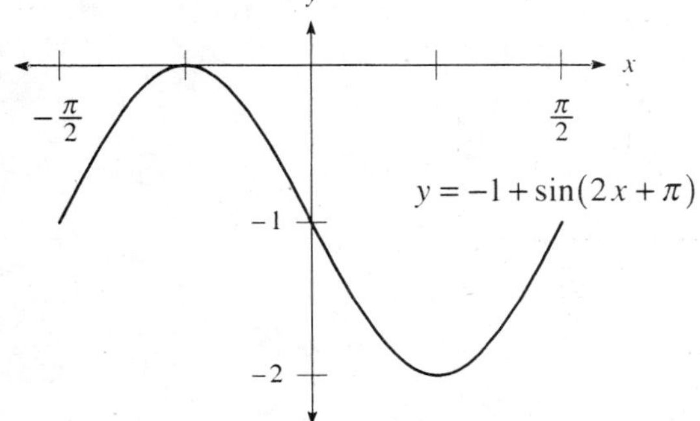

26.   Displacing the graph from Problem 14 by 3 units:

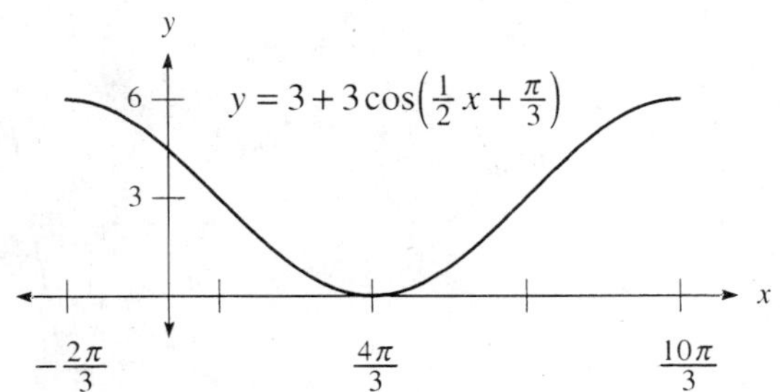

28.   Displacing the graph from Problem 16 by −2 units:

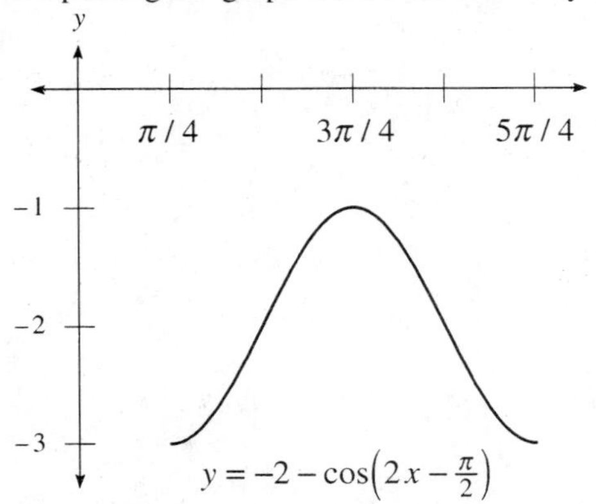

30.   Displacing the graph from Problem 18 by 3 units:

32.   Displacing the graph from Problem 20 by $\frac{2}{3}$ units:

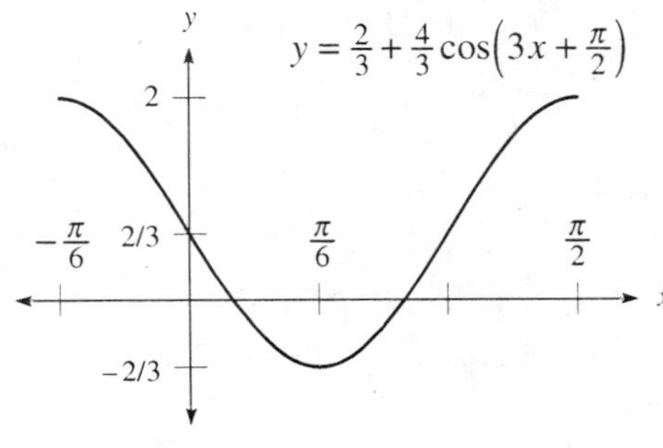

34.   The amplitude is 3, the period is $\dfrac{2\pi}{2} = \pi$, and the phase shift is $\dfrac{\pi/3}{2} = \dfrac{\pi}{6}$:

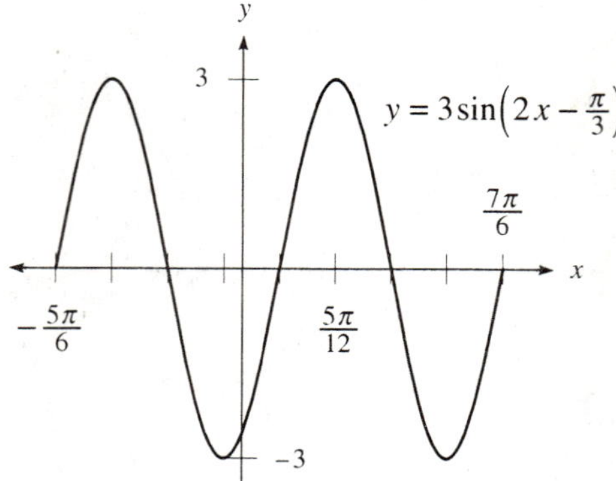

36.   Reflecting the graph from Problem 34 across the x-axis:

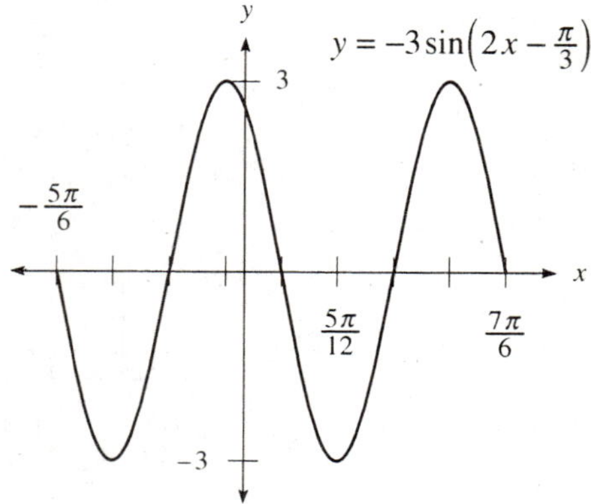

38.   The amplitude is $\frac{3}{4}$, the period is $\dfrac{2\pi}{3}$, and the phase shift is $\dfrac{\pi/2}{3} = \dfrac{\pi}{6}$:

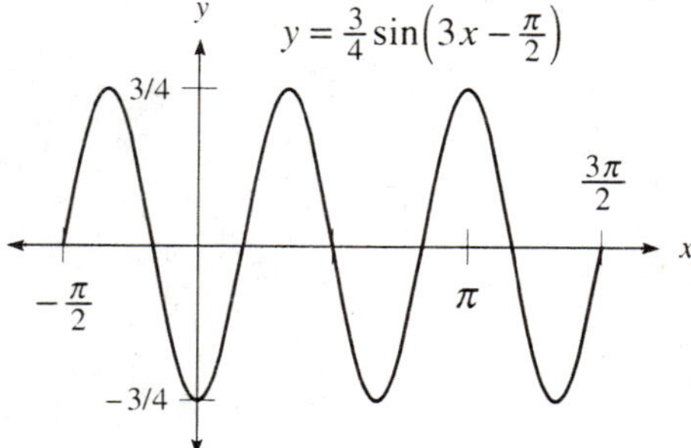

40. Reflecting the graph from Problem 38 across the x-axis:

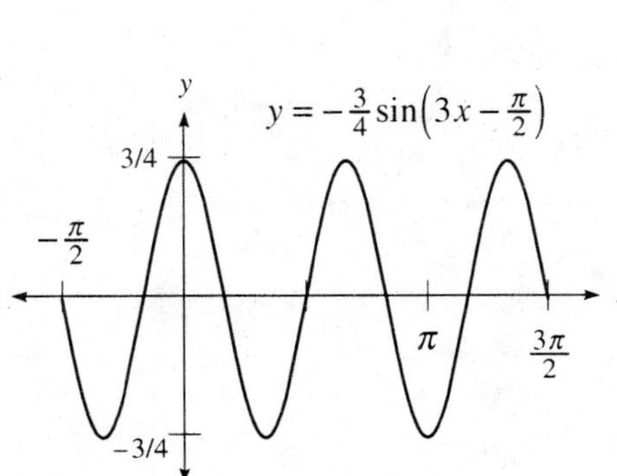

42. Sketching the graph by first graphing $y = \cos\left(x + \frac{\pi}{4}\right)$:

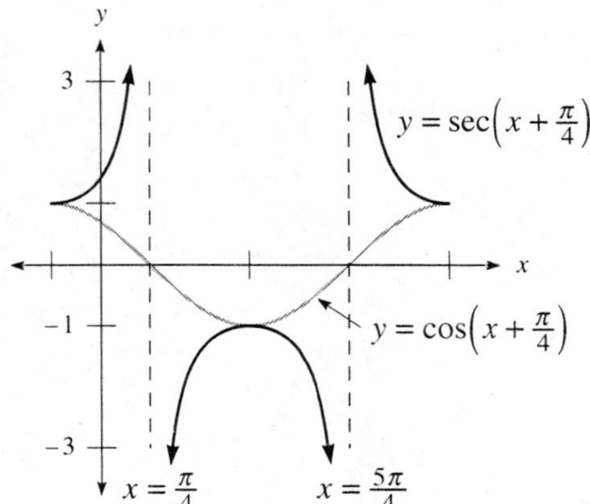

44. Sketching the graph by first graphing $y = 2\sin\left(2x - \frac{\pi}{2}\right)$:

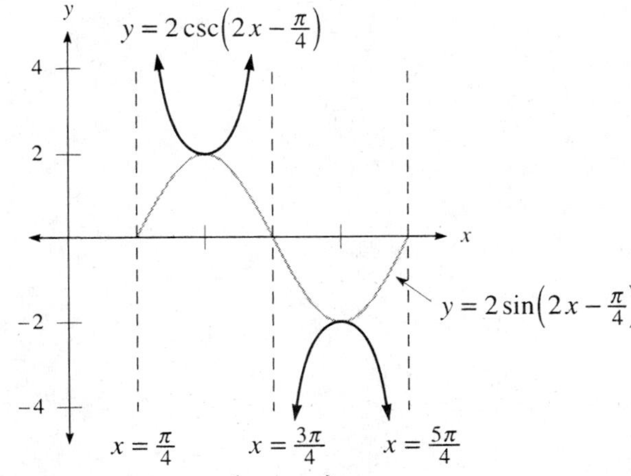

46. Sketching the graph by first graphing $y = 3\cos\left(2x - \frac{\pi}{3}\right)$:

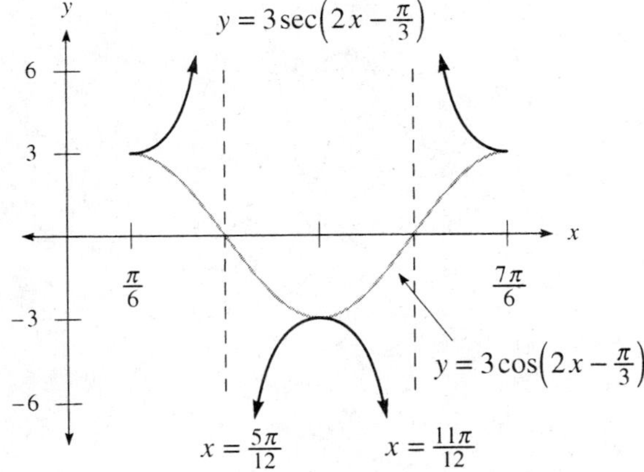

48.   The phase shift is $\dfrac{\pi}{4}$:

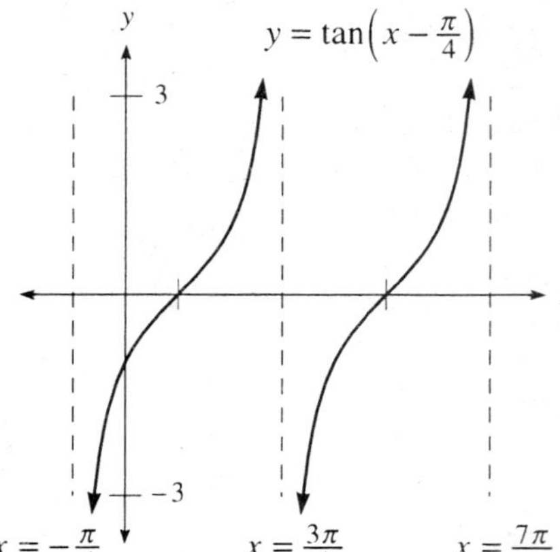

50.   The phase shift is $-\dfrac{\pi}{4}$:

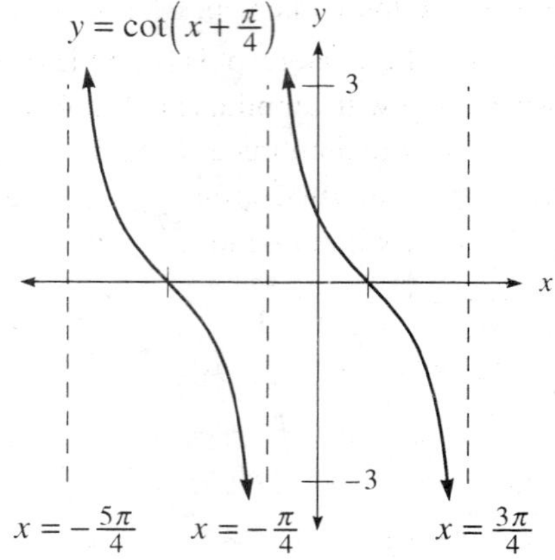

52.   The period is $\dfrac{\pi}{2}$ and the phase shift is $\dfrac{-\pi/2}{2} = -\dfrac{\pi}{4}$:

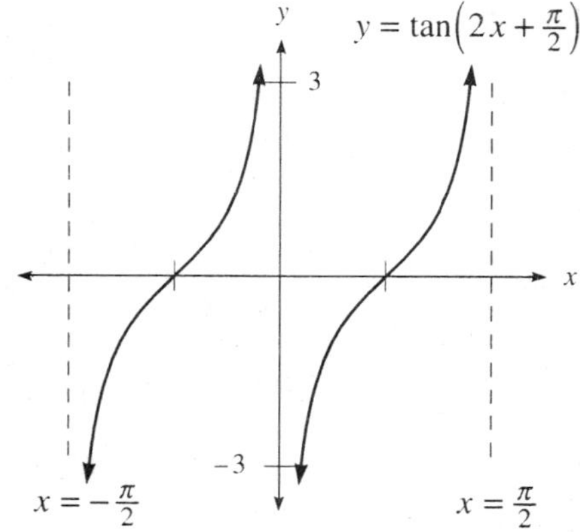

54.   First convert 90° to radians: $90° = 90 \cdot \dfrac{\pi}{180} = \dfrac{\pi}{2}$ radians

The arc length is given by: $s = r\theta = 22 \cdot \dfrac{\pi}{2} = 11\pi\,$cm

56.   In 5 hours the hour hand has moved $\frac{5}{12}$ revolutions, so the central angle is given by $\frac{5}{12} \cdot 2\pi = \dfrac{5\pi}{6}$ radians. So the arc

length is given by: $s = r\theta = 3 \cdot \dfrac{5\pi}{6} = \dfrac{5\pi}{2}$ inches

58.   First convert 135° to radians: $135° = 135 \cdot \dfrac{\pi}{180} = \dfrac{3\pi}{4}$ radians. Since $s = r\theta$:

$$75 = r \cdot \dfrac{3\pi}{4}$$
$$300 = 3\pi r$$
$$r = \dfrac{100}{\pi}\,\text{m}$$

The radius of the circle is $\dfrac{100}{\pi}$ meters.

4.4 Finding an Equation from Its Graph

2. The slope is $-\frac{1}{2}$ and the y-intercept is 1, so the equation of this line is $y = -\frac{1}{2}x + 1$.

4. The slope is -2 and the y-intercept is 3, so the equation of this line is $y = -2x + 3$.

6. This is a cosine curve with amplitude $= 1$ and period $= 2\pi$. There is no phase shift, so the equation is $y = \cos x$.

8. This is a sine curve with amplitude $= 2$ and period $= 2\pi$. There is no phase shift, so the equation is $y = 2\sin x$.

10. This is a reflection of the curve from Problem 8 across the x-axis, so the equation is $y = -2\sin x$.

12. This is a cosine curve with amplitude $= 1$ and period $= \pi$, therefore:

$$\frac{2\pi}{B} = \pi$$
$$2\pi = \pi B$$
$$B = 2$$

There is no phase shift, so the equation is $y = \cos 2x$.

14. This is a cosine curve with amplitude $= 1$ and period $= 4\pi$, therefore:

$$\frac{2\pi}{B} = 4\pi$$
$$2\pi = 4\pi B$$
$$B = \tfrac{1}{2}$$

There is no phase shift, so the equation is $y = \cos \frac{1}{2}x$.

16. This is a sine curve with amplitude $= 3$ and period $= \pi$, therefore:

$$\frac{2\pi}{B} = \pi$$
$$2\pi = \pi B$$
$$B = 2$$

There is no phase shift, so the equation is $y = 3\sin 2x$.

18. This is a cosine curve with amplitude $= 2$ and period $= 2$, therefore:

$$\frac{2\pi}{B} = 2$$
$$2\pi = 2B$$
$$B = \pi$$

There is no phase shift, so the equation is $y = 2\cos \pi x$.

20. This is a reflection of the curve from Problem 18 across the x-axis, so the equation is $y = -2\cos \pi x$.

22. This is a vertical translation of the curve from Problem 20 of 3 units, so its equation is $y = -2\cos \pi x + 3$, or $y = 3 - 2\cos \pi x$.

24. This is a sine curve with amplitude $= 2$ and period $= \dfrac{2\pi}{3}$, therefore:

$$\frac{2\pi}{B} = \frac{2\pi}{3}$$
$$6\pi = 2\pi B$$
$$B = 3$$

The phase shift is $\dfrac{\pi}{6}$, therefore:

$$-\frac{C}{B} = \frac{\pi}{6}$$
$$-\frac{C}{3} = \frac{\pi}{6}$$
$$C = -\frac{\pi}{2}$$

The equation is $y = 2\sin\left(3x - \dfrac{\pi}{2}\right)$.

26.   This is a sine curve with amplitude = 3 and period = π, therefore:

$$\frac{2\pi}{B} = \pi$$
$$2\pi = \pi B$$
$$B = 2$$

The phase shift is $\dfrac{\pi}{4}$, therefore:

$$-\frac{C}{B} = \frac{\pi}{4}$$
$$-\frac{C}{2} = \frac{\pi}{4}$$
$$C = -\frac{\pi}{2}$$

The curve is a reflection across the x-axis, so its equation is $y = -3\sin\left(2x - \dfrac{\pi}{2}\right)$.

28.   This is a vertical translation of the curve from Problem 24 of -2 units, so its equation is $y = -2 + 2\sin\left(3x - \dfrac{\pi}{2}\right)$.

30.   This is a vertical translation of the curve from Problem 26 of 3 units, so its equation is $y = 3 - 3\sin\left(2x - \dfrac{\pi}{2}\right)$.

32.   The curve will be a cosine curve reflected across the x-axis. The highest point is $165 + 9 = 174$, and the lowest point is 9, so the amplitude is $\frac{1}{2}(174 - 9) = 82.5$. There is no phase shift, so the curve will have the form $h = K - 82.5\cos(Bt)$. When $t = 0$, $h = 9$, so:

$$9 = K - 82.5$$
$$K = 91.5$$

Since the wheel revolves at $1.5\frac{\text{revolutions}}{\text{min}}$, it requires $\frac{2}{3}$ min to complete 1 revolution, so the period is $\frac{2}{3}$. Therefore:

$$\frac{2\pi}{B} = \frac{2}{3}$$
$$6\pi = 2B$$
$$B = 3\pi$$

Therefore, the equation is given by $h = 91.5 - 82.5\cos(3\pi t)$.

34.   The reference angle is $180° - 148° = 32°$.

36.   The reference angle is $-125° + 180° = 55°$.

38.   Since $450°$ has the same trigonometric values as $90°$, the reference angle is not defined.

40.   Since $\cos\theta = 0.7455$, the reference angle is given by $\hat{\theta} = \cos^{-1}(0.7455) \approx 41.8°$. Since θ terminates in quadrant IV, $\theta = 360° - 41.8° = 318.2°$.

42.   Since $\sec\theta = -3.1416$, $\cos\theta = -\dfrac{1}{3.1416}$, so the reference angle is given by $\hat{\theta} = \cos^{-1}\left(\dfrac{1}{3.1416}\right) \approx 71.4°$. Since θ terminates in quadrant III, $\theta = 180° + 71.4° = 251.4°$.

44.   Since $\tan\theta = 0.8156$, the reference angle is given by $\hat{\theta} = \tan^{-1}(0.8156) \approx 39.2°$. Since θ terminates in quadrant III, $\theta = 180° + 39.2° = 219.2°$.

4.5 Graphing Combinations of Functions

2. Graphing $y_1 = 1$, $y_2 = \cos x$, and $y = 1 + \cos x$:

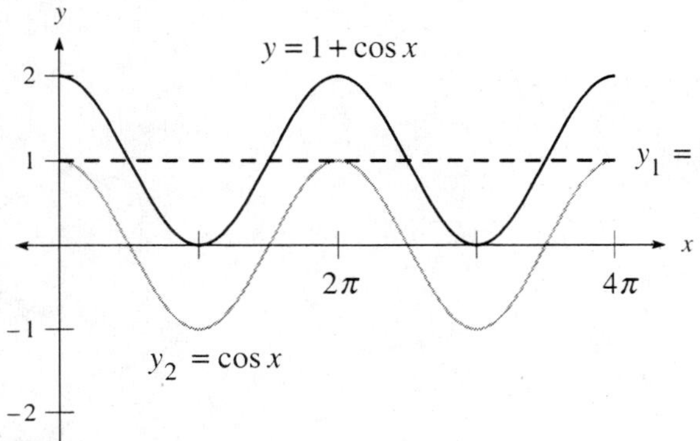

4. Graphing $y_1 = 2$, $y_2 = -\sin x$, and $y = 2 - \sin x$:

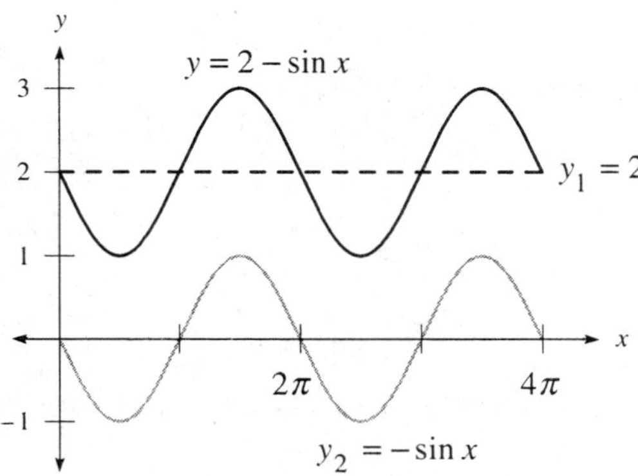

6. Graphing $y_1 = 4$, $y_2 = 2\cos x$, and $y = 4 + 2\cos x$:

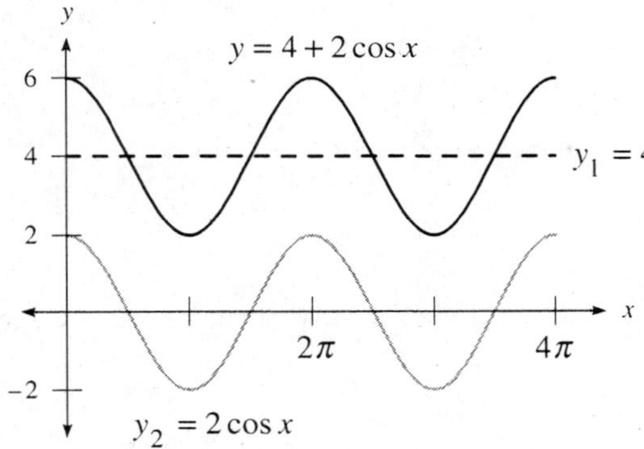

8. Graphing $y_1 = \frac{1}{2}x$, $y_2 = -\sin x$, and $y = \frac{1}{2}x - \sin x$:

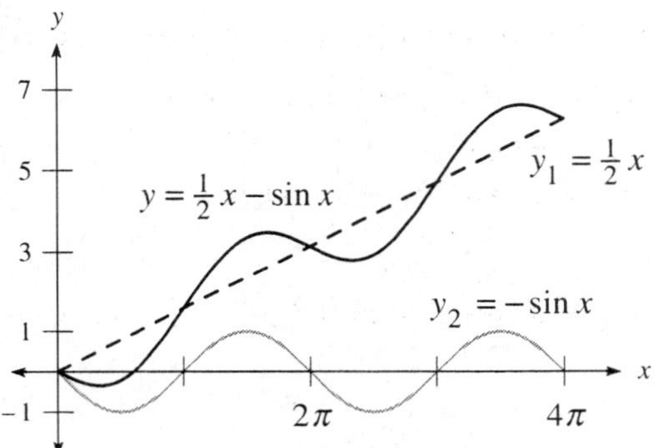

10. Graphing $y_1 = \frac{1}{3}x$, $y_2 = \cos x$, and $y = \frac{1}{3}x + \cos x$:

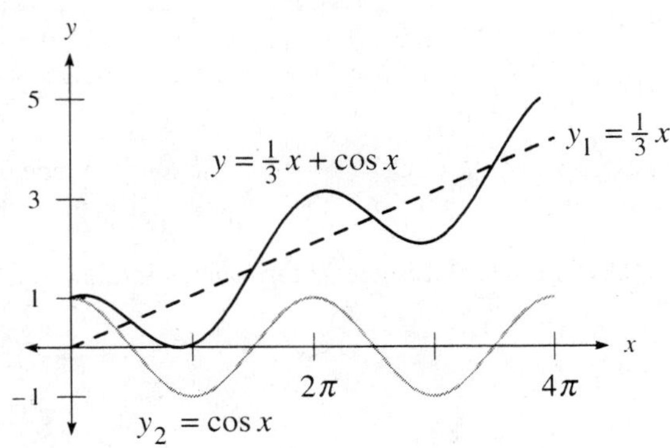

12. Graphing $y_1 = x$, $y_2 = \cos \pi x$, and $y = x + \cos \pi x$:

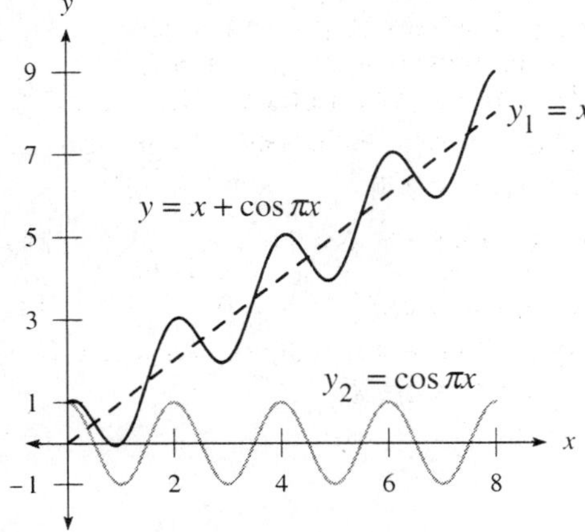

14.   Graphing $y_1 = 3\cos x$, $y_2 = \sin 2x$, and $y = 3\cos x + \sin 2x$:

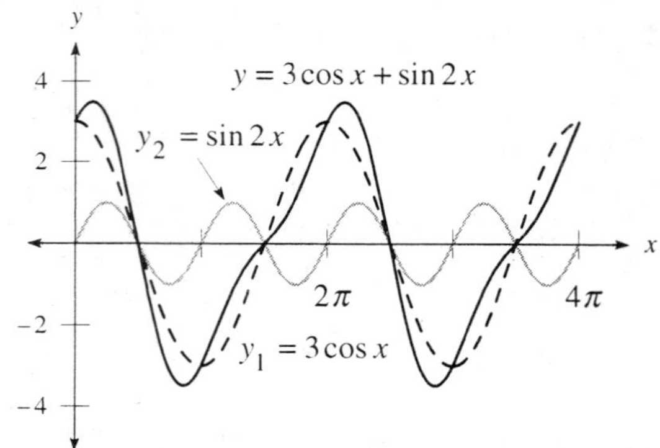

16.   Graphing $y_1 = 2\cos x$, $y_2 = -\sin 2x$, and $y = 2\cos x - \sin 2x$:

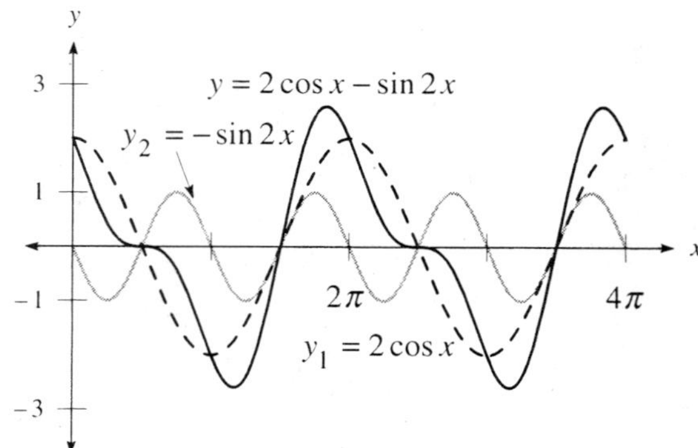

18.   Graphing $y_1 = \cos x$, $y_2 = \cos \dfrac{x}{2}$, and $y = \cos x + \cos \dfrac{x}{2}$:

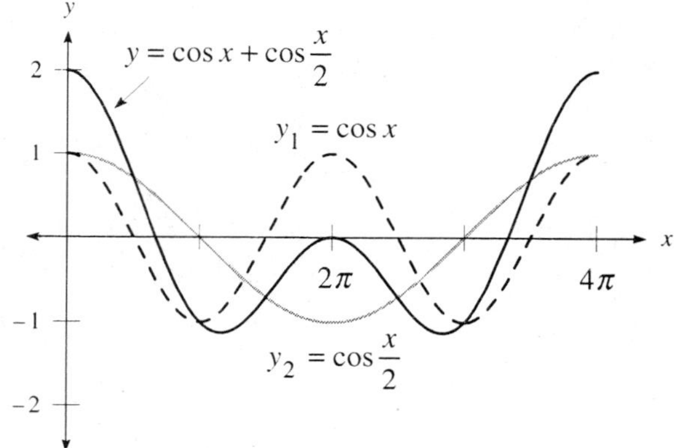

20. Graphing $y_1 = \cos x$, $y_2 = \cos 2x$, and $y = \cos x + \cos 2x$:

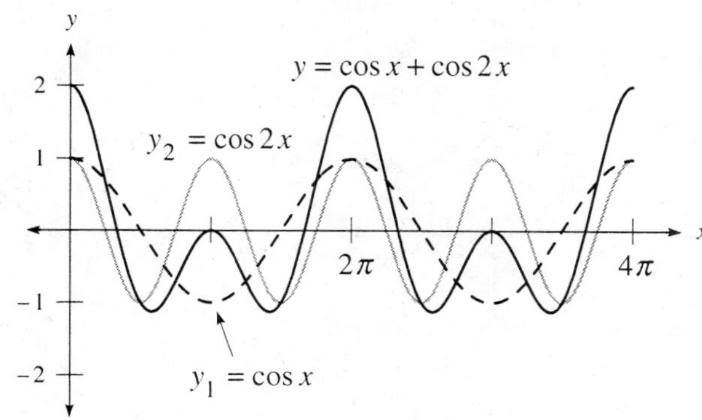

22. Graphing $y_1 = \sin x$, $y_2 = \frac{1}{2}\cos 2x$, and $y = \sin x + \frac{1}{2}\cos 2x$:

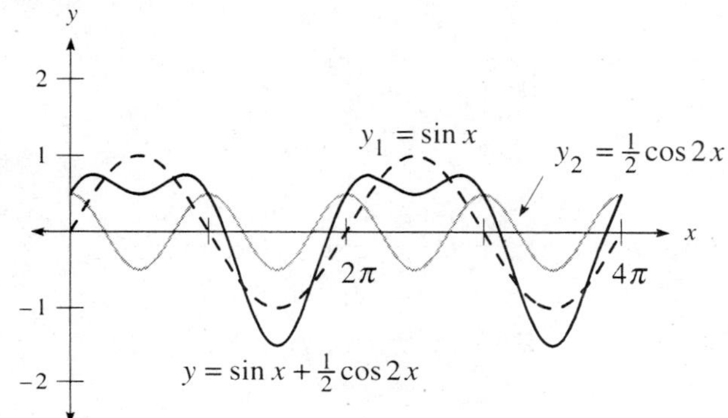

24. Graphing $y_1 = \cos x$, $y_2 = -\sin x$, and $y = \cos x - \sin x$:

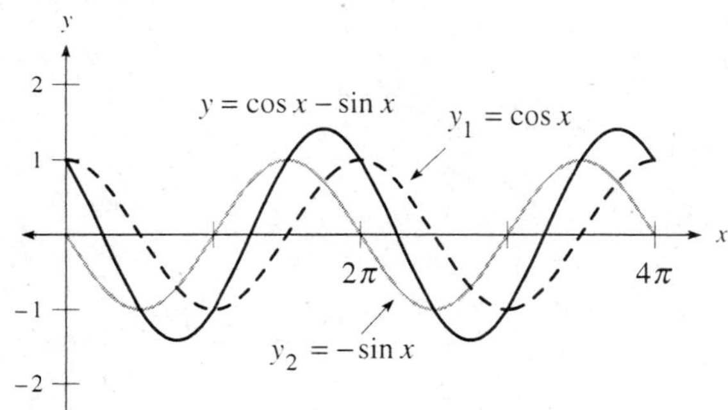

26. Sketching the graph of $y = x\cos x$:

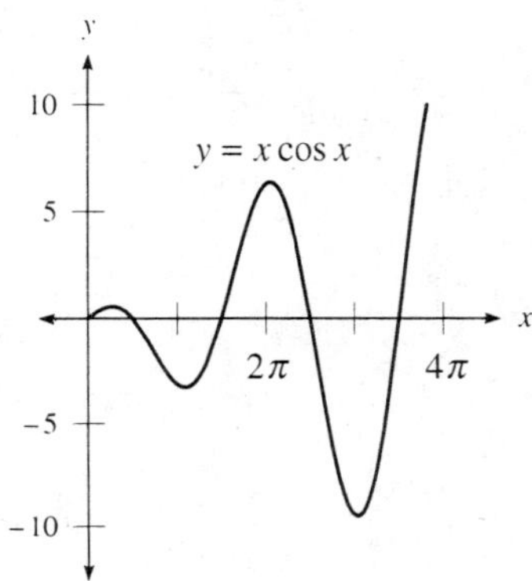

28.   Graphing $y_1 = 2$, $y_2 = \cos x$, and $y = y_1 + y_2 = 2 + \cos x$:

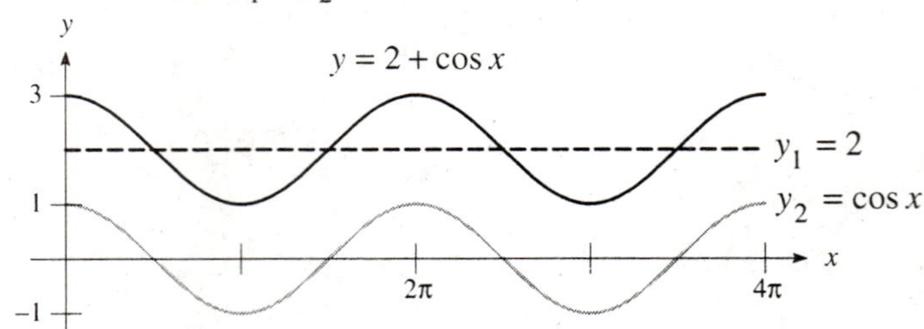

30.   Graphing $y_1 = x$, $y_2 = -\sin x$, and $y = y_1 + y_2 = x - \sin x$:

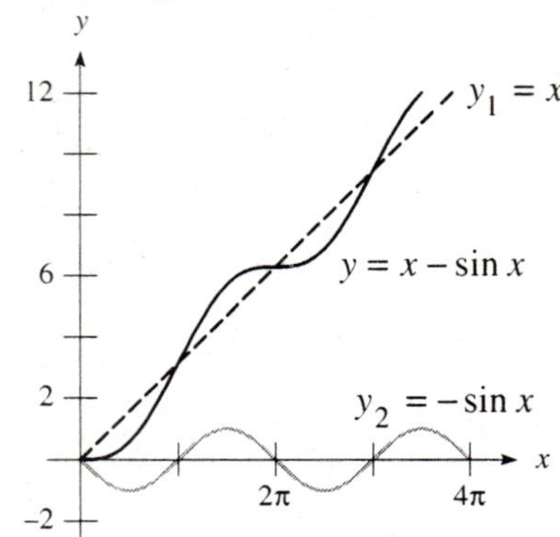

32.   Graphing $y_1 = \cos x$, $y_2 = 2\sin x$, and $y = y_1 + y_2 = \cos x + 2\sin x$:

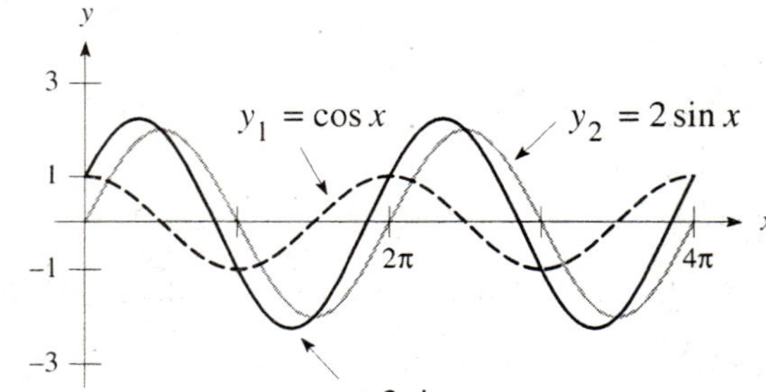

34.   Graphing $y_1 = \sin 2x$, $y_2 = -2\sin 3x$, and $y = y_1 + y_2 = \sin 2x - 2\sin 3x$:

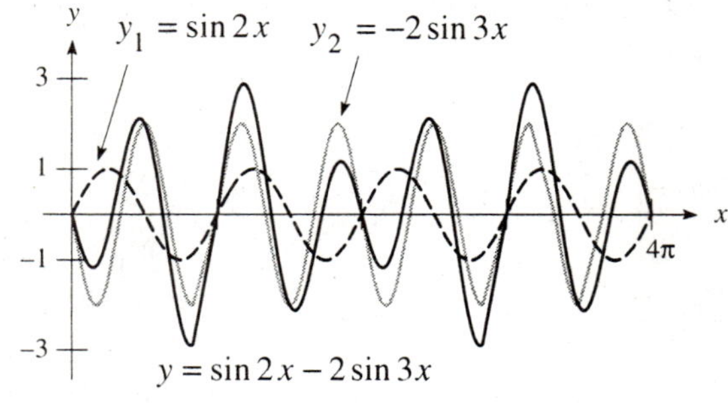

36. **a.** Sketching the graph:

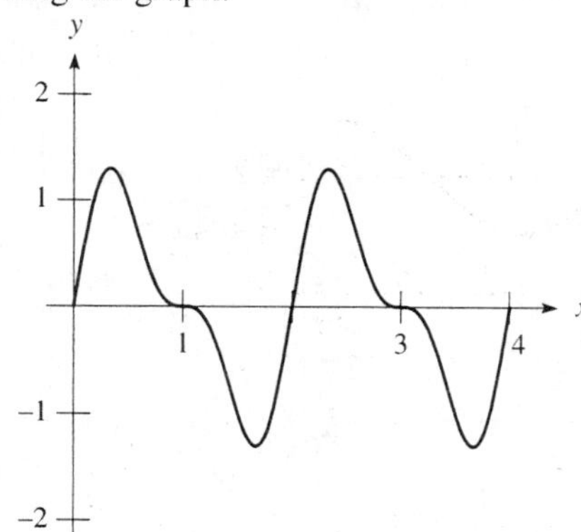

b. Sketching the graph:

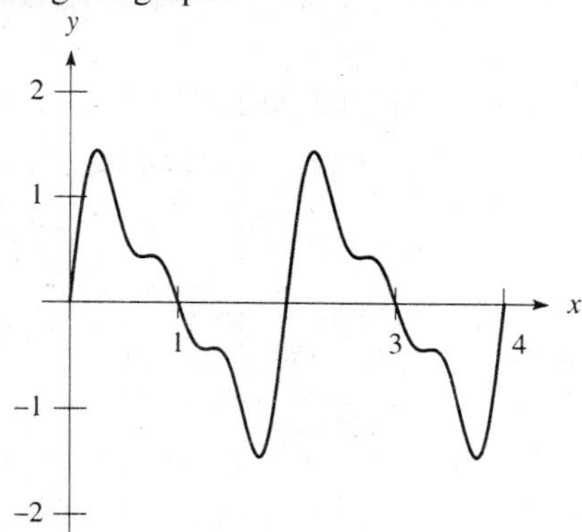

c. Sketching the graph:

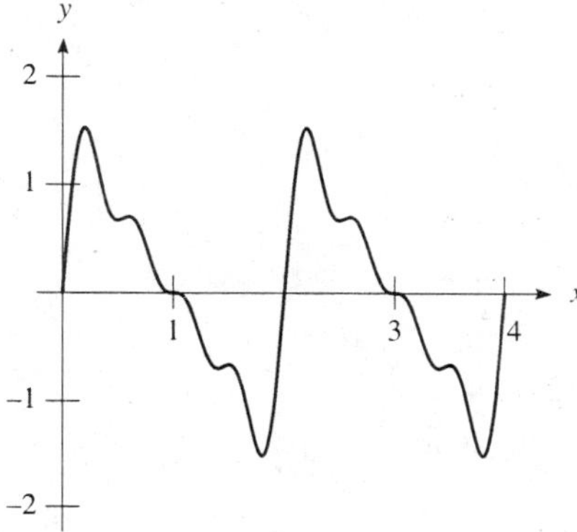

d. Sketching the graph:

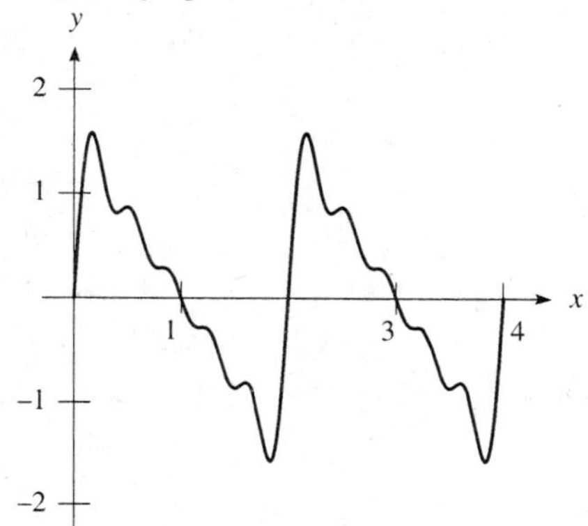

38. The linear velocity is given by: $v = \dfrac{s}{t} = \dfrac{15 \text{ cm}}{30 \text{ sec}} = 0.5 \dfrac{\text{cm}}{\text{sec}}$

40. The distance the point travels is given by: $s = vt = 65 \dfrac{\text{m}}{\text{sec}} \cdot 60 \text{ sec} = 3900 \text{ m}$

42. Converting to angular velocity: $30 \dfrac{\text{rev}}{\text{min}} \cdot 2\pi \dfrac{\text{radians}}{\text{rev}} = 60\pi \dfrac{\text{radians}}{\text{min}} \cdot \dfrac{1 \text{ min}}{60 \text{ sec}} = \pi \dfrac{\text{radians}}{\text{sec}}$

44. First convert to angular velocity: $\omega = 5 \dfrac{\text{rev}}{\text{min}} \cdot 2\pi \dfrac{\text{radians}}{\text{rev}} = 10\pi \dfrac{\text{radians}}{\text{min}}$

The linear velocity is given by: $v = r\omega = 6 \text{ in.} \cdot 10\pi \dfrac{\text{radians}}{\text{min}} = 60\pi \dfrac{\text{in.}}{\text{min}}$

46. In 1 day the hour hand moves 2 revolutions $\cdot 2\pi \dfrac{\text{radians}}{\text{rev}} = 4\pi$ radians. Therefore: $s = r\theta = 8 \text{ cm} \cdot 4\pi = 32\pi \text{ cm}$

4.6 Inverse Trigonometric Functions

2.   Sketching the graphs of $y = \sin x$ and $y = \sin^{-1} x$ between $-\dfrac{\pi}{2}$ and $\dfrac{\pi}{2}$:

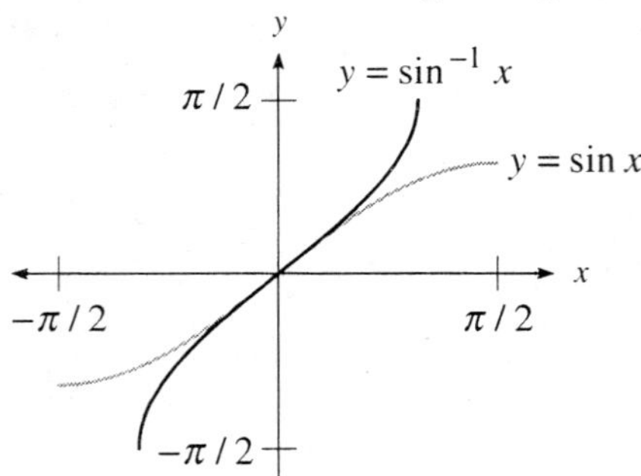

4.   Sketching the graphs of $y = \cot x$ and $x = \cot y$ between 0 and 2π:

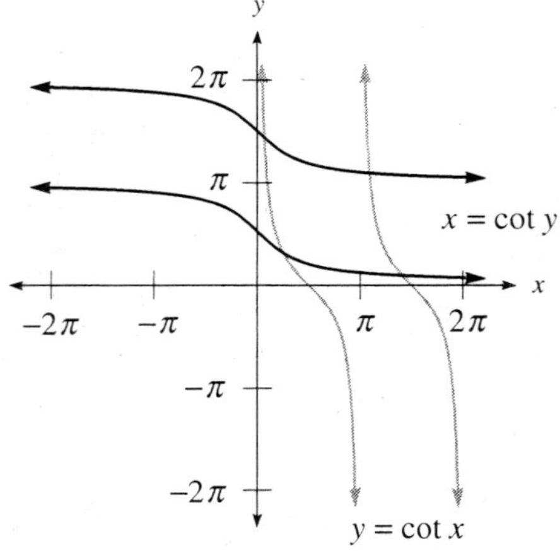

6.   Let $x = \cos^{-1}\left(\tfrac{1}{2}\right)$, so $\cos x = \tfrac{1}{2}$ and $0 \le x \le \pi$. Thus $x = \dfrac{\pi}{3}$, so $\cos^{-1}\left(\tfrac{1}{2}\right) = \dfrac{\pi}{3}$.

8.   Let $x = \cos^{-1}(0)$, so $\cos x = 0$ and $0 \le x \le \pi$. Thus $x = \dfrac{\pi}{2}$, so $\cos^{-1}(0) = \dfrac{\pi}{2}$.

10.   Let $x = \tan^{-1}(0)$, so $\tan x = 0$ and $-\dfrac{\pi}{2} < x < \dfrac{\pi}{2}$. Thus $x = 0$, so $\tan^{-1}(0) = 0$.

12.   Let $x = \arccos(1)$, so $\cos x = 1$ and $0 \le x \le \pi$. Thus $x = 0$, so $\arccos(1) = 0$.

14.   Let $x = \sin^{-1}\left(\dfrac{1}{\sqrt{2}}\right)$, so $\sin x = \dfrac{1}{\sqrt{2}}$ and $-\dfrac{\pi}{2} \le x \le \dfrac{\pi}{2}$. Thus $x = \dfrac{\pi}{4}$, so $\sin^{-1}\left(\dfrac{1}{\sqrt{2}}\right) = \dfrac{\pi}{4}$.

16.   Let $x = \arctan\left(\dfrac{1}{\sqrt{3}}\right)$, so $\tan x = \dfrac{1}{\sqrt{3}}$ and $-\dfrac{\pi}{2} < x < \dfrac{\pi}{2}$. Thus $x = \dfrac{\pi}{6}$, so $\arctan\left(\dfrac{1}{\sqrt{3}}\right) = \dfrac{\pi}{6}$.

18.   Let $x = \arcsin\left(-\dfrac{\sqrt{3}}{2}\right)$, so $\sin x = -\dfrac{\sqrt{3}}{2}$ and $-\dfrac{\pi}{2} \le x \le \dfrac{\pi}{2}$. Thus $x = -\dfrac{\pi}{3}$, so $\arcsin\left(-\dfrac{\sqrt{3}}{2}\right) = -\dfrac{\pi}{3}$.

20.   Let $x = \tan^{-1}\left(-\sqrt{3}\right)$, so $\tan x = -\sqrt{3}$ and $-\dfrac{\pi}{2} < x < \dfrac{\pi}{2}$. Thus $x = -\dfrac{\pi}{3}$, so $\tan^{-1}\left(-\sqrt{3}\right) = -\dfrac{\pi}{3}$.

22.   Let $x = \sin^{-1}(1)$, so $\sin x = 1$ and $-\dfrac{\pi}{2} \le x \le \dfrac{\pi}{2}$. Thus $x = \dfrac{\pi}{2}$, so $\sin^{-1}(1) = \dfrac{\pi}{2}$.

24.   Let $x = \arcsin(-1)$, so $\sin x = -1$ and $-\dfrac{\pi}{2} \le x \le \dfrac{\pi}{2}$. Thus $x = -\dfrac{\pi}{2}$, so $\arcsin(-1) = -\dfrac{\pi}{2}$.

26.   Using a calculator: $\sin^{-1}(-0.1702) \approx -9.8°$ **28.**   Using a calculator: $\cos^{-1}(0.8425) \approx 32.6°$

30.   Using a calculator: $\tan^{-1}(-0.3799) \approx -20.8°$ **32.**   Using a calculator: $\arccos(0.9627) \approx 15.7°$

34.   Using a calculator: $\sin^{-1}(-0.4664) \approx -27.8°$ **36.**   Using a calculator: $\arctan(-0.3640) \approx -20.0°$

38.   Using a calculator: $\cos^{-1}(-0.7660) \approx 140.0°$

40.   **a.**   The value is $\cos^{-1}\left(\frac{1}{2}\right) = 60°$.   **b.**   The value is $\cos^{-1}\left(-\frac{\sqrt{3}}{2}\right) = 150°$.

       c.   The value is $\arccos\left(\frac{1}{\sqrt{2}}\right) = 45°$.

42.   Let $\theta = \sin^{-1}\frac{x}{4}$, so $-\frac{\pi}{2} < \theta < \frac{\pi}{2}$. Then $\cos\theta \geq 0$, and so $4|\cos\theta| = 4\cos\theta$.

44.   Let $\theta = \tan^{-1}\frac{x}{5}$, so $-\frac{\pi}{2} < \theta < \frac{\pi}{2}$. Then $\sec\theta > 0$, and so $5|\sec\theta| = 5\sec\theta$.

46.   Since they are inverse functions: $\cos\left(\cos^{-1}\frac{3}{5}\right) = \frac{3}{5}$ **48.**   Since they are inverse functions: $\sin\left(\sin^{-1}\frac{1}{\sqrt{2}}\right) = \frac{1}{\sqrt{2}}$

50.   Since they are inverse functions: $\tan\left(\tan^{-1}\frac{3}{4}\right) = \frac{3}{4}$

52.   Evaluating: $\sin^{-1}(\sin 330°) = \sin^{-1}\left(-\frac{1}{2}\right) = -30°$ or $-\frac{\pi}{6}$

54.   Evaluating: $\sin^{-1}\left(\sin\frac{\pi}{4}\right) = \sin^{-1}\left(\frac{1}{\sqrt{2}}\right) = \frac{\pi}{4}$ **56.**   Evaluating: $\cos^{-1}(\cos 45°) = \cos^{-1}\left(\frac{1}{\sqrt{2}}\right) = 45°$ or $\frac{\pi}{4}$

58.   Evaluating: $\cos^{-1}\left(\cos\frac{7\pi}{6}\right) = \cos^{-1}\left(-\frac{\sqrt{3}}{2}\right) = \frac{5\pi}{6}$ **60.**   Evaluating: $\tan^{-1}(\tan 60°) = \tan^{-1}\left(\sqrt{3}\right) = 60°$ or $\frac{\pi}{3}$

62.   Evaluating: $\tan^{-1}\left(\tan\frac{2\pi}{3}\right) = \tan^{-1}\left(-\sqrt{3}\right) = -\frac{\pi}{3}$

64.   Let $\theta = \tan^{-1}\frac{3}{4}$, so $\tan\theta = \frac{3}{4}$. Drawing the triangle:

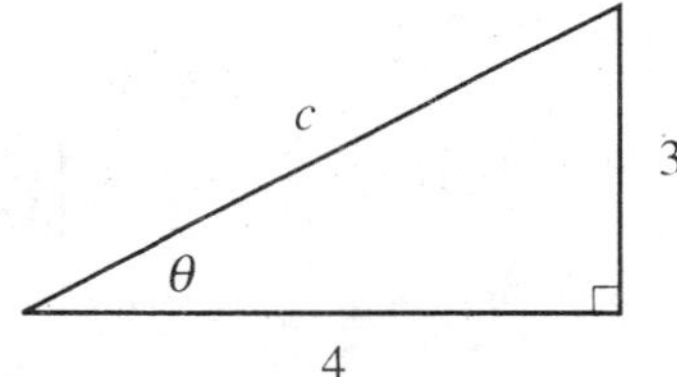

Using the Pythagorean Theorem: $c = \sqrt{4^2 + 3^2} = \sqrt{16+9} = \sqrt{25} = 5$

Therefore: $\csc\left(\tan^{-1}\frac{3}{4}\right) = \csc\theta = \frac{1}{\sin\theta} = \frac{1}{3/5} = \frac{5}{3}$

66.   Let $\theta = \cos^{-1}\frac{3}{5}$, so $\cos\theta = \frac{3}{5}$. Drawing the triangle:

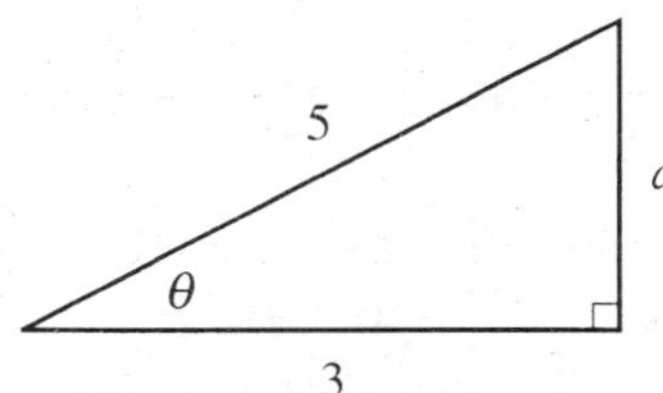

Using the Pythagorean Theorem: $a = \sqrt{5^2 - 3^2} = \sqrt{25-9} = \sqrt{16} = 4$. Therefore: $\tan\left(\cos^{-1}\frac{3}{5}\right) = \tan\theta = \frac{4}{3}$

68. Let $\theta = \cos^{-1} \dfrac{1}{\sqrt{5}}$, so $\cos\theta = \dfrac{1}{\sqrt{5}}$. Drawing the triangle:

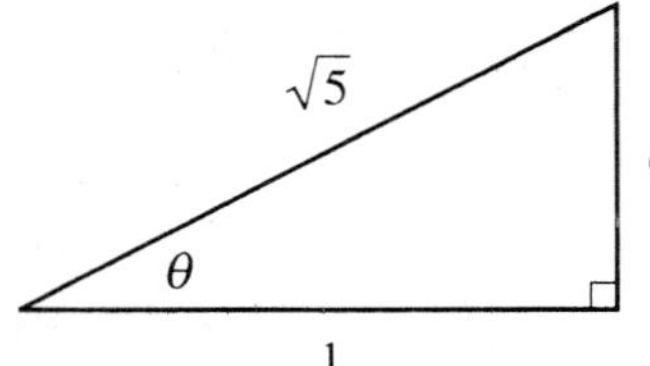

Using the Pythagorean Theorem: $a = \sqrt{\left(\sqrt{5}\right)^2 - 1^2} = \sqrt{5-1} = \sqrt{4} = 2$. Therefore: $\sin\left(\cos^{-1} \dfrac{1}{\sqrt{5}}\right) = \sin\theta = \dfrac{2}{\sqrt{5}}$

70. Let $\theta = \sin^{-1} \tfrac{1}{2}$, so $\sin\theta = \tfrac{1}{2}$ and thus $\theta = \dfrac{\pi}{6}$. Therefore: $\cos\left(\sin^{-1} \tfrac{1}{2}\right) = \cos\dfrac{\pi}{6} = \dfrac{\sqrt{3}}{2}$

72. Let $\theta = \tan^{-1} \tfrac{1}{3}$, so $\tan\theta = \tfrac{1}{3}$. Therefore: $\cot\left(\tan^{-1} \tfrac{1}{3}\right) = \cot\theta = \dfrac{1}{\tan\theta} = \dfrac{1}{\frac{1}{3}} = 3$

74. $\cos^{-1}(\cos x) = x$ as long as $0 \le x \le \pi$.

76. $\sin\left(\sin^{-1} x\right) = x$ as long as $-1 \le x \le 1$.

78. Let $\theta = \cos^{-1} x$, so $\cos\theta = x$. Drawing the triangle:

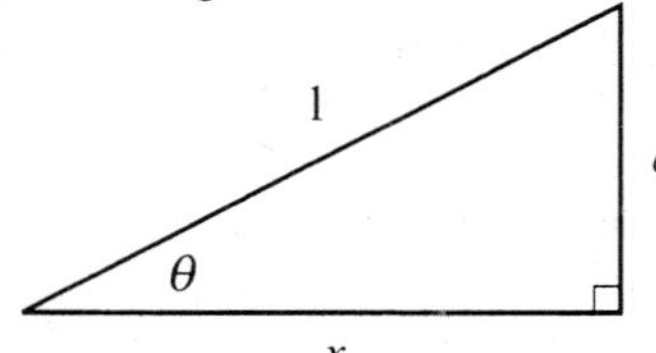

Using the Pythagorean Theorem: $a = \sqrt{1 - x^2}$. Therefore: $\tan\left(\cos^{-1} x\right) = \tan\theta = \dfrac{\sqrt{1 - x^2}}{x}$

80. Let $\theta = \tan^{-1} x$, so $\tan\theta = x$. Drawing the triangle:

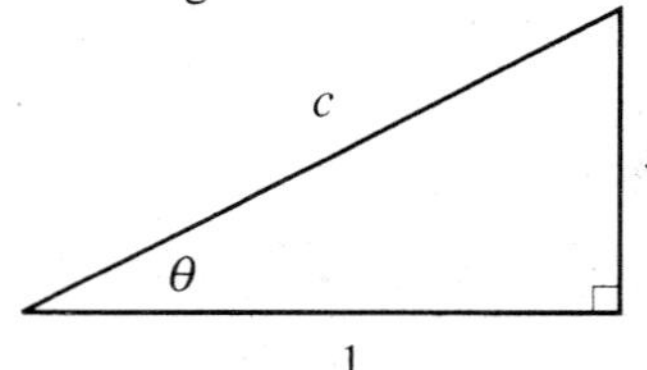

Using the Pythagorean Theorem: $c = \sqrt{x^2 + 1}$. Therefore: $\cos\left(\tan^{-1} x\right) = \cos\theta = \dfrac{1}{\sqrt{x^2 + 1}}$

82. Let $\theta = \sin^{-1} \dfrac{1}{x}$, so $\sin\theta = \dfrac{1}{x}$. Drawing the triangle:

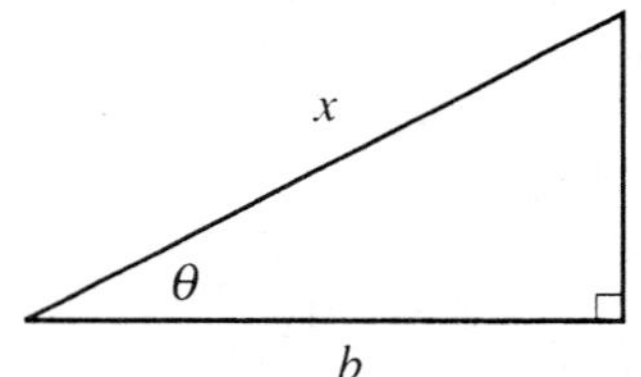

Using the Pythagorean Theorem: $b = \sqrt{x^2 - 1}$. Therefore: $\cos\left(\sin^{-1} \dfrac{1}{x}\right) = \cos\theta = \dfrac{\sqrt{x^2 - 1}}{x}$

84. From Problem 82: $\csc\left(\sin^{-1} \dfrac{1}{x}\right) = \csc\theta = \dfrac{1}{\sin\theta} = \dfrac{1}{\frac{1}{x}} = x$

86. The amplitude is 2 and the period is $\dfrac{2\pi}{4} = \dfrac{\pi}{2}$:

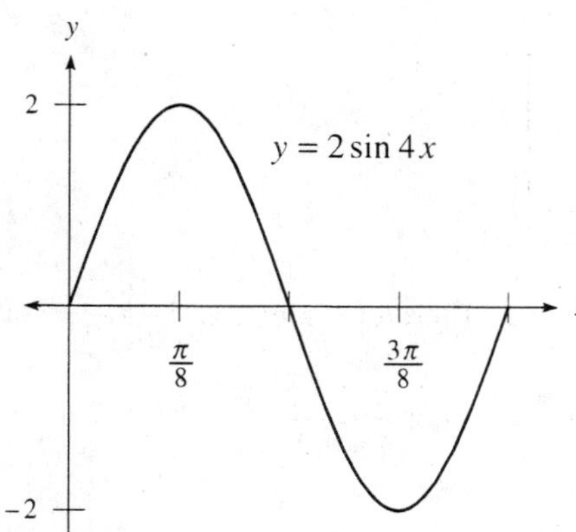

88. The amplitude is 3 and the period is $\dfrac{2\pi}{\pi} = 2$:

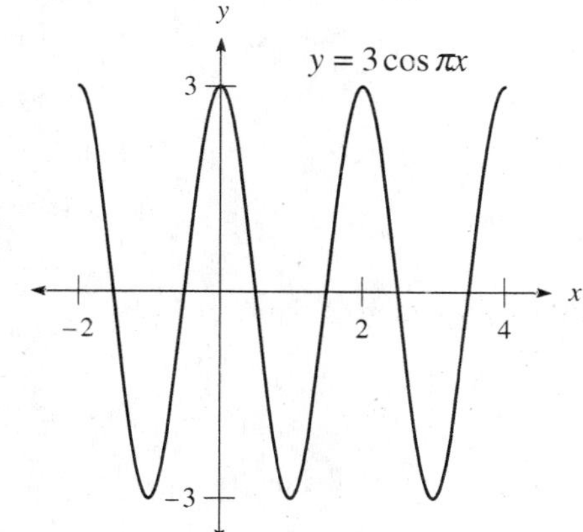

90. The amplitude is 3 and the period is $\dfrac{2\pi}{2} = \pi$. Note this curve is a reflection of $y = 3\sin 2x$ across the x-axis:

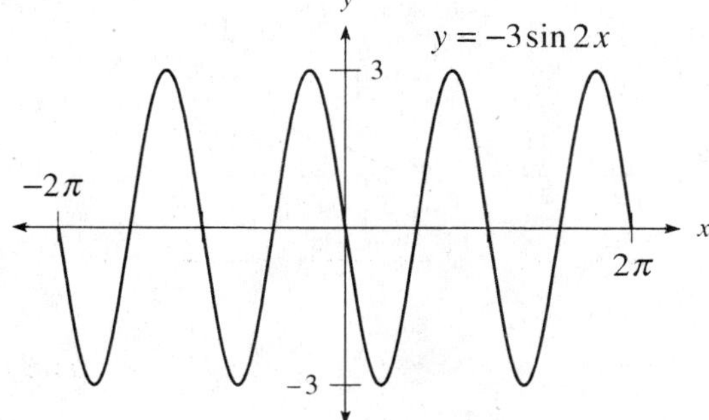

92. The amplitude is 1, the period is 2π, and the phase shift is $-\dfrac{\pi}{6}$:

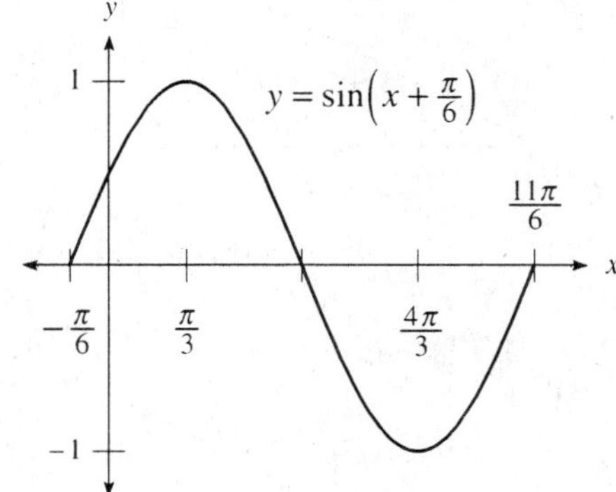

94.   The amplitude is 1, the period is $\dfrac{2\pi}{2} = \pi$, and the phase shift is $-\dfrac{\pi}{2}$:

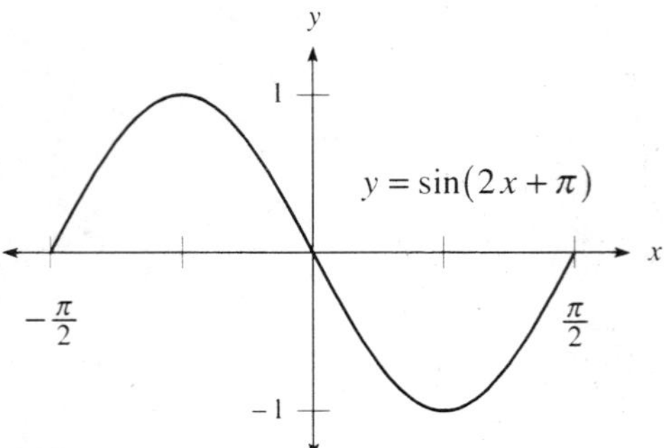

96.   The amplitude is 3, the period is $\dfrac{2\pi}{2} = \pi$, and the phase shift is $\dfrac{\pi/3}{2} = \dfrac{\pi}{6}$:

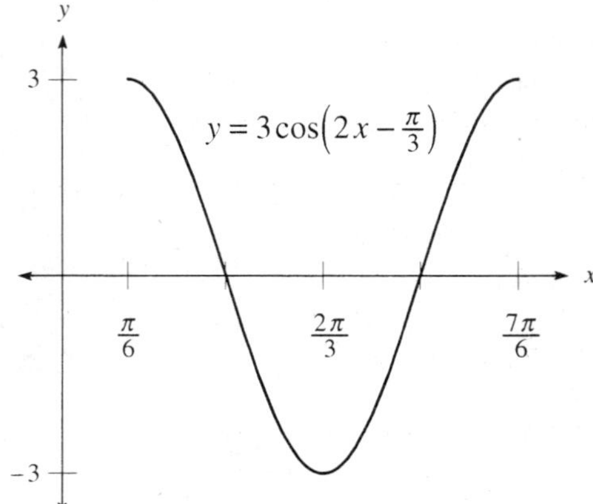

Chapter 4 Test

1.   Sketching the graph:

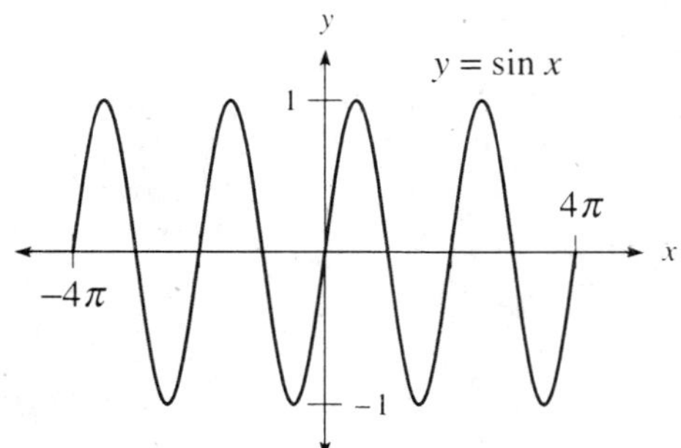

2.   Sketching the graph:

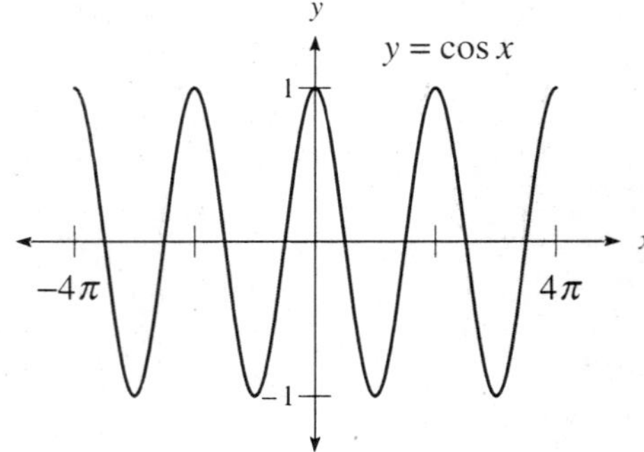

3.   Sketching the graph:

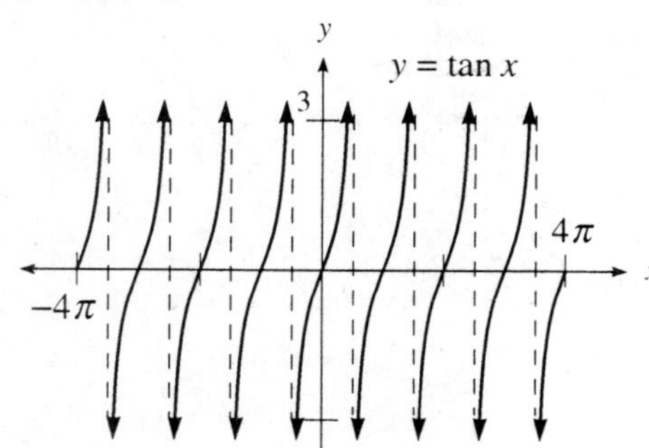

4.   Sketching the graph:

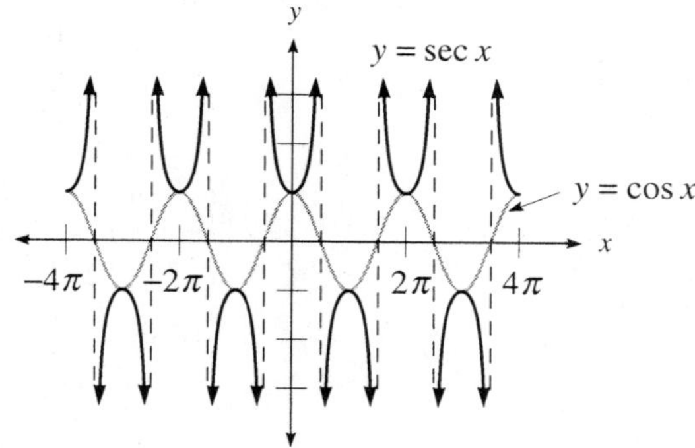

5.   Since 1 cycle is completed every 2π radians, and the graph from Problem 1 is graphed for $4\pi - (-4\pi) = 8\pi$ radians, there are $\dfrac{8\pi}{2\pi} = 4$ cycles completed in the graph.

6.   Since 1 cycle is completed every π radians, and the graph from Problem 3 is graphed for $4\pi - (-4\pi) = 8\pi$ radians, there are $\dfrac{8\pi}{\pi} = 8$ cycles completed in the graph.

7. $\sec x = -1$ when $x = -3\pi, -\pi, \pi, 3\pi$.

8. $\sec x$ is undefined whenever $\cos x = 0$, which occurs when $x = -\dfrac{7\pi}{2}, -\dfrac{5\pi}{2}, -\dfrac{3\pi}{2}, -\dfrac{\pi}{2}, \dfrac{\pi}{2}, \dfrac{3\pi}{2}, \dfrac{5\pi}{2}, \dfrac{7\pi}{2}$.

9.   The amplitude is 1 and the period is $\dfrac{2\pi}{\pi} = 2$:

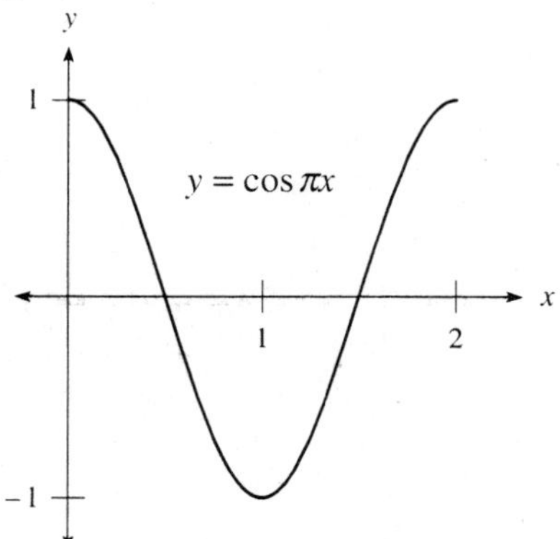

10.   The amplitude is 3 and the period is 2π. Note this curve is a reflection of $y = 3\cos x$ across the x-axis:

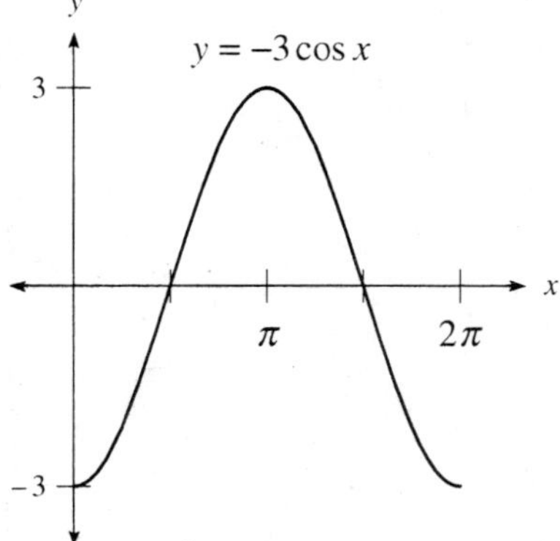

11.   The amplitude is 3 and the period is $\dfrac{2\pi}{2} = \pi$. Note this curve is a vertical translation of $y = 3\sin 2x$ by 2 units:

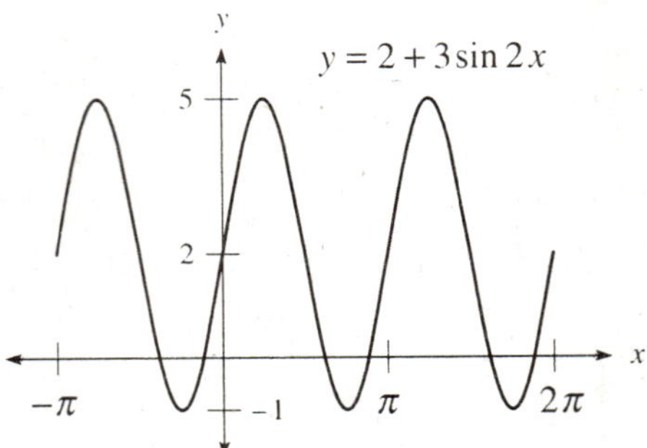

12.   The amplitude is 2 and the period is $\dfrac{2\pi}{\pi} = 2$:

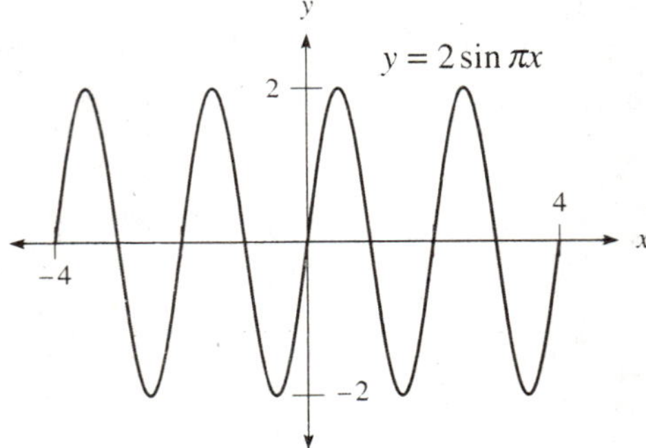

13.   The amplitude is 1, the period is 2π, and the phase shift is $-\dfrac{\pi}{4}$:

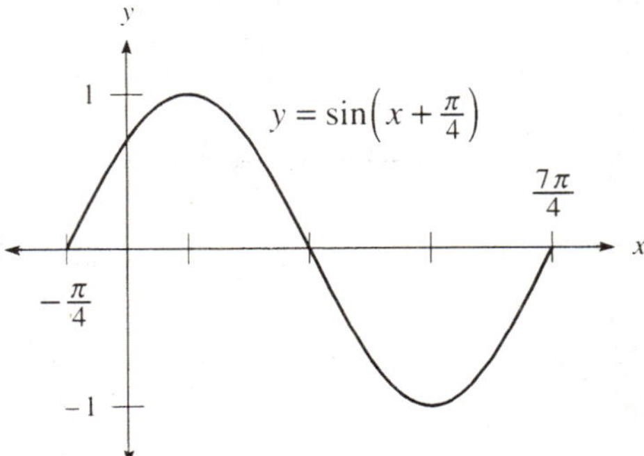

14.   The amplitude is 1, the period is 2π, and the phase shift is $\dfrac{\pi}{2}$:

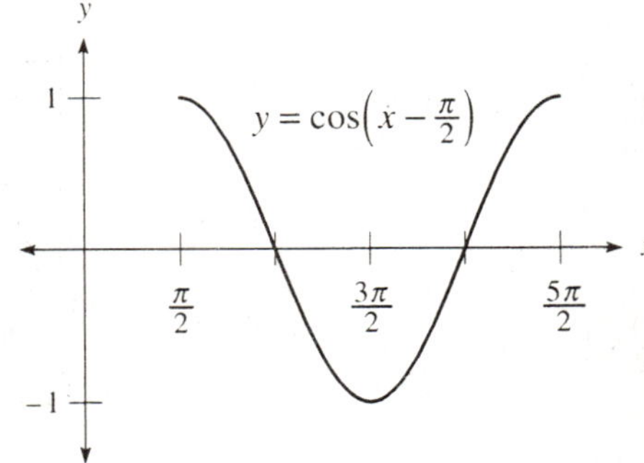

15.   The amplitude is 3, the period is $\dfrac{2\pi}{2} = \pi$, and the phase shift is $\dfrac{\pi/3}{2} = \dfrac{\pi}{6}$:

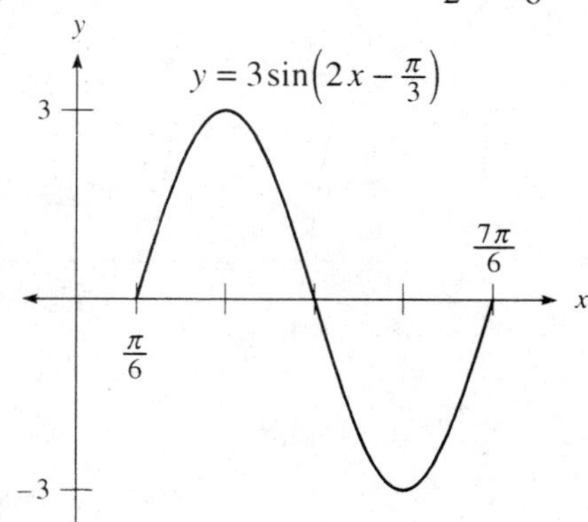

16.   The amplitude is 3, the period is $\dfrac{2\pi}{\pi/3} = 6$, and the phase shift is $\dfrac{\pi/3}{\pi/3} = 1$. Note the curve is a vertical translation of

$y = 3\sin\left(\dfrac{\pi}{3}x - \dfrac{\pi}{3}\right)$ by -3 units:

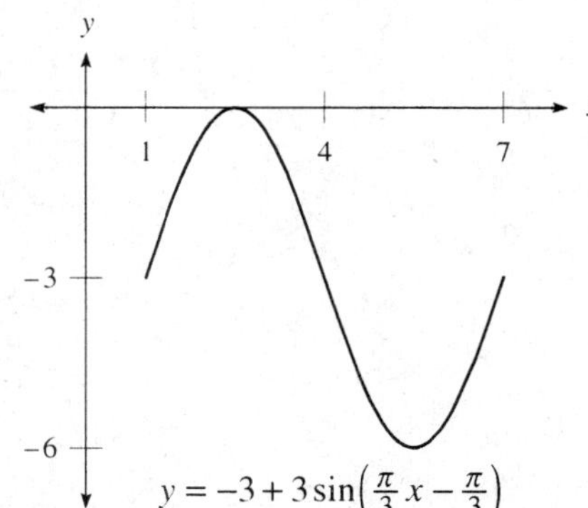

17.   The period is 2π and the phase shift is $-\dfrac{\pi}{4}$:

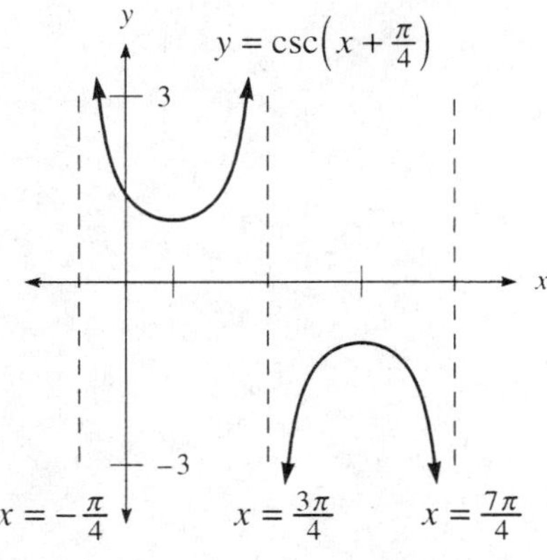

18.   The period is $\dfrac{\pi}{2}$ and the phase shift is $\dfrac{\pi/2}{2} = \dfrac{\pi}{4}$:

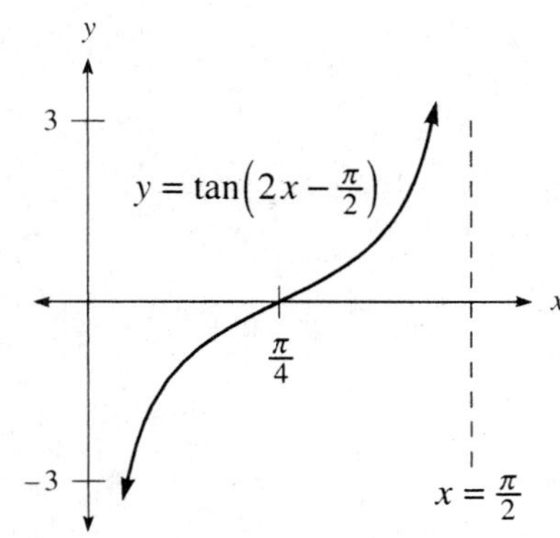

19. The amplitude is 2, the period is $\dfrac{2\pi}{3}$, and the phase shift is $\dfrac{\pi}{3}$:

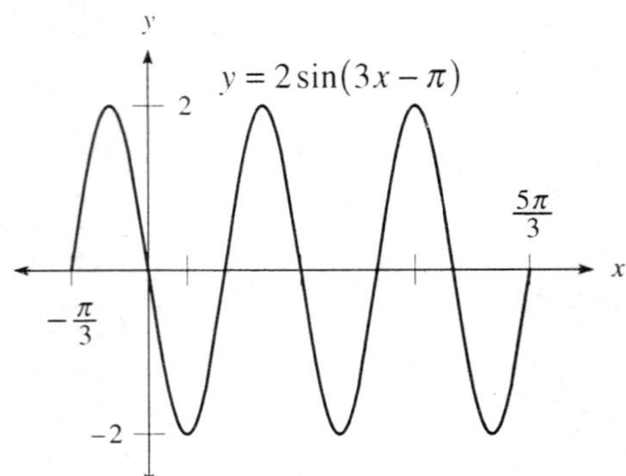

20. The amplitude is 2, the period is $\dfrac{2\pi}{\pi/2} = 4$, and the phase shift is $\dfrac{\pi/4}{\pi/2} = \dfrac{1}{2}$:

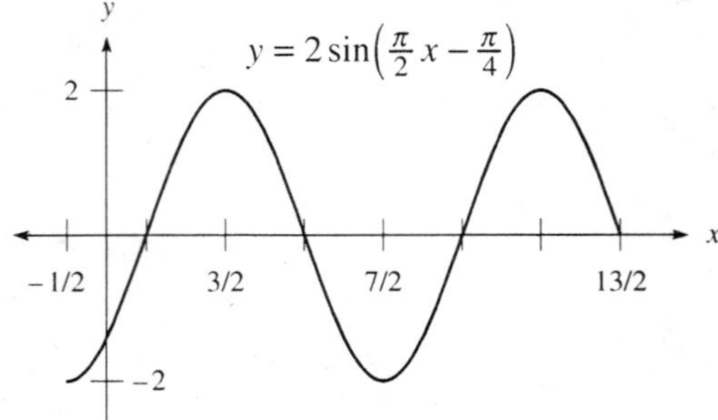

21. This is a sine curve with amplitude $= 2$ and period $= 4\pi$, therefore:

$$\frac{2\pi}{B} = 4\pi$$
$$2\pi = 4\pi B$$
$$B = \tfrac{1}{2}$$

The phase shift is $-\pi$, therefore:

$$-\frac{C}{B} = -\pi$$
$$\frac{-C}{1/2} = -\pi$$
$$C = \frac{\pi}{2}$$

Thus the equation of the curve is $y = 2\sin\left(\tfrac{1}{2}x + \dfrac{\pi}{2}\right)$.

22. This is a sine curve with amplitude $= \tfrac{1}{2}(1-0) = \tfrac{1}{2}$ and a vertical translation of $\tfrac{1}{2}$ unit. The period is 4, therefore:

$$\frac{2\pi}{B} = 4$$
$$2\pi = 4B$$
$$B = \frac{\pi}{2}$$

There is no phase shift, thus the equation of the curve is $y = \tfrac{1}{2} + \tfrac{1}{2}\sin\left(\dfrac{\pi}{2}x\right)$.

23. Graphing $y_1 = \frac{1}{2}x$, $y_2 = -\sin x$, and $y = \frac{1}{2}x - \sin x$:

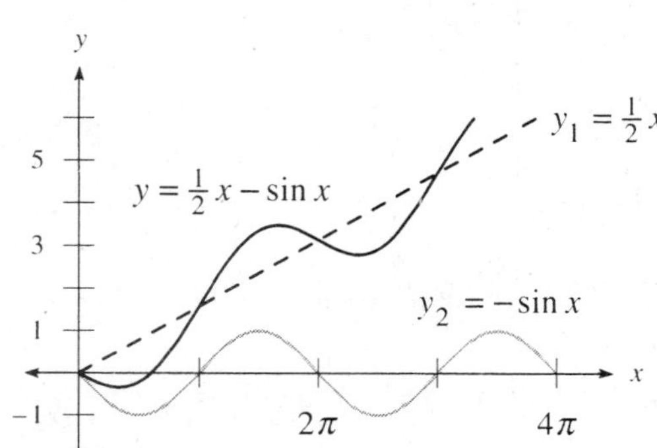

24. Graphing $y_1 = \sin x$, $y_2 = \cos 2x$, $y = \sin x + \cos 2x$:

25. Graphing $y = \cos^{-1} x$:

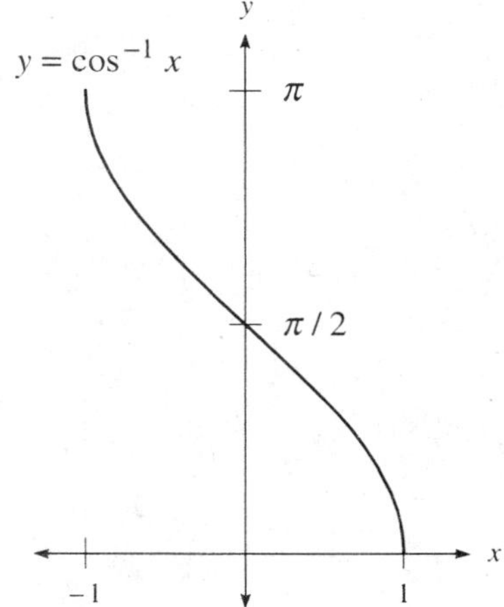

26. Graphing $y = \arcsin x$:

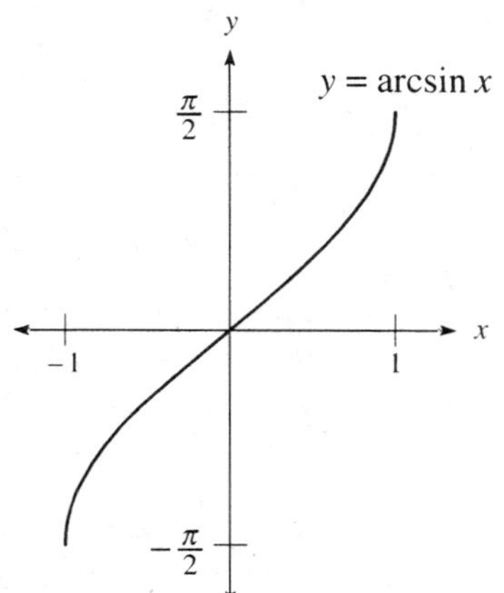

27. Let $x = \sin^{-1}\left(\frac{1}{2}\right)$, so $\sin x = \frac{1}{2}$ and $-\frac{\pi}{2} \le x \le \frac{\pi}{2}$. Thus $x = \frac{\pi}{6}$, so $\sin^{-1}\left(\frac{1}{2}\right) = \frac{\pi}{6}$.

28. Let $x = \cos^{-1}\left(-\frac{\sqrt{3}}{2}\right)$, so $\cos x = -\frac{\sqrt{3}}{2}$ and $0 \le x \le \pi$. Thus $x = \frac{5\pi}{6}$, so $\cos^{-1}\left(-\frac{\sqrt{3}}{2}\right) = \frac{5\pi}{6}$.

29. Let $x = \arctan(-1)$, so $\tan x = -1$ and $-\frac{\pi}{2} < x < \frac{\pi}{2}$. Thus $x = -\frac{\pi}{4}$, so $\arctan(-1) = -\frac{\pi}{4}$.

30. Let $x = \arcsin(1)$, so $\sin x = 1$ and $-\frac{\pi}{2} \le x \le \frac{\pi}{2}$. Thus $x = \frac{\pi}{2}$, so $\arcsin(1) = \frac{\pi}{2}$.

31. Using a calculator: $\arcsin(0.5934) \approx 36.4°$

32. Using a calculator: $\arctan(-0.8302) \approx -39.7°$

33. Using a calculator: $\arccos(-0.6981) \approx 134.3°$

34. Using a calculator: $\arcsin(-0.2164) \approx -12.5°$

35. Let $\theta = \cos^{-1}\frac{2}{3}$, so $\cos\theta = \frac{2}{3}$. Drawing the triangle:

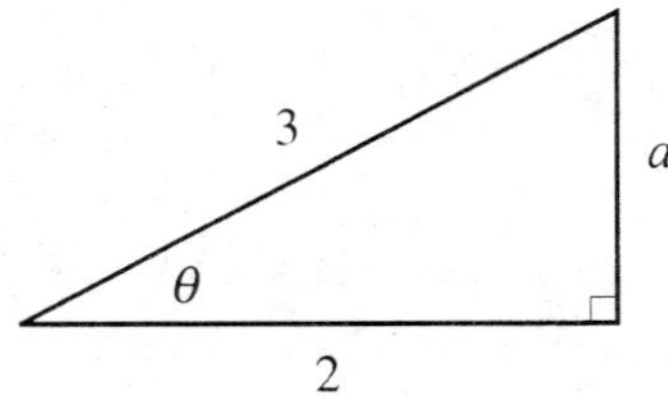

Using the Pythagorean Theorem: $a = \sqrt{3^2 - 2^2} = \sqrt{9-4} = \sqrt{5}$. Therefore: $\tan\left(\cos^{-1}\frac{2}{3}\right) = \tan\theta = \frac{\sqrt{5}}{2}$

36.   Let $\theta = \tan^{-1} \frac{2}{3}$, so $\tan \theta = \frac{2}{3}$. Drawing the triangle:

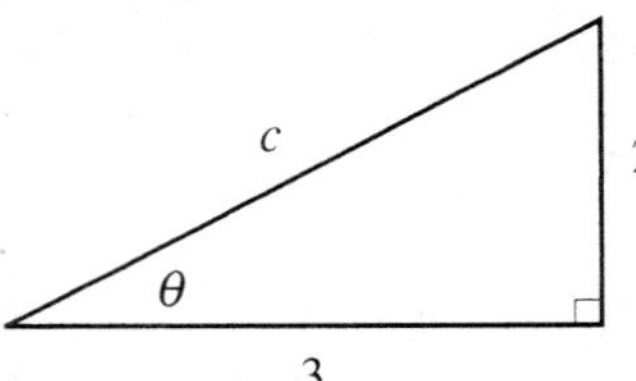

Using the Pythagorean Theorem: $c = \sqrt{3^2 + 2^2} = \sqrt{9 + 4} = \sqrt{13}$. Therefore: $\cos\left(\tan^{-1} \frac{2}{3}\right) = \cos \theta = \dfrac{3}{\sqrt{13}}$

37.   Evaluating: $\cos^{-1}(\cos 30°) = \cos^{-1}\left(\dfrac{\sqrt{3}}{2}\right) = 30°$ or $\dfrac{\pi}{6}$ **38.**   Evaluating: $\tan^{-1}\left(\tan \dfrac{7\pi}{6}\right) = \tan^{-1}\left(\dfrac{1}{\sqrt{3}}\right) = \dfrac{\pi}{6}$

39.   Let $\theta = \cos^{-1} x$, so $\cos \theta = x$. Drawing the triangle:

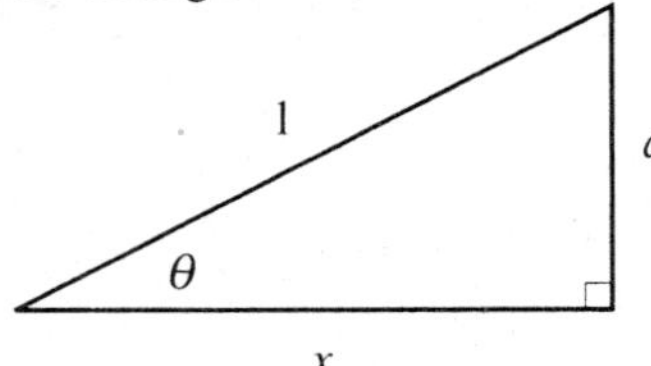

Using the Pythagorean Theorem: $a = \sqrt{1 - x^2}$. Therefore: $\sin\left(\cos^{-1} x\right) = \sin \theta = \dfrac{\sqrt{1 - x^2}}{1} = \sqrt{1 - x^2}$

40.   Let $\theta = \sin^{-1} x$, so $\sin \theta = x$. Drawing the triangle:

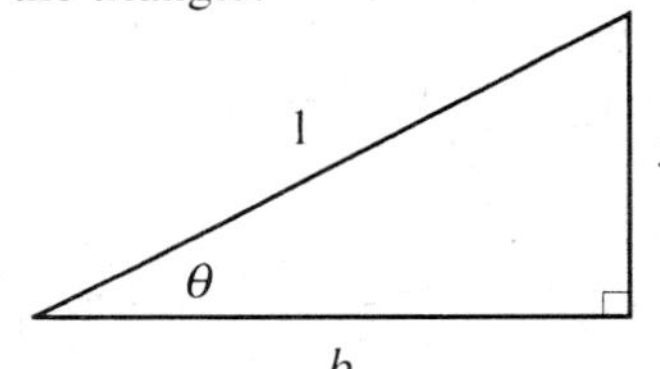

Using the Pythagorean Theorem: $b = \sqrt{1 - x^2}$. Therefore: $\tan\left(\sin^{-1} x\right) = \tan \theta = \dfrac{x}{\sqrt{1 - x^2}}$

Chapter 5
Identities and Formulas

5.1 Proving Identities

2. Working from the left side: $\sec\theta\cot\theta = \dfrac{1}{\cos\theta}\cdot\dfrac{\cos\theta}{\sin\theta} = \dfrac{1}{\sin\theta} = \csc\theta$

4. Working from the left side: $\tan\theta\cot\theta = \tan\theta\cdot\dfrac{1}{\tan\theta} = 1$

6. Working from the left side: $\dfrac{\cot A}{\csc A} = \dfrac{\dfrac{\cos A}{\sin A}}{\dfrac{1}{\sin A}} = \dfrac{\cos A}{\sin A}\cdot\dfrac{\sin A}{1} = \cos A$

8. Working from the left side: $\tan\theta\csc\theta\cos\theta = \dfrac{\sin\theta}{\cos\theta}\cdot\dfrac{1}{\sin\theta}\cdot\dfrac{\cos\theta}{1} = \dfrac{\sin\theta\cos\theta}{\cos\theta\sin\theta} = 1$

10. Working from the left side: $\sin x(\sec x + \csc x) = \sin x\left(\dfrac{1}{\cos x} + \dfrac{1}{\sin x}\right) = \dfrac{\sin x}{\cos x} + \dfrac{\sin x}{\sin x} = \tan x + 1$

12. Working from the left side: $\tan x(\cos x + \cot x) = \dfrac{\sin x}{\cos x}\left(\cos x + \dfrac{\cos x}{\sin x}\right) = \sin x + 1$

14. Working from the left side: $\sin^2 x\left(\cot^2 x + 1\right) = \sin^2 x\left(\dfrac{\cos^2 x}{\sin^2 x} + 1\right) = \cos^2 x + \sin^2 x = 1$

16. Working from the left side: $(1-\cos x)(1+\cos x) = 1 - \cos x + \cos x - \cos^2 x = 1 - \cos^2 x = \sin^2 x$

18. Working from the left side:

$$\dfrac{\sin^4 t - \cos^4 t}{\sin^2 t\cos^2 t} = \dfrac{\left(\sin^2 t + \cos^2 t\right)\left(\sin^2 t - \cos^2 t\right)}{\sin^2 t\cos^2 t}$$

$$= \dfrac{\sin^2 t - \cos^2 t}{\sin^2 t\cos^2 t}$$

$$= \dfrac{\sin^2 t}{\sin^2 t\cos^2 t} - \dfrac{\cos^2 t}{\sin^2 t\cos^2 t}$$

$$= \dfrac{1}{\cos^2 t} - \dfrac{1}{\sin^2 t}$$

$$= \sec^2 t - \csc^2 t$$

20. Working from the right side: $\dfrac{\cos^2\theta}{1+\sin\theta} = \dfrac{\cos^2\theta}{1+\sin\theta}\cdot\dfrac{1-\sin\theta}{1-\sin\theta} = \dfrac{\cos^2\theta(1-\sin\theta)}{1-\sin^2\theta} = \dfrac{\cos^2\theta(1-\sin\theta)}{\cos^2\theta} = 1-\sin\theta$

22. Working from the left side: $\dfrac{1-\cos^4\theta}{1+\cos^2\theta} = \dfrac{\left(1+\cos^2\theta\right)\left(1-\cos^2\theta\right)}{1+\cos^2\theta} = 1-\cos^2\theta = \sin^2\theta$

24. Working from the left side: $\csc^2\theta - \cot^2\theta = \dfrac{1}{\sin^2\theta} - \dfrac{\cos^2\theta}{\sin^2\theta} = \dfrac{1-\cos^2\theta}{\sin^2\theta} = \dfrac{\sin^2\theta}{\sin^2\theta} = 1$

26. Working from the left side (and using the result from Problem 24):

$$\csc^4\theta - \cot^4\theta = \left(\csc^2\theta - \cot^2\theta\right)\left(\csc^2\theta + \cot^2\theta\right) = 1\cdot\left(\dfrac{1}{\sin^2\theta} + \dfrac{\cos^2\theta}{\sin^2\theta}\right) = \dfrac{1+\cos^2\theta}{\sin^2\theta}$$

28. Working from the left side: $\sec\theta - \csc\theta = \dfrac{1}{\cos\theta} - \dfrac{1}{\sin\theta} = \dfrac{\sin\theta}{\sin\theta\cos\theta} - \dfrac{\cos\theta}{\sin\theta\cos\theta} = \dfrac{\sin\theta-\cos\theta}{\sin\theta\cos\theta}$

Note that this could also be proved working from the right side:

$$\dfrac{\sin\theta-\cos\theta}{\sin\theta\cos\theta} = \dfrac{\sin\theta}{\sin\theta\cos\theta} - \dfrac{\cos\theta}{\sin\theta\cos\theta} = \dfrac{1}{\cos\theta} - \dfrac{1}{\sin\theta} = \sec\theta - \csc\theta$$

30. Working from the left side: $\sec B - \cos B = \dfrac{1}{\cos B} - \cos B = \dfrac{1-\cos^2 B}{\cos B} = \dfrac{\sin^2 B}{\cos B} = \dfrac{\sin B}{\cos B}\cdot\sin B = \tan B\sin B$

32. Working from the left side: $\tan\theta\sin\theta + \cos\theta = \dfrac{\sin\theta}{\cos\theta}\cdot\sin\theta + \cos\theta = \dfrac{\sin^2\theta}{\cos\theta} + \dfrac{\cos^2\theta}{\cos\theta} = \dfrac{1}{\cos\theta} = \sec\theta$

34. Working from the left side:

$$\dfrac{\cos x}{1+\sin x} - \dfrac{1-\sin x}{\cos x} = \dfrac{\cos^2 x}{\cos x(1+\sin x)} - \dfrac{1-\sin^2 x}{\cos x(1+\sin x)} = \dfrac{\cos^2 x}{\cos x(1+\sin x)} - \dfrac{\cos^2 x}{\cos x(1+\sin x)} = 0$$

36. Working from the left side: $\dfrac{1}{1-\sin x} + \dfrac{1}{1+\sin x} = \dfrac{1+\sin x}{1-\sin^2 x} + \dfrac{1-\sin x}{1-\sin^2 x} = \dfrac{2}{1-\sin^2 x} = \dfrac{2}{\cos^2 x} = 2\sec^2 x$

38. Working from the left side: $\dfrac{\csc x - 1}{\csc x + 1} = \dfrac{\dfrac{1}{\sin x} - 1}{\dfrac{1}{\sin x} + 1}\cdot\dfrac{\sin x}{\sin x} = \dfrac{1-\sin x}{1+\sin x}$

40. Working from the left side: $\dfrac{\sin t}{1+\cos t} = \dfrac{\sin t}{1+\cos t}\cdot\dfrac{1-\cos t}{1-\cos t} = \dfrac{\sin t(1-\cos t)}{1-\cos^2 t} = \dfrac{\sin t(1-\cos t)}{\sin^2 t} = \dfrac{1-\cos t}{\sin t}$

Note that this could also be proved working from the right side:

$$\dfrac{1-\cos t}{\sin t} = \dfrac{1-\cos t}{\sin t}\cdot\dfrac{1+\cos t}{1+\cos t} = \dfrac{1-\cos^2 t}{\sin t(1+\cos t)} = \dfrac{\sin^2 t}{\sin t(1+\cos t)} = \dfrac{\sin t}{1+\cos t}$$

42. Working from the left side: $\dfrac{\sin^2 t}{(1-\cos t)^2} = \dfrac{1-\cos^2 t}{(1-\cos t)^2} = \dfrac{(1+\cos t)(1-\cos t)}{(1-\cos t)^2} = \dfrac{1+\cos t}{1-\cos t}$

44. Working from the left side: $\dfrac{\csc\theta-1}{\cot\theta} = \dfrac{\csc\theta-1}{\cot\theta}\cdot\dfrac{\csc\theta+1}{\csc\theta+1} = \dfrac{\csc^2\theta-1}{\cot\theta(\csc\theta+1)} = \dfrac{\cot^2\theta}{\cot\theta(\csc\theta+1)} = \dfrac{\cot\theta}{\csc\theta+1}$

46. Working from the right side:

$$(\csc x + \cot x)^2 = \left(\dfrac{1}{\sin x} + \dfrac{\cos x}{\sin x}\right)^2 = \dfrac{(1+\cos x)^2}{\sin^2 x} = \dfrac{(1+\cos x)^2}{1-\cos^2 x} = \dfrac{(1+\cos x)^2}{(1+\cos x)(1-\cos x)} = \dfrac{1+\cos x}{1-\cos x}$$

48. Working from the left side:

$$\dfrac{1}{\csc x - \cot x} = \dfrac{1}{\csc x - \cot x}\cdot\dfrac{\csc x + \cot x}{\csc x + \cot x} = \dfrac{\csc x + \cot x}{\csc^2 x - \cot^2 x} = \dfrac{\csc x + \cot x}{1} = \csc x + \cot x$$

50. Working from the left side: $\dfrac{\cos x + 1}{\cot x} = \dfrac{\cos x + 1}{\dfrac{\cos x}{\sin x}} = \dfrac{\sin x\cos x + \sin x}{\cos x} = \sin x + \dfrac{\sin x}{\cos x} = \sin x + \tan x$

52. Working from the left side:
$$\cos^4 A - \sin^4 A = \left(\cos^2 A + \sin^2 A\right)\left(\cos^2 A - \sin^2 A\right) = 1 \cdot \left(1 - \sin^2 A - \sin^2 A\right) = 1 - 2\sin^2 A$$

54. Working from the left side:
$$\frac{\cot^2 B - \cos^2 B}{\csc^2 B - 1} = \frac{\dfrac{\cos^2 B}{\sin^2 B} - \cos^2 B}{\dfrac{1}{\sin^2 B} - 1} \cdot \frac{\sin^2 B}{\sin^2 B} = \frac{\cos^2 B - \sin^2 B \cos^2 B}{1 - \sin^2 B} = \frac{\cos^2 B\left(1 - \sin^2 B\right)}{1 - \sin^2 B} = \cos^2 B$$

56. Working from the left side: $\dfrac{\csc^2 y + \cot^2 y}{\csc^4 y - \cot^4 y} = \dfrac{\csc^2 y + \cot^2 y}{\left(\csc^2 y + \cot^2 y\right)\left(\csc^2 y - \cot^2 y\right)} = \dfrac{1}{\csc^2 y - \cot^2 y} = \dfrac{1}{1} = 1$

58. Working from the left side (factor the numerator as a difference of cubes):
$$\frac{1 - \cos^3 A}{1 - \cos A} = \frac{\left(1 - \cos A\right)\left(1 + \cos A + \cos^2 A\right)}{1 - \cos A} = 1 + \cos A + \cos^2 A$$

60. Working from the left side (factor the numerator as a sum of cubes):
$$\frac{1 + \cot^3 t}{1 + \cot t} = \frac{\left(1 + \cot t\right)\left(1 - \cot t + \cot^2 t\right)}{1 + \cot t} = 1 - \cot t + \cot^2 t = \csc^2 t - \cot t$$

62. Working from the left side:
$$\frac{\cot^2 x}{\sin x + \cos x} = \frac{\cos^2 x}{\sin^2 x(\sin x + \cos x)} \cdot \frac{\sin x - \cos x}{\sin x - \cos x}$$
$$= \frac{\cos^2 x \sin x - \cos^3 x}{\sin^2 x\left(\sin^2 x - \cos^2 x\right)}$$
$$= \frac{\cos^2 x \sin x - \cos^3 x}{\sin^2 x\left(\sin^2 x - 1 + \sin^2 x\right)}$$
$$= \frac{\cos^2 x \sin x - \cos^3 x}{\sin^2 x\left(2\sin^2 x - 1\right)}$$
$$= \frac{\cos^2 x \sin x - \cos^3 x}{2\sin^4 x - \sin^2 x}$$

64. Working from the left side:
$$\frac{\tan^2 \psi + 2}{1 + \tan^2 \psi} = \frac{\dfrac{\sin^2 \psi}{\cos^2 \psi} + 2}{1 + \dfrac{\sin^2 \psi}{\cos^2 \psi}} = \frac{\sin^2 \psi + 2\cos^2 \psi}{\cos^2 \psi + \sin^2 \psi} = \sin^2 \psi + 2\cos^2 \psi = 1 - \cos^2 \psi + 2\cos^2 \psi = 1 + \cos^2 \psi$$

66. Working from the left side:
$$\frac{\cos \beta}{1 - \tan \beta} + \frac{\sin \beta}{1 - \cot \beta} = \frac{\cos \beta}{1 - \dfrac{\sin \beta}{\cos \beta}} + \frac{\sin \beta}{1 - \dfrac{\cos \beta}{\sin \beta}}$$
$$= \frac{\cos^2 \beta}{\cos \beta - \sin \beta} + \frac{\sin^2 \beta}{\sin \beta - \cos \beta}$$
$$= \frac{\sin^2 \beta - \cos^2 \beta}{\sin \beta - \cos \beta}$$
$$= \frac{\left(\sin \beta + \cos \beta\right)\left(\sin \beta - \cos \beta\right)}{\sin \beta - \cos \beta}$$
$$= \sin \beta + \cos \beta$$

68.   Graphing both curves:

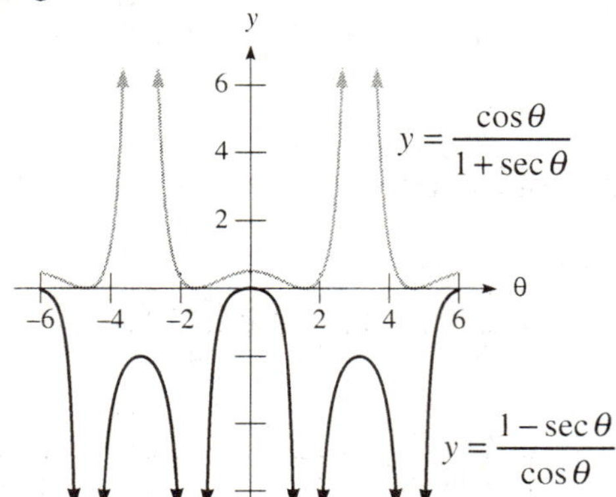

The equation does not appear to be an identity.

70.   Graphing both curves:

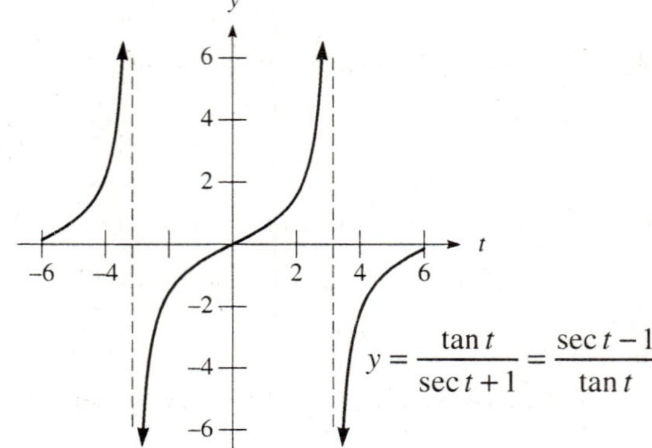

The equation appears to be an identity.

72.   Graphing both curves:

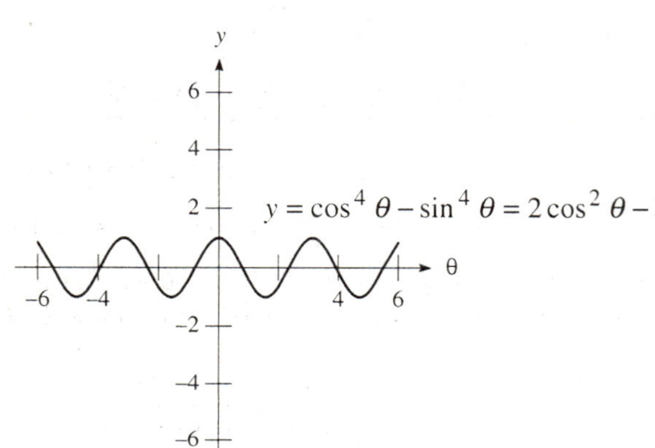

The equation appears to be an identity.

74.   Graphing both curves:

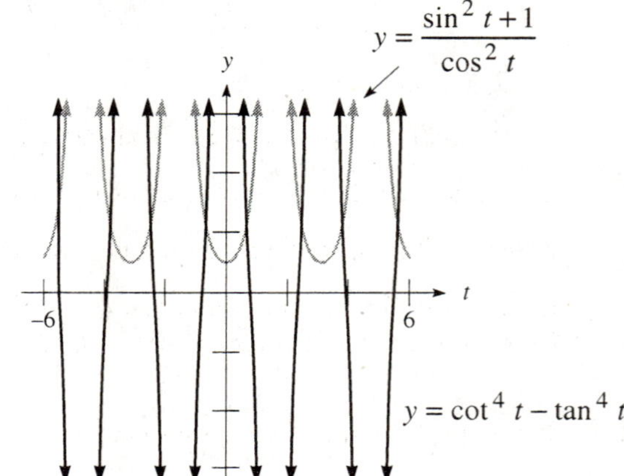

The equation does not appear to be an identity.

76.   Substituting $\theta = \dfrac{\pi}{6}$: $\sin\dfrac{\pi}{6} + \cos\dfrac{\pi}{6} = \dfrac{1}{2} + \dfrac{\sqrt{3}}{2} = \dfrac{1+\sqrt{3}}{2} \neq 1$

78.   Substituting $\theta = \dfrac{\pi}{4}$: $\tan^2\dfrac{\pi}{4} + \cot^2\dfrac{\pi}{4} = 1^2 + 1^2 = 2 \neq 1$

80.   Substituting $\theta = 0$: $\sin 0\cos 0 = 0 \cdot 1 = 0 \neq 1$

82.   Substituting $x = 30°$ into both expressions:

$$\sin 2x = \sin(2 \cdot 30°) = \sin 60° = \dfrac{\sqrt{3}}{2}$$

$$2\sin x = 2\sin 30° = 2 \cdot \dfrac{1}{2} = 1$$

So $\sin 2x \neq 2\sin x$ for all values of x.

84.   Since B terminates in quadrant III, $\sin B < 0$, thus:

$$\sin B = -\sqrt{1 - \cos^2 B} = -\sqrt{1 - \left(-\tfrac{5}{13}\right)^2} = -\sqrt{1 - \tfrac{25}{169}} = -\sqrt{\tfrac{144}{169}} = -\tfrac{12}{13}$$

Therefore: $\tan B = \dfrac{\sin B}{\cos B} = \dfrac{-\tfrac{12}{13}}{-\tfrac{5}{13}} = \dfrac{12}{5}$

86.   The exact value is given by: $\cos\dfrac{\pi}{3} = \dfrac{1}{2}$

88.   The exact value is given by: $\sin\dfrac{\pi}{6} = \dfrac{1}{2}$

90. Converting to degrees: $\dfrac{5\pi}{12}$ radians $= \dfrac{5\pi}{12} \cdot \dfrac{180°}{\pi} = 75°$

92. Converting to degrees: $\dfrac{11\pi}{12}$ radians $= \dfrac{11\pi}{12} \cdot \dfrac{180°}{\pi} = 165°$

5.2 Sum and Difference Formulas

2. Using the sum formula for sine:
$$\sin 75° = \sin(30° + 45°) = \sin 30° \cos 45° + \cos 30° \sin 45° = \frac{1}{2} \cdot \frac{\sqrt{2}}{2} + \frac{\sqrt{3}}{2} \cdot \frac{\sqrt{2}}{2} = \frac{\sqrt{2} + \sqrt{6}}{4}$$

4. Using the sum formula for tangent: $\tan 75° = \tan(30° + 45°) = \dfrac{\tan 30° + \tan 45°}{1 - \tan 30° \tan 45°} = \dfrac{\dfrac{1}{\sqrt{3}} + 1}{1 - \dfrac{1}{\sqrt{3}} \cdot 1} \cdot \dfrac{\sqrt{3}}{\sqrt{3}} = \dfrac{\sqrt{3} + 1}{\sqrt{3} - 1}$

6. Using the sum formula for cosine:
$$\cos \frac{7\pi}{12} = \cos\left(\frac{\pi}{4} + \frac{\pi}{3}\right) = \cos \frac{\pi}{4} \cos \frac{\pi}{3} - \sin \frac{\pi}{4} \sin \frac{\pi}{3} = \frac{\sqrt{2}}{2} \cdot \frac{1}{2} - \frac{\sqrt{2}}{2} \cdot \frac{\sqrt{3}}{2} = \frac{\sqrt{2} - \sqrt{6}}{4}$$

8. Using the sum formula for sine:
$$\sin 105° = \sin(60° + 45°) = \sin 60° \cos 45° + \cos 60° \sin 45° = \frac{\sqrt{3}}{2} \cdot \frac{\sqrt{2}}{2} + \frac{1}{2} \cdot \frac{\sqrt{2}}{2} = \frac{\sqrt{2} + \sqrt{6}}{4}$$

10. Using the difference formula for cosine: $\cos(x - 2\pi) = \cos x \cos 2\pi + \sin x \sin 2\pi = \cos x \cdot 1 + \sin x \cdot 0 = \cos x$

12. Using the difference formula for sine: $\sin\left(x - \dfrac{\pi}{2}\right) = \sin x \cos \dfrac{\pi}{2} - \cos x \sin \dfrac{\pi}{2} = \sin x \cdot 0 - \cos x \cdot 1 = -\cos x$

14. Using the difference formula for sine: $\sin(180° - \theta) = \sin 180° \cos \theta - \cos 180° \sin \theta = 0 \cdot \cos \theta - (-1) \cdot \sin \theta = \sin \theta$

16. Using the sum formula for cosine: $\cos(90° + \theta) = \cos 90° \cos \theta - \sin 90° \sin \theta = 0 \cdot \cos \theta - 1 \cdot \sin \theta = -\sin \theta$

18. Using the difference formula for tangent: $\tan\left(x - \dfrac{\pi}{4}\right) = \dfrac{\tan x - \tan \dfrac{\pi}{4}}{1 + \tan x \tan \dfrac{\pi}{4}} = \dfrac{\tan x - 1}{1 + \tan x \cdot 1} = \dfrac{\tan x - 1}{\tan x + 1}$

20. Using the difference formula for cosine: $\cos\left(x - \dfrac{3\pi}{2}\right) = \cos x \cos \dfrac{3\pi}{2} + \sin x \sin \dfrac{3\pi}{2} = \cos x \cdot 0 + \sin x \cdot (-1) = -\sin x$

22. Using the difference formula for cosine: $\cos 3x \cos 2x + \sin 3x \sin 2x = \cos(3x - 2x) = \cos x$

24. Using the difference formula for sine: $\sin 8x \cos x - \cos 8x \sin x = \sin(8x - x) = \sin 7x$

26. Using the difference formula for cosine: $\cos 15° \cos 75° + \sin 15° \sin 75° = \cos(15° - 75°) = \cos(-60°) = \cos 60° = \frac{1}{2}$

28. First use the sum formula for sine: $y = \sin x \cos 2x + \cos x \sin 2x = \sin(x + 2x) = \sin 3x$

The amplitude is 1 and the period is $\dfrac{2\pi}{3}$. Graphing the curve:

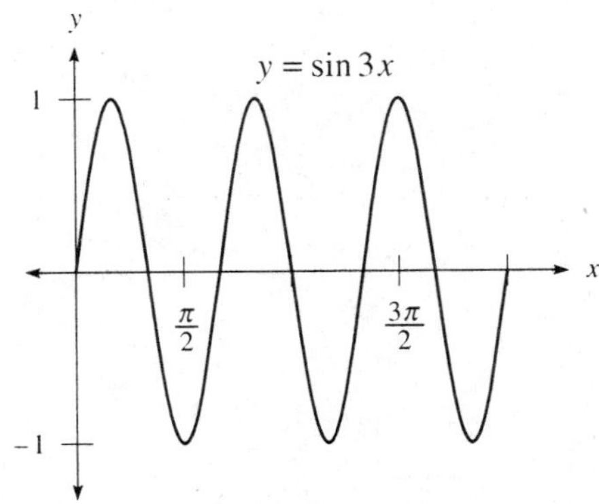

30.   First use the difference formula for cosine: $y = 2(\cos 4x \cos x + \sin 4x \sin x) = 2\cos(4x - x) = 2\cos 3x$

The amplitude is 2 and the period is $\dfrac{2\pi}{3}$. Graphing the curve:

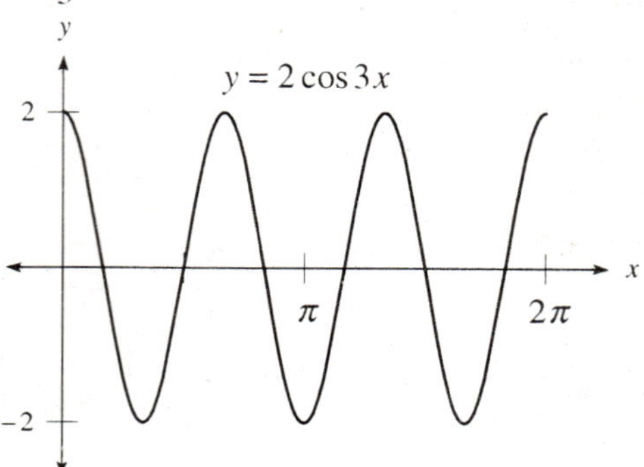

32.   Using the difference formula for sine: $y = \sin x \cos \dfrac{\pi}{6} - \cos x \sin \dfrac{\pi}{6} = \sin\left(x - \dfrac{\pi}{6}\right)$

The amplitude is 1, the period is 2π, and the phase shift is $\dfrac{\pi}{6}$. Graphing the curve:

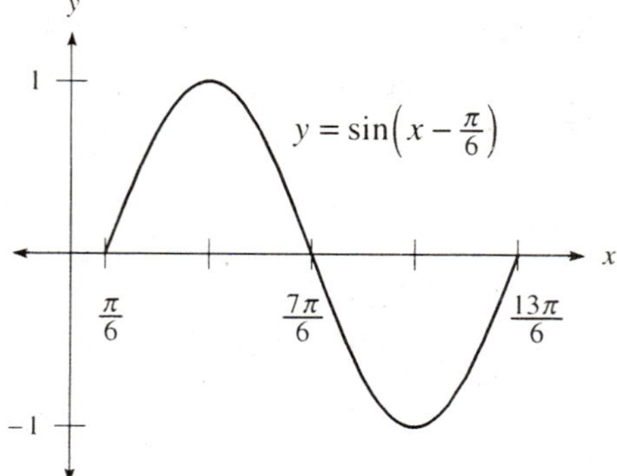

34.   Using the difference formula for sine: $y = 2\left(\sin x \cos \dfrac{\pi}{3} - \cos x \sin \dfrac{\pi}{3}\right) = 2\sin\left(x - \dfrac{\pi}{3}\right)$

The amplitude is 2, the period is 2π, and the phase shift is $\dfrac{\pi}{3}$. Graphing the curve:

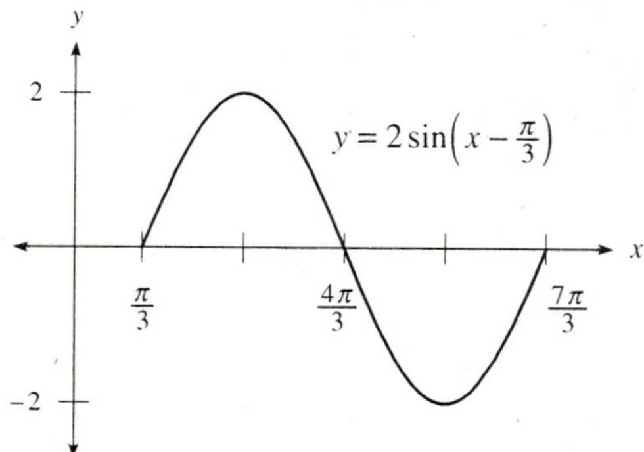

36. Since $\cos A = -\frac{5}{13}$ and A terminates in quadrant II, $\sin A > 0$ and thus:

$$\sin A = \sqrt{1 - \cos^2 A} = \sqrt{1 - \left(-\frac{5}{13}\right)^2} = \sqrt{1 - \frac{25}{169}} = \sqrt{\frac{144}{169}} = \frac{12}{13}$$

Since $\sin B = \frac{3}{5}$ and B terminates in quadrant I, $\cos B > 0$ and thus:

$$\cos B = \sqrt{1 - \sin^2 B} = \sqrt{1 - \left(\frac{3}{5}\right)^2} = \sqrt{1 - \frac{9}{25}} = \sqrt{\frac{16}{25}} = \frac{4}{5}$$

Now use the difference formulas for sine and cosine:

$$\sin(A - B) = \sin A \cos B - \cos A \sin B = \left(\frac{12}{13}\right)\left(\frac{4}{5}\right) - \left(-\frac{5}{13}\right)\left(\frac{3}{5}\right) = \frac{48}{65} + \frac{15}{65} = \frac{63}{65}$$

$$\cos(A - B) = \cos A \cos B + \sin A \sin B = \left(-\frac{5}{13}\right)\left(\frac{4}{5}\right) + \left(\frac{12}{13}\right)\left(\frac{3}{5}\right) = -\frac{20}{65} + \frac{36}{65} = \frac{16}{65}$$

Now note that $\tan A = \dfrac{\sin A}{\cos A} = \dfrac{^{12}/_{13}}{-^5/_{13}} = -\dfrac{12}{5}$ and $\tan B = \dfrac{\sin B}{\cos B} = \dfrac{^3/_5}{^4/_5} = \dfrac{3}{4}$, therefore:

$$\tan(A - B) = \frac{\tan A - \tan B}{1 + \tan A \tan B} = \frac{-\frac{12}{5} - \frac{3}{4}}{1 + \left(-\frac{12}{5}\right)\left(\frac{3}{4}\right)} = \frac{-\frac{63}{20}}{1 - \frac{36}{20}} = \frac{-\frac{63}{20}}{-\frac{16}{20}} = \frac{63}{16}$$

Note that we could have also found $\tan(A - B)$ directly: $\tan(A - B) = \dfrac{\sin(A - B)}{\cos(A - B)} = \dfrac{^{63}/_{65}}{^{16}/_{65}} = \dfrac{63}{16}$

Since $\sin(A - B) > 0$ and $\cos(A - B) > 0$, $A - B$ must terminate in quadrant I.

38. Since $\sec A = \sqrt{5}$, $\cos A = \dfrac{1}{\sqrt{5}}$. Since A terminates in quadrant I, $\sin A > 0$ and thus:

$$\sin A = \sqrt{1 - \cos^2 A} = \sqrt{1 - \left(\frac{1}{\sqrt{5}}\right)^2} = \sqrt{1 - \frac{1}{5}} = \sqrt{\frac{4}{5}} = \frac{2}{\sqrt{5}}$$

Since $\sec B = \sqrt{10}$, $\cos B = \dfrac{1}{\sqrt{10}}$. Since B terminates in quadrant I, $\sin B > 0$ and thus:

$$\sin B = \sqrt{1 - \cos^2 B} = \sqrt{1 - \left(\frac{1}{\sqrt{10}}\right)^2} = \sqrt{1 - \frac{1}{10}} = \sqrt{\frac{9}{10}} = \frac{3}{\sqrt{10}}$$

Now use the sum formula for cosine:

$$\cos(A + B) = \cos A \cos B - \sin A \sin B = \left(\frac{1}{\sqrt{5}}\right)\left(\frac{1}{\sqrt{10}}\right) - \left(\frac{2}{\sqrt{5}}\right)\left(\frac{3}{\sqrt{10}}\right) = \frac{1}{\sqrt{50}} - \frac{6}{\sqrt{50}} = -\frac{5}{\sqrt{50}} = -\frac{5}{5\sqrt{2}} = -\frac{1}{\sqrt{2}}$$

Therefore: $\sec(A + B) = \dfrac{1}{\cos(A + B)} = \dfrac{1}{-^1/_{\sqrt{2}}} = -\sqrt{2}$

40. Using the identity $\tan(A + B) = \dfrac{\tan A + \tan B}{1 - \tan A \tan B}$:

$$2 = \frac{\tan A + \frac{1}{3}}{1 - \tan A \cdot \frac{1}{3}}$$

$$2 = \frac{3 \tan A + 1}{3 - \tan A}$$

$$2(3 - \tan A) = 3 \tan A + 1$$

$$6 - 2 \tan A = 3 \tan A + 1$$

$$5 = 5 \tan A$$

$$\tan A = 1$$

42. Using the addition formula for cosine: $\cos 2x = \cos(x + x) = \cos x \cos x - \sin x \sin x = \cos^2 x - \sin^2 x$

44.   Working from the left side:
$$\sin(90° + x) - \sin(90° - x) = (\sin 90° \cos x + \cos 90° \sin x) - (\sin 90° \cos x - \cos 90° \sin x)$$
$$= (1 \cdot \cos x + 0 \cdot \sin x) - (1 \cdot \cos x - 0 \cdot \sin x)$$
$$= \cos x - \cos x$$
$$= 0$$

46.   Working from the left side:
$$\cos(x + 90°) + \cos(x - 90°) = (\cos x \cos 90° - \sin x \sin 90°) + (\cos x \cos 90° + \sin x \sin 90°)$$
$$= (\cos x \cdot 0 - \sin x \cdot 1) + (\cos x \cdot 0 + \sin x \cdot 1)$$
$$= -\sin x + \sin x$$
$$= 0$$

48.   Working from the left side:
$$\cos\left(\frac{\pi}{3} + x\right) + \cos\left(\frac{\pi}{3} - x\right) = \left(\cos\frac{\pi}{3}\cos x - \sin\frac{\pi}{3}\sin x\right) + \left(\cos\frac{\pi}{3}\cos x + \sin\frac{\pi}{3}\sin x\right)$$
$$= \left(\frac{1}{2}\cos x - \frac{\sqrt{3}}{2}\sin x\right) + \left(\frac{1}{2}\cos x + \frac{\sqrt{3}}{2}\sin x\right)$$
$$= \frac{1}{2}\cos x + \frac{1}{2}\cos x$$
$$= \cos x$$

50.   Working from the left side:
$$\sin\left(\frac{\pi}{4} + x\right) + \sin\left(\frac{\pi}{4} - x\right) = \left(\sin\frac{\pi}{4}\cos x + \cos\frac{\pi}{4}\sin x\right) + \left(\sin\frac{\pi}{4}\cos x - \cos\frac{\pi}{4}\sin x\right)$$
$$= \left(\frac{\sqrt{2}}{2}\cos x + \frac{\sqrt{2}}{2}\sin x\right) + \left(\frac{\sqrt{2}}{2}\cos x - \frac{\sqrt{2}}{2}\sin x\right)$$
$$= \frac{\sqrt{2}}{2}\cos x + \frac{\sqrt{2}}{2}\cos x$$
$$= \sqrt{2}\cos x$$

52.   Working from the left side:
$$\cos\left(x + \frac{3\pi}{2}\right) + \cos\left(x - \frac{3\pi}{2}\right) = \left(\cos x \cos\frac{3\pi}{2} - \sin x \sin\frac{3\pi}{2}\right) + \left(\cos x \cos\frac{3\pi}{2} + \sin x \sin\frac{3\pi}{2}\right)$$
$$= (0 \cdot \cos x - (-1) \cdot \sin x) + (0 \cdot \cos x + (-1) \cdot \sin x)$$
$$= \sin x - \sin x$$
$$= 0$$

54.   Working from the left side:
$$\cos(A + B) + \cos(A - B) = (\cos A \cos B - \sin A \sin B) + (\cos A \cos B + \sin A \sin B)$$
$$= \cos A \cos B + \cos A \cos B$$
$$= 2\cos A \cos B$$

56.   Working from the left side:
$$\frac{\cos(A + B)}{\sin A \cos B} = \frac{\cos A \cos B - \sin A \sin B}{\sin A \cos B} = \frac{\cos A \cos B}{\sin A \cos B} - \frac{\sin A \sin B}{\sin A \cos B} = \frac{\cos A}{\sin A} - \frac{\sin B}{\cos B} = \cot A - \tan B$$

58.   Working from the left side:

$$\sec(A - B) = \frac{1}{\cos(A - B)}$$

$$= \frac{1}{\cos A \cos B + \sin A \sin B}$$

$$= \frac{1}{\cos A \cos B + \sin A \sin B} \cdot \frac{\cos A \cos B - \sin A \sin B}{\cos A \cos B - \sin A \sin B}$$

$$= \frac{\cos A \cos B - \sin A \sin B}{\cos^2 A \cos^2 B - \sin^2 A \sin^2 B}$$

$$= \frac{\cos(A + B)}{\cos^2 A \left(1 - \sin^2 B\right) - \left(1 - \cos^2 A\right)\sin^2 B}$$

$$= \frac{\cos(A + B)}{\cos^2 A - \cos^2 A \sin^2 B - \sin^2 B + \cos^2 A \sin^2 B}$$

$$= \frac{\cos(A + B)}{\cos^2 A - \sin^2 B}$$

60.   Graphing each side:

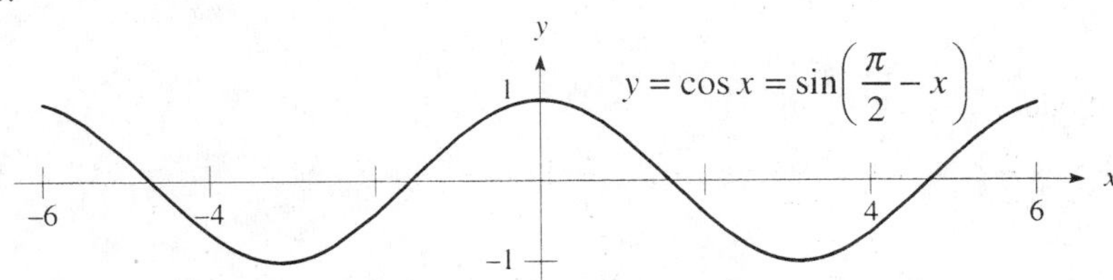

This appears to be an identity. Working from the right side:

$$\sin\left(\frac{\pi}{2} - x\right) = \sin\frac{\pi}{2}\cos x - \cos\frac{\pi}{2}\sin x = 1 \cdot \cos x - 0 \cdot \sin x = \cos x$$

62.   Graphing each side:

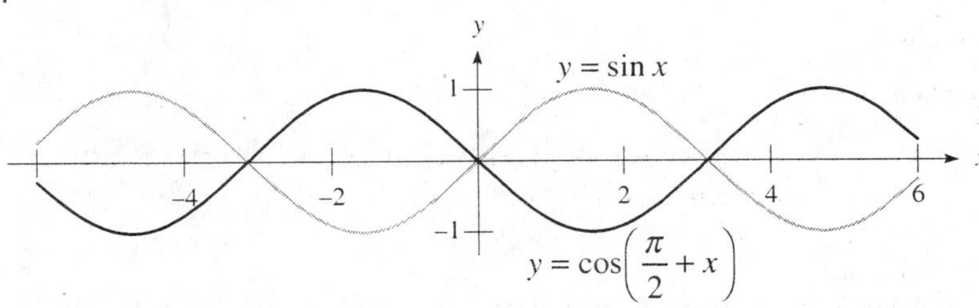

The two graphs are different, so it does not appear the equation is an identity.

64.   Graphing each side:

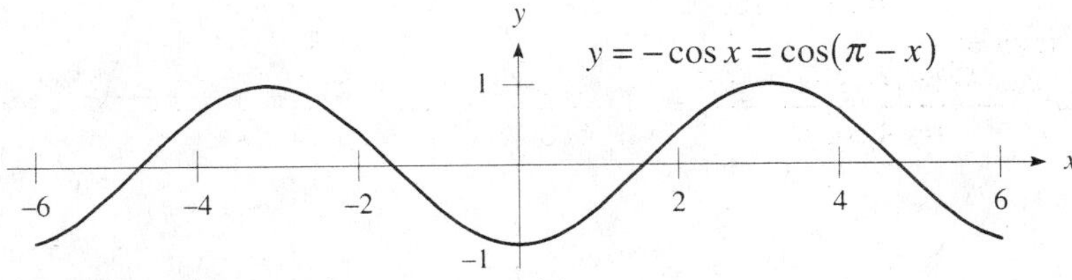

This appears to be an identity. Working from the right side:

$$\cos(\pi - x) = \cos\pi \cos x + \sin\pi \sin x = -1 \cdot \cos x + 0 \cdot \sin x = -\cos x$$

66.   The amplitude is 2 and the period is $\dfrac{2\pi}{4} = \dfrac{\pi}{2}$:

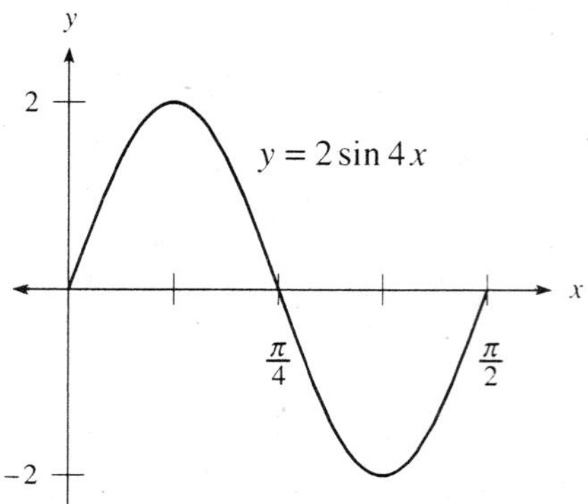

68.   The amplitude is 5 and the period is $\dfrac{2\pi}{\frac{1}{3}} = 6\pi$:

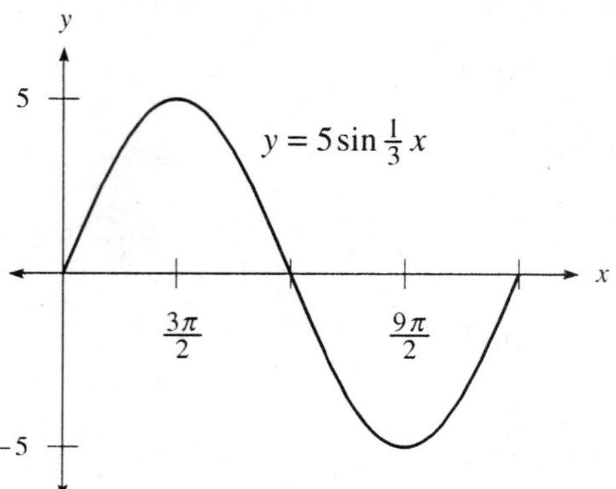

70.   The amplitude is 1 and the period is $\dfrac{2\pi}{2\pi} = 1$:

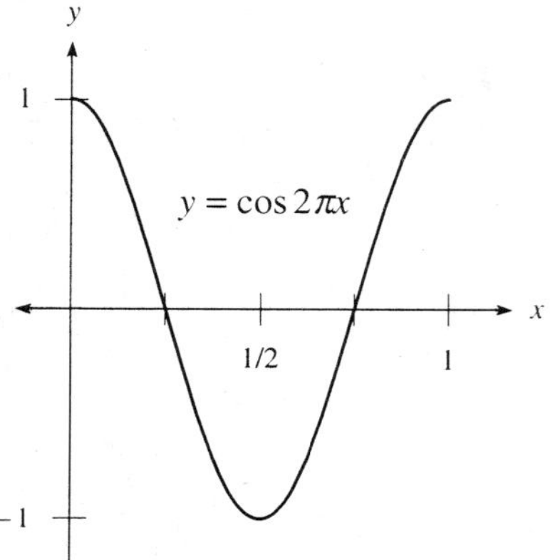

72.   The period is $\dfrac{2\pi}{3}$:

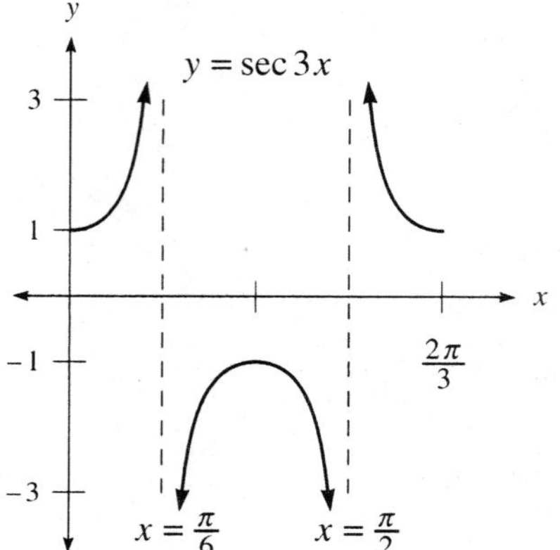

74.   The amplitude is $\dfrac{1}{2}$ and the period is $\dfrac{2\pi}{3}$:

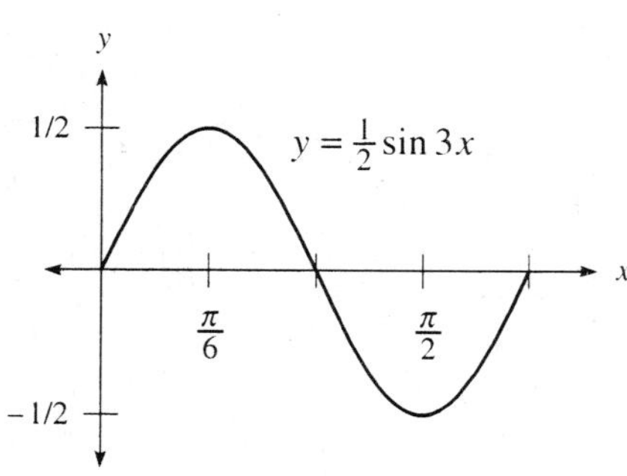

76.   The amplitude is 2 and the period is $\dfrac{2\pi}{\pi/2} = 4$:

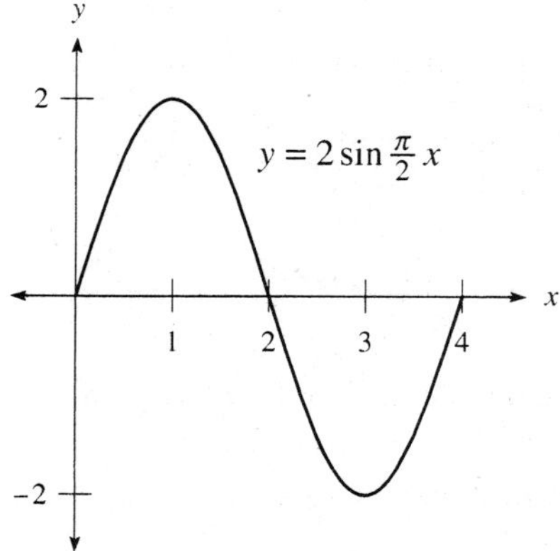

5.3 Double-Angle Formulas

2. Since A terminates in quadrant III, $\cos A < 0$ and thus:

$$\cos A = -\sqrt{1 - \sin^2 A} = -\sqrt{1 - \left(-\tfrac{3}{5}\right)^2} = -\sqrt{1 - \tfrac{9}{25}} = -\sqrt{\tfrac{16}{25}} = -\tfrac{4}{5}$$

Using the double-angle formula for cosine: $\cos 2A = \cos^2 A - \sin^2 A = \left(-\tfrac{4}{5}\right)^2 - \left(-\tfrac{3}{5}\right)^2 = \tfrac{16}{25} - \tfrac{9}{25} = \tfrac{7}{25}$

4. Since $\tan A = \dfrac{\sin A}{\cos A} = \dfrac{-\tfrac{3}{5}}{-\tfrac{4}{5}} = \tfrac{3}{4}$, use the double-angle formula for tangent:

$$\tan 2A = \frac{2\tan A}{1 - \tan^2 A} = \frac{2 \cdot \tfrac{3}{4}}{1 - \left(\tfrac{3}{4}\right)^2} = \frac{\tfrac{3}{2}}{1 - \tfrac{9}{16}} = \frac{\tfrac{3}{2}}{\tfrac{7}{16}} = \tfrac{3}{2} \cdot \tfrac{16}{7} = \tfrac{24}{7}$$

Therefore $\cot 2A = \dfrac{1}{\tan 2A} = \dfrac{1}{\tfrac{24}{7}} = \tfrac{7}{24}$.

6. Since x terminates in quadrant IV, $\sin x < 0$ and thus:

$$\sin x = -\sqrt{1 - \cos^2 x} = -\sqrt{1 - \left(\frac{1}{\sqrt{10}}\right)^2} = -\sqrt{1 - \tfrac{1}{10}} = -\sqrt{\tfrac{9}{10}} = -\frac{3}{\sqrt{10}}$$

Using the double-angle formula for sine: $\sin 2x = 2\sin x \cos x = 2\left(-\dfrac{3}{\sqrt{10}}\right)\left(\dfrac{1}{\sqrt{10}}\right) = -\tfrac{6}{10} = -\tfrac{3}{5}$

8. Since $\tan x = \dfrac{\sin x}{\cos x} = \dfrac{-\tfrac{3}{\sqrt{10}}}{\tfrac{1}{\sqrt{10}}} = -3$, use the double-angle formula for tangent:

$$\tan 2x = \frac{2\tan x}{1 - \tan^2 x} = \frac{2 \cdot (-3)}{1 - (-3)^2} = \frac{-6}{1 - 9} = \frac{-6}{-8} = \tfrac{3}{4}$$

10. Since θ terminates in quadrant I, $\sec\theta > 0$ and thus: $\sec\theta = \sqrt{\tan^2\theta + 1} = \sqrt{\left(\tfrac{5}{12}\right)^2 + 1} = \sqrt{\tfrac{25}{144} + 1} = \sqrt{\tfrac{169}{144}} = \tfrac{13}{12}$

So $\cos\theta = \dfrac{1}{\tfrac{13}{12}} = \tfrac{12}{13}$, therefore: $\sin\theta = \sqrt{1 - \cos^2\theta} = \sqrt{1 - \left(\tfrac{12}{13}\right)^2} = \sqrt{1 - \tfrac{144}{169}} = \sqrt{\tfrac{25}{169}} = \tfrac{5}{13}$

Using the double-angle formula for cosine: $\cos 2\theta = \cos^2\theta - \sin^2\theta = \left(\tfrac{12}{13}\right)^2 - \left(\tfrac{5}{13}\right)^2 = \tfrac{144}{169} - \tfrac{25}{169} = \tfrac{119}{169}$

12. Since, from Problem 10, $\cos 2\theta = \tfrac{119}{169}$, $\sec 2\theta = \dfrac{1}{\cos 2\theta} = \dfrac{1}{\tfrac{119}{169}} = \tfrac{169}{119}$.

14. Since $\csc t = \sqrt{5}$, $\sin t = \dfrac{1}{\sqrt{5}}$. Since t terminates in quadrant II, $\cos t < 0$ and thus:

$$\cos t = -\sqrt{1 - \sin^2 t} = -\sqrt{1 - \left(\frac{1}{\sqrt{5}}\right)^2} = -\sqrt{1 - \tfrac{1}{5}} = -\sqrt{\tfrac{4}{5}} = -\frac{2}{\sqrt{5}}$$

Using the double-angle formula for sine: $\sin 2t = 2\sin t \cos t = 2\left(\dfrac{1}{\sqrt{5}}\right)\left(-\dfrac{2}{\sqrt{5}}\right) = -\tfrac{4}{5}$

16. Since, from Problem 14, $\sin 2t = -\tfrac{4}{5}$, $\csc 2t = \dfrac{1}{\sin 2t} = \dfrac{1}{-\tfrac{4}{5}} = -\tfrac{5}{4}$.

18.   Using the double-angle formula for cosine: $y = 2 - 4\sin^2 x = 2\left(1 - 2\sin^2 x\right) = 2\cos 2x$

The amplitude is 2 and the period is $\dfrac{2\pi}{2} = \pi$. Sketching the graph:

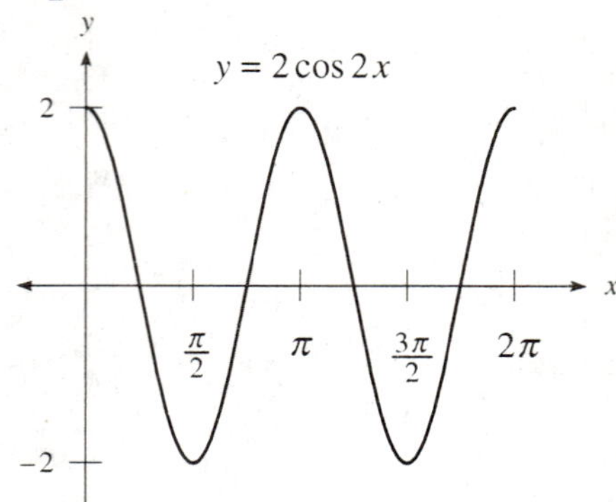

20.   Using the double-angle formula for cosine: $y = 4\cos^2 x - 2 = 2\left(2\cos^2 x - 1\right) = 2\cos 2x$

The amplitude is 2 and the period is $\dfrac{2\pi}{2} = \pi$. Sketching the graph:

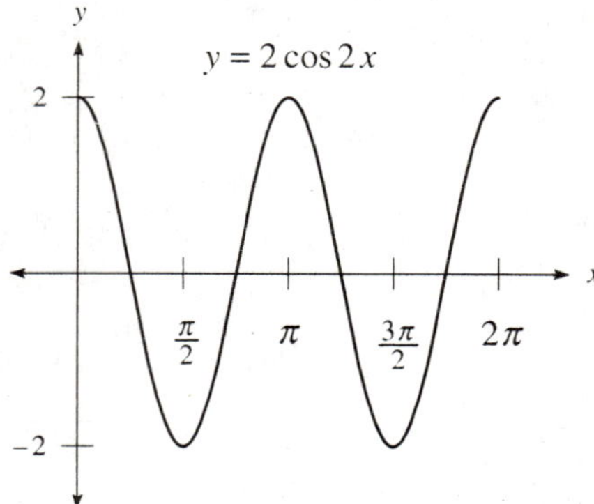

22.   Using the double-angle formula for cosine: $y = 2\cos^2 2x - 1 = \cos 4x$

The amplitude is 1 and the period is $\dfrac{2\pi}{4} = \dfrac{\pi}{2}$. Sketching the graph:

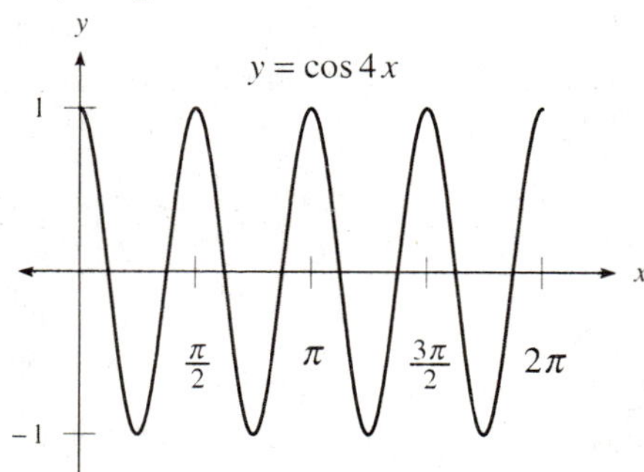

24.   Computing each side:

$$\cos 60° = \tfrac{1}{2}$$

$$1 - 2\sin^2 60° = 1 - 2\left(\tfrac{1}{2}\right)^2 = 1 - 2 \cdot \tfrac{1}{4} = 1 - \tfrac{1}{2} = \tfrac{1}{2}$$

Therefore $\cos 60° = 1 - 2\sin^2 60°$.

26.   Computing each side:

$$\sin 90° = 1$$

$$2\sin 45° \cos 45° = 2\left(\frac{1}{\sqrt{2}}\right)\left(\frac{1}{\sqrt{2}}\right) = 2 \cdot \tfrac{1}{2} = 1$$

Therefore $\sin 90° = 2\sin 45° \cos 45°$.

28. Using the double-angle formula for tangent: $\tan 2A = \dfrac{2\tan A}{1-\tan^2 A} = \dfrac{2(-\sqrt{3})}{1-(-\sqrt{3})^2} = \dfrac{-2\sqrt{3}}{1-3} = \dfrac{-2\sqrt{3}}{-2} = \sqrt{3}$

30. Using the double-angle formula for cosine: $\cos^2 15° - \sin^2 15° = \cos(2 \cdot 15°) = \cos 30° = \dfrac{\sqrt{3}}{2}$

32. Using the double-angle formula for cosine: $2\cos^2 105° - 1 = \cos(2 \cdot 105°) = \cos 210° = -\dfrac{\sqrt{3}}{2}$

34. Using the double-angle formula for sine: $\sin\dfrac{\pi}{8}\cos\dfrac{\pi}{8} = \tfrac{1}{2}\sin\left(2 \cdot \dfrac{\pi}{8}\right) = \tfrac{1}{2}\sin\dfrac{\pi}{4} = \tfrac{1}{2} \cdot \dfrac{\sqrt{2}}{2} = \dfrac{\sqrt{2}}{4}$

36. Using the double-angle formula for tangent: $\dfrac{\tan\dfrac{3\pi}{8}}{1-\tan^2\dfrac{3\pi}{8}} = \tfrac{1}{2}\tan\left(2 \cdot \dfrac{3\pi}{8}\right) = \tfrac{1}{2}\tan\dfrac{3\pi}{4} = \tfrac{1}{2}(-1) = -\tfrac{1}{2}$

38. Working from the left side:
$$(\cos x - \sin x)(\cos x + \sin x) = \cos^2 x - \sin x\cos x + \sin x\cos x - \sin^2 x = \cos^2 x - \sin^2 x = \cos 2x$$

40. Working from the right side: $\dfrac{1-\cos 2\theta}{2} = \dfrac{1-(1-2\sin^2\theta)}{2} = \dfrac{1-1+2\sin^2\theta}{2} = \dfrac{2\sin^2\theta}{2} = \sin^2\theta$

42. Working from the right side: $\dfrac{1-\tan^2\theta}{1+\tan^2\theta} = \dfrac{1-\dfrac{\sin^2\theta}{\cos^2\theta}}{1+\dfrac{\sin^2\theta}{\cos^2\theta}} = \dfrac{\cos^2\theta - \sin^2\theta}{\cos^2\theta + \sin^2\theta} = \dfrac{\cos 2\theta}{1} = \cos 2\theta$

44. Working from the right side: $\cot x - \tan x = \dfrac{\cos x}{\sin x} - \dfrac{\sin x}{\cos x} = \dfrac{\cos^2 x - \sin^2 x}{\sin x\cos x} = \dfrac{\cos 2x}{\tfrac{1}{2}\sin 2x} = 2 \cdot \dfrac{\cos 2x}{\sin 2x} = 2\cot 2x$

46. Working from the left side:
$$\cos 3\theta = \cos(\theta + 2\theta)$$
$$= \cos\theta\cos 2\theta - \sin\theta\sin 2\theta$$
$$= \cos\theta(2\cos^2\theta - 1) - \sin\theta \cdot 2\sin\theta\cos\theta$$
$$= 2\cos^3\theta - \cos\theta - 2\cos\theta(1-\cos^2\theta)$$
$$= 2\cos^3\theta - \cos\theta - 2\cos\theta + 2\cos^3\theta$$
$$= 4\cos^3\theta - 3\cos\theta$$

48. Working from the left side: $2\sin^4 x + 2\sin^2 x\cos^2 x = 2\sin^2 x(\sin^2 x + \cos^2 x) = 2\sin^2 x$

Working from the right side: $1 - \cos 2x = 1 - (1 - 2\sin^2 x) = 1 - 1 + 2\sin^2 x = 2\sin^2 x$

Since both sides simplify to the same expression, they are equal.

50. Working from the left side: $\csc\theta - 2\sin\theta = \dfrac{1}{\sin\theta} - 2\sin\theta = \dfrac{1-2\sin^2\theta}{\sin\theta} = \dfrac{\cos 2\theta}{\sin\theta}$

52. Working from the left side:
$$\cos 4A = \cos(2A + 2A)$$
$$= \cos 2A\cos 2A - \sin 2A\sin 2A$$
$$= (\cos^2 A - \sin^2 A)(\cos^2 A - \sin^2 A) - (2\sin A\cos A)(2\sin A\cos A)$$
$$= \cos^4 A - 2\sin^2 A\cos^2 A + \sin^4 A - 4\sin^2 A\cos^2 A$$
$$= \cos^4 A - 6\sin^2 A\cos^2 A + \sin^4 A$$

54.   Working from the right side:

$$\sec x \csc x - \cot x + \tan x = \frac{1}{\cos x} \cdot \frac{1}{\sin x} - \frac{\cos x}{\sin x} + \frac{\sin x}{\cos x}$$

$$= \frac{1}{\sin x \cos x} - \frac{\cos^2 x}{\sin x \cos x} + \frac{\sin^2 x}{\sin x \cos x}$$

$$= \frac{1 - \cos^2 x + \sin^2 x}{\sin x \cos x}$$

$$= \frac{1 - \left(\cos^2 x - \sin^2 x\right)}{\sin x \cos x}$$

$$= \frac{1 - \cos 2x}{\frac{1}{2}\sin 2x}$$

$$= \frac{2 - 2\cos 2x}{\sin 2x}$$

56.   Graphing the two curves:

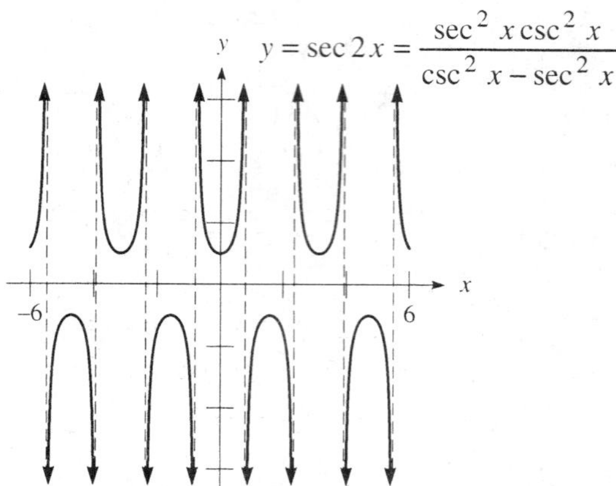

The equation appears to be an identity. Working from the right side:

$$\frac{\sec^2 x \csc^2 x}{\csc^2 x - \sec^2 x} = \frac{\dfrac{1}{\cos^2 x} \cdot \dfrac{1}{\sin^2 x}}{\dfrac{1}{\sin^2 x} - \dfrac{1}{\cos^2 x}} \cdot \frac{\sin^2 x \cos^2 x}{\sin^2 x \cos^2 x} = \frac{1}{\cos^2 x - \sin^2 x} = \frac{1}{\cos 2x} = \sec 2x$$

58.   Graphing the two curves:

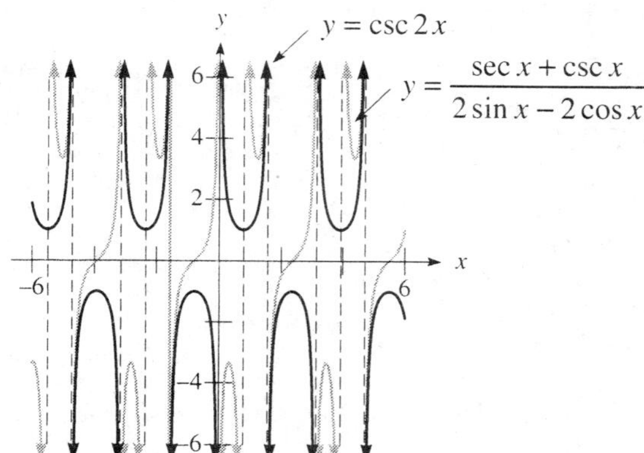

The equation does not appear to be an identity.

60.   Graphing the two curves:

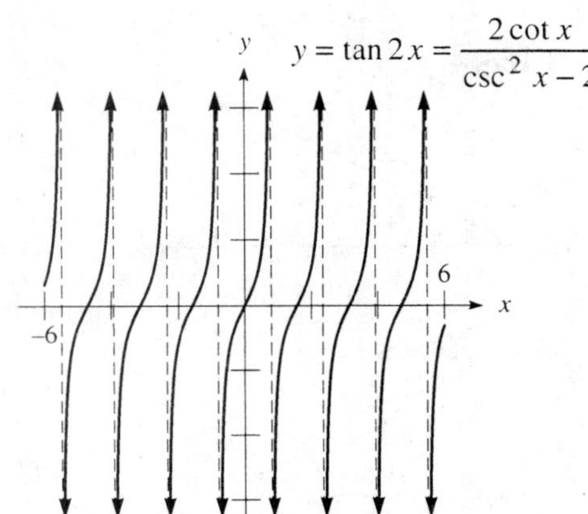

The equation appears to be an identity. Working from the right side:

$$\frac{2\cot x}{\csc^2 x - 2} = \frac{\dfrac{2\cos x}{\sin x}}{\dfrac{1}{\sin^2 x} - 2} \cdot \frac{\sin^2 x}{\sin^2 x} = \frac{2\sin x \cos x}{1 - 2\sin^2 x} = \frac{\sin 2x}{\cos 2x} = \tan 2x$$

62.   Since $x = 4\sin\theta$, $\sin\theta = \dfrac{x}{4}$ and thus $\theta = \sin^{-1}\dfrac{x}{4}$. Draw the triangle:

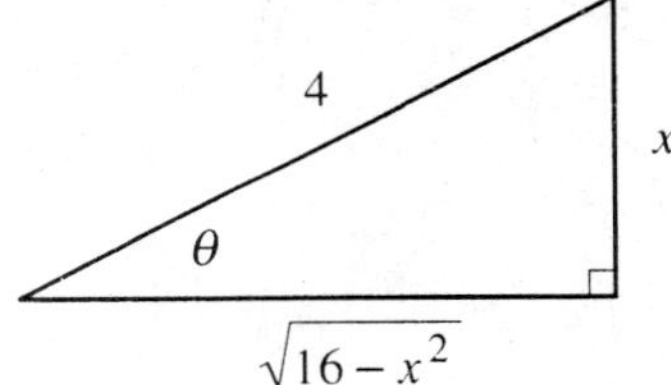

So $\cos\theta = \dfrac{\sqrt{16 - x^2}}{4}$. Thus the expression becomes:

$$\frac{\theta}{2} - \frac{\sin 2\theta}{4} = \tfrac{1}{2}\theta - \tfrac{1}{4}(2\sin\theta\cos\theta)$$

$$= \tfrac{1}{2}\theta - \tfrac{1}{2}\sin\theta\cos\theta$$

$$= \tfrac{1}{2}\sin^{-1}\frac{x}{4} - \tfrac{1}{2}\cdot\frac{x}{4}\cdot\frac{\sqrt{16 - x^2}}{4}$$

$$= \tfrac{1}{2}\sin^{-1}\frac{x}{4} - \tfrac{1}{32}x\sqrt{16 - x^2}$$

64.   Since $x = 2\sin\theta$, $\sin\theta = \dfrac{x}{2}$ and thus $\theta = \sin^{-1}\dfrac{x}{2}$. Draw the triangle:

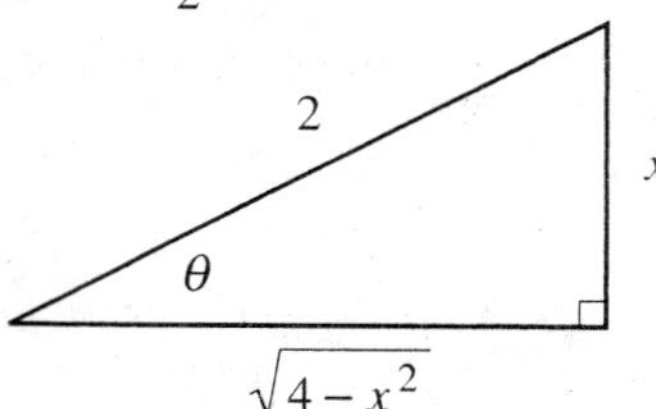

So $\tan\theta = \dfrac{x}{\sqrt{4-x^2}}$. Thus the expression becomes:

$$2\theta - \tan 2\theta = 2\theta - \frac{2\tan\theta}{1-\tan^2\theta} = 2\sin^{-1}\frac{x}{2} - \frac{2\cdot\dfrac{x}{\sqrt{4-x^2}}}{1-\dfrac{x^2}{4-x^2}} = 2\sin^{-1}\frac{x}{2} - \frac{2x\sqrt{4-x^2}}{4-2x^2} = 2\sin^{-1}\frac{x}{2} - \frac{x\sqrt{4-x^2}}{2-x^2}$$

66.   Graphing the curve:

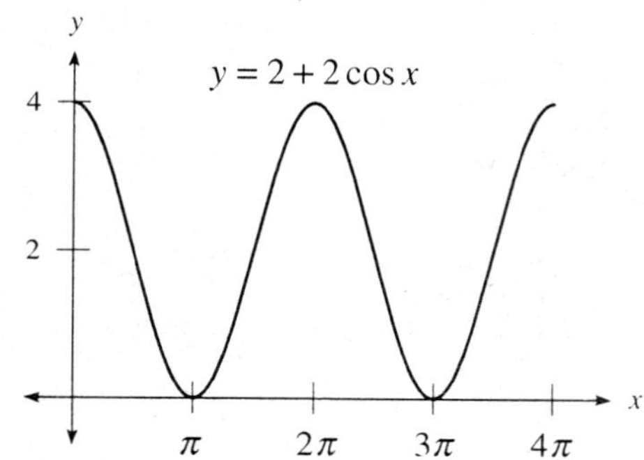

68.   Graphing the curve:

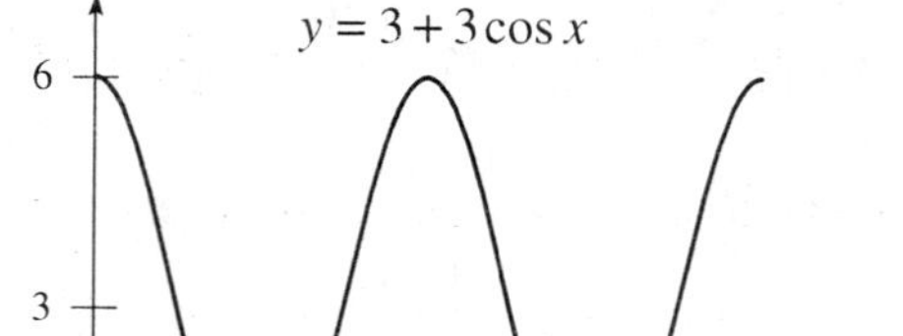

70.   Graphing the curve:

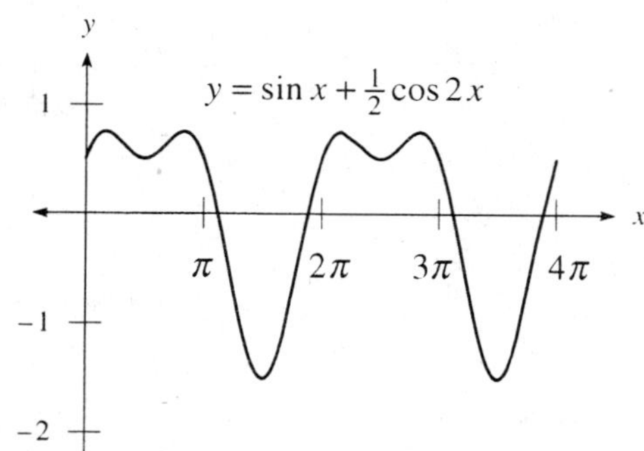

72.   Graphing the curve:

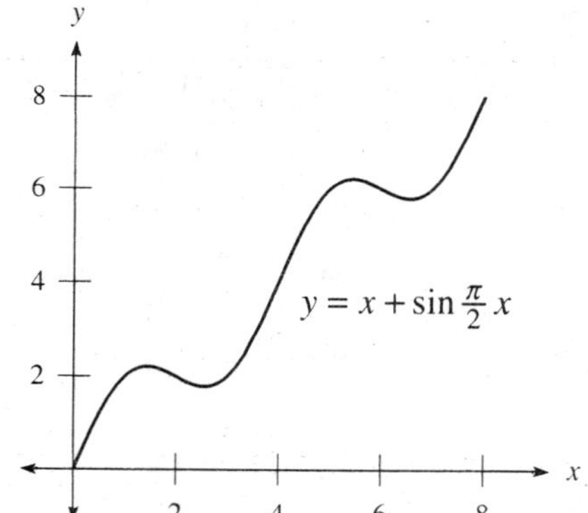

5.4 Half-Angle Formulas

2. Since $270° < A < 360°$, $135° < \dfrac{A}{2} < 180°$. Thus $\dfrac{A}{2}$ terminates in quadrant II, so $\cos\dfrac{A}{2} < 0$ and therefore:

$$\cos\frac{A}{2} = -\sqrt{\frac{1+\cos A}{2}} = -\sqrt{\frac{1+\frac{1}{2}}{2}} = -\sqrt{\frac{3}{4}} = -\frac{\sqrt{3}}{2}$$

4. Since, from Problem 2, $\cos\dfrac{A}{2} = -\dfrac{\sqrt{3}}{2}$, $\sec\dfrac{A}{2} = \dfrac{1}{\cos\dfrac{A}{2}} = \dfrac{1}{-\sqrt{3}/2} = -\dfrac{2}{\sqrt{3}}$.

6. Since A terminates in quadrant III, $\cos A < 0$ and thus:

$$\cos A = -\sqrt{1-\sin^2 A} = -\sqrt{1-\left(-\tfrac{3}{5}\right)^2} = -\sqrt{1-\tfrac{9}{25}} = -\sqrt{\tfrac{16}{25}} = -\tfrac{4}{5}$$

Since $180° < A < 270°$, $90° < \dfrac{A}{2} < 135°$. Thus $\dfrac{A}{2}$ terminates in quadrant II, so $\sin\dfrac{A}{2} > 0$ and therefore:

$$\sin\frac{A}{2} = \sqrt{\frac{1-\cos A}{2}} = \sqrt{\frac{1-\left(-\tfrac{4}{5}\right)}{2}} = \sqrt{\tfrac{9}{10}} = \frac{3}{\sqrt{10}}$$

8. Since, from Problem 6, $\sin\dfrac{A}{2} = \dfrac{3}{\sqrt{10}}$, $\csc\dfrac{A}{2} = \dfrac{1}{\sin\dfrac{A}{2}} = \dfrac{1}{3/\sqrt{10}} = \dfrac{\sqrt{10}}{3}$.

10. Since B terminates in quadrant III, $\cos B < 0$ and thus:

$$\cos B = -\sqrt{1-\sin^2 B} = -\sqrt{1-\left(-\tfrac{1}{3}\right)^2} = -\sqrt{1-\tfrac{1}{9}} = -\sqrt{\tfrac{8}{9}} = -\frac{2\sqrt{2}}{3}$$

Since $180° < B < 270°$, $90° < \dfrac{B}{2} < 135°$. Thus $\dfrac{B}{2}$ terminates in quadrant II, so $\sin\dfrac{B}{2} > 0$ and therefore:

$$\sin\frac{B}{2} = \sqrt{\frac{1-\cos B}{2}} = \sqrt{\frac{1+\frac{2\sqrt{2}}{3}}{2}} = \sqrt{\frac{3+2\sqrt{2}}{6}}$$

Thus $\csc\dfrac{B}{2} = \dfrac{1}{\sin\dfrac{B}{2}} = \dfrac{1}{\sqrt{\frac{3+2\sqrt{2}}{6}}} = \sqrt{\dfrac{6}{3+2\sqrt{2}}}$.

12. First find $\cos\dfrac{B}{2}$ (note that $\dfrac{B}{2}$ terminates in quadrant II, thus $\cos\dfrac{B}{2} < 0$):

$$\cos\frac{B}{2} = -\sqrt{\frac{1+\cos B}{2}} = -\sqrt{\frac{1-\frac{2\sqrt{2}}{3}}{2}} = -\sqrt{\frac{3-2\sqrt{2}}{6}}$$

Thus $\sec\dfrac{B}{2} = \dfrac{1}{\cos\dfrac{B}{2}} = \dfrac{1}{-\sqrt{\frac{3-2\sqrt{2}}{6}}} = -\sqrt{\dfrac{6}{3-2\sqrt{2}}}$.

14. Using the results from Problems 10 and 12: $\cot\dfrac{B}{2} = \dfrac{\cos\dfrac{B}{2}}{\sin\dfrac{B}{2}} = \dfrac{-\sqrt{\frac{3-2\sqrt{2}}{6}}}{\sqrt{\frac{3+2\sqrt{2}}{6}}} = -\sqrt{\dfrac{3-2\sqrt{2}}{3+2\sqrt{2}}} \cdot \dfrac{\sqrt{3+2\sqrt{2}}}{\sqrt{3+2\sqrt{2}}} = -\dfrac{1}{3+2\sqrt{2}}$

16. Since A terminates in quadrant II, $\cos A < 0$ and thus: $\cos A = -\sqrt{1 - \sin^2 A} = -\sqrt{1 - \left(\frac{4}{5}\right)^2} = -\sqrt{1 - \frac{16}{25}} = -\sqrt{\frac{9}{25}} = -\frac{3}{5}$

Also note that $\dfrac{A}{2}$ terminates in quadrant I, thus $\cos\dfrac{A}{2} > 0$ and therefore: $\cos\dfrac{A}{2} = \sqrt{\dfrac{1 + \cos A}{2}} = \sqrt{\dfrac{1 - \frac{3}{5}}{2}} = \sqrt{\dfrac{1}{5}} = \dfrac{1}{\sqrt{5}}$

18. Using the double-angle formula for sine: $\sin 2A = 2\sin A\cos A = 2\left(\frac{4}{5}\right)\left(-\frac{3}{5}\right) = -\frac{24}{25}$

20. Using the result from Problem 18: $\csc 2A = \dfrac{1}{\sin 2A} = \dfrac{1}{-24/25} = -\dfrac{25}{24}$

22. Since B terminates in quadrant I, $\cos B > 0$ and thus: $\cos B = \sqrt{1 - \sin^2 B} = \sqrt{1 - \left(\frac{3}{5}\right)^2} = \sqrt{1 - \frac{9}{25}} = \sqrt{\frac{16}{25}} = \frac{4}{5}$

Note that $\dfrac{B}{2}$ terminates in quadrant I, thus $\sin\dfrac{B}{2} > 0$ and therefore: $\sin\dfrac{B}{2} = \sqrt{\dfrac{1 - \cos B}{2}} = \sqrt{\dfrac{1 - \frac{4}{5}}{2}} = \sqrt{\dfrac{1}{10}} = \dfrac{1}{\sqrt{10}}$

24. Using the results of Problems 16 and 22 and the addition formula for cosine:
$$\cos(A + B) = \cos A\cos B - \sin A\sin B = \left(-\frac{3}{5}\right)\left(\frac{4}{5}\right) - \left(\frac{4}{5}\right)\left(\frac{3}{5}\right) = -\frac{12}{25} - \frac{12}{25} = -\frac{24}{25}$$

26. Using the results of Problems 16 and 22 and the subtraction formula for sine:
$$\sin(A - B) = \sin A\cos B - \cos A\sin B = \left(\frac{4}{5}\right)\left(\frac{4}{5}\right) - \left(-\frac{3}{5}\right)\left(\frac{3}{5}\right) = \frac{16}{25} + \frac{9}{25} = 1$$

28. Using the half-angle formula for cosine: $y = 6\cos^2\dfrac{x}{2} = 6\left(\dfrac{1 + \cos x}{2}\right) = 3(1 + \cos x) = 3 + 3\cos x$

The amplitude is 3, the period is 2π, and the vertical translation is 3 units. Now graphing the curve:

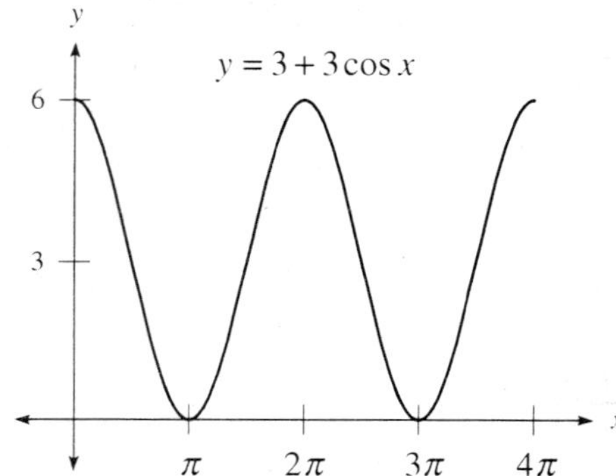

30. Using the half-angle formula for cosine: $y = 2\sin^2\dfrac{x}{2} = 2\left(\dfrac{1 - \cos x}{2}\right) = 1 - \cos x$

Now graphing the curve:

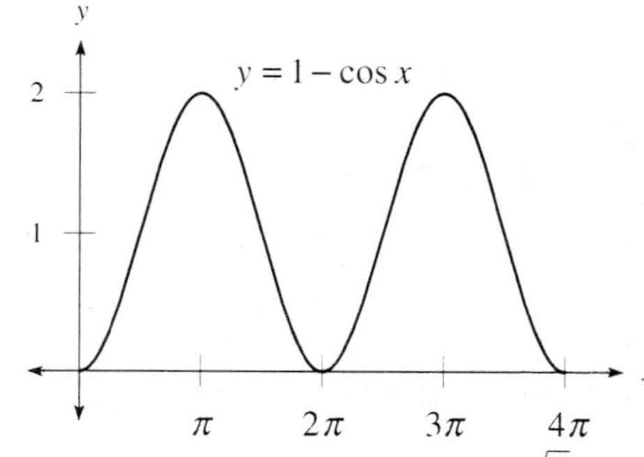

32. Using the half-angle formula for tangent: $\tan 15° = \dfrac{1 - \cos 30°}{\sin 30°} = \dfrac{1 - \frac{\sqrt{3}}{2}}{\frac{1}{2}} = 2 - \sqrt{3}$

34. Using the half-angle formula for cosine (note that $\cos 75° > 0$):

$$\cos 75° = \sqrt{\frac{1 + \cos 150°}{2}} = \sqrt{\frac{1 - \frac{\sqrt{3}}{2}}{2}} = \sqrt{\frac{2 - \sqrt{3}}{4}} = \frac{\sqrt{2 - \sqrt{3}}}{2}$$

36. Using the half-angle formula for sine (note that $\sin 105° > 0$):

$$\sin 105° = \sqrt{\frac{1 - \cos 210°}{2}} = \sqrt{\frac{1 + \frac{\sqrt{3}}{2}}{2}} = \sqrt{\frac{2 + \sqrt{3}}{4}} = \frac{\sqrt{2 + \sqrt{3}}}{2}$$

38. Working from the left side: $2\cos^2\frac{\theta}{2} = 2\left(\frac{1 + \cos\theta}{2}\right) = 1 + \cos\theta$

Working from the right side: $\dfrac{\sin^2\theta}{1 - \cos\theta} = \dfrac{1 - \cos^2\theta}{1 - \cos\theta} = \dfrac{(1 + \cos\theta)(1 - \cos\theta)}{1 - \cos\theta} = 1 + \cos\theta$

Since both sides simplify to the same expression, they are equal.

40. Working from the left side: $\csc^2\dfrac{A}{2} = \dfrac{1}{\sin^2\dfrac{A}{2}} = \dfrac{1}{\dfrac{1 - \cos A}{2}} = \dfrac{2}{1 - \cos A}$

Working from the right side: $\dfrac{2\sec A}{\sec A - 1} = \dfrac{\dfrac{2}{\cos A}}{\dfrac{1}{\cos A} - 1} \cdot \dfrac{\cos A}{\cos A} = \dfrac{2}{1 - \cos A}$

Since both sides simplify to the same expression, they are equal.

42. Working from the right side: $\dfrac{\sec B}{\sec B\csc B + \csc B} = \dfrac{\dfrac{1}{\cos B}}{\dfrac{1}{\cos B} \cdot \dfrac{1}{\sin B} + \dfrac{1}{\sin B}} \cdot \dfrac{\sin B\cos B}{\sin B\cos B} = \dfrac{\sin B}{1 + \cos B} = \tan\dfrac{B}{2}$

44. Working from the left side:

$$\tan\frac{x}{2} - \cot\frac{x}{2} = \tan\frac{x}{2} - \frac{1}{\tan\dfrac{x}{2}}$$

$$= \frac{\sin x}{1 + \cos x} - \frac{\sin x}{1 - \cos x}$$

$$= \frac{\sin x(1 - \cos x) - \sin x(1 + \cos x)}{1 - \cos^2 x}$$

$$= \frac{\sin x(1 - \cos x - 1 - \cos x)}{\sin^2 x}$$

$$= \frac{-2\cos x}{\sin x}$$

$$= -2\cot x$$

46. Working from the left side: $2\sin^2\frac{\theta}{2} = 2\left(\frac{1 - \cos\theta}{2}\right) = 1 - \cos\theta$

Working from the right side: $\dfrac{\sin^2\theta}{1 + \cos\theta} \cdot \dfrac{1 - \cos\theta}{1 - \cos\theta} = \dfrac{\sin^2\theta(1 - \cos\theta)}{1 - \cos^2\theta} = \dfrac{\sin^2\theta(1 - \cos\theta)}{\sin^2\theta} = 1 - \cos\theta$

Since both sides simplify to the same expression, they are equal.

48. Working from the right side: $1 - 2\cos 2\theta + \cos^2 2\theta = (1 - \cos 2\theta)^2 = \left(2\sin^2\theta\right)^2 = 4\sin^4\theta$

50.   Let $\theta = \arcsin\frac{3}{5}$, so $\sin\theta = \frac{3}{5}$. Draw the triangle:

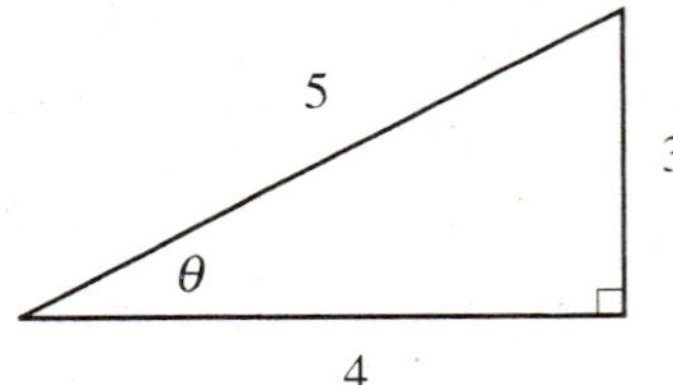

Therefore: $\cos\left(\arcsin\frac{3}{5}\right) = \cos\theta = \frac{4}{5}$

52.   Let $\theta = \arctan 2$, so $\tan\theta = 2$. Draw the triangle:

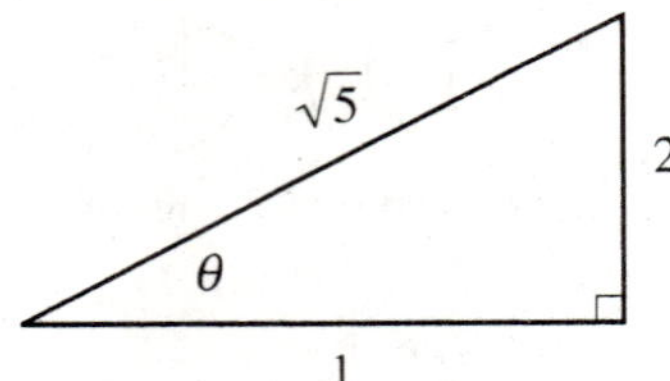

Therefore: $\sin(\arctan 2) = \sin\theta = \dfrac{2}{\sqrt{5}}$

54.   Let $\theta = \tan^{-1} x$, so $\tan\theta = x$. Draw the triangle:

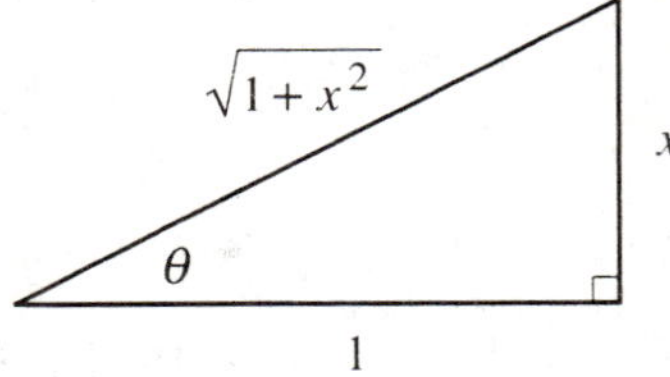

Therefore: $\cos\left(\tan^{-1} x\right) = \cos\theta = \dfrac{1}{\sqrt{1+x^2}}$

56.   Let $\theta = \cos^{-1} x$, so $\cos\theta = x$. Draw the triangle:

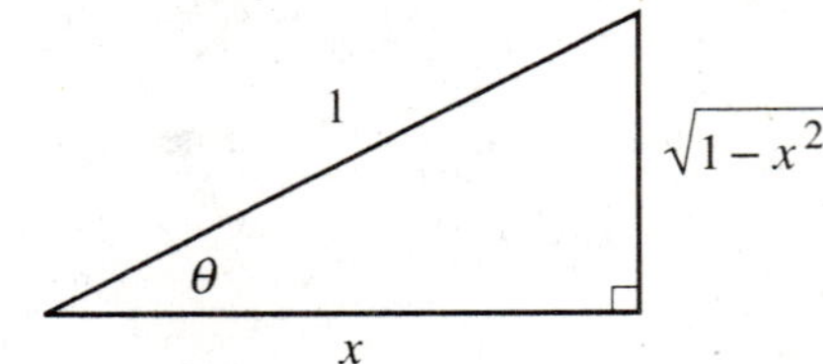

Therefore: $\tan\left(\cos^{-1} x\right) = \tan\theta = \dfrac{\sqrt{1-x^2}}{x}$

58.   Sketching the graph:

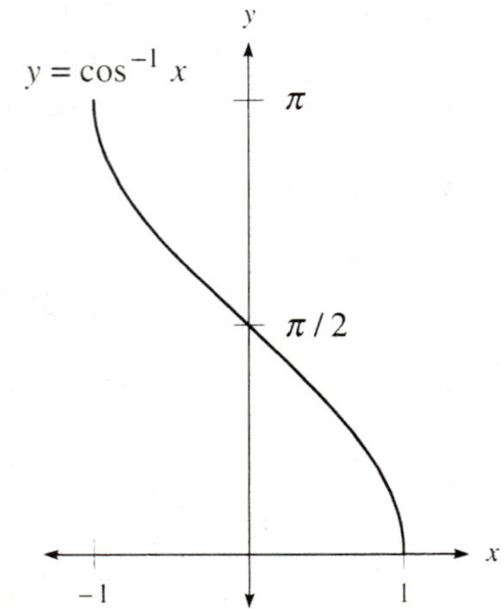

5.5 Additional Identities

2.   Let $\alpha = \arcsin\frac{3}{5}$, so $\sin\alpha = \frac{3}{5}$. Let $\beta = \arctan 2$, so $\tan\beta = 2$. Draw the triangles:

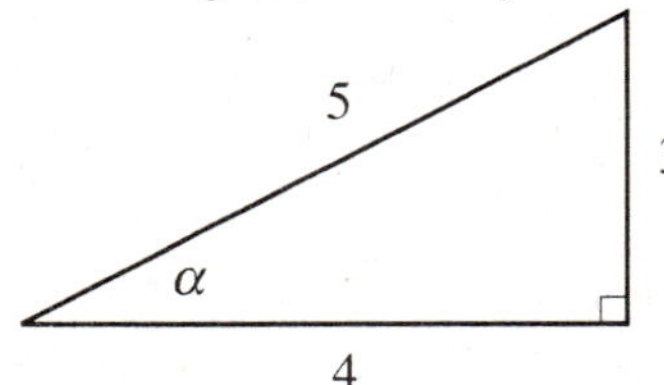
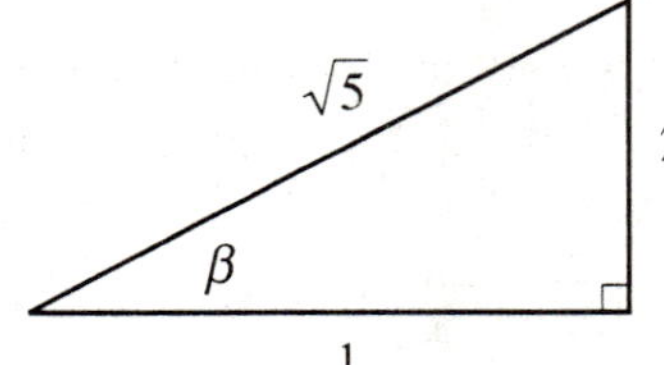

Therefore: $\cos\left(\arcsin\frac{3}{5} - \arctan 2\right) = \cos(\alpha - \beta) = \cos\alpha\cos\beta + \sin\alpha\sin\beta = \left(\frac{4}{5}\right)\left(\frac{1}{\sqrt{5}}\right) + \left(\frac{3}{5}\right)\left(\frac{2}{\sqrt{5}}\right) = \frac{10}{5\sqrt{5}} = \frac{2}{\sqrt{5}}$

4. Let $\alpha = \tan^{-1}\frac{1}{2}$, so $\tan\alpha = \frac{1}{2}$. Let $\beta = \sin^{-1}\frac{1}{2}$, so $\sin\beta = \frac{1}{2}$. Draw the triangles:

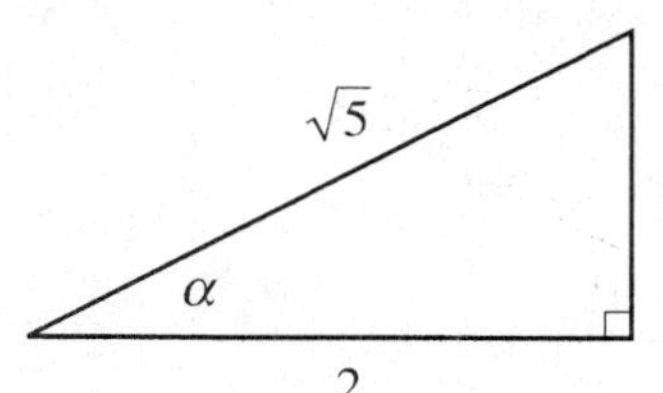
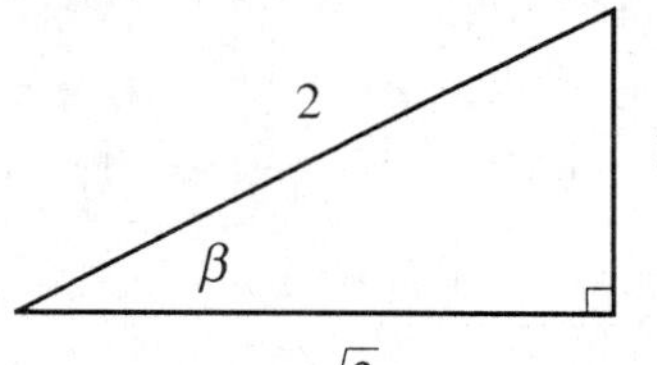

Therefore: $\sin\left(\tan^{-1}\frac{1}{2} - \sin^{-1}\frac{1}{2}\right) = \sin(\alpha - \beta) = \sin\alpha\cos\beta - \cos\alpha\sin\beta = \left(\frac{1}{\sqrt{5}}\right)\left(\frac{\sqrt{3}}{2}\right) - \left(\frac{2}{\sqrt{5}}\right)\left(\frac{1}{2}\right) = \frac{\sqrt{3}-2}{2\sqrt{5}}$

6. Let $\alpha = \tan^{-1}\frac{3}{4}$, so $\tan\alpha = \frac{3}{4}$. Draw the triangle:

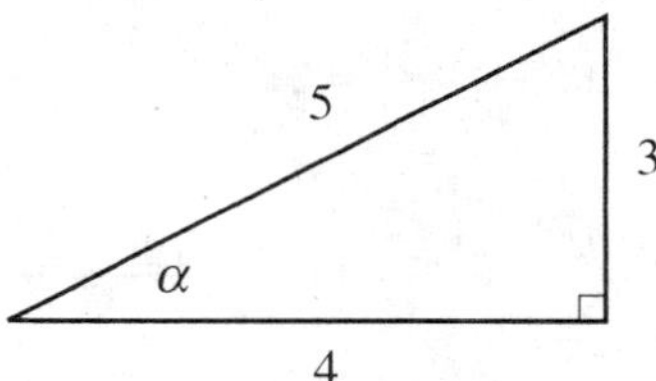

Therefore: $\sin\left(2\tan^{-1}\frac{3}{4}\right) = \sin(2\alpha) = 2\sin\alpha\cos\alpha = 2\left(\frac{3}{5}\right)\left(\frac{4}{5}\right) = \frac{24}{25}$

8. Let $\alpha = \cos^{-1}x$, so $\cos\alpha = x$. Draw the triangle:

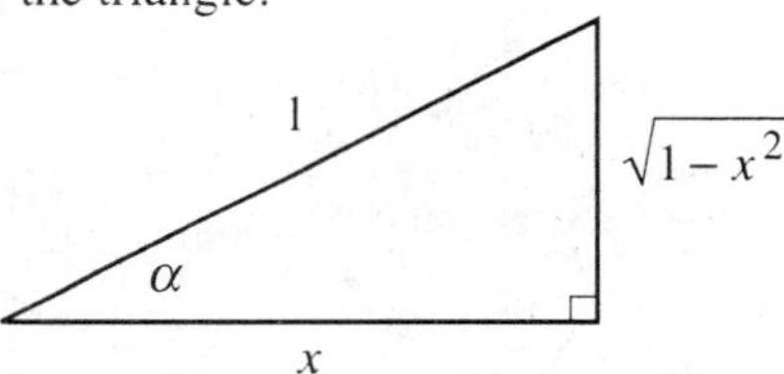

Therefore: $\tan\left(\cos^{-1}x\right) = \tan\alpha = \frac{\sqrt{1-x^2}}{x}$

10. Using the triangle from Problem 8: $\sin\left(2\cos^{-1}x\right) = \sin(2\alpha) = 2\sin\alpha\cos\alpha = 2\cdot\sqrt{1-x^2}\cdot x = 2x\sqrt{1-x^2}$

12. Let $\alpha = \sin^{-1}x$, so $\sin\alpha = x$. Draw the triangle:

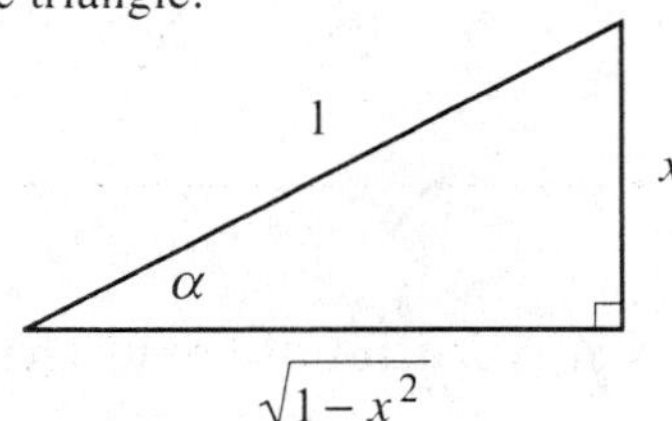

Therefore: $\cos\left(2\sin^{-1}x\right) = \cos(2\alpha) = \cos^2\alpha - \sin^2\alpha = \left(\sqrt{1-x^2}\right)^2 - x^2 = 1 - x^2 - x^2 = 1 - 2x^2$

14. Substituting $A = 120°$ and $B = 30°$ into product formula (1):
$$\frac{1}{2}\left[\sin(A+B) + \sin(A-B)\right] = \frac{1}{2}\left(\sin 150° + \sin 90°\right) = \frac{1}{2}\left(\frac{1}{2}+1\right) = \frac{1}{2}\cdot\frac{3}{2} = \frac{3}{4}$$
$$\sin A\cos B = \left(\frac{\sqrt{3}}{2}\right)\left(\frac{\sqrt{3}}{2}\right) = \frac{3}{4}$$

Both sides of the formula evaluate to the same number.

16. Using product formula (4): $10\sin 5x\sin 3x = 10\cdot\frac{1}{2}\left[\cos(5x-3x) - \cos(5x+3x)\right] = 5(\cos 2x - \cos 8x)$

18. Using product formula (2):
$$\cos 2x\sin 8x = \frac{1}{2}\left[\sin(2x+8x) - \sin(2x-8x)\right] = \frac{1}{2}\left[\sin 10x - \sin(-6x)\right] = \frac{1}{2}\left(\sin 10x + \sin 6x\right)$$

20. Using product formula (3):
$$\cos 90° \cos 180° = \tfrac{1}{2}\left[\cos(90° + 180°) + \cos(90° - 180°)\right] = \tfrac{1}{2}\left[\cos 270° + \cos(-90°)\right] = \tfrac{1}{2}(0 + 0) = 0$$

22. Using product formula (2): $\cos 3\pi \sin \pi = \tfrac{1}{2}\left[\sin(3\pi + \pi) - \sin(3\pi - \pi)\right] = \tfrac{1}{2}\left[\sin 4\pi - \sin 2\pi\right] = \tfrac{1}{2}(0 - 0) = 0$

24. Substituting $\alpha = 90°$ and $\beta = 30°$ into sum formula (8):

$$\cos \alpha - \cos \beta = \cos 90° - \cos 30° = 0 - \frac{\sqrt{3}}{2} = -\frac{\sqrt{3}}{2}$$

$$-2 \sin \frac{\alpha + \beta}{2} \sin \frac{\alpha - \beta}{2} = -2 \sin \frac{120°}{2} \sin \frac{60°}{2} = -2 \sin 60° \sin 30° = -2 \cdot \frac{\sqrt{3}}{2} \cdot \frac{1}{2} = -\frac{\sqrt{3}}{2}$$

26. Using sum formula (8): $\cos 5x - \cos 3x = -2 \sin \dfrac{5x + 3x}{2} \sin \dfrac{5x - 3x}{2} = -2 \sin \dfrac{8x}{2} \sin \dfrac{2x}{2} = -2 \sin 4x \sin x$

28. Using sum formula (6):

$$\sin 75° - \sin 15° = 2 \cos \frac{75° + 15°}{2} \sin \frac{75° - 15°}{2} = 2 \cos \frac{90°}{2} \sin \frac{60°}{2} = 2 \cos 45° \sin 30° = 2 \cdot \frac{1}{\sqrt{2}} \cdot \frac{1}{2} = \frac{1}{\sqrt{2}}$$

30. Using sum formula (7): $\cos \dfrac{\pi}{12} + \cos \dfrac{7\pi}{12} = 2 \cos \dfrac{\dfrac{\pi}{12} + \dfrac{7\pi}{12}}{2} \cos \dfrac{\dfrac{\pi}{12} - \dfrac{7\pi}{12}}{2} = 2 \cos \dfrac{\pi}{3} \cos\left(-\dfrac{\pi}{4}\right) = 2 \cdot \dfrac{1}{2} \cdot \dfrac{1}{\sqrt{2}} = \dfrac{1}{\sqrt{2}}$

32. Working from the right side: $\dfrac{\cos 3x + \cos x}{\sin 3x - \sin x} = \dfrac{2 \cos \dfrac{3x + x}{2} \cos \dfrac{3x - x}{2}}{2 \cos \dfrac{3x + x}{2} \sin \dfrac{3x - x}{2}} = \dfrac{2 \cos 2x \cos x}{2 \cos 2x \sin x} = \dfrac{\cos x}{\sin x} = \cot x$

34. Working from the right side: $\dfrac{\cos 3x - \cos 5x}{\sin 3x - \sin 5x} = \dfrac{-2 \sin \dfrac{3x + 5x}{2} \sin \dfrac{3x - 5x}{2}}{2 \cos \dfrac{3x + 5x}{2} \sin \dfrac{3x - 5x}{2}} = \dfrac{-2 \sin 4x \sin(-x)}{2 \cos 4x \sin(-x)} = -\dfrac{\sin 4x}{\cos 4x} = -\tan 4x$

36. Working from the right side:

$$\frac{\sin 3x - \sin x}{\cos x - \cos 3x} = \frac{2 \cos \dfrac{3x + x}{2} \sin \dfrac{3x - x}{2}}{-2 \sin \dfrac{x + 3x}{2} \sin \dfrac{x - 3x}{2}} = \frac{2 \cos 2x \sin x}{-2 \sin 2x \sin(-x)} = \frac{2 \cos 2x \sin x}{2 \sin 2x \sin x} = \frac{\cos 2x}{\sin 2x} = \cot 2x$$

38. The amplitude is 1, the period is 2π , and the phase shift is $\dfrac{\pi}{4}$:

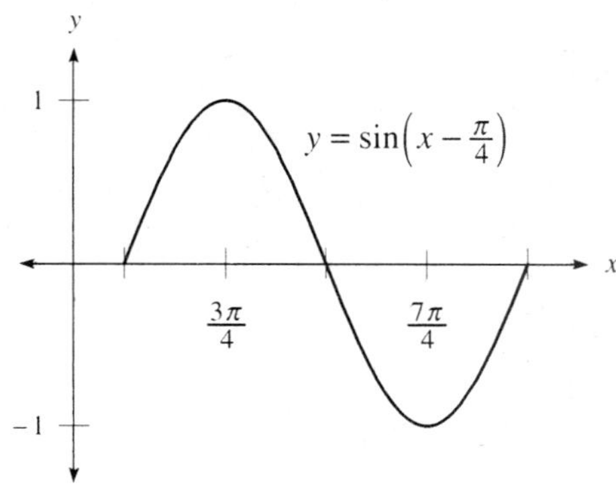

40.   The amplitude is 1, the period is 2π, and the phase shift is $-\dfrac{\pi}{3}$:

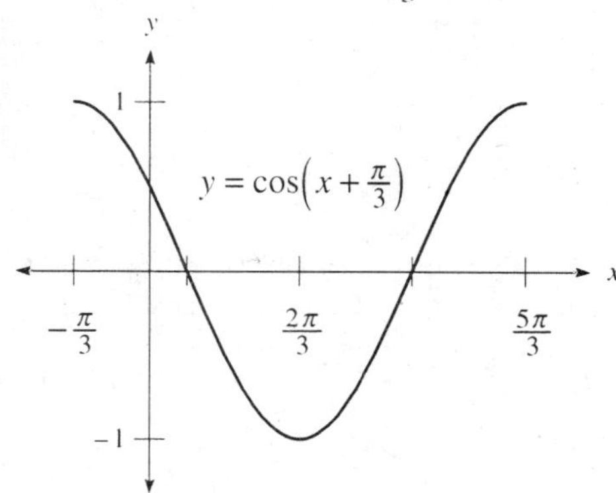

42.   The amplitude is 1, the period is $\dfrac{2\pi}{2} = \pi$, and the phase shift is $\dfrac{\pi/3}{2} = \dfrac{\pi}{6}$:

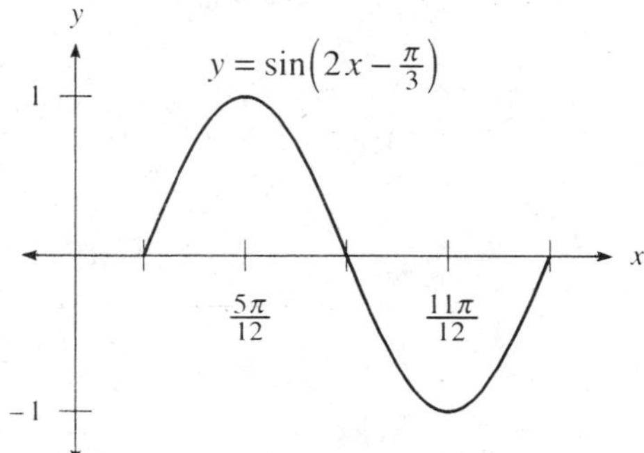

44.   The amplitude is $\dfrac{4}{3}$, the period is $\dfrac{2\pi}{3}$, and the phase shift is $\dfrac{\pi/2}{3} = \dfrac{\pi}{6}$:

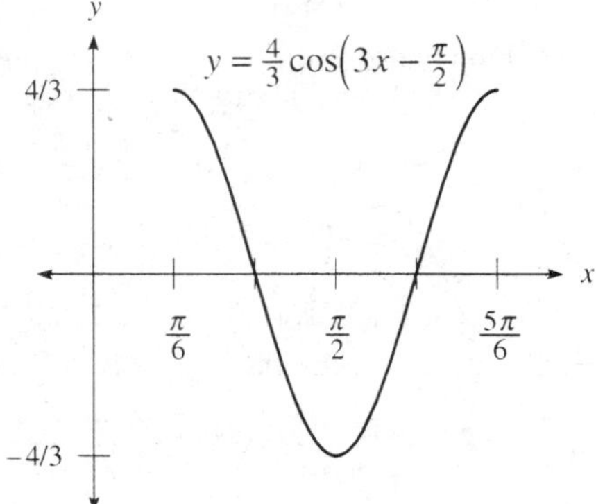

46. The amplitude is 4, the period is $\dfrac{2\pi}{2\pi} = 1$, and the phase shift is $\dfrac{\pi/2}{2\pi} = \tfrac{1}{4}$:

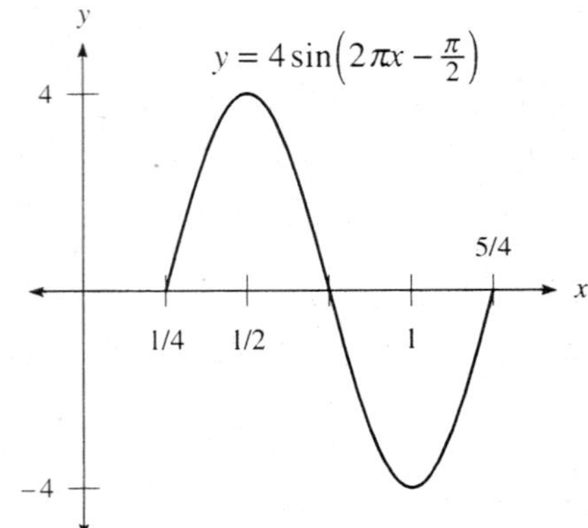

Chapter 5 Test

1. Working from the right side: $\sin\theta\sec\theta = \sin\theta \cdot \dfrac{1}{\cos\theta} = \dfrac{\sin\theta}{\cos\theta} = \tan\theta$

2. Working from the left side: $\dfrac{\cot\theta}{\csc\theta} = \dfrac{\dfrac{\cos\theta}{\sin\theta}}{\dfrac{1}{\sin\theta}} = \dfrac{\cos\theta}{\sin\theta} \cdot \dfrac{\sin\theta}{1} = \cos\theta$

3. Working from the left side: $(\sec x - 1)(\sec x + 1) = \sec^2 x - \sec x + \sec x - 1 = \sec^2 x - 1 = \tan^2 x$

4. Working from the left side: $\sec\theta - \cos\theta = \dfrac{1}{\cos\theta} - \cos\theta \cdot \dfrac{\cos\theta}{\cos\theta} = \dfrac{1 - \cos^2\theta}{\cos\theta} = \dfrac{\sin^2\theta}{\cos\theta} = \dfrac{\sin\theta}{\cos\theta} \cdot \sin\theta = \tan\theta\sin\theta$

5. Working from the left side: $\dfrac{\cos t}{1 - \sin t} = \dfrac{\cos t}{1 - \sin t} \cdot \dfrac{1 + \sin t}{1 + \sin t} = \dfrac{\cos t(1 + \sin t)}{1 - \sin^2 t} = \dfrac{\cos t(1 + \sin t)}{\cos^2 t} = \dfrac{1 + \sin t}{\cos t}$

6. Working from the left side: $\dfrac{1}{1 - \sin t} + \dfrac{1}{1 + \sin t} = \dfrac{1 + \sin t}{1 - \sin^2 t} + \dfrac{1 - \sin t}{1 - \sin^2 t} = \dfrac{2}{1 - \sin^2 t} = \dfrac{2}{\cos^2 t} = 2\sec^2 t$

7. Using the difference formula for sine: $\sin(\theta - 90°) = \sin\theta\cos 90° - \cos\theta\sin 90° = \sin\theta \cdot 0 - \cos\theta \cdot 1 = -\cos\theta$

8. Using the addition formula for cosine: $\cos\left(\dfrac{\pi}{2} + \theta\right) = \cos\dfrac{\pi}{2}\cos\theta - \sin\dfrac{\pi}{2}\sin\theta = 0 \cdot \cos\theta - 1 \cdot \sin\theta = -\sin\theta$

9. Working from the left side: $\cos^4 A - \sin^4 A = \left(\cos^2 A + \sin^2 A\right)\left(\cos^2 A - \sin^2 A\right) = 1 \cdot \cos 2A = \cos 2A$

10. Working from the right side: $\dfrac{\sin 2A}{1 - \cos 2A} = \dfrac{2\sin A\cos A}{1 - \left(1 - 2\sin^2 A\right)} = \dfrac{2\sin A\cos A}{2\sin^2 A} = \dfrac{\cos A}{\sin A} = \cot A$

11. Working from the left side: $\cot x - \tan x = \dfrac{\cos x}{\sin x} - \dfrac{\sin x}{\cos x} = \dfrac{\cos^2 x - \sin^2 x}{\sin x\cos x} = \dfrac{\cos 2x}{\sin x\cos x}$

12. Working from the right side: $\dfrac{\tan x}{\sec x + 1} = \dfrac{\dfrac{\sin x}{\cos x}}{\dfrac{1}{\cos x} + 1} \cdot \dfrac{\cos x}{\cos x} = \dfrac{\sin x}{1 + \cos x} = \tan\dfrac{x}{2}$

13.   Sketching the graph:

$$y = \frac{\sec^2 \alpha}{\sin^2 \alpha} = \csc^2 \alpha + \sec^2 \alpha$$

The equation appears to be an identity. Working from the right side:

$$\csc^2 \alpha + \sec^2 \alpha = \frac{1}{\sin^2 \alpha} + \frac{1}{\cos^2 \alpha}$$

$$= \frac{\cos^2 \alpha}{\sin^2 \alpha \cos^2 \alpha} + \frac{\sin^2 \alpha}{\sin^2 \alpha \cos^2 \alpha}$$

$$= \frac{\cos^2 \alpha + \sin^2 \alpha}{\sin^2 \alpha \cos^2 \alpha}$$

$$= \frac{1}{\sin^2 \alpha \cos^2 \alpha}$$

$$= \frac{1}{\sin^2 \alpha} \cdot \frac{1}{\cos^2 \alpha}$$

$$= \frac{\sec^2 \alpha}{\sin^2 \alpha}$$

14.   Sketching the graph:

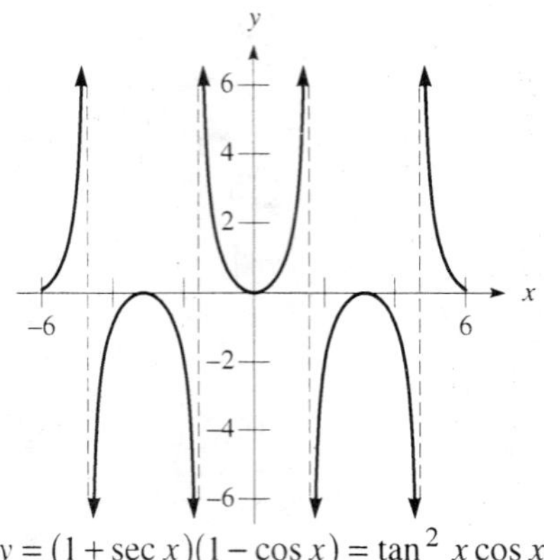

$$y = (1 + \sec x)(1 - \cos x) = \tan^2 x \cos x$$

The equation appears to be an identity. Working from the left side:

$$(1 + \sec x)(1 - \cos x) = 1 + \sec x - \cos x - 1$$

$$= \sec x - \cos x$$

$$= \frac{1}{\cos x} - \cos x$$

$$= \frac{1 - \cos^2 x}{\cos x}$$

$$= \frac{\sin^2 x}{\cos x}$$

$$= \frac{\sin^2 x}{\cos^2 x} \cdot \cos x$$

$$= \tan^2 x \cos x$$

15.   Sketching the graph:

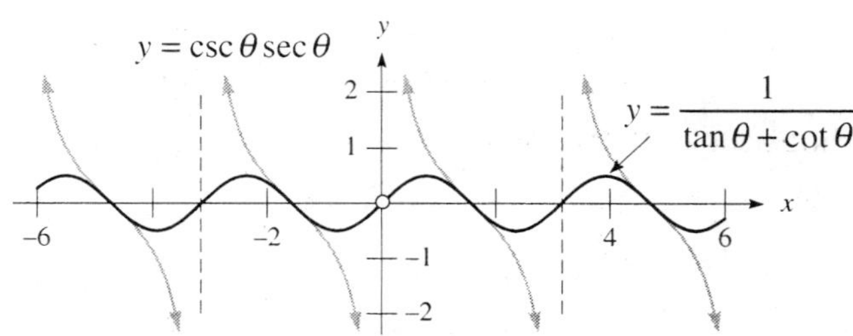

The equation does not appear to be an identity.

16.   Sketching the graph:

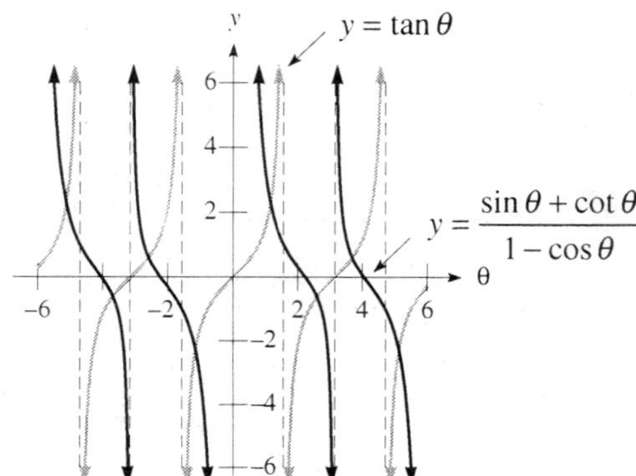

The equation does not appear to be an identity.

17.   Sketching the graph:

$$y = \cot^2 \theta - \cos^2 \theta = \cot^2 \theta \cos^2 \theta$$

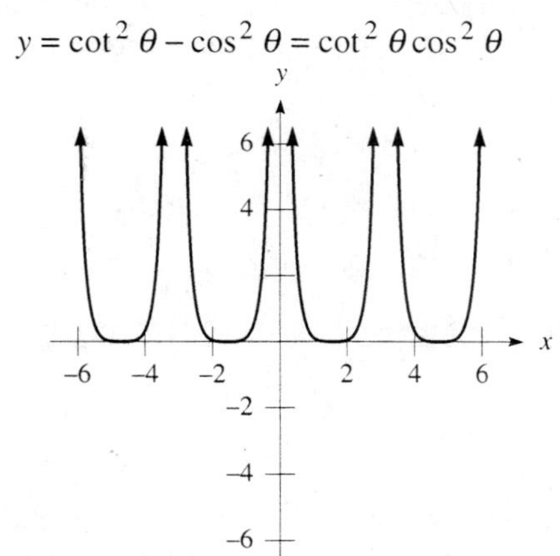

The equation appears to be an identity. Working from the left side:

$$\cot^2 \theta - \cos^2 \theta = \frac{\cos^2 \theta}{\sin^2 \theta} - \cos^2 \theta$$

$$= \frac{\cos^2 \theta - \cos^2 \theta \sin^2 \theta}{\sin^2 \theta}$$

$$= \frac{\cos^2 \theta \left(1 - \sin^2 \theta\right)}{\sin^2 \theta}$$

$$= \frac{\cos^2 \theta}{\sin^2 \theta} \cdot \left(1 - \sin^2 \theta\right)$$

$$= \cot^2 \theta \cos^2 \theta$$

18.   Sketching the graph:

$$y = \sec^2 x \csc^2 x = \sec^2 x + \csc^2 x$$

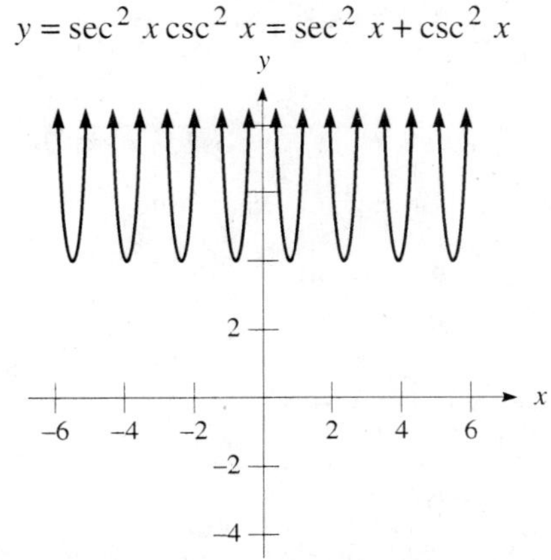

The equation appears to be an identity. Working from the right side:

$$\sec^2 x + \csc^2 x = \frac{1}{\cos^2 x} + \frac{1}{\sin^2 x} = \frac{\sin^2 x + \cos^2 x}{\sin^2 x \cos^2 x} = \frac{1}{\sin^2 x \cos^2 x} = \frac{1}{\cos^2 x} \cdot \frac{1}{\sin^2 x} = \sec^2 x \csc^2 x$$

19.   Since A terminates in quadrant IV, $\cos A > 0$ and thus: $\cos A = \sqrt{1 - \sin^2 A} = \sqrt{1 - \left(-\frac{3}{5}\right)^2} = \sqrt{1 - \frac{9}{25}} = \sqrt{\frac{16}{25}} = \frac{4}{5}$

Since B terminates in quadrant II, $\cos B < 0$ and thus:

$$\cos B = -\sqrt{1 - \sin^2 B} = -\sqrt{1 - \left(\frac{12}{13}\right)^2} = -\sqrt{1 - \frac{144}{169}} = -\sqrt{\frac{25}{169}} = -\frac{5}{13}$$

Using the addition formula for sine: $\sin(A + B) = \sin A \cos B + \cos A \sin B = \left(-\frac{3}{5}\right)\left(-\frac{5}{13}\right) + \left(\frac{4}{5}\right)\left(\frac{12}{13}\right) = \frac{15}{65} + \frac{48}{65} = \frac{63}{65}$

20. Using the subtraction formula for cosine and the results from Problem 19:
$$\cos(A - B) = \cos A \cos B + \sin A \sin B = \left(\tfrac{4}{5}\right)\left(-\tfrac{5}{13}\right) + \left(-\tfrac{3}{5}\right)\left(\tfrac{12}{13}\right) = -\tfrac{20}{65} - \tfrac{36}{65} = -\tfrac{56}{65}$$

21. Using the double-angle formula for cosine and the results from Problem 19:
$$\cos 2B = \cos^2 B - \sin^2 B = \left(-\tfrac{5}{13}\right)^2 - \left(\tfrac{12}{13}\right)^2 = \tfrac{25}{169} - \tfrac{144}{169} = -\tfrac{119}{169}$$

22. Using the double-angle formula for sine and the results from Problem 19:
$$\sin 2B = 2 \sin B \cos B = 2 \cdot \tfrac{12}{13} \cdot \left(-\tfrac{5}{13}\right) = -\tfrac{120}{169}$$

23. Since $270° \le A \le 360°$, $135° \le \dfrac{A}{2} \le 180°$. So $\sin \dfrac{A}{2} > 0$, therefore (using the results from Problem 19):
$$\sin \frac{A}{2} = \sqrt{\frac{1 - \cos A}{2}} = \sqrt{\frac{1 - \frac{4}{5}}{2}} = \sqrt{\tfrac{1}{10}} = \frac{1}{\sqrt{10}}$$

24. Since $135° \le \dfrac{A}{2} \le 180°$, $\cos \dfrac{A}{2} < 0$, therefore (using the results from Problem 19):
$$\cos \frac{A}{2} = -\sqrt{\frac{1 + \cos A}{2}} = -\sqrt{\frac{1 + \frac{4}{5}}{2}} = -\sqrt{\tfrac{9}{10}} = -\frac{3}{\sqrt{10}}$$

25. Using the half-angle formula for sine (note that $\sin 75° > 0$):
$$\sin 75° = \sqrt{\frac{1 - \cos 150°}{2}} = \sqrt{\frac{1 + \frac{\sqrt{3}}{2}}{2}} = \sqrt{\frac{2 + \sqrt{3}}{4}} = \frac{\sqrt{2 + \sqrt{3}}}{2}$$
We could also solve this using the addition formula for sine:
$$\sin 75° = \sin(45° + 30°) = \sin 45° \cos 30° + \cos 45° \sin 30° = \frac{\sqrt{2}}{2} \cdot \frac{\sqrt{3}}{2} + \frac{\sqrt{2}}{2} \cdot \frac{1}{2} = \frac{\sqrt{6} + \sqrt{2}}{4}$$

26. Using the half-angle formula for cosine (note that $\cos 15° > 0$):
$$\cos 15° = \sqrt{\frac{1 + \cos 30°}{2}} = \sqrt{\frac{1 + \frac{\sqrt{3}}{2}}{2}} = \sqrt{\frac{2 + \sqrt{3}}{4}} = \frac{\sqrt{2 + \sqrt{3}}}{2}$$
Also note that we could use the result from Problem 25: $\cos 15° = \sin(90° - 15°) = \sin 75° = \dfrac{\sqrt{2 + \sqrt{3}}}{2}$

We could also solve this using the subtraction formula for cosine:
$$\cos 15° = \cos(45° - 30°) = \cos 45° \cos 30° + \sin 45° \sin 30° = \frac{\sqrt{2}}{2} \cdot \frac{\sqrt{3}}{2} + \frac{\sqrt{2}}{2} \cdot \frac{1}{2} = \frac{\sqrt{6} + \sqrt{2}}{4}$$

27. Using the half-angle formula for tangent: $\tan \dfrac{\pi}{12} = \dfrac{1 - \cos \frac{\pi}{6}}{\sin \frac{\pi}{6}} = \dfrac{1 - \frac{\sqrt{3}}{2}}{\frac{1}{2}} = 2 - \sqrt{3}$

28. Using the result from Problem 28: $\cot \dfrac{\pi}{12} = \dfrac{1}{\tan \frac{\pi}{12}} = \dfrac{1}{2 - \sqrt{3}}$

29. Using the addition formula for cosine: $\cos 4x \cos 5x - \sin 4x \sin 5x = \cos(4x + 5x) = \cos 9x$

30. Using the addition formula for sine: $\sin 15° \cos 75° + \cos 15° \sin 75° = \sin(15° + 75°) = \sin 90° = 1$

31. Since $180° \le A \le 270°$, $\cos A < 0$ and thus: $\cos A = -\sqrt{1 - \sin^2 A} = -\sqrt{1 - \left(-\dfrac{1}{\sqrt{5}}\right)^2} = -\sqrt{1 - \tfrac{1}{5}} = -\sqrt{\tfrac{4}{5}} = -\dfrac{2}{\sqrt{5}}$

Using the double-angle formula for cosine: $\cos 2A = \cos^2 A - \sin^2 A = \left(-\dfrac{2}{\sqrt{5}}\right)^2 - \left(-\dfrac{1}{\sqrt{5}}\right)^2 = \tfrac{4}{5} - \tfrac{1}{5} = \tfrac{3}{5}$

Since $180° \le A \le 270°$, $90° \le \dfrac{A}{2} \le 135°$ and thus $\cos\dfrac{A}{2} < 0$. Therefore:

$$\cos\frac{A}{2} = -\sqrt{\frac{1 + \cos A}{2}} = -\sqrt{\frac{1 - \dfrac{2}{\sqrt{5}}}{2}} = -\sqrt{\frac{\sqrt{5} - 2}{2\sqrt{5}}} = -\sqrt{\frac{5 - 2\sqrt{5}}{10}}$$

32. Since $\sec A = \sqrt{10}$, $\cos A = \dfrac{1}{\sqrt{10}}$. Since $0° \le A \le 90°$, $\sin A > 0$ and thus:

$$\sin A = \sqrt{1 - \cos^2 A} = \sqrt{1 - \left(\frac{1}{\sqrt{10}}\right)^2} = \sqrt{1 - \tfrac{1}{10}} = \sqrt{\tfrac{9}{10}} = \frac{3}{\sqrt{10}}$$

Using the double-angle formula for sine: $\sin 2A = 2\sin A \cos A = 2\left(\dfrac{3}{\sqrt{10}}\right)\left(\dfrac{1}{\sqrt{10}}\right) = \tfrac{6}{10} = \tfrac{3}{5}$

Since $0° \le A \le 90°$, $0° \le \dfrac{A}{2} \le 45°$, thus $\sin\dfrac{A}{2} > 0$. Therefore:

$$\sin\frac{A}{2} = \sqrt{\frac{1 - \cos A}{2}} = \sqrt{\frac{1 - \dfrac{1}{\sqrt{10}}}{2}} = \sqrt{\frac{\sqrt{10} - 1}{2\sqrt{10}}} = \sqrt{\frac{10 - \sqrt{10}}{20}}$$

33. Using the addition formula for tangent:

$$\tan(A + B) = \frac{\tan A + \tan B}{1 - \tan A \tan B}$$

$$3 = \frac{\tan A + \tfrac{1}{2}}{1 - \tan A \cdot \tfrac{1}{2}}$$

$$3 = \frac{2\tan A + 1}{2 - \tan A}$$

$$3(2 - \tan A) = 2\tan A + 1$$

$$6 - 3\tan A = 2\tan A + 1$$

$$5 = 5\tan A$$

$$\tan A = 1$$

34. Using the double-angle formula for cosine:

$$\cos 2x = 2\cos^2 x - 1$$

$$\tfrac{1}{2} = 2\cos^2 x - 1$$

$$2\cos^2 x = \tfrac{3}{2}$$

$$\cos^2 x = \tfrac{3}{4}$$

$$\cos x = \pm\frac{\sqrt{3}}{2}$$

35. Let $\alpha = \arcsin\tfrac{4}{5}$, so $\sin\alpha = \tfrac{4}{5}$. Let $\beta = \arctan 2$, so $\tan\beta = 2$. Draw the triangles:

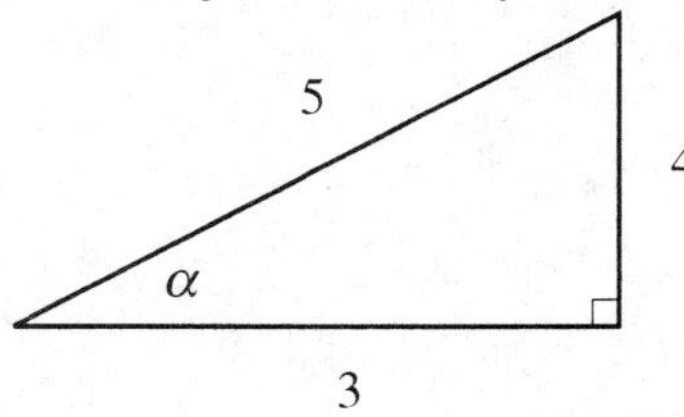

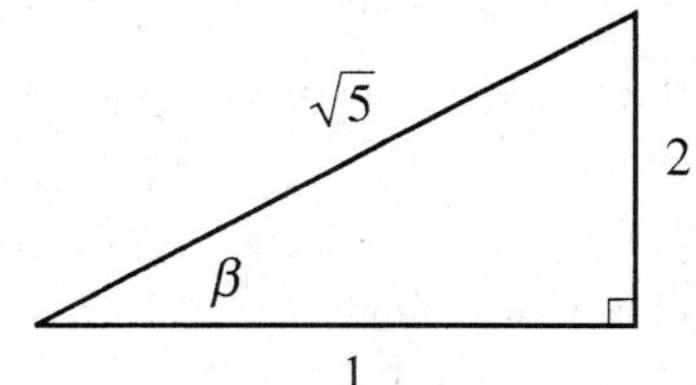

Therefore: $\cos\left(\arcsin\tfrac{4}{5} - \arctan 2\right) = \cos(\alpha - \beta) = \cos\alpha \cos\beta + \sin\alpha \sin\beta = \left(\tfrac{3}{5}\right)\left(\dfrac{1}{\sqrt{5}}\right) + \left(\tfrac{4}{5}\right)\left(\dfrac{2}{\sqrt{5}}\right) = \dfrac{11}{5\sqrt{5}}$

36.   Let $\alpha = \arccos \frac{4}{5}$, so $\cos \alpha = \frac{4}{5}$. Let $\beta = \arctan 2$, so $\tan \beta = 2$. Draw the triangles:

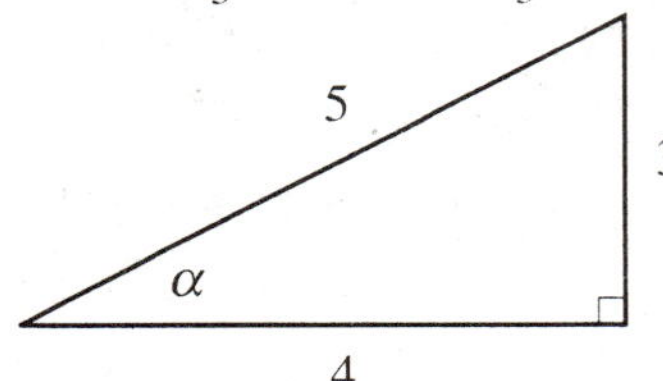

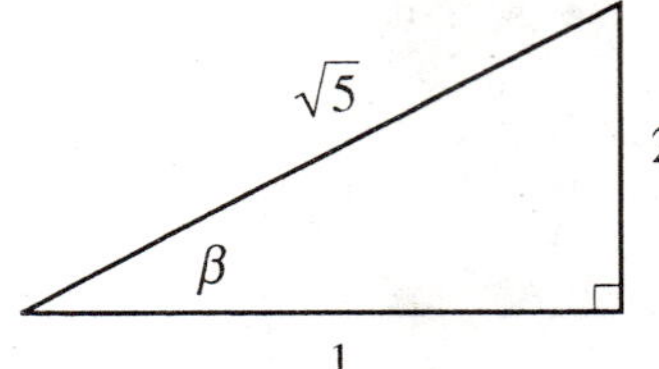

Therefore: $\sin\left(\arccos \frac{4}{5} + \arctan 2\right) = \sin(\alpha + \beta) = \sin \alpha \cos \beta + \cos \alpha \sin \beta = \left(\frac{3}{5}\right)\left(\frac{1}{\sqrt{5}}\right) + \left(\frac{4}{5}\right)\left(\frac{2}{\sqrt{5}}\right) = \frac{11}{5\sqrt{5}}$

37.   Let $\alpha = \sin^{-1} x$, so $\sin \alpha = x$. Draw the triangle:

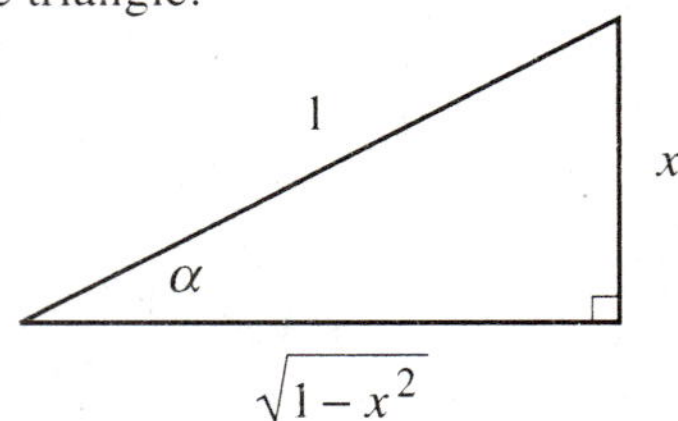

Therefore: $\cos\left(2 \sin^{-1} x\right) = \cos 2\alpha = \cos^2 \alpha - \sin^2 \alpha = \left(\sqrt{1-x^2}\right)^2 - x^2 = 1 - x^2 - x^2 = 1 - 2x^2$

38.   Let $\alpha = \cos^{-1} x$, so $\cos \alpha = x$. Draw the triangle:

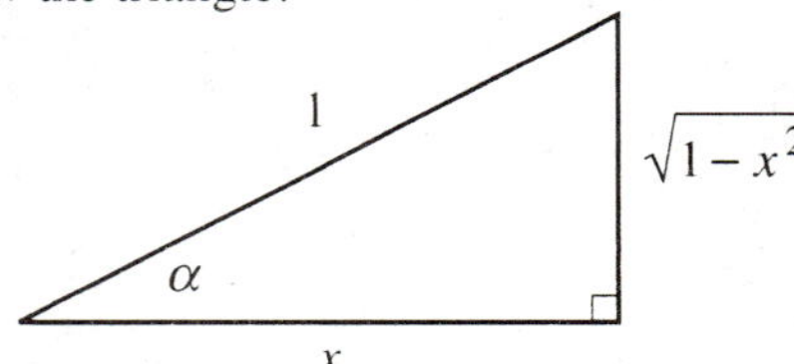

Therefore: $\sin\left(2 \cos^{-1} x\right) = \sin 2\alpha = 2 \sin \alpha \cos \alpha = 2 \bullet x \bullet \sqrt{1-x^2} = 2x\sqrt{1-x^2}$

39.   Using the product-to-sum formula: $\sin 6x \sin 4x = \frac{1}{2}\left[\cos(6x - 4x) - \cos(6x + 4x)\right] = \frac{1}{2}(\cos 2x - \cos 10x)$

40.   Using the sum-to-product formula:

$$\cos 15° + \cos 75° = 2 \cos \frac{15° + 75°}{2} \cos \frac{15° - 75°}{2} = 2 \cos 45° \cos(-30°) = 2 \bullet \frac{\sqrt{2}}{2} \bullet \frac{\sqrt{3}}{2} = \frac{\sqrt{6}}{2}$$

Chapter 6
Equations

6.1 Solving Trigonometric Equations

2. Solving the equation:

$$2\cos\theta = 1$$
$$\cos\theta = \tfrac{1}{2}$$
$$\theta = 60°, 300°$$

4. Solving the equation:

$$2\cos\theta + \sqrt{3} = 0$$
$$2\cos\theta = -\sqrt{3}$$
$$\cos\theta = -\frac{\sqrt{3}}{2}$$
$$\theta = 150°, 210°$$

6. Solving the equation:

$$\sqrt{3}\cot\theta - 1 = 0$$
$$\sqrt{3}\cot\theta = 1$$
$$\cot\theta = \frac{1}{\sqrt{3}}$$
$$\theta = 60°, 240°$$

8. Solving the equation:

$$\sqrt{3} + 5\sin t = 3\sin t$$
$$2\sin t = -\sqrt{3}$$
$$\sin t = -\frac{\sqrt{3}}{2}$$
$$t = \frac{4\pi}{3}, \frac{5\pi}{3}$$

10. Solving the equation:

$$5\cos t + \sqrt{12} = \cos t$$
$$4\cos t = -\sqrt{12}$$
$$4\cos t = -2\sqrt{3}$$
$$\cos t = -\frac{\sqrt{3}}{2}$$
$$t = \frac{5\pi}{6}, \frac{7\pi}{6}$$

12. Solving the equation:

$$3\sin t + 4 = 4$$
$$3\sin t = 0$$
$$\sin t = 0$$
$$t = 0, \pi$$

14. Solving the equation:

$$4\sin\theta + 3 = 0$$
$$4\sin\theta = -3$$
$$\sin\theta = -\tfrac{3}{4}$$
$$\theta \approx 180° + 48.6° = 228.6°$$
$$\theta \approx 360° - 48.6° = 311.4°$$

16. Solving the equation:
$$4\cos\theta - 1 = 3\cos\theta + 4$$
$$\cos\theta = 5$$
Since $-1 \le \cos\theta \le 1$, $\cos\theta = 5$ is impossible. This equation has no solution ($\varnothing$).

18. Solving the equation:
$$\sin\theta - 4 = -2\sin\theta$$
$$3\sin\theta = 4$$
$$\sin\theta = \tfrac{4}{3}$$
Since $-1 \le \sin\theta \le 1$, $\sin\theta = \tfrac{4}{3}$ is impossible. This equation has no solution ($\varnothing$).

20. Setting each factor equal to 0, $\cos x - 1 = 0$ when $\cos x = 1$, and $2\cos x - 1 = 0$ when $\cos x = \tfrac{1}{2}$. Thus the solutions are $x = 0, \dfrac{\pi}{3}, \dfrac{5\pi}{3}$.

22. Setting each factor equal to 0, $\tan x = 0$, or $\tan x + 1 = 0$ when $\tan x = -1$. Thus the solutions are $x = 0, \dfrac{3\pi}{4}, \pi, \dfrac{7\pi}{4}$.

24. Factoring the equation:
$$\cos x - 2\sin x \cos x = 0$$
$$\cos x(1 - 2\sin x) = 0$$
Thus either $\cos x = 0$, or $1 - 2\sin x = 0$, which occurs when $\sin x = \tfrac{1}{2}$. Thus the solutions are $x = \dfrac{\pi}{6}, \dfrac{\pi}{2}, \dfrac{5\pi}{6}, \dfrac{3\pi}{2}$.

26. Factoring the equation:
$$2\cos^2 x + \cos x - 1 = 0$$
$$(2\cos x - 1)(\cos x + 1) = 0$$
Thus either $2\cos x - 1 = 0$, so $\cos x = \tfrac{1}{2}$, or $\cos x + 1 = 0$, so $\cos x = -1$. Thus the solutions are $x = \dfrac{\pi}{3}, \pi, \dfrac{5\pi}{3}$.

28. Setting each factor equal to 0, $2\sin\theta - \sqrt{3} = 0$ when $\sin\theta = \dfrac{\sqrt{3}}{2}$, or $2\sin\theta - 1 = 0$ when $\sin\theta = \tfrac{1}{2}$. Thus the solutions are $\theta = 30°, 60°, 120°, 150°$.

30. Factoring the equation:
$$\tan\theta - 2\cos\theta\tan\theta = 0$$
$$\tan\theta(1 - 2\cos\theta) = 0$$
Thus either $\tan\theta = 0$ or $1 - 2\cos\theta = 0$, so $\cos\theta = \tfrac{1}{2}$. Thus the solutions are $\theta = 0°, 60°, 180°, 300°$.

32. Factoring the equation:
$$2\sin^2\theta - 7\sin\theta = -3$$
$$2\sin^2\theta - 7\sin\theta + 3 = 0$$
$$(2\sin\theta - 1)(\sin\theta - 3) = 0$$
Thus either $2\sin\theta - 1 = 0$, so $\sin\theta = \tfrac{1}{2}$, or $\sin\theta - 3 = 0$, so $\sin\theta = 3$, which is impossible. Thus the solutions are $\theta = 30°, 150°$.

34. Using the quadratic formula with $a = 2$, $b = 2$, and $c = -1$:
$$\cos\theta = \frac{-2 \pm \sqrt{2^2 - 4(2)(-1)}}{2(2)} = \frac{-2 \pm \sqrt{4 + 8}}{4} = \frac{-2 \pm 2\sqrt{3}}{4} = \frac{-1 \pm \sqrt{3}}{2} \approx -1.37, 0.37$$
Since $-1 \le \cos\theta \le 1$, $\cos\theta = -1.37$ is impossible. Thus $\theta = 68.5°$ or $\theta = 360° - 68.5° = 291.5°$.

36. Using the quadratic formula with $a = 1$, $b = -1$, and $c = -1$:
$$\sin\theta = \frac{1 \pm \sqrt{(-1)^2 - 4(1)(-1)}}{2(1)} = \frac{1 \pm \sqrt{1 + 4}}{2} = \frac{1 \pm \sqrt{5}}{2} \approx -0.62, 1.62$$
Since $-1 \le \sin\theta \le 1$, $\sin\theta = 1.62$ is impossible. Thus $\theta = 180° + 38.2° = 218.2°$ or $\theta = 360° - 38.2° = 321.8°$.

38. First write the equation as $2\cos^2\theta - 4\cos\theta + 1 = 0$. Using the quadratic formula with $a = 2$, $b = -4$, and $c = 1$:

$$\cos\theta = \frac{4 \pm \sqrt{(-4)^2 - 4(2)(1)}}{2(2)} = \frac{4 \pm \sqrt{16-8}}{4} = \frac{4 \pm 2\sqrt{2}}{4} = \frac{2 \pm \sqrt{2}}{2} \approx 0.29, 1.71$$

Since $-1 \le \cos\theta \le 1$, $\cos\theta = 1.71$ is impossible. Thus $\theta = 73.0°$ or $\theta = 360° - 73.0° = 287.0°$.

40. From the results of Problem 2, $\theta = 60° + 360°k$ or $\theta = 300° + 360°k$, where k is any integer.

42. From the results of Problem 8, $t = \dfrac{4\pi}{3} + 2k\pi$ or $t = \dfrac{5\pi}{3} + 2k\pi$, where k is any integer.

44. From the results of Problem 12, $t = k\pi$, where k is any integer.

46. From the results of Problem 14, $\theta = 228.6° + 360°k$ or $\theta = 311.4° + 360°k$.

48. The principal solutions are $\theta = 60°$ and $\theta = 120°$. Therefore:

$$2A + 50° = 60° + 360°k \qquad\qquad 2A + 50° = 120° + 360°k$$
$$2A = 10° + 360°k \qquad\qquad 2A = 70° + 360°k$$
$$A = 5° + 180°k \qquad\qquad A = 35° + 180°k$$

50. The principal solutions are $\theta = 60°$ and $\theta = 300°$. Therefore:

$$3A + 30° = 60° + 360°k \qquad\qquad 3A + 30° = 300° + 360°k$$
$$3A = 30° + 360°k \qquad\qquad 3A = 270° + 360°k$$
$$A = 10° + 120°k \qquad\qquad A = 90° + 120°k$$

52. The principal solutions are $\theta = 210°$ and $\theta = 330°$. Therefore:

$$4A - 20° = 210° + 360°k \qquad\qquad 4A - 20° = 330° + 360°k$$
$$4A = 230° + 360°k \qquad\qquad 4A = 350° + 360°k$$
$$A = 57.5° + 90°k \qquad\qquad A = 87.5° + 90°k$$

54. The principal solutions are $\theta = 135°$ and $\theta = 225°$. Therefore:

$$5A + 15° = 135° + 360°k \qquad\qquad 5A + 15° = 225° + 360°k$$
$$5A = 120° + 360°k \qquad\qquad 5A = 210° + 360°k$$
$$A = 24° + 72°k \qquad\qquad A = 42° + 72°k$$

56. Graphing the equation:

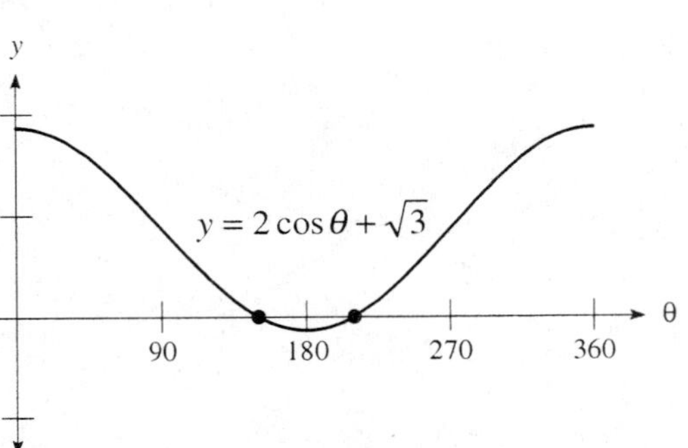

The zeros are $\theta = 150°, 210°$.

58. Graphing the equation:

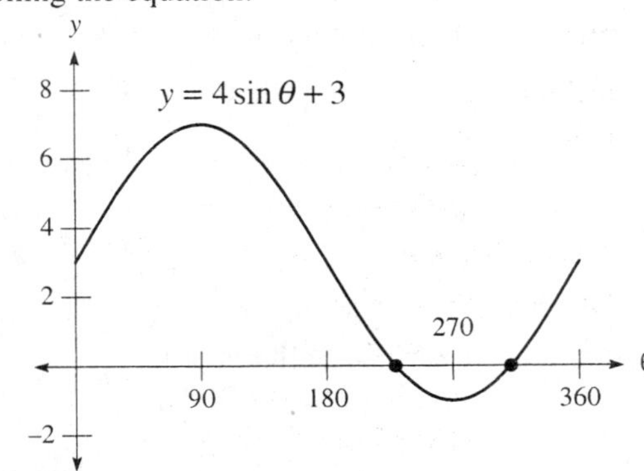

The zeros are $\theta \approx 228.6°, 311.4°$.

60. Graphing the equation:

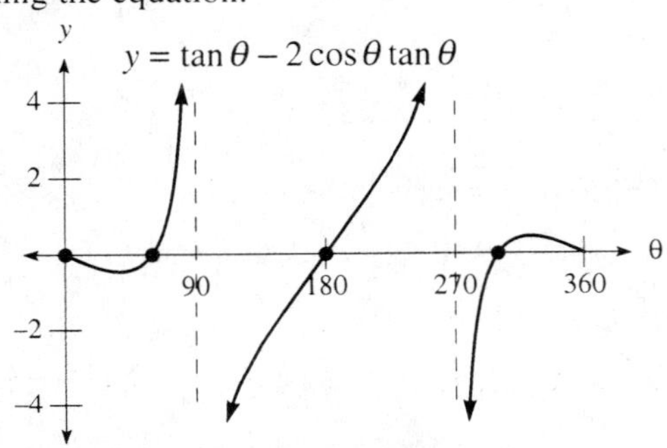

The zeros are $\theta = 0°, 180°, 60°, 300°$.

62. Graphing the equation:

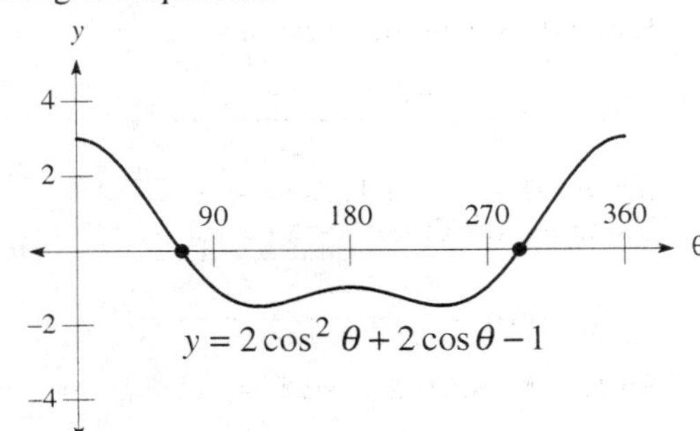

The zeros are $\theta \approx 68.5°, 291.5°$.

64.   Graphing each side of the equation:

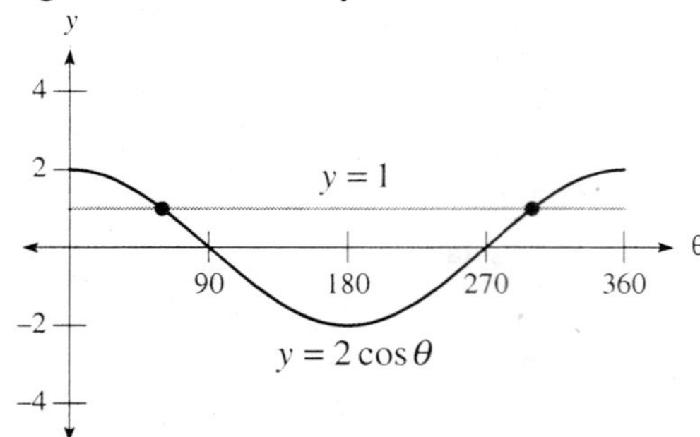

The intersection points are $\theta = 60°, 300°$.

66.   Graphing each side of the equation:

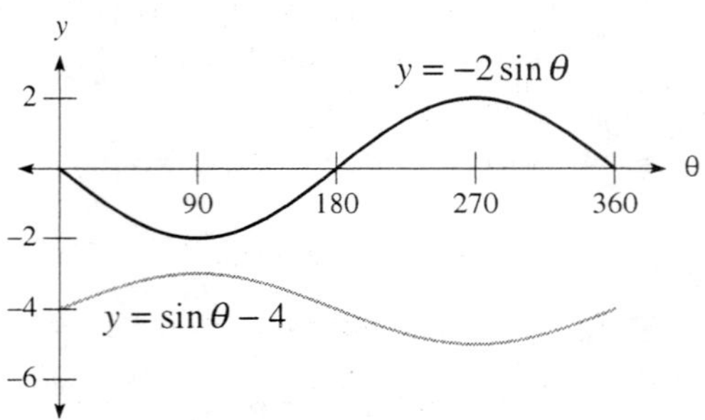

There are no intersection points.

68.   Graphing each side of the equation:

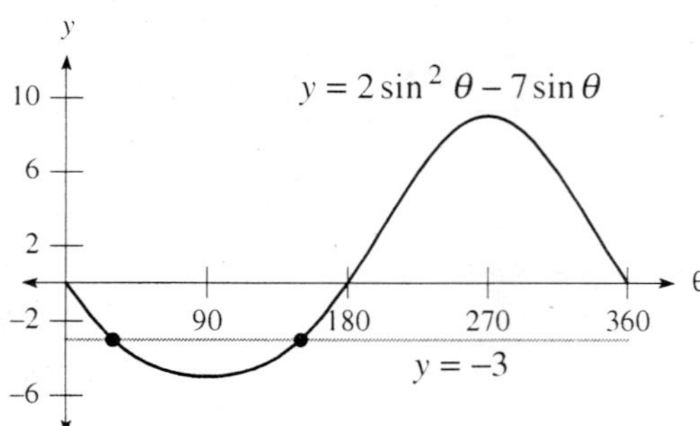

The intersection points are $\theta = 30°, 150°$.

70.   Graphing each side of the equation:

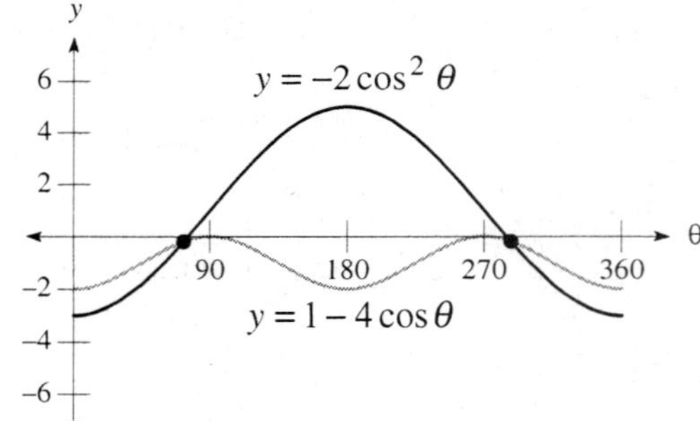

The intersection points are $\theta \approx 73.0°, 287.0°$.

72.   Graphing each side of the equation:

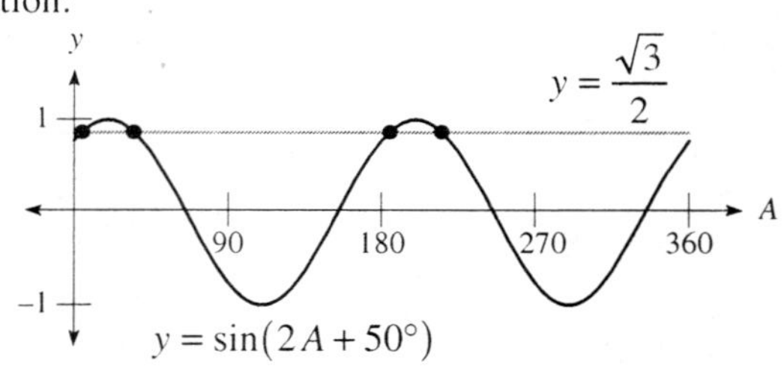

The intersection points are $A = 5°, 35°, 185°, 215°$.

74.   Graphing each side of the equation:

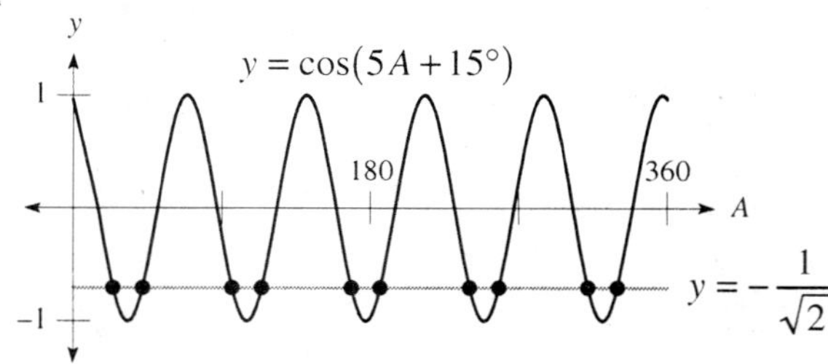

The intersection points are $A = 24°, 42°, 96°, 114°, 168°, 186°, 240°, 258°, 312°, 330°$.

76.   Substituting $v = 600$ and $\theta = 45°$: $h = -16t^2 + 600t \sin 45° = -16t^2 + 600t \cdot \dfrac{\sqrt{2}}{2} = -16t^2 + 300\sqrt{2}\, t$

78.   Let $t = \sqrt{3}$: $h = -16\left(\sqrt{3}\right)^2 + 300\sqrt{2} \cdot \sqrt{3} = -48 + 300\sqrt{6} \approx 686.8$ ft

80.   Substituting $v = 1500$, $t = 3$, and $h = 750$:

$$750 = -16(3)^2 + 1500(3)\sin\theta$$
$$750 = -144 + 4500\sin\theta$$
$$894 = 4500\sin\theta$$
$$\sin\theta = \frac{894}{4500} \approx 0.1987$$
$$\theta \approx 11.5°$$

82.   Using the double-angle formula for cosine: $\cos 2A = \cos^2 A - \sin^2 A = \left(1 - \sin^2 A\right) - \sin^2 A = 1 - 2\sin^2 A$

84.   Using the double-angle formula for cosine: $\cos 2A = \cos^2 A - \sin^2 A$

86.   Using the addition formula for sine: $\sin(\theta + 30°) = \sin\theta\cos 30° + \cos\theta\sin 30° = \dfrac{\sqrt{3}}{2}\sin\theta + \tfrac{1}{2}\cos\theta$

88.   Using the addition formula for cosine:

$$\cos 105° = \cos(45° + 60°) = \cos 45°\cos 60° - \sin 45°\sin 60° = \left(\frac{\sqrt{2}}{2}\right)\left(\frac{1}{2}\right) - \left(\frac{\sqrt{2}}{2}\right)\left(\frac{\sqrt{3}}{2}\right) = \frac{\sqrt{2} - \sqrt{6}}{4}$$

90.   Working from the right side: $\dfrac{2}{\tan x + \cot x} = \dfrac{2}{\dfrac{\sin x}{\cos x} + \dfrac{\cos x}{\sin x}} \cdot \dfrac{\sin x \cos x}{\sin x \cos x} = \dfrac{2\sin x \cos x}{\sin^2 x + \cos^2 x} = \dfrac{\sin 2x}{1} = \sin 2x$

6.2 More on Trigonometric Equations

2.   Since $\csc\theta = \dfrac{1}{\sin\theta}$, the equation becomes:

$$\sqrt{2} \cdot \frac{1}{\sin\theta} = 2$$
$$\sqrt{2} = 2\sin\theta$$
$$\sin\theta = \frac{\sqrt{2}}{2}$$
$$\theta = 45°, 135°$$

4.   Since $\sec\theta = \dfrac{1}{\cos\theta}$, the equation becomes:

$$2\sqrt{3} \cdot \frac{1}{\cos\theta} + 7 = 3$$
$$\frac{2\sqrt{3}}{\cos\theta} = -4$$
$$2\sqrt{3} = -4\cos\theta$$
$$\cos\theta = -\frac{\sqrt{3}}{2}$$
$$\theta = 150°, 210°$$

6.   Since $\sec\theta = \dfrac{1}{\cos\theta}$, the equation becomes:

$$4\cos\theta - 3 \cdot \frac{1}{\cos\theta} = 0$$
$$4\cos^2\theta - 3 = 0$$
$$4\cos^2\theta = 3$$
$$\cos^2\theta = \tfrac{3}{4}$$
$$\cos\theta = \pm\frac{\sqrt{3}}{2}$$
$$\theta = 30°, 150°, 210°, 330°$$

8.   Since $\csc\theta = \dfrac{1}{\sin\theta}$ and $\cot\theta = \dfrac{\cos\theta}{\sin\theta}$, the equation becomes:

$$\frac{1}{\sin\theta} + 2 \cdot \frac{\cos\theta}{\sin\theta} = 0$$
$$1 + 2\cos\theta = 0$$
$$2\cos\theta = -1$$
$$\cos\theta = -\tfrac{1}{2}$$
$$\theta = 120°, 240°$$

10. Using the double-angle formula for sine:
$$2\sin\theta + \sin 2\theta = 0$$
$$2\sin\theta + 2\sin\theta\cos\theta = 0$$
$$2\sin\theta(1+\cos\theta) = 0$$
Thus either $\sin\theta = 0$ or $\cos\theta = -1$. Thus $\theta = 0°, 180°$.

12. Since $\csc\theta = \dfrac{1}{\sin\theta}$, the equation becomes:
$$2\sin\theta - 1 = \frac{1}{\sin\theta}$$
$$2\sin^2\theta - \sin\theta = 1$$
$$2\sin^2\theta - \sin\theta - 1 = 0$$
$$(2\sin\theta + 1)(\sin\theta - 1) = 0$$
$$\sin\theta = -\tfrac{1}{2}, 1$$
$$\theta = 90°, 210°, 330°$$

14. Using the double-angle formula for cosine:
$$\cos 2x - \cos x - 2 = 0$$
$$2\cos^2 x - 1 - \cos x - 2 = 0$$
$$2\cos^2 x - \cos x - 3 = 0$$
$$(2\cos x - 3)(\cos x + 1) = 0$$
$$\cos x = \tfrac{3}{2}, -1$$
$$x = \pi$$

Note that $\cos x = \tfrac{3}{2}$ has no solution.

16. Using the double-angle formula for cosine:
$$\sin x = -\cos 2x$$
$$\sin x = -\left(1 - 2\sin^2 x\right)$$
$$\sin x = -1 + 2\sin^2 x$$
$$2\sin^2 x - \sin x - 1 = 0$$
$$(2\sin x + 1)(\sin x - 1) = 0$$
$$\sin x = -\tfrac{1}{2}, 1$$
$$x = \frac{\pi}{2}, \frac{7\pi}{6}, \frac{11\pi}{6}$$

18. Substituting $\sin^2 x = 1 - \cos^2 x$ into the equation:
$$2\sin^2 x - \cos x - 1 = 0$$
$$2\left(1 - \cos^2 x\right) - \cos x - 1 = 0$$
$$2 - 2\cos^2 x - \cos x - 1 = 0$$
$$2\cos^2 x + \cos x - 1 = 0$$
$$(2\cos x - 1)(\cos x + 1) = 0$$
$$\cos x = \tfrac{1}{2}, -1$$
$$x = \frac{\pi}{3}, \pi, \frac{5\pi}{3}$$

20. Substituting $\cos^2 x = 1 - \sin^2 x$ into the equation:
$$4\cos^2 x - 4\sin x - 5 = 0$$
$$4\left(1 - \sin^2 x\right) - 4\sin x - 5 = 0$$
$$4 - 4\sin^2 x - 4\sin x - 5 = 0$$
$$-4\sin^2 x - 4\sin x - 1 = 0$$
$$4\sin^2 x + 4\sin x + 1 = 0$$
$$(2\sin x + 1)^2 = 0$$
$$\sin x = -\tfrac{1}{2}$$
$$x = \frac{7\pi}{6}, \frac{11\pi}{6}$$

22. Since $\tan x = \dfrac{\sin x}{\cos x}$ and $\sec x = \dfrac{1}{\cos x}$, the equation becomes:

$$2\cos x + \frac{\sin x}{\cos x} = \frac{1}{\cos x}$$

$$2\cos^2 x + \sin x = 1$$

$$2\left(1 - \sin^2 x\right) + \sin x = 1$$

$$2 - 2\sin^2 x + \sin x - 1 = 0$$

$$-2\sin^2 x + \sin x + 1 = 0$$

$$2\sin^2 x - \sin x - 1 = 0$$

$$(2\sin x + 1)(\sin x - 1) = 0$$

$$\sin x = -\tfrac{1}{2}, 1$$

$$x = \frac{7\pi}{6}, \frac{11\pi}{6}, \frac{\pi}{2}$$

Note that $x = \dfrac{\pi}{2}$ is impossible, since $\tan \dfrac{\pi}{2}$ is undefined. So the solutions are $x = \dfrac{7\pi}{6}, \dfrac{11\pi}{6}$.

24. Isolating $\sin x$ and then squaring:

$$\sin x - \cos x = \sqrt{2}$$

$$\sin x = \cos x + \sqrt{2}$$

$$\sin^2 x = \left(\cos x + \sqrt{2}\right)^2$$

$$\sin^2 x = \cos^2 x + 2\sqrt{2}\cos x + 2$$

$$1 - \cos^2 x = \cos^2 x + 2\sqrt{2}\cos x + 2$$

$$2\cos^2 x + 2\sqrt{2}\cos x + 1 = 0$$

$$\left(\sqrt{2}\cos x + 1\right)^2 = 0$$

$$\sqrt{2}\cos x = -1$$

$$\cos x = -\frac{1}{\sqrt{2}}$$

$$x = \frac{3\pi}{4}, \frac{5\pi}{4}$$

Since this equation was solved by squaring, each solution must be checked:

$$\sin \frac{3\pi}{4} - \cos \frac{3\pi}{4} = \frac{\sqrt{2}}{2} - \left(-\frac{\sqrt{2}}{2}\right) = \frac{\sqrt{2}}{2} + \frac{\sqrt{2}}{2} = \sqrt{2} \quad \text{(checks)}$$

$$\sin \frac{5\pi}{4} - \cos \frac{5\pi}{4} = -\frac{\sqrt{2}}{2} - \left(-\frac{\sqrt{2}}{2}\right) = -\frac{\sqrt{2}}{2} + \frac{\sqrt{2}}{2} = 0 \quad \text{(doesn't check)}$$

Only $x = \dfrac{3\pi}{4}$ checks in the original equation.

26. Isolating $\sin\theta$ and then squaring:

$$\sin\theta - \sqrt{3}\cos\theta = \sqrt{3}$$
$$\sin\theta = \sqrt{3} + \sqrt{3}\cos\theta$$
$$\sin^2\theta = \left(\sqrt{3} + \sqrt{3}\cos\theta\right)^2$$
$$\sin^2\theta = 3 + 6\cos\theta + 3\cos^2\theta$$
$$1 - \cos^2\theta = 3 + 6\cos\theta + 3\cos^2\theta$$
$$4\cos^2\theta + 6\cos\theta + 2 = 0$$
$$2\cos^2\theta + 3\cos\theta + 1 = 0$$
$$(2\cos\theta + 1)(\cos\theta + 1) = 0$$
$$\cos\theta = -\tfrac{1}{2}, -1$$
$$\theta = 120°, 240°, 180°$$

Since this equation was solved by squaring, each solution must be checked:

$$\sin 120° - \sqrt{3}\cos 120° = \frac{\sqrt{3}}{2} - \sqrt{3}\left(-\tfrac{1}{2}\right) = \frac{\sqrt{3}}{2} + \frac{\sqrt{3}}{2} = \sqrt{3} \ \ (\text{checks})$$

$$\sin 240° - \sqrt{3}\cos 240° = -\frac{\sqrt{3}}{2} - \sqrt{3}\left(-\tfrac{1}{2}\right) = -\frac{\sqrt{3}}{2} + \frac{\sqrt{3}}{2} = 0 \ \ (\text{doesn't check})$$

$$\sin 180° - \sqrt{3}\cos 180° = 0 - \sqrt{3}(-1) = 0 + \sqrt{3} = \sqrt{3} \ \ (\text{checks})$$

Thus $\theta = 120°, 180°$ are the solutions.

28. Isolating $\sin\theta$ and then squaring:

$$\sin\theta - \sqrt{3}\cos\theta = 1$$
$$\sin\theta = 1 + \sqrt{3}\cos\theta$$
$$\sin^2\theta = \left(1 + \sqrt{3}\cos\theta\right)^2$$
$$\sin^2\theta = 1 + 2\sqrt{3}\cos\theta + 3\cos^2\theta$$
$$1 - \cos^2\theta = 1 + 2\sqrt{3}\cos\theta + 3\cos^2\theta$$
$$4\cos^2\theta + 2\sqrt{3}\cos\theta = 0$$
$$2\cos\theta(2\cos\theta + \sqrt{3}) = 0$$
$$\cos\theta = 0, -\frac{\sqrt{3}}{2}$$
$$\theta = 90°, 270°, 150°, 210°$$

Since this equation was solved by squaring, each solution must be checked:

$$\sin 90° - \sqrt{3}\cos 90° = 1 - \sqrt{3}\cdot 0 = 1 \ \ (\text{checks})$$
$$\sin 270° - \sqrt{3}\cos 270° = -1 - \sqrt{3}\cdot 0 = -1 \ \ (\text{doesn't check})$$

$$\sin 150° - \sqrt{3}\cos 150° = \tfrac{1}{2} - \sqrt{3}\cdot\left(-\frac{\sqrt{3}}{2}\right) = \tfrac{1}{2} + \tfrac{3}{2} = 2 \ \ (\text{doesn't check})$$

$$\sin 210° - \sqrt{3}\cos 210° = -\tfrac{1}{2} - \sqrt{3}\cdot\left(-\frac{\sqrt{3}}{2}\right) = -\tfrac{1}{2} + \tfrac{3}{2} = 1 \ \ (\text{checks})$$

Thus $\theta = 90°, 210°$ are the solutions.

30. Isolating $\sin\dfrac{\theta}{2}$ and then squaring:

$$\sin\frac{\theta}{2}+\cos\theta=1$$

$$\sin\frac{\theta}{2}=1-\cos\theta$$

$$\sin^{2}\frac{\theta}{2}=(1-\cos\theta)^{2}$$

$$\frac{1-\cos\theta}{2}=1-2\cos\theta+\cos^{2}\theta$$

$$1-\cos\theta=2-4\cos\theta+2\cos^{2}\theta$$

$$2\cos^{2}\theta-3\cos\theta+1=0$$

$$(2\cos\theta-1)(\cos\theta-1)=0$$

$$\cos\theta=\tfrac{1}{2},1$$

$$\theta=60°,300°,0°$$

Since this equation was solved by squaring, each solution must be checked:

$$\sin\frac{60°}{2}+\cos60°=\sin30°+\cos60°=\tfrac{1}{2}+\tfrac{1}{2}=1 \quad\text{(checks)}$$

$$\sin\frac{300°}{2}+\cos300°=\sin150°+\cos300°=\tfrac{1}{2}+\tfrac{1}{2}=1 \quad\text{(checks)}$$

$$\sin\frac{0°}{2}+\cos0°=\sin0°+\cos0°=0+1=1 \quad\text{(checks)}$$

Thus $\theta=0°,60°,300°$ are the solutions.

32. Isolating $\cos\dfrac{\theta}{2}$ and then squaring:

$$\cos\frac{\theta}{2}-\cos\theta=0$$

$$\cos\frac{\theta}{2}=\cos\theta$$

$$\cos^{2}\frac{\theta}{2}=\cos^{2}\theta$$

$$\frac{1+\cos\theta}{2}=\cos^{2}\theta$$

$$1+\cos\theta=2\cos^{2}\theta$$

$$2\cos^{2}\theta-\cos\theta-1=0$$

$$(2\cos\theta+1)(\cos\theta-1)=0$$

$$\cos\theta=-\tfrac{1}{2},1$$

$$\theta=120°,240°,0°$$

Checking each solution in the original equation:

$$\cos\frac{120°}{2}-\cos120°=\cos60°-\cos120°=\tfrac{1}{2}-\left(-\tfrac{1}{2}\right)=\tfrac{1}{2}+\tfrac{1}{2}=1 \quad\text{(doesn't check)}$$

$$\cos\frac{240°}{2}-\cos240°=\cos120°-\cos240°=-\tfrac{1}{2}-\left(-\tfrac{1}{2}\right)=-\tfrac{1}{2}+\tfrac{1}{2}=0 \quad\text{(checks)}$$

$$\cos\frac{0°}{2}-\cos0°=\cos0°-\cos0°=1-1=0 \quad\text{(checks)}$$

Thus $\theta=0°,240°$ are the solutions.

34. Since $\cot\theta = \dfrac{\cos\theta}{\sin\theta}$ and $\csc\theta = \dfrac{1}{\sin\theta}$, the equation becomes:

$$13\cdot\frac{\cos\theta}{\sin\theta}+\frac{11}{\sin\theta}=6\sin\theta$$

$$13\cos\theta+11=6\sin^2\theta$$

$$13\cos\theta+11=6\left(1-\cos^2\theta\right)$$

$$13\cos\theta+11=6-6\cos^2\theta$$

$$6\cos^2\theta+13\cos\theta+5=0$$

$$(3\cos\theta+5)(2\cos\theta+1)=0$$

$$\cos\theta=-\tfrac{5}{3},-\tfrac{1}{2}$$

$$\theta=120°,240°$$

Note that $\cos\theta=-\tfrac{5}{3}$ has no solutions. These solutions are verified graphically.

36. Since $\sec\theta = \dfrac{1}{\cos\theta}$ and $\tan\theta = \dfrac{\sin\theta}{\cos\theta}$, the equation becomes:

$$\frac{18}{\cos^2\theta}-17\cdot\frac{\sin\theta}{\cos\theta}\cdot\frac{1}{\cos\theta}-12=0$$

$$18-17\sin\theta-12\cos^2\theta=0$$

$$18-17\sin\theta-12\left(1-\sin^2\theta\right)=0$$

$$18-17\sin\theta-12+12\sin^2\theta=0$$

$$12\sin^2\theta-17\sin\theta+6=0$$

$$(3\sin\theta-2)(4\sin\theta-3)=0$$

$$\sin\theta=\tfrac{2}{3},\tfrac{3}{4}$$

$$\theta\approx41.8°,180°-41.8°=138.2°$$

$$\theta\approx48.6°,180°-48.6°=131.4°$$

These solutions are verified graphically.

38. Using the double-angle formula for cosine:

$$16\cos 2\theta-18\sin^2\theta=0$$

$$16\left(1-2\sin^2\theta\right)-18\sin^2\theta=0$$

$$16-32\sin^2\theta-18\sin^2\theta=0$$

$$16-50\sin^2\theta=0$$

$$50\sin^2\theta=16$$

$$\sin^2\theta=\tfrac{8}{25}$$

$$\sin\theta=\pm\frac{2\sqrt{2}}{5}$$

$$\theta\approx34.4°,180°-34.4°=145.6°$$

$$\theta\approx180°+34.4°=214.4°,360°-34.4°=325.6°$$

These solutions are verified graphically.

40. Using the results of Problem 4, $\theta=150°+360°k$ or $\theta=210°+360°k$, where k is any integer.

42. Using the results of Problem 24, $x=\dfrac{3\pi}{4}+2k\pi$, where k is any integer.

44. Using the results of Problem 32, $\theta=360°k$ or $\theta=240°+360°k$, where k is any integer.

46.   Substituting $r = 2$ and $R = 4$, then set the expression equal to 0:

$$2^4 \csc^2 \theta - 4^4 \csc\theta \cot\theta = 0$$

$$\frac{16}{\sin^2 \theta} - 256 \cdot \frac{1}{\sin\theta} \cdot \frac{\cos\theta}{\sin\theta} = 0$$

$$16 - 256\cos\theta = 0$$

$$256\cos\theta = 16$$

$$\cos\theta = \tfrac{1}{16}$$

$$\theta \approx 86.4°$$

48.   Substituting $\cos^2\theta = 1 - \sin^2\theta$ into the equation:

$$2\left(1 - \sin^2\theta\right) + 2\sin\theta - 1 = 0$$

$$2 - 2\sin^2\theta + 2\sin\theta - 1 = 0$$

$$-2\sin^2\theta + 2\sin\theta + 1 = 0$$

$$2\sin^2\theta - 2\sin\theta - 1 = 0$$

Using the quadratic formula with $a = 2$, $b = -2$, and $c = -1$:

$$\sin\theta = \frac{-(-2) \pm \sqrt{(-2)^2 - 4(2)(-1)}}{2(2)} = \frac{2 \pm \sqrt{4+8}}{4} = \frac{2 \pm 2\sqrt{3}}{4} = \frac{1 \pm \sqrt{3}}{2}$$

Either $\sin\theta = \dfrac{1+\sqrt{3}}{2} \approx 1.37$, which is impossible, or $\sin\theta = \dfrac{1-\sqrt{3}}{2} \approx -0.37$. Thus $\theta = 180° + 21.5° = 201.5°$
or $\theta = 360° - 21.5° = 338.5°$.

50.   Substituting $\sin^2\theta = 1 - \cos^2\theta$ into the equation:

$$1 - \cos^2\theta = \cos\theta$$

$$\cos^2\theta + \cos\theta - 1 = 0$$

Using the quadratic formula with $a = 1$, $b = 1$, and $c = -1$: $\cos\theta = \dfrac{-1 \pm \sqrt{1^2 - 4(1)(-1)}}{2(1)} = \dfrac{-1 \pm \sqrt{1+4}}{2} = \dfrac{-1 \pm \sqrt{5}}{2}$

Either $\cos\theta = \dfrac{-1-\sqrt{5}}{2} \approx -1.62$, which is impossible, or $\cos\theta = \dfrac{-1+\sqrt{5}}{2} \approx 0.62$. Thus $\theta \approx 51.8°$ or
$\theta \approx 360° - 51.8° = 308.2°$.

52.   Substituting $\cos^2\theta = 1 - \sin^2\theta$ into the equation:

$$4\sin\theta = 3 - 2\left(1 - \sin^2\theta\right)$$

$$4\sin\theta = 3 - 2 + 2\sin^2\theta$$

$$2\sin^2\theta - 4\sin\theta + 1 = 0$$

Using the quadratic formula with $a = 2$, $b = -4$, and $c = 1$:

$$\sin\theta = \frac{-(-4) \pm \sqrt{(-4)^2 - 4(2)(1)}}{2(2)} = \frac{4 \pm \sqrt{16-8}}{4} = \frac{4 \pm 2\sqrt{2}}{4} = \frac{2 \pm \sqrt{2}}{2}$$

Either $\sin\theta = \dfrac{2+\sqrt{2}}{2} \approx 1.71$, which is impossible, or $\sin\theta = \dfrac{2-\sqrt{2}}{2} \approx 0.29$. Thus $\theta \approx 17.0°$ or
$\theta \approx 180° - 17.0° = 163.0°$.

54. Sketching the graph:

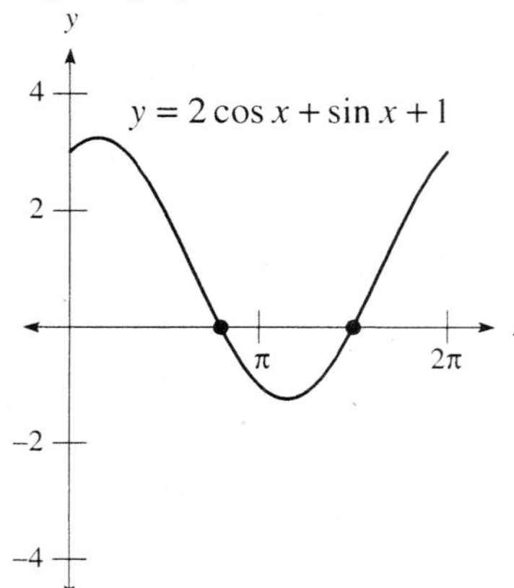

The solutions are $x \approx 2.4981, 4.7124$.

56. Sketching the graph:

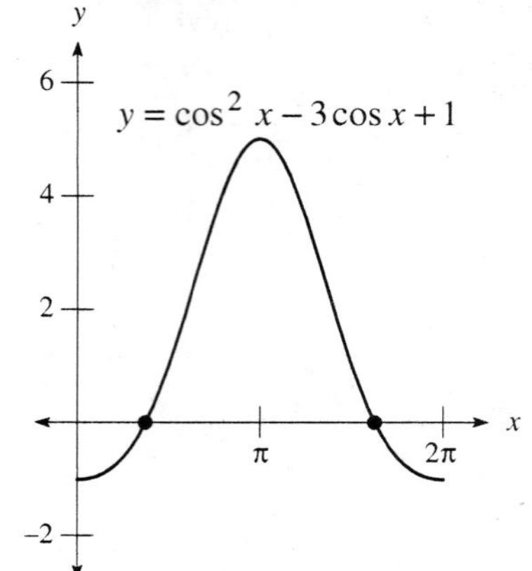

The solutions are $x \approx 1.1789, 5.1043$.

58. Write the equation as $\csc x - \tan x - 3 = 0$. Sketching the graph:

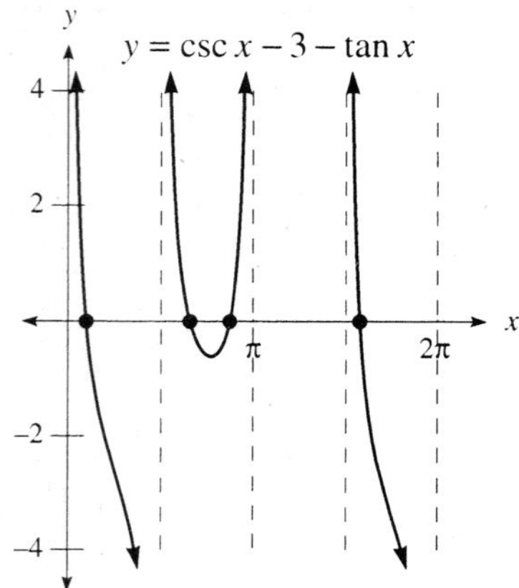

The solutions are $x \approx 0.3063, 2.0629, 2.7435, 4.9556$.

60. Since $0° \le A \le 90°$, $\cos A > 0$ and thus: $\cos A = \sqrt{1 - \sin^2 A} = \sqrt{1 - \left(\frac{2}{3}\right)^2} = \sqrt{1 - \frac{4}{9}} = \sqrt{\frac{5}{9}} = \frac{\sqrt{5}}{3}$

Since $0° \le A \le 90°$, $0° \le \frac{A}{2} \le 45°$ and thus $\cos \frac{A}{2} > 0$. Therefore: $\cos \frac{A}{2} = \sqrt{\frac{1 + \cos A}{2}} = \sqrt{\frac{1 + \frac{\sqrt{5}}{3}}{2}} = \sqrt{\frac{3 + \sqrt{5}}{6}}$

62. Since $\cos \frac{A}{2} = \sqrt{\frac{3 + \sqrt{5}}{6}}$, $\sec \frac{A}{2} = \frac{1}{\cos \frac{A}{2}} = \sqrt{\frac{6}{3 + \sqrt{5}}}$.

64. First find $\sin \frac{A}{2}$ (since $0° \le \frac{A}{2} \le 45°$, $\sin \frac{A}{2} > 0$): $\sin \frac{A}{2} = \sqrt{\frac{1 - \cos A}{2}} = \sqrt{\frac{1 - \frac{\sqrt{5}}{3}}{2}} = \sqrt{\frac{3 - \sqrt{5}}{6}}$

Therefore: $\cot \frac{A}{2} = \frac{\cos \frac{A}{2}}{\sin \frac{A}{2}} = \frac{\sqrt{\frac{3 + \sqrt{5}}{6}}}{\sqrt{\frac{3 - \sqrt{5}}{6}}} = \sqrt{\frac{3 + \sqrt{5}}{3 - \sqrt{5}}}$

66. Using the half-angle formula for cosine: $y = 6\cos^2 \dfrac{x}{2} = 6\left(\dfrac{1+\cos x}{2}\right) = 3(1+\cos x) = 3 + 3\cos x$

The amplitude is 3, the period is 2π, and the vertical translation is 3 units. Now graphing the curve:

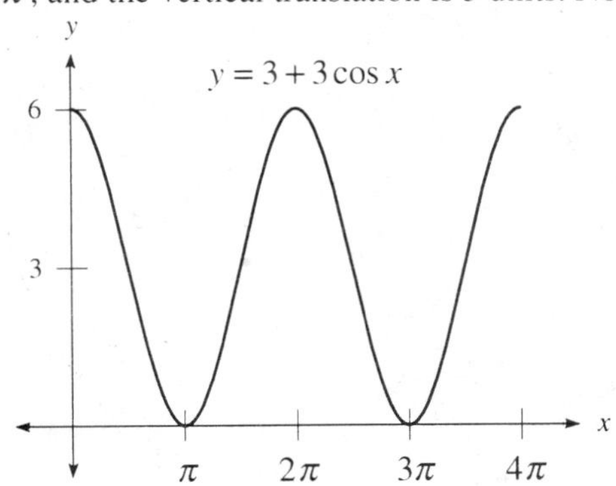

68. Since $\cos 15° > 0$, use the half-angle formula for cosine: $\cos 15° = \sqrt{\dfrac{1+\cos 30°}{2}} = \sqrt{\dfrac{1+\dfrac{\sqrt{3}}{2}}{2}} = \sqrt{\dfrac{2+\sqrt{3}}{4}} = \dfrac{\sqrt{2+\sqrt{3}}}{2}$

6.3 Trigonometric Equations Involving Multiple Angles

2. For $\sin 2\theta = -\dfrac{\sqrt{3}}{2}$, all values will take the form:

$$2\theta = 240° + 360°k \qquad\qquad 2\theta = 300° + 360°k$$
$$\theta = 120° + 180°k \qquad\qquad \theta = 150° + 180°k$$

Substituting $k = 0$ and $k = 1$ results in the solutions $\theta = 120°, 150°, 300°, 330°$. These solutions are verified graphically.

4. For $\cot 2\theta = 1$, all values will take the form:

$$2\theta = 45° + 180°k$$
$$\theta = 22.5° + 90°k$$

Substituting $k = 0, 1, 2,$ and 3 results in the solutions $\theta = 22.5°, 112.5°, 202.5°, 292.5°$. These solutions are verified graphically.

6. For $\sin 3\theta = -1$, all values will take the form:

$$3\theta = 270° + 360°k$$
$$\theta = 90° + 120°k$$

Substituting $k = 0, 1,$ and 2 results in the solutions $\theta = 90°, 210°, 330°$. These solutions are verified graphically.

8. For $\cos 2x = \dfrac{1}{\sqrt{2}}$, all values will take the form:

$$2x = \dfrac{\pi}{4} + 2\pi k \qquad\qquad 2x = \dfrac{7\pi}{4} + 2\pi k$$
$$x = \dfrac{\pi}{8} + \pi k \qquad\qquad x = \dfrac{7\pi}{8} + \pi k$$

Substituting $k = 0$ and $k = 1$ results in the solutions $x = \dfrac{\pi}{8}, \dfrac{7\pi}{8}, \dfrac{9\pi}{8}, \dfrac{15\pi}{8}$. These solutions are verified graphically.

10. For $\csc 3x = 1$, all values will take the form:

$$3x = \dfrac{\pi}{2} + 2\pi k$$
$$x = \dfrac{\pi}{6} + \dfrac{2\pi}{3} k$$

Substituting $k = 0, 1,$ and 2 results in the solutions $x = \dfrac{\pi}{6}, \dfrac{5\pi}{6}, \dfrac{3\pi}{2}$. These solutions are verified graphically.

12.   For $\tan 2x = -\sqrt{3}$, all values will take the form:
$$2x = \frac{2\pi}{3} + \pi k$$
$$x = \frac{\pi}{3} + \frac{\pi}{2}k$$

Substituting $k = 0$, 1, 2, and 3 results in the solutions $x = \frac{\pi}{3}, \frac{5\pi}{6}, \frac{4\pi}{3}, \frac{11\pi}{6}$. These solutions are verified graphically.

14.   For $\sin 2\theta = -\frac{\sqrt{3}}{2}$, all values will take the form:

$$2\theta = 240° + 360°k \qquad\qquad 2\theta = 300° + 360°k$$
$$\theta = 120° + 180°k \qquad\qquad \theta = 150° + 180°k$$

16.   For $\cos 3\theta = -1$, all values will take the form:
$$3\theta = 180° + 360°k$$
$$\theta = 60° + 120°k$$

18.   For $\cos 8\theta = \frac{1}{2}$, all values will take the form:
$$8\theta = 60° + 360°k \qquad\qquad 8\theta = 300° + 360°k$$
$$\theta = 7.5° + 45°k \qquad\qquad \theta = 37.5° + 45°k$$

20.   Graphing each side of the equation:

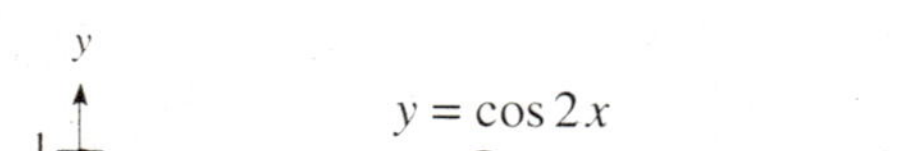

The solutions are $x = 60°, 120°, 240°, 300°$.

22.   Graphing each side of the equation:

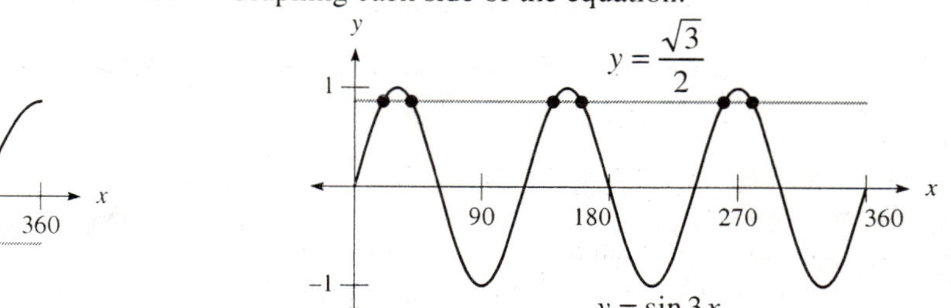

The solutions are $x = 20°, 40°, 140°, 160°, 260°, 280°$.

24.   Graphing each side of the equation:

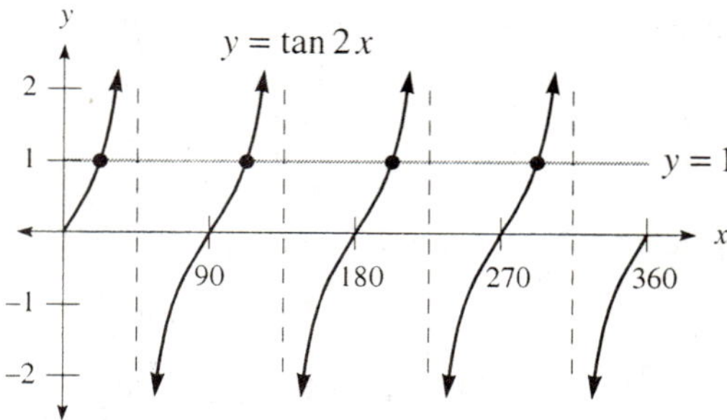

The solutions are $x = 22.5°, 112.5°, 202.5°, 292.5°$.

26.   Since $\sin 2x \cos x + \cos 2x \sin x = \sin(2x + x) = \sin 3x$, solving $\sin 3x = -\frac{1}{2}$ results in all values of the form:

$$3x = \frac{7\pi}{6} + 2\pi k \qquad\qquad 3x = \frac{11\pi}{6} + 2\pi k$$
$$x = \frac{7\pi}{18} + \frac{2\pi}{3}k \qquad\qquad x = \frac{11\pi}{18} + \frac{2\pi}{3}k$$

Substituting $k = 0$, 1, and 2 results in the solutions $x = \frac{7\pi}{18}, \frac{11\pi}{18}, \frac{19\pi}{18}, \frac{23\pi}{18}, \frac{31\pi}{18}, \frac{35\pi}{18}$.

28. Since $\cos 2x \cos x - \sin 2x \sin x = \cos(2x + x) = \cos 3x$, solving $\cos 3x = \dfrac{1}{\sqrt{2}}$ results in all values of the form:

$$3x = \frac{\pi}{4} + 2\pi k \qquad\qquad\qquad 3x = \frac{7\pi}{4} + 2\pi k$$
$$x = \frac{\pi}{12} + \frac{2\pi}{3}k \qquad\qquad\qquad x = \frac{7\pi}{12} + \frac{2\pi}{3}k$$

Substituting $k = 0$, 1, and 2 results in the solutions $x = \dfrac{\pi}{12}, \dfrac{7\pi}{12}, \dfrac{3\pi}{4}, \dfrac{5\pi}{4}, \dfrac{17\pi}{12}, \dfrac{23\pi}{12}$.

30. Since $\sin 2x \cos 3x + \cos 2x \sin 3x = \sin(2x + 3x) = \sin 5x$, solving $\sin 5x = -1$ results in all values of the form:

$$5x = \frac{3\pi}{2} + 2k\pi$$
$$x = \frac{3\pi}{10} + \frac{2k\pi}{5}$$

32. Since $\cos^2 4x = 1$, $\cos 4x = \pm 1$. All values will take the form:

$$4x = 0 + 2k\pi \qquad\qquad\qquad 4x = \pi + 2k\pi$$
$$x = \frac{k\pi}{2} \qquad\qquad\qquad\qquad x = \frac{\pi}{4} + \frac{k\pi}{2}$$

Note that these solutions can be combined as $x = \dfrac{k\pi}{4}$, where k is any integer.

34. Since $\sin^3 5x = -1$, $\sin 5x = -1$. All values will take the form:

$$5x = \frac{3\pi}{2} + 2k\pi$$
$$x = \frac{3\pi}{10} + \frac{2k\pi}{5}$$

36. First factor the equation:

$$2\sin^2 3\theta + 3\sin 3\theta + 1 = 0$$
$$(2\sin 3\theta + 1)(\sin 3\theta + 1) = 0$$
$$\sin 3\theta = -\tfrac{1}{2}, -1$$

For $\sin 3\theta = -\tfrac{1}{2}$, all values will take the form:

$$3\theta = 210° + 360°k \qquad\qquad\qquad 3\theta = 330° + 360°k$$
$$\theta = 70° + 120°k \qquad\qquad\qquad\quad \theta = 110° + 120°k$$

For $\sin 3\theta = -1$, all values will take the form:

$$3\theta = 270° + 360°k$$
$$\theta = 90° + 120°k$$

38. First factor the equation:

$$2\cos^2 2\theta - \cos 2\theta - 1 = 0$$
$$(2\cos 2\theta + 1)(\cos 2\theta - 1) = 0$$
$$\cos 2\theta = -\tfrac{1}{2}, 1$$

For $\cos 2\theta = -\tfrac{1}{2}$, all values will take the form:

$$2\theta = 120° + 360°k \qquad\qquad\qquad 2\theta = 240° + 360°k$$
$$\theta = 60° + 180°k \qquad\qquad\qquad\quad \theta = 120° + 180°k$$

For $\cos 2\theta = 1$, all values will take the form:

$$2\theta = 0° + 360°k$$
$$\theta = 180°k$$

Note that these solutions can be combined as $\theta = 60°k$, where k is any integer.

40. Since $\cot^2 3\theta = 1$, $\cot 3\theta = \pm 1$. For $\cot 3\theta = 1$, all values will take the form:
$$3\theta = 45° + 180°k$$
$$\theta = 15° + 60°k$$
For $\cot 3\theta = -1$, all values will take the form:
$$3\theta = 135° + 180°k$$
$$\theta = 45° + 60°k$$
Note that these solutions can be combined as $\theta = 15° + 30°k$, where k is any integer.

42. Squaring each side of the equation:
$$\sin\theta - \cos\theta = 1$$
$$(\sin\theta - \cos\theta)^2 = 1$$
$$\sin^2\theta - 2\sin\theta\cos\theta + \cos^2\theta = 1$$
$$1 - \sin 2\theta = 1$$
$$\sin 2\theta = 0$$
All solutions will take the form:
$$2\theta = 0° + 360°k \qquad\qquad 2\theta = 180° + 360°k$$
$$\theta = 180°k \qquad\qquad\qquad \theta = 90° + 180°k$$
Substituting $k = 0$ and $k = 1$ results in the possible solutions $\theta = 0°, 90°, 180°, 270°$. Since this equation was solved by squaring, these possible solutions must be checked:
$$\sin 0° - \cos 0° = 0 - 1 = -1 \text{ (doesn' t check)}$$
$$\sin 90° - \cos 90° = 1 - 0 = 1 \text{ (checks)}$$
$$\sin 180° - \cos 180° = 0 - (-1) = 1 \text{ (checks)}$$
$$\sin 270° - \cos 270° = -1 - 0 = -1 \text{ (doesn' t check)}$$
Only $\theta = 90°, 180°$ are solutions.

44. Squaring each side of the equation:
$$\cos\theta - \sin\theta = -1$$
$$(\cos\theta - \sin\theta)^2 = 1$$
$$\cos^2\theta - 2\sin\theta\cos\theta + \sin^2\theta = 1$$
$$1 - \sin 2\theta = 1$$
$$\sin 2\theta = 0$$
As in Problem 42, the possible solutions $\theta = 0°, 90°, 180°, 270°$. Checking these solutions:
$$\cos 0° - \sin 0° = 1 - 0 = 1 \text{ (doesn' t check)}$$
$$\cos 90° - \sin 90° = 0 - 1 = -1 \text{ (checks)}$$
$$\cos 180° - \sin 180° = -1 - 0 = -1 \text{ (checks)}$$
$$\cos 270° - \sin 270° = 0 - (-1) = 1 \text{ (doesn' t check)}$$
Only $\theta = 90°, 180°$ are solutions.

46. Using $a = 1$, $b = -6$, and $c = 4$ in the quadratic formula:
$$\cos 3\theta = \frac{-(-6) \pm \sqrt{(-6)^2 - 4(1)(4)}}{2(1)} = \frac{6 \pm \sqrt{36 - 16}}{2} = \frac{6 \pm 2\sqrt{5}}{2} = 3 \pm \sqrt{5}$$
So either $\cos 3\theta = 3 + \sqrt{5} \approx 5.24$, which is impossible, or $\cos 3\theta = 3 - \sqrt{5} \approx 0.764$. All solutions will take the form:
$$3\theta = 40.2° + 360°k \qquad\qquad 3\theta = 319.8° + 360°k$$
$$\theta = 13.4° + 120°k \qquad\qquad \theta = 106.6° + 120°k$$
Substituting $k = 0$, 1, and 2 results in the solutions $\theta = 13.4°, 106.6°, 133.4°, 226.6°, 253.4°, 346.6°$.

48. Using $a = 2$, $b = -6$, and $c = 3$ in the quadratic formula:
$$\sin 2\theta = \frac{-(-6) \pm \sqrt{(-6)^2 - 4(2)(3)}}{2(2)} = \frac{6 \pm \sqrt{36 - 24}}{4} = \frac{6 \pm 2\sqrt{3}}{4} = \frac{3 \pm \sqrt{3}}{2}$$
So either $\sin 2\theta = \dfrac{3 + \sqrt{3}}{2} \approx 2.37$, which is impossible, or $\sin 2\theta = \dfrac{3 - \sqrt{3}}{2} \approx 0.63$. All solutions will take the form:
$$2\theta = 39.3° + 360°k \qquad\qquad 2\theta = 140.7° + 360°k$$
$$\theta \approx 19.7° + 180°k \qquad\qquad \theta \approx 70.3° + 180°k$$
Substituting $k = 0$ and $k = 1$ results in the solutions $\theta = 19.7°, 70.3°, 199.7°, 250.3°$.

50. Using $\sin^2 4\theta = 1 - \cos^2 4\theta$:

$$2\left(1 - \cos^2 4\theta\right) - 2\cos 4\theta = 1$$
$$2 - 2\cos^2 4\theta - 2\cos 4\theta = 1$$
$$-2\cos^2 4\theta - 2\cos 4\theta + 1 = 0$$
$$2\cos^2 4\theta + 2\cos 4\theta - 1 = 0$$

Using $a = 2$, $b = 2$, and $c = -1$ in the quadratic formula:

$$\cos 4\theta = \frac{-2 \pm \sqrt{(2)^2 - 4(2)(-1)}}{2(2)} = \frac{-2 \pm \sqrt{4 + 8}}{4} = \frac{-2 \pm 2\sqrt{3}}{4} = \frac{-1 \pm \sqrt{3}}{2}$$

So either $\cos 4\theta = \dfrac{-1 - \sqrt{3}}{2} \approx -1.37$, which is impossible, or $\cos 4\theta = \dfrac{-1 + \sqrt{3}}{2} \approx 0.37$. All solutions will take

the form:

$$4\theta = 68.5° + 360°k \qquad\qquad 4\theta = 291.5° + 360°k$$
$$\theta = 17.1° + 90°k \qquad\qquad \theta = 72.9° + 90°k$$

Substituting $k = 0$, 1, 2, and 3 results in the solutions $\theta = 17.1°, 72.9°, 107.1°, 162.9°, 197.1°, 252.9°, 287.1°, 342.9°$.

52. Substituting $h = 100$ ft:

$$100 = 110.5 - 98.5\cos\frac{2\pi}{15}t$$
$$-10.5 = -98.5\cos\frac{2\pi}{15}t$$
$$\cos\frac{2\pi}{15}t \approx 0.1066$$
$$\frac{2\pi}{15}t = \cos^{-1}(0.1066)$$

Using radian mode (since t is a real number) yields:

$$\frac{2\pi}{15}t = 1.46 \qquad\qquad \frac{2\pi}{15}t = 2\pi - 1.46 = 4.82$$
$$t \approx 3.5 \qquad\qquad t \approx 11.5$$

The time is either 3.5 min or 11.5 min. This solution is verified graphically.

54. Substituting $c = \sqrt{3}\,r$:

$$2r\sin\frac{\theta}{2} = \sqrt{3}\,r$$
$$\sin\frac{\theta}{2} = \frac{\sqrt{3}}{2}$$
$$\frac{\theta}{2} = 60°, 120°$$
$$\theta = 120°, 240°$$

56. Substituting $d = 100$ ft:

$$100\tan\tfrac{1}{2}\pi t = 100$$
$$\tan\tfrac{1}{2}\pi t = 1$$

Using radian mode (since t is a real number) yields:

$$\tfrac{1}{2}\pi t = \frac{\pi}{4}$$
$$\pi t = \frac{\pi}{2}$$
$$t = \tfrac{1}{2}$$

The other will see the light after $\tfrac{1}{2}$ sec if the person opposite the lighthouse first sees the light. Since the light makes a complete revolution every 2 seconds, the other will see the light after $2 - \tfrac{1}{2} = \tfrac{3}{2}$ sec if the second person sees the light first.

58.   Using radian mode:

$$\sin 2\pi t = \frac{1}{\sqrt{2}}$$

$$2\pi t = \frac{\pi}{4}$$

$$t = \frac{1}{8}$$

60.   Working from the left side: $\dfrac{\sin^2 x}{(1-\cos x)^2} = \dfrac{1-\cos^2 x}{(1-\cos x)^2} = \dfrac{(1+\cos x)(1-\cos x)}{(1-\cos x)^2} = \dfrac{1+\cos x}{1-\cos x}$

62.   Working from the left side: $\dfrac{1}{1-\sin t} + \dfrac{1}{1+\sin t} = \dfrac{1+\sin t + 1 - \sin t}{(1-\sin t)(1+\sin t)} = \dfrac{2}{1-\sin^2 t} = \dfrac{2}{\cos^2 t} = 2\sec^2 t$

64.   Using the half-angle formulas for sine and cosine:

$$\tan \frac{A}{2} = \frac{\sin \frac{A}{2}}{\cos \frac{A}{2}} = \frac{\sqrt{\dfrac{1-\cos A}{2}}}{\sqrt{\dfrac{1+\cos A}{2}}} = \sqrt{\frac{1-\cos A}{1+\cos A}} \cdot \sqrt{\frac{1-\cos A}{1-\cos A}} = \sqrt{\frac{(1-\cos A)^2}{1-\cos^2 A}} = \sqrt{\frac{(1-\cos A)^2}{\sin^2 A}} = \frac{1-\cos A}{\sin A}$$

66.   Since $0° \le B \le 90°$, $\cos B > 0$ and thus: $\cos B = \sqrt{1-\sin^2 B} = \sqrt{1-\left(\frac{3}{5}\right)^2} = \sqrt{1-\frac{9}{25}} = \sqrt{\frac{16}{25}} = \frac{4}{5}$

Using the double-angle formula for cosine: $\cos 2B = \cos^2 B - \sin^2 B = \left(\frac{4}{5}\right)^2 - \left(\frac{3}{5}\right)^2 = \frac{16}{25} - \frac{9}{25} = \frac{7}{25}$

68.   Since $0° \le B \le 90°$, $0° \le \frac{B}{2} \le 45°$, so $\sin \frac{B}{2} > 0$. Therefore (using the results from Problem 66):

$$\sin \frac{B}{2} = \sqrt{\frac{1-\cos B}{2}} = \sqrt{\frac{1-\frac{4}{5}}{2}} = \sqrt{\frac{1}{10}} = \frac{1}{\sqrt{10}}$$

70.   Since $90° \le A \le 180°$, $\cos A < 0$ and thus: $\cos A = -\sqrt{1-\sin^2 A} = -\sqrt{1-\left(\frac{1}{3}\right)^2} = -\sqrt{1-\frac{1}{9}} = -\sqrt{\frac{8}{9}} = -\frac{2\sqrt{2}}{3}$

Using the difference formula for cosine and the results from Problem 66:

$$\cos(A-B) = \cos A \cos B + \sin A \sin B = \left(-\frac{2\sqrt{2}}{3}\right)\left(\frac{4}{5}\right) + \left(\frac{1}{3}\right)\left(\frac{3}{5}\right) = \frac{3-8\sqrt{2}}{15}$$

72.   Using the result from Problem 70: $\sec(A-B) = \dfrac{1}{\cos(A-B)} = \dfrac{1}{\dfrac{3-8\sqrt{2}}{15}} = \dfrac{15}{3-8\sqrt{2}}$

6.4 Parametric Equations and Further Graphing

2. Since $\sin t = -x$ and $\cos t = y$:

$$\sin^2 t + \cos^2 t = 1$$
$$(-x)^2 + y^2 = 1$$
$$x^2 + y^2 = 1$$

The graph is a circle with center = (0,0) and radius = 1:

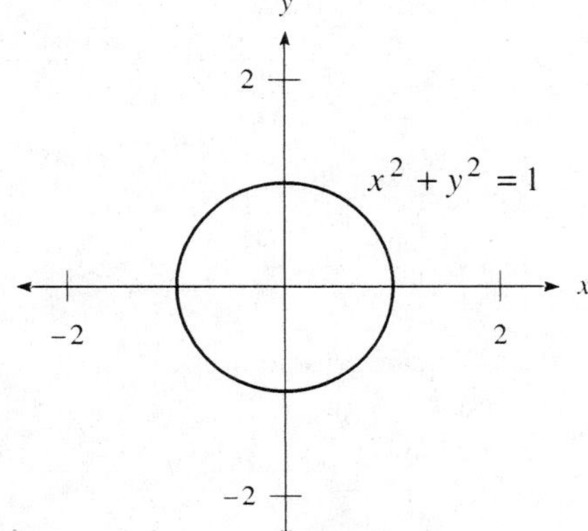

4. Since $\cos t = \dfrac{x}{2}$ and $\sin t = \dfrac{y}{2}$:

$$\sin^2 t + \cos^2 t = 1$$
$$\left(\frac{y}{2}\right)^2 + \left(\frac{x}{2}\right)^2 = 1$$
$$\frac{y^2}{4} + \frac{x^2}{4} = 1$$
$$x^2 + y^2 = 4$$

The graph is a circle with center = (0,0) and radius = 2:

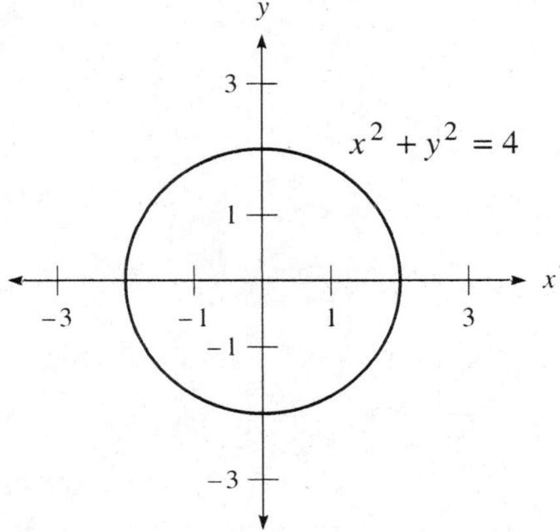

6. Since $\sin t = \dfrac{x}{3}$ and $\cos t = \dfrac{y}{4}$:

$$\sin^2 t + \cos^2 t = 1$$
$$\left(\frac{x}{3}\right)^2 + \left(\frac{y}{4}\right)^2 = 1$$
$$\frac{x^2}{9} + \frac{y^2}{16} = 1$$

The graph is an ellipse:

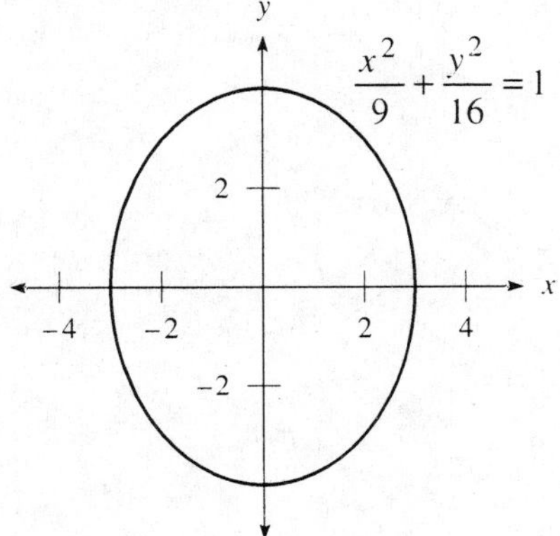

8. Since $\sin t = x - 3$ and $\cos t = y - 2$:

$$\sin^2 t + \cos^2 t = 1$$
$$(x-3)^2 + (y-2)^2 = 1$$

The graph is a circle with center = (3,2) and radius = 1:

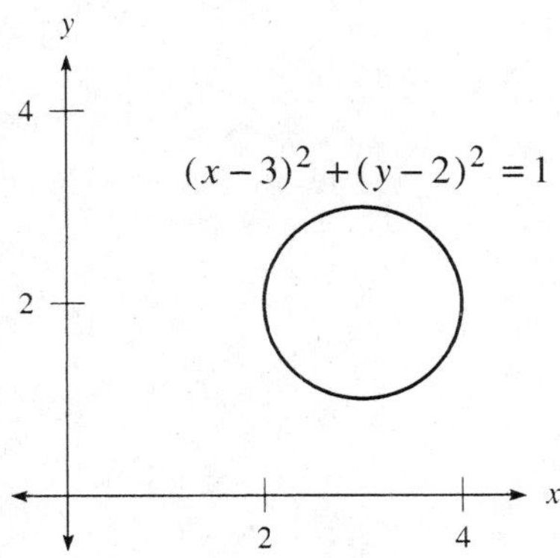

10.   Since $\cos t = x + 3$ and $\sin t = y - 2$:

$$\sin^2 t + \cos^2 t = 1$$
$$(y-2)^2 + (x+3)^2 = 1$$
$$(x+3)^2 + (y-2)^2 = 1$$

The graph is a circle with center $= (-3, 2)$ and radius $= 1$:

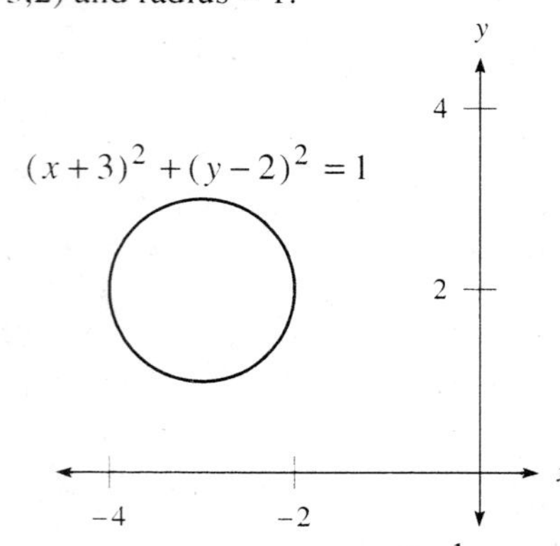

12.   Since $x - 2 = 3\sin t$, $\sin t = \dfrac{x-2}{3}$. Since $y - 1 = 3\cos t$, $\cos t = \dfrac{y-1}{3}$. Therefore:

$$\sin^2 t + \cos^2 t = 1$$
$$\left(\frac{x-2}{3}\right)^2 + \left(\frac{y-1}{3}\right)^2 = 1$$
$$\frac{(x-2)^2}{9} + \frac{(y-1)^2}{9} = 1$$
$$(x-2)^2 + (y-1)^2 = 9$$

The graph is a circle with center $= (2, 1)$ and radius $= 3$:

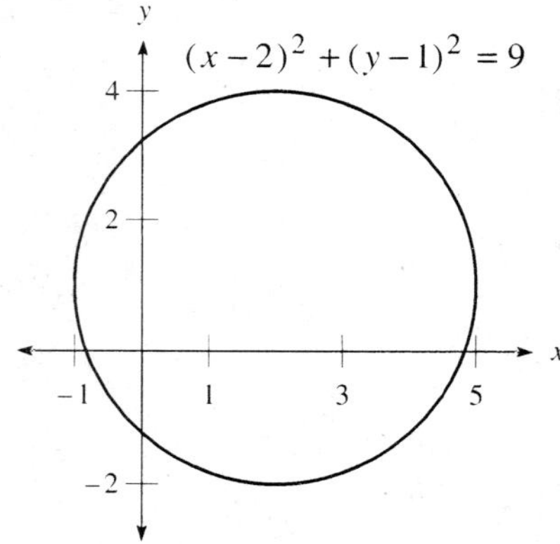

14. Since $x + 5 = 4\sin t$, $\sin t = \dfrac{x+5}{4}$. Since $y + 3 = 4\cos t$, $\cos t = \dfrac{y+3}{4}$. Therefore:

$$\sin^2 t + \cos^2 t = 1$$
$$\left(\frac{x+5}{4}\right)^2 + \left(\frac{y+3}{4}\right)^2 = 1$$
$$\frac{(x+5)^2}{16} + \frac{(y+3)^2}{16} = 1$$
$$(x+5)^2 + (y+3)^2 = 16$$

The graph is a circle with center = $(-5,-3)$ and radius = 4:

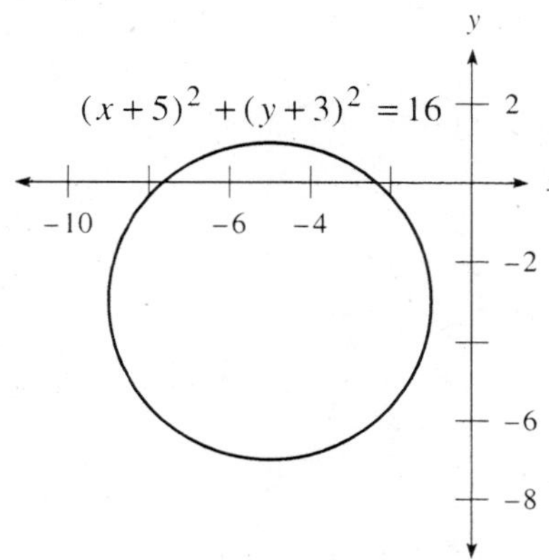

16. Since $\tan t = x$ and $\sec t = y$:

$$1 + \tan^2 t = \sec^2 t$$
$$1 + x^2 = y^2$$
$$y^2 - x^2 = 1$$

The graph would be a hyperbola.

18. Since $\cot t = \dfrac{x}{3}$ and $\csc t = \dfrac{y}{3}$:

$$1 + \cot^2 t = \csc^2 t$$
$$1 + \left(\frac{x}{3}\right)^2 = \left(\frac{y}{3}\right)^2$$
$$1 + \frac{x^2}{9} = \frac{y^2}{9}$$
$$\frac{y^2}{9} - \frac{x^2}{9} = 1$$

The graph would be a hyperbola.

20. Since $x - 3 = 5\tan t$, $\tan t = \dfrac{x-3}{5}$. Since $y - 2 = 5\sec t$, $\sec t = \dfrac{y-2}{5}$. Therefore:

$$1 + \tan^2 t = \sec^2 t$$
$$1 + \left(\frac{x-3}{5}\right)^2 = \left(\frac{y-2}{5}\right)^2$$
$$1 + \frac{(x-3)^2}{25} = \frac{(y-2)^2}{25}$$
$$\frac{(y-2)^2}{25} - \frac{(x-3)^2}{25} = 1$$

The graph would be a hyperbola.

22. Since $x = \cos 2t = 2\cos^2 t - 1$:

$$\cos 2t = 2\cos^2 t - 1$$
$$x = 2y^2 - 1$$

The graph would be a portion of a parabola.

24. Since $\cos t = \cos t$, the equation is $y = x$. This graph would be a line segment.

26. Since $\dfrac{x}{2} = \sin t$ and $\dfrac{y}{3} = \sin t$:

$$\frac{x}{2} = \frac{y}{3}$$
$$y = \frac{3}{2}x$$

The graph would be a line segment.

28. Let t represent the central angle in radians. Since the radius of the wheel is 82.5 feet, we have:

$$x = 82.5\cos t$$
$$y = 82.5\sin t$$

To shift the wheel 100 feet to the right and 91.5 feet up ($82.5 + 9 = 91.5$), the parametric equations are:

$$x = 100 + 82.5\cos t$$
$$y = 91.5 + 82.5\sin t$$

To shift the wheel so that the rider starts at the bottom, subtract $\dfrac{\pi}{2}$ from t to obtain the equations:

$$x = 100 + 82.5\cos\left(t - \frac{\pi}{2}\right)$$
$$y = 91.5 + 82.5\sin\left(t - \frac{\pi}{2}\right)$$

We now need to find a connection between t and the parameter T. Since it takes 1.5 min = 90 sec to make 1 revolution, then:

$$\frac{t}{T} = \frac{2\pi}{90}$$
$$t = \frac{\pi}{45}T$$

Therefore the parametric equations are:

$$x = 100 + 82.5\cos\left(\frac{\pi}{45}T - \frac{\pi}{2}\right)$$
$$y = 91.5 + 82.5\sin\left(\frac{\pi}{45}T - \frac{\pi}{2}\right)$$

Graphing these equations:

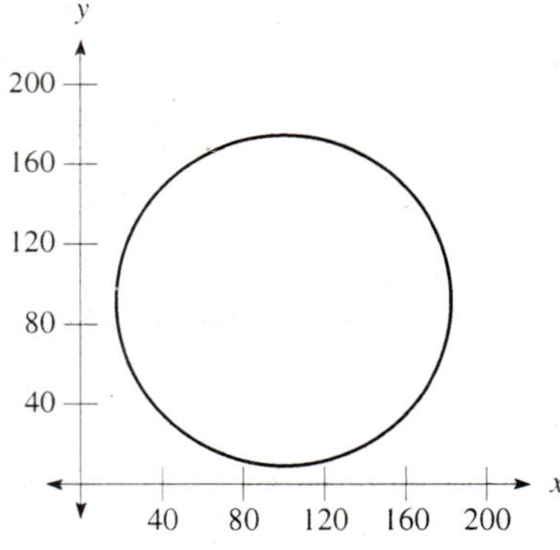

30. Let $\alpha = \sin^{-1} \frac{1}{3}$, so $\sin \alpha = \frac{1}{3}$. Drawing the triangle:

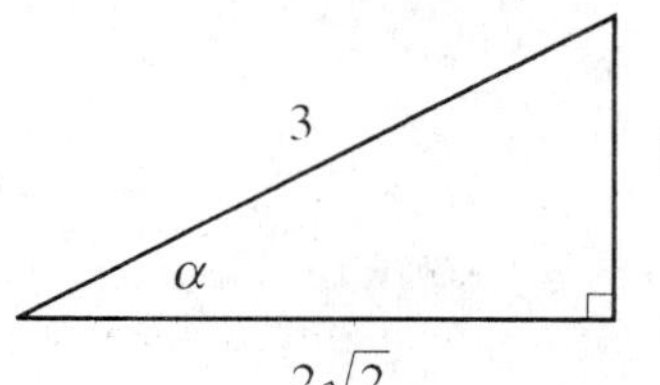

Therefore: $\tan\left(\sin^{-1} \frac{1}{3}\right) = \tan \alpha = \dfrac{1}{2\sqrt{2}}$

32. Let $\alpha = \tan^{-1} \frac{2}{3}$, so $\tan \alpha = \frac{2}{3}$, and $\beta = \cos^{-1} \frac{1}{3}$, so $\cos \beta = \frac{1}{3}$. Drawing the triangles:

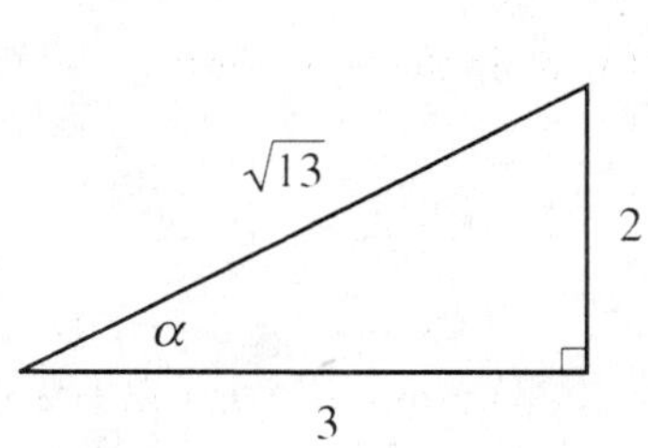 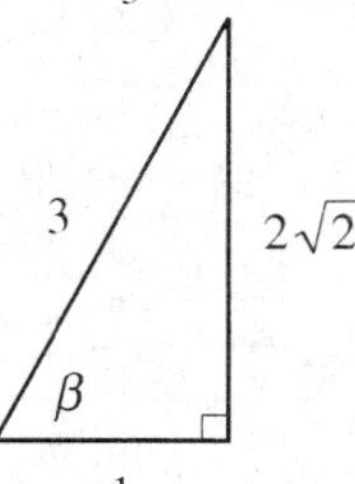

Therefore: $\cos\left(\tan^{-1} \frac{2}{3} + \cos^{-1} \frac{1}{3}\right) = \cos(\alpha + \beta) = \cos \alpha \cos \beta - \sin \alpha \sin \beta = \left(\dfrac{3}{\sqrt{13}}\right)\left(\dfrac{1}{3}\right) - \left(\dfrac{2}{\sqrt{13}}\right)\left(\dfrac{2\sqrt{2}}{3}\right) = \dfrac{3 - 4\sqrt{2}}{3\sqrt{13}}$

34. Let $\alpha = \cos^{-1} x$, so $\cos \alpha = x$. Drawing the triangle:

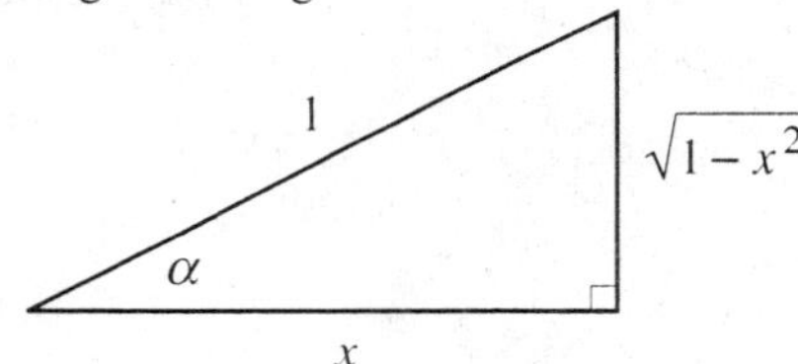

Therefore: $\sin\left(\cos^{-1} x\right) = \sin \alpha = \sqrt{1 - x^2}$

36. Let $\alpha = \tan^{-1} x$, so $\tan \alpha = x$. Drawing the triangle:

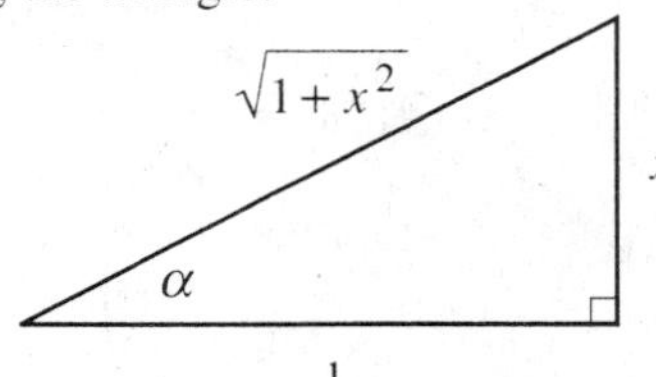

Therefore: $\sin\left(2 \tan^{-1} x\right) = \sin 2\alpha = 2 \sin \alpha \cos \alpha = 2 \cdot \dfrac{x}{\sqrt{1 + x^2}} \cdot \dfrac{1}{\sqrt{1 + x^2}} = \dfrac{2x}{x^2 + 1}$

38. Using the sum-to-product formula: $\sin 8x + \sin 4x = 2 \sin \dfrac{8x + 4x}{2} \cos \dfrac{8x - 4x}{2} = 2 \sin 6x \cos 2x$

40.   First convert the initial speed 50 mi/hr to ft/sec: $\dfrac{50 \text{ mi}}{1 \text{ hr}} \cdot \dfrac{5280 \text{ ft}}{1 \text{ mi}} \cdot \dfrac{1 \text{ hr}}{3600 \text{ sec}} \approx 73.3 \text{ ft / sec}$

The parametric equations for the path are then:
$$x = (73.3\cos\theta)t$$
$$y = (73.3\sin\theta)t - 16t^2$$

Now substituting $\theta = 20°, 30°, 40°, 50°, 60°, 70°, 80°$ results in the graphs:

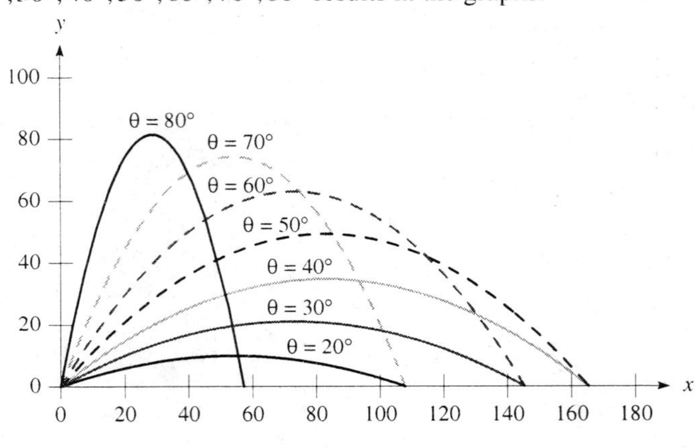

Chapter 6 Test

1.   Solving the equation:
$$2\sin\theta - 1 = 0$$
$$2\sin\theta = 1$$
$$\sin\theta = \tfrac{1}{2}$$
$$\theta = 30°, 150°$$

2.   Solving the equation:
$$\sqrt{3}\tan\theta + 1 = 0$$
$$\sqrt{3}\tan\theta = -1$$
$$\tan\theta = -\frac{1}{\sqrt{3}}$$
$$\theta = 150°, 330°$$

3.   Factoring the equation:
$$\cos\theta - 2\sin\theta\cos\theta = 0$$
$$\cos\theta(1 - 2\sin\theta) = 0$$
Thus either $\cos\theta = 0$ or $1 - 2\sin\theta = 0$, which occurs when $\sin\theta = \tfrac{1}{2}$. Thus the solutions are $\theta = 30°, 90°, 150°, 270°$.

4.   Factoring the equation:
$$\tan\theta - 2\cos\theta\tan\theta = 0$$
$$\tan\theta(1 - 2\cos\theta) = 0$$
Thus either $\tan\theta = 0$ or $1 - 2\cos\theta = 0$, which occurs when $\cos\theta = \tfrac{1}{2}$. Thus the solutions are $\theta = 0°, 60°, 180°, 300°$.

5.   Since $\sec\theta = \dfrac{1}{\cos\theta}$, the equation becomes:
$$4\cos\theta - \frac{2}{\cos\theta} = 0$$
$$4\cos^2\theta - 2 = 0$$
$$4\cos^2\theta = 2$$
$$\cos^2\theta = \tfrac{1}{2}$$
$$\cos\theta = \pm\frac{1}{\sqrt{2}}$$
$$\theta = 45°, 135°, 225°, 315°$$

6.   Since $\csc\theta = \dfrac{1}{\sin\theta}$, the equation becomes:
$$2\sin\theta - \frac{1}{\sin\theta} = 1$$
$$2\sin^2\theta - 1 = \sin\theta$$
$$2\sin^2\theta - \sin\theta - 1 = 0$$
$$(2\sin\theta + 1)(\sin\theta - 1) = 0$$
$$\sin\theta = -\tfrac{1}{2}, 1$$
$$\theta = 90°, 210°, 330°$$

7. Isolating $\sin\dfrac{\theta}{2}$ and then squaring:

$$\sin\frac{\theta}{2}+\cos\theta=0$$

$$\sin\frac{\theta}{2}=-\cos\theta$$

$$\sin^2\frac{\theta}{2}=(-\cos\theta)^2$$

$$\frac{1-\cos\theta}{2}=\cos^2\theta$$

$$1-\cos\theta=2\cos^2\theta$$

$$2\cos^2\theta+\cos\theta-1=0$$

$$(2\cos\theta-1)(\cos\theta+1)=0$$

$$\cos\theta=\tfrac{1}{2},-1$$

$$\theta=60°,180°,300°$$

Since this equation was solved by squaring, each solution must be checked:

$$\sin\frac{60°}{2}+\cos 60°=\sin 30°+\cos 60°=\tfrac{1}{2}+\tfrac{1}{2}=1 \text{ (doesn' t check)}$$

$$\sin\frac{180°}{2}+\cos 180°=\sin 90°+\cos 180°=1+(-1)=0 \text{ (checks)}$$

$$\sin\frac{300°}{2}+\cos 300°=\sin 150°+\cos 300°=\tfrac{1}{2}+\tfrac{1}{2}=1 \text{ (doesn' t check)}$$

Thus $\theta=180°$ is the only solution.

8. Isolating $\cos\dfrac{\theta}{2}$ and then squaring:

$$\cos\frac{\theta}{2}-\cos\theta=0$$

$$\cos\frac{\theta}{2}=\cos\theta$$

$$\cos^2\frac{\theta}{2}=\cos^2\theta$$

$$\frac{1+\cos\theta}{2}=\cos^2\theta$$

$$1+\cos\theta=2\cos^2\theta$$

$$2\cos^2\theta-\cos\theta-1=0$$

$$(2\cos\theta+1)(\cos\theta-1)=0$$

$$\cos\theta=-\tfrac{1}{2},1$$

$$\theta=0°,120°,240°$$

Since this equation was solved by squaring, each solution must be checked:

$$\cos\frac{0°}{2}-\cos 0°=\cos 0°-\cos 0°=1-1=0 \text{ (checks)}$$

$$\cos\frac{120°}{2}-\cos 120°=\cos 60°-\cos 120°=\tfrac{1}{2}-\left(-\tfrac{1}{2}\right)=1 \text{ (doesn' t check)}$$

$$\cos\frac{240°}{2}-\cos 240°=\cos 120°-\cos 240°=-\tfrac{1}{2}-\left(-\tfrac{1}{2}\right)=0 \text{ (checks)}$$

Thus $\theta=0°,240°$ are the solutions.

9. Using the double-angle formula for cosine:
$$4\cos 2\theta + 2\sin\theta = 1$$
$$4\left(1 - 2\sin^2\theta\right) + 2\sin\theta = 1$$
$$4 - 8\sin^2\theta + 2\sin\theta = 1$$
$$-8\sin^2\theta + 2\sin\theta + 3 = 0$$
$$8\sin^2\theta - 2\sin\theta - 3 = 0$$
$$(4\sin\theta - 3)(2\sin\theta + 1) = 0$$
$$\sin\theta = -\tfrac{1}{2}, \tfrac{3}{4}$$

If $\sin\theta = -\tfrac{1}{2}$, $\theta = 210°, 330°$, and if $\sin\theta = \tfrac{3}{4}$, $\theta \approx 48.6°, 131.4°$.

10. For $\sin(3\theta - 45°) = -\dfrac{\sqrt{3}}{2}$, all values will take the form:

$$3\theta - 45° = 240° + 360°k \qquad\qquad 3\theta - 45° = 300° + 360°k$$
$$3\theta = 285° + 360°k \qquad\qquad 3\theta = 345° + 360°k$$
$$\theta = 95° + 120°k \qquad\qquad \theta = 115° + 120°k$$

Substituting $k = 0$, 1, and 2 results in the solutions $\theta = 95°, 115°, 215°, 235°, 335°, 355°$.

11. Squaring each side of the equation:
$$(\sin\theta + \cos\theta)^2 = 1$$
$$\sin^2\theta + 2\sin\theta\cos\theta + \cos^2\theta = 1$$
$$1 + \sin 2\theta = 1$$
$$\sin 2\theta = 0$$

All values will take the form:
$$2\theta = 0° + 360°k \qquad\qquad 2\theta = 180° + 360°k$$
$$\theta = 180°k \qquad\qquad \theta = 90° + 180°k$$

Substituting $k = 0$ and $k = 1$ results in the possible solutions $\theta = 0°, 90°, 180°, 270°$. Since this equation was solved by squaring, each solution must be checked:
$$\sin 0° + \cos 0° = 0 + 1 = 1 \text{ (checks)}$$
$$\sin 90° + \cos 90° = 1 + 0 = 1 \text{ (checks)}$$
$$\sin 180° + \cos 180° = 0 - 1 = -1 \text{ (doesn't check)}$$
$$\sin 270° + \cos 270° = -1 + 0 = -1 \text{ (doesn't check)}$$

Thus $\theta = 0°, 90°$ are the solutions.

12. Squaring each side of the equation:
$$(\sin\theta - \cos\theta)^2 = 1$$
$$\sin^2\theta - 2\sin\theta\cos\theta + \cos^2\theta = 1$$
$$1 - \sin 2\theta = 1$$
$$\sin 2\theta = 0$$

All values will take the form:
$$2\theta = 0° + 360°k \qquad\qquad 2\theta = 180° + 360°k$$
$$\theta = 180°k \qquad\qquad \theta = 90° + 180°k$$

Substituting $k = 0$ and $k = 1$ results in the possible solutions $\theta = 0°, 90°, 180°, 270°$. Since this equation was solved by squaring, each solution must be checked:
$$\sin 0° - \cos 0° = 0 - 1 = -1 \text{ (doesn't check)}$$
$$\sin 90° - \cos 90° = 1 - 0 = 1 \text{ (checks)}$$
$$\sin 180° - \cos 180° = 0 - (-1) = 1 \text{ (checks)}$$
$$\sin 270° - \cos 270° = -1 - 0 = -1 \text{ (doesn't check)}$$

Thus $\theta = 90°, 180°$ are the solutions.

13. All values will take the form:
$$3\theta = 120° + 360°k \qquad\qquad 3\theta = 240° + 360°k$$
$$\theta = 40° + 120°k \qquad\qquad \theta = 80° + 120°k$$

Substituting $k = 0$, $k = 1$, and $k = 2$ results in the solutions $\theta = 40°, 80°, 160°, 200°, 280°, 320°$.

14. All values will take the form:
$$2\theta = 45^\circ + 180^\circ k$$
$$\theta = 22.5^\circ + 90^\circ k$$
Substituting $k = 0$, $k = 1$, $k = 2$, and $k = 3$ results in the solutions $\theta = 22.5^\circ, 112.5^\circ, 202.5^\circ, 292.5^\circ$.

15. Using the double-angle formula for cosine:
$$\cos 2x - 3\cos x = -2$$
$$\left(2\cos^2 x - 1\right) - 3\cos x = -2$$
$$2\cos^2 x - 3\cos x + 1 = 0$$
$$(2\cos x - 1)(\cos x - 1) = 0$$
$$\cos x = \tfrac{1}{2}, 1$$

All solutions will take the form $x = 2k\pi$, $x = \dfrac{\pi}{3} + 2k\pi$, or $x = \dfrac{5\pi}{3} + 2k\pi$.

16. Solving the equation:
$$\sqrt{3}\sin x - \cos x = 0$$
$$\sqrt{3}\sin x = \cos x$$
$$\frac{\sqrt{3}\sin x}{\cos x} = \frac{\cos x}{\cos x}$$
$$\sqrt{3}\tan x = 1$$
$$\tan x = \frac{1}{\sqrt{3}}$$
$$x = \frac{\pi}{6}, \frac{7\pi}{6}$$

All solutions will take the form $x = \dfrac{\pi}{6} + 2k\pi$ or $x = \dfrac{7\pi}{6} + 2k\pi$. Note that these solutions combine as $x = \dfrac{\pi}{6} + k\pi$.

17. Since $\sin 2x \cos x + \cos 2x \sin x = \sin(2x + x) = \sin 3x$, the equation is $\sin 3x = -1$. All solutions will take the form:
$$x = \frac{\pi}{2} + \frac{2k\pi}{3}$$

18. Solving by taking cube roots:
$$\sin^3 4x = 1$$
$$\sin 4x = 1$$
$$4x = \frac{\pi}{2} + 2\pi k$$
$$x = \frac{\pi}{8} + \frac{k\pi}{2}$$

19. Solving by factoring:
$$5\sin^2 \theta - 3\sin \theta = 2$$
$$5\sin^2 \theta - 3\sin \theta - 2 = 0$$
$$(5\sin \theta + 2)(\sin \theta - 1) = 0$$
$$\sin \theta = -\tfrac{2}{5}, 1$$

For $\sin \theta = 1$, $\theta = 90^\circ$. For $\sin \theta = -\tfrac{2}{5}$, $\theta = 203.6^\circ, 336.4^\circ$.

20. Write the equation as $4\cos^2 \theta - 4\cos \theta - 2 = 0$, or $2\cos^2 \theta - 2\cos \theta - 1 = 0$. Using $a = 2$, $b = -2$, and $c = -1$ in the quadratic formula: $\cos \theta = \dfrac{-(-2) \pm \sqrt{(-2)^2 - 4(2)(-1)}}{2(2)} = \dfrac{2 \pm \sqrt{4 + 8}}{4} = \dfrac{2 \pm \sqrt{12}}{4} = \dfrac{2 \pm 2\sqrt{3}}{4} = \dfrac{1 \pm \sqrt{3}}{2}$

Either $\cos \theta = \dfrac{1 + \sqrt{3}}{2} \approx 1.37$, which is impossible, or $\cos \theta = \dfrac{1 - \sqrt{3}}{2} \approx -0.367$, which occurs when $\theta \approx 111.5^\circ, 248.5^\circ$.

21.   Sketching the graph:

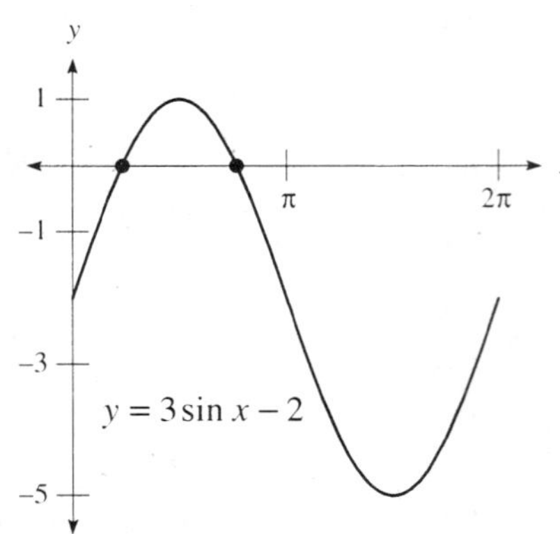

The solutions are $x \approx 0.7297, 2.4119$.

22.   Sketching the graph:

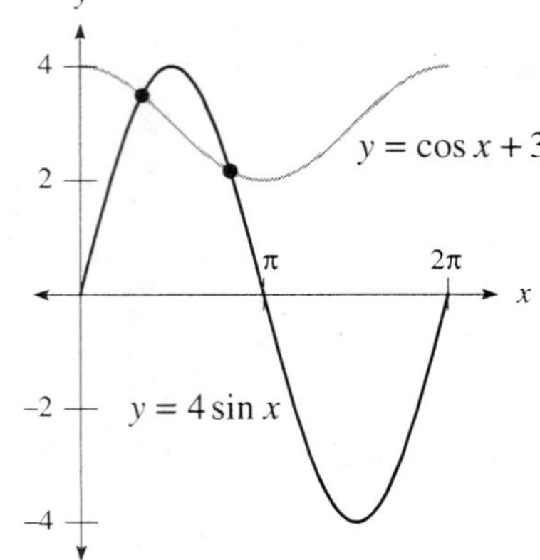

The solutions are $x \approx 1.0598, 2.5717$.

23.   Sketching the graph:

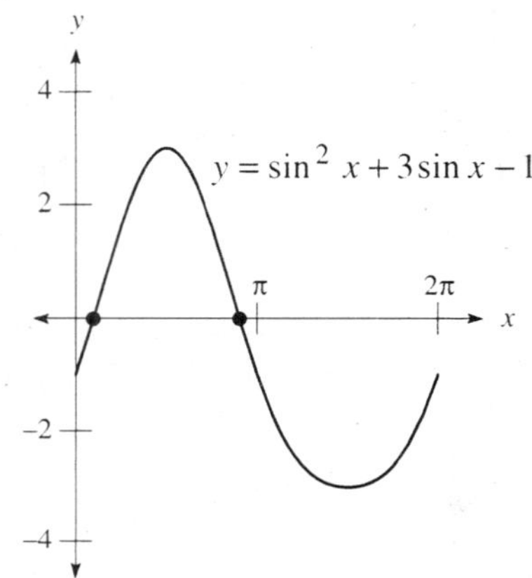

The solutions are $x \approx 0.3076, 2.8340$.

24.   Sketching the graph:

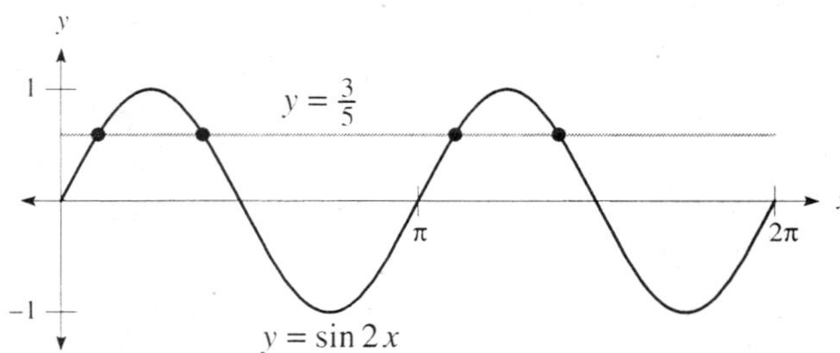

The solutions are $x \approx 0.3218, 1.2490, 3.4633, 4.3906$.

25.   Solving the equation:

$$139 - 125\cos\frac{\pi}{10}t = 150$$

$$-125\cos\frac{\pi}{10}t = 11$$

$$\cos\frac{\pi}{10}t = -\frac{11}{125}$$

Then either:

$$\frac{\pi}{10}t = \cos^{-1}\left(-\frac{11}{125}\right) \qquad\qquad \frac{\pi}{10}t = \pi + \cos^{-1}\left(\frac{11}{125}\right)$$

$$t = \frac{10}{\pi}\cos^{-1}\left(-\frac{11}{125}\right) \approx 5.3 \qquad\qquad t = \frac{10}{\pi}\left(\pi + \cos^{-1}\left(\frac{11}{125}\right)\right) \approx 14.7$$

The time will be either 5.3 minutes or 14.7 minutes.

26.   Since $\cos t = \dfrac{x}{3}$ and $\sin t = \dfrac{y}{3}$, the equation becomes: **27.**   Since $\sec t = x$ and $\tan t = y$, the equation becomes:

$$\cos^2 t + \sin^2 t = 1 \qquad\qquad\qquad\qquad\qquad 1 + \tan^2 t = \sec^2 t$$

$$\left(\frac{x}{3}\right)^2 + \left(\frac{y}{3}\right)^2 = 1 \qquad\qquad\qquad\qquad 1 + y^2 = x^2$$

$$\frac{x^2}{9} + \frac{y^2}{9} = 1 \qquad\qquad\qquad\qquad\qquad x^2 - y^2 = 1$$

$$x^2 + y^2 = 9$$

The graph is a circle with center $= (0,0)$ and radius $= 3$:   The graph is a hyperbola:

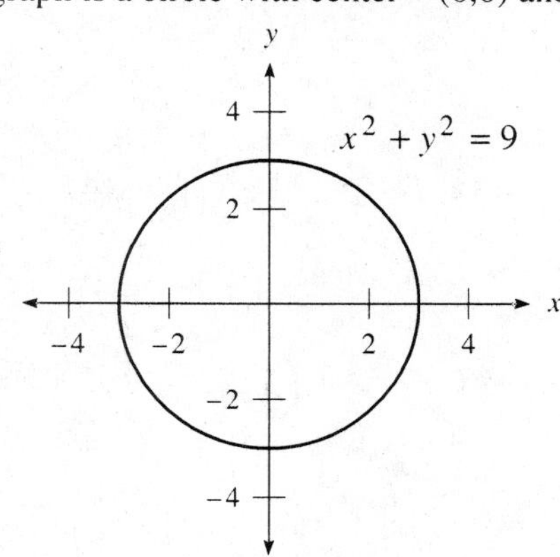

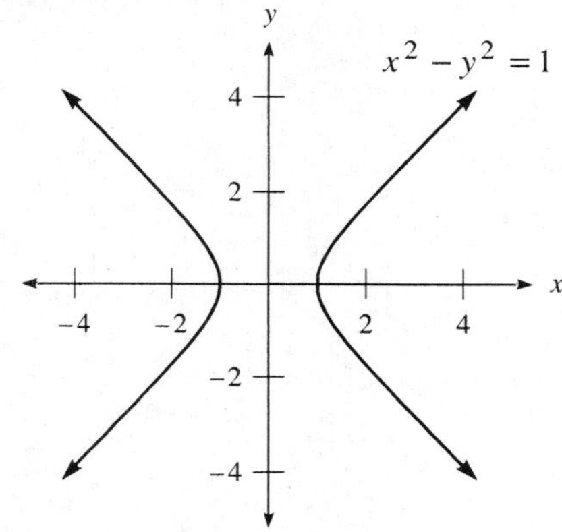

28. Since $x - 3 = 2\sin t$, $\sin t = \dfrac{x-3}{2}$. Since $y - 1 = 2\cos t$, $\cos t = \dfrac{y-1}{2}$. Therefore:

$$\sin^2 t + \cos^2 t = 1$$
$$\left(\frac{x-3}{2}\right)^2 + \left(\frac{y-1}{2}\right)^2 = 1$$
$$\frac{(x-3)^2}{4} + \frac{(y-1)^2}{4} = 1$$
$$(x-3)^2 + (y-1)^2 = 4$$

The graph is a circle with center $= (3,1)$ and radius $= 2$:

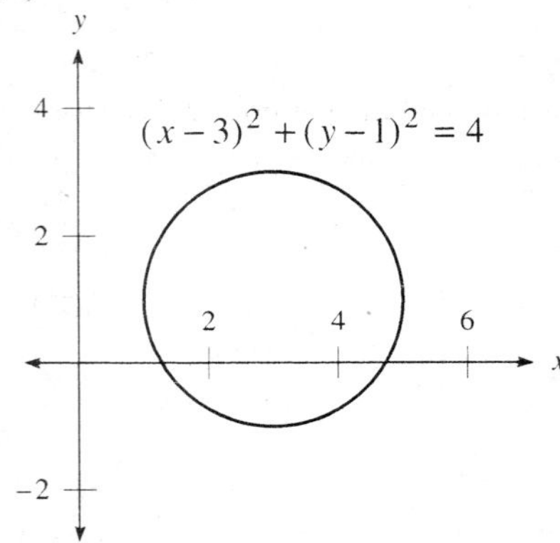

29. Since $x + 3 = 3\cos t$, $\cos t = \dfrac{x+3}{3}$. Since $y - 1 = 3\sin t$, $\sin t = \dfrac{y-1}{3}$. Therefore:

$$\sin^2 t + \cos^2 t = 1$$
$$\left(\frac{y-1}{3}\right)^2 + \left(\frac{x+3}{3}\right)^2 = 1$$
$$\frac{(x+3)^2}{9} + \frac{(y-1)^2}{9} = 1$$
$$(x+3)^2 + (y-1)^2 = 9$$

The graph is a circle with center $= (-3,1)$ and radius $= 3$:

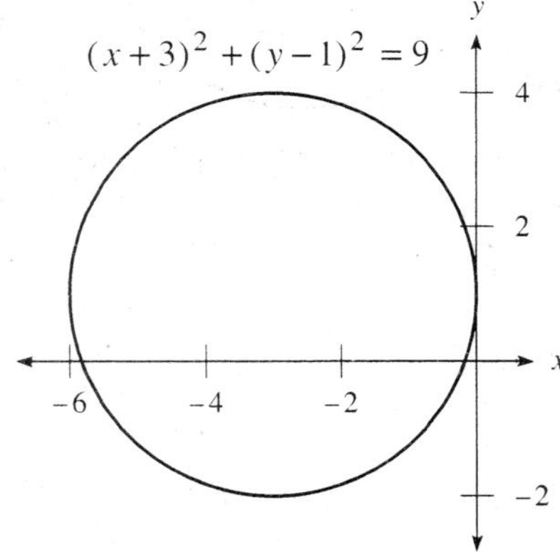

30.   Let t represent the central angle in radians. Since the radius of the wheel is 90 feet, we have:

$$x = 90\cos t$$
$$y = 90\sin t$$

To shift the wheel 98 feet up $(90 + 8 = 98)$, the parametric equations are:

$$x = 90\cos t$$
$$y = 98 + 90\sin t$$

To shift the wheel so that the rider starts at the bottom, subtract $\dfrac{\pi}{2}$ from t to obtain the equations:

$$x = 90\cos\left(t - \frac{\pi}{2}\right)$$
$$y = 98 + 90\sin\left(t - \frac{\pi}{2}\right)$$

We now need to find a connection between t and the parameter T. Since it takes 3 min to make 1 revolution, then:

$$\frac{t}{T} = \frac{2\pi}{3}$$
$$t = \frac{2\pi}{3}T$$

Therefore the parametric equations are:

$$x = 90\cos\left(\frac{2\pi}{3}T - \frac{\pi}{2}\right)$$
$$y = 98 + 90\sin\left(\frac{2\pi}{3}T - \frac{\pi}{2}\right)$$

Graphing these equations:

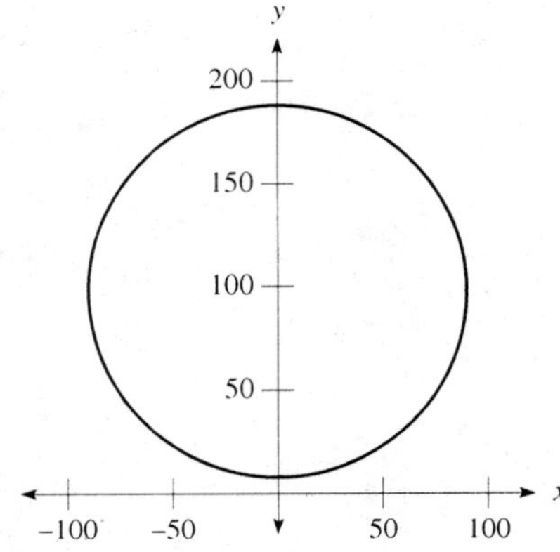

Chapter 7
Triangles

7.1 The Law of Sines

2. Using the law of sines:

$$\frac{a}{\sin A} = \frac{b}{\sin B}$$

$$\frac{a}{\sin 80°} = \frac{14}{\sin 30°}$$

$$a = \frac{14 \sin 80°}{\sin 30°} \approx 28 \text{ cm}$$

4. Using the law of sines:

$$\frac{c}{\sin C} = \frac{b}{\sin B}$$

$$\frac{c}{\sin 40°} = \frac{18}{\sin 110°}$$

$$c = \frac{18 \sin 40°}{\sin 110°} \approx 12 \text{ inches}$$

6. Using the law of sines:

$$\frac{a}{\sin A} = \frac{c}{\sin C}$$

$$\frac{a}{\sin 5°} = \frac{510}{\sin 125°}$$

$$a = \frac{510 \sin 5°}{\sin 125°} \approx 54 \text{ yd}$$

8. First note that $A = 180° - (B + C) = 180° - (40° + 70°) = 70°$. Using the law of sines:

$$\frac{a}{\sin A} = \frac{c}{\sin C}$$

$$\frac{a}{\sin 70°} = \frac{42}{\sin 70°}$$

$$a = \frac{42 \sin 70°}{\sin 70°} = 42 \text{ km}$$

10. First note that $B = 180° - (A + C) = 180° - (33° + 82°) = 65°$. Using the law of sines:

$$\frac{c}{\sin C} = \frac{b}{\sin B}$$

$$\frac{c}{\sin 82°} = \frac{18}{\sin 65°}$$

$$c = \frac{18 \sin 82°}{\sin 65°} \approx 20 \text{ cm}$$

12. Note that $B = 180° - (A + C) = 180° - (110.4° + 21.8°) = 47.8°$. Find a using the law of sines:

$$\frac{a}{\sin A} = \frac{c}{\sin C}$$

$$\frac{a}{\sin 110.4°} = \frac{246}{\sin 21.8°}$$

$$a = \frac{246 \sin 110.4°}{\sin 21.8°} \approx 621 \text{ inches}$$

Find b using the law of sines:

$$\frac{b}{\sin B} = \frac{c}{\sin C}$$

$$\frac{b}{\sin 47.8°} = \frac{246}{\sin 21.8°}$$

$$b = \frac{246 \sin 47.8°}{\sin 21.8°} \approx 491 \text{ inches}$$

14. Note that $A = 180° - (B + C) = 180° - (57° + 31°) = 92°$. Find b using the law of sines:

$$\frac{b}{\sin B} = \frac{a}{\sin A}$$

$$\frac{b}{\sin 57°} = \frac{7.3}{\sin 92°}$$

$$b = \frac{7.3 \sin 57°}{\sin 92°} \approx 6.1 \text{ m}$$

Find c using the law of sines:

$$\frac{c}{\sin C} = \frac{a}{\sin A}$$

$$\frac{c}{\sin 31°} = \frac{7.3}{\sin 92°}$$

$$c = \frac{7.3 \sin 31°}{\sin 92°} \approx 3.8 \text{ m}$$

16. Note that $A = 180° - (B + C) = 180° - (14°20' + 75°40') = 90°$. Find a using the law of sines:

$$\frac{a}{\sin A} = \frac{b}{\sin B}$$

$$\frac{a}{\sin 90°} = \frac{2.72}{\sin 14°20'}$$

$$a = \frac{2.72 \sin 90°}{\sin 14°20'} \approx 11.0 \text{ ft}$$

Find c using the law of sines:

$$\frac{c}{\sin C} = \frac{b}{\sin B}$$

$$\frac{c}{\sin 75°40'} = \frac{2.72}{\sin 14°20'}$$

$$c = \frac{2.72 \sin 75°40'}{\sin 14°20'} \approx 10.6 \text{ ft}$$

18.   Note that $C = 180° - (A + B) = 180° - (105° + 45°) = 30°$. Finding a using the law of sines:

$$\frac{a}{\sin A} = \frac{c}{\sin C}$$

$$\frac{a}{\sin 105°} = \frac{630}{\sin 30°}$$

$$a = \frac{630 \sin 105°}{\sin 30°} \approx 1200 \text{ cm}$$

Finding b using the law of sines:

$$\frac{b}{\sin B} = \frac{c}{\sin C}$$

$$\frac{b}{\sin 45°} = \frac{630}{\sin 30°}$$

$$b = \frac{630 \sin 45°}{\sin 30°} \approx 890 \text{ cm}$$

20.   Using the law of sines:

$$\frac{b}{\sin B} = \frac{a}{\sin A}$$

$$\frac{20}{\sin B} = \frac{18}{\sin 40°}$$

$$\sin B = \frac{20 \sin 40°}{18} \approx 0.7142$$

$$B \approx 46°, 134°$$

22.   Re-drawing the figure:

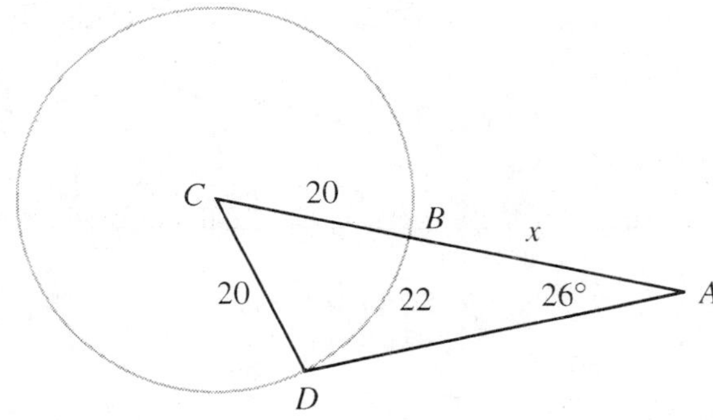

First find C using the arc length: $C = \dfrac{s}{r} = \dfrac{22}{20} = 1.1$ radians

Converting to degrees: 1.1 radians $= 1.1 \cdot \dfrac{180°}{\pi} \approx 63°$

Thus $D = 180° - (A + C) = 180° - (26° + 63°) = 91°$. Using the law of sines:

$$\frac{CA}{\sin D} = \frac{CD}{\sin A}$$

$$\frac{20 + x}{\sin 91°} = \frac{20}{\sin 26°}$$

$$20 + x = \frac{20 \sin 91°}{\sin 26°}$$

$$x = \frac{20 \sin 91°}{\sin 26°} - 20 \approx 26$$

24.   Re-drawing the figure:

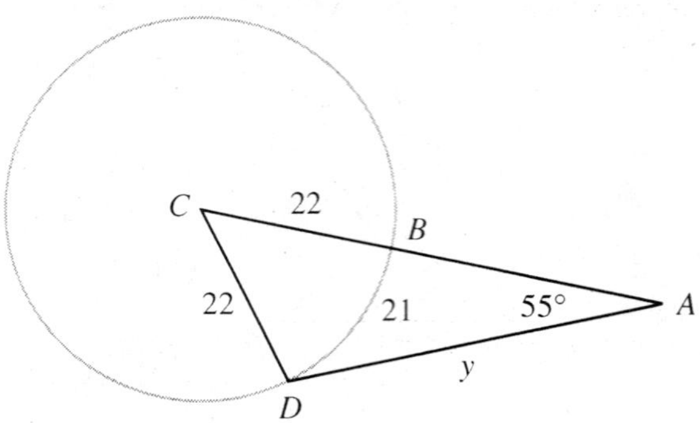

First find C using the arc length: $C = \dfrac{s}{r} = \dfrac{21}{22}$ radians

Converting to degrees: $\dfrac{21}{22}$ radians $= \dfrac{21}{22} \cdot \dfrac{180°}{\pi} \approx 54.7°$

Using the law of sines:
$$\frac{DA}{\sin C} = \frac{DC}{\sin A}$$
$$\frac{y}{\sin 54.7°} = \frac{22}{\sin 55°}$$
$$y = \frac{22 \sin 54.7°}{\sin 55°} \approx 22$$

26.   Note that this problem can be solved using right triangles. To use the law of sines, first label missing information:

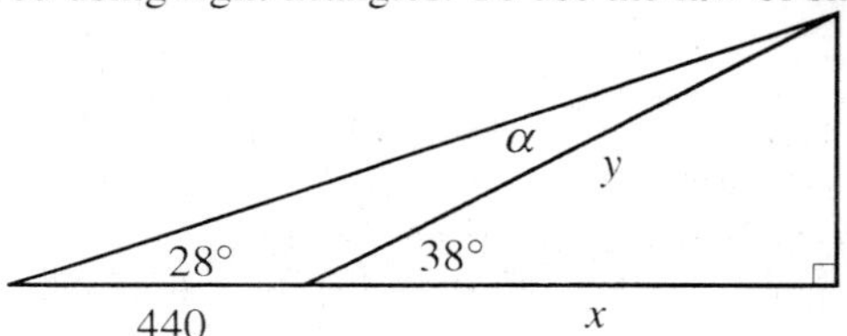

Note that $\alpha = 180° - (28° + 142°) = 10°$. Find y using the law of sines:
$$\frac{y}{\sin 28°} = \frac{440}{\sin 10°}$$
$$y = \frac{440 \sin 28°}{\sin 10°} \approx 1190 \text{ feet}$$

Therefore (using the smaller right triangle):
$$\cos 38° = \frac{x}{1190}$$
$$x = 1190 \cos 38° \approx 937 \text{ feet}$$

The distance from the building on the second observation is therefore $937 + 440 \approx 1,400$ feet.

28.   First draw the figure, where h represents the height of the building:

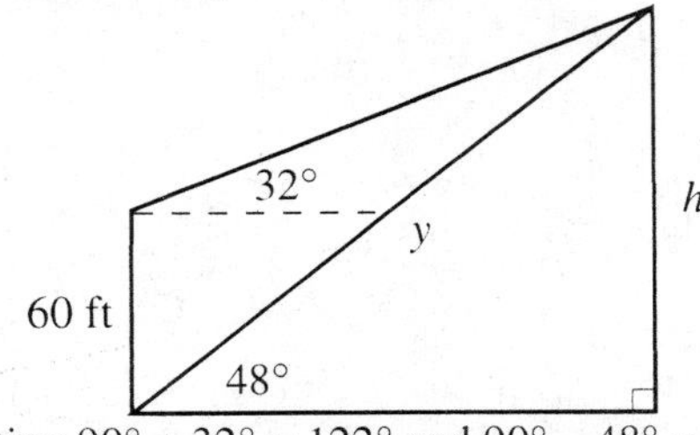

The upper triangle can be drawn, noting $90° + 32° = 122°$ and $90° - 48° = 42°$:

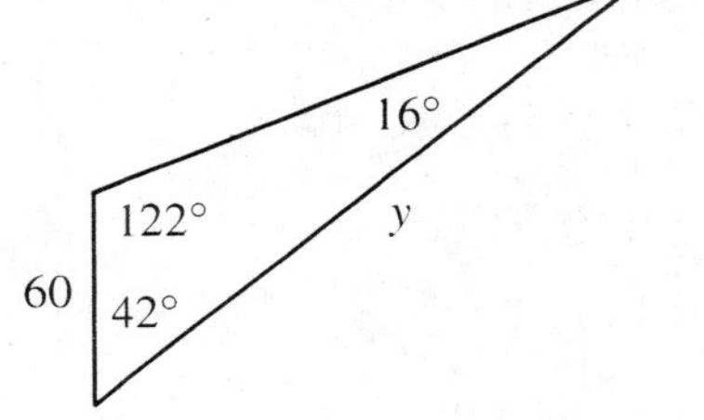

Using the law of sines:
$$\frac{y}{\sin 122°} = \frac{60}{\sin 16°}$$
$$y = \frac{60 \sin 122°}{\sin 16°} \approx 185 \text{ feet}$$

Therefore (using the smaller right triangle):
$$\sin 48° = \frac{h}{185}$$
$$b = 185 \sin 48° \approx 137 \text{ feet}$$

The height of the building is approximately 137 feet.

30.   First draw the figure, where h represents the height of the building:

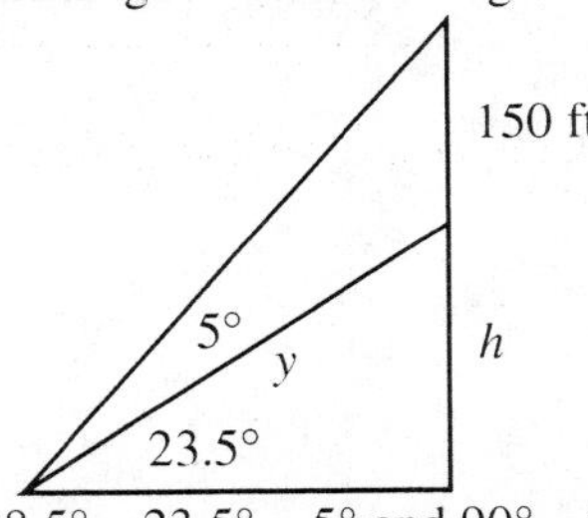

The upper triangle can be drawn, noting $28.5° - 23.5° = 5°$ and $90° - 28.5° = 61.5°$:

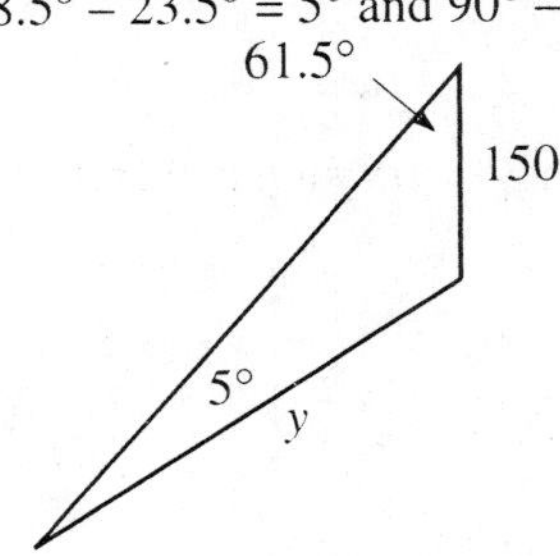

Using the law of sines:

$$\frac{y}{\sin 61.5^\circ} = \frac{150}{\sin 5^\circ}$$

$$y = \frac{150\sin 61.5^\circ}{\sin 5^\circ} \approx 1512 \text{ feet}$$

Therefore (using the smaller right triangle):

$$\sin 23.5^\circ = \frac{h}{1512}$$

$$h = 1512\sin 23.5^\circ \approx 603 \text{ feet}$$

The building is approximately 603 feet tall.

32.   First find CB using right triangle CBD (let x represent length CB):

$$\tan 22.5^\circ = \frac{1050}{x}$$

$$x\tan 22.5^\circ = 1050$$

$$x = \frac{1050}{\tan 22.5^\circ} \approx 2535 \text{ feet}$$

Now consider triangle ABC, where y is the desired distance:

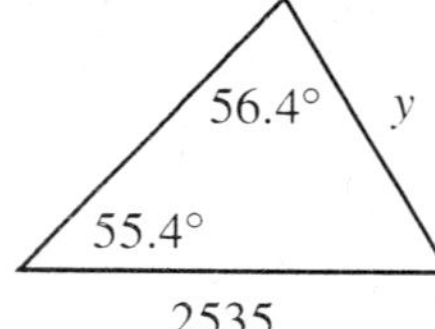

Using the law of sines:

$$\frac{y}{\sin 55.4^\circ} = \frac{2535}{\sin 56.4^\circ}$$

$$y = \frac{2535\sin 55.4^\circ}{\sin 56.4^\circ} \approx 2510 \text{ feet}$$

The rescue boat will need to travel approximately 2,510 feet.

34.   First draw the figure, where x and y represent the required distances. Note that the angle at the rocket is $180^\circ - (75^\circ + 65^\circ) = 40^\circ$:

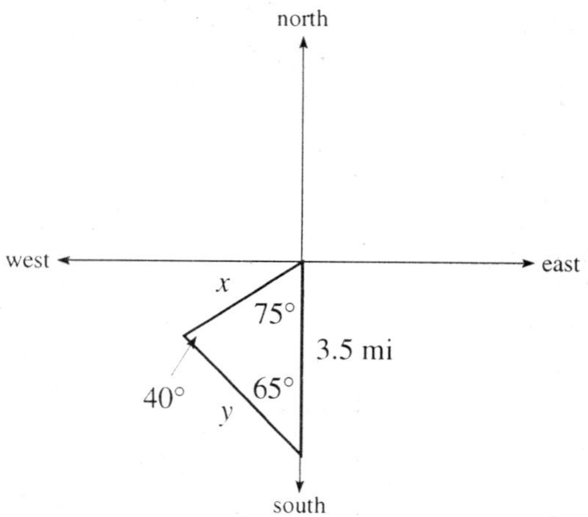

Find x using the law of sines:

$$\frac{x}{\sin 65^\circ} = \frac{3.5}{\sin 40^\circ}$$

$$x = \frac{3.5\sin 65^\circ}{\sin 40^\circ} \approx 4.9 \text{ mi}$$

Find y using the law of sines:

$$\frac{y}{\sin 75^\circ} = \frac{3.5}{\sin 40^\circ}$$

$$y = \frac{3.5\sin 75^\circ}{\sin 40^\circ} \approx 5.3 \text{ mi}$$

Tom is approximately 4.9 mi from the rocket, and Fred is approximately 5.3 mi from the rocket.

36.   First note that 14.5 inches $\approx$ 1.21 feet. Since the tightrope walker is standing in the center of the rope, we can construct the figure:

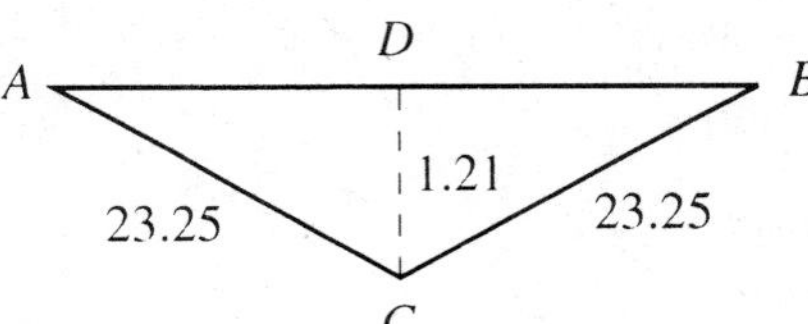

Finding $\angle DCB$ using the right triangle:

$$\cos\angle DCB = \frac{1.21}{23.25}$$

$$\angle DCB = \cos^{-1}\left(\frac{1.21}{23.25}\right) \approx 87.0°$$

Now re-drawing the tension vectors:

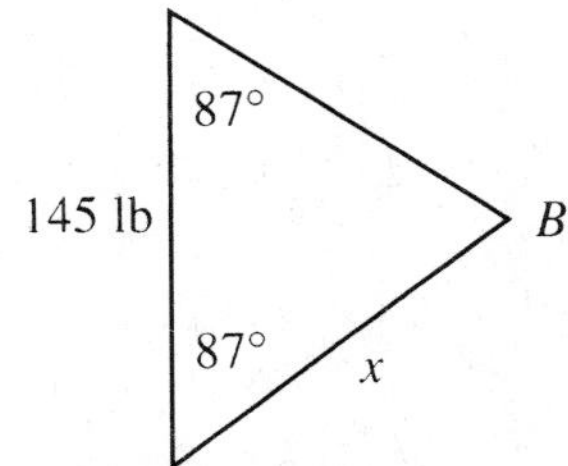

Note that $B = 180° - (87° + 87°) = 6°$. Using the law of sines:

$$\frac{x}{\sin 87°} = \frac{145}{\sin 6°}$$

$$x = \frac{145\sin 87°}{\sin 6°} \approx 1390 \text{ pounds}$$

The tension is approximately 1,390 pounds.

38.   Since the chair lift is in the center of the cable, construct the figure:

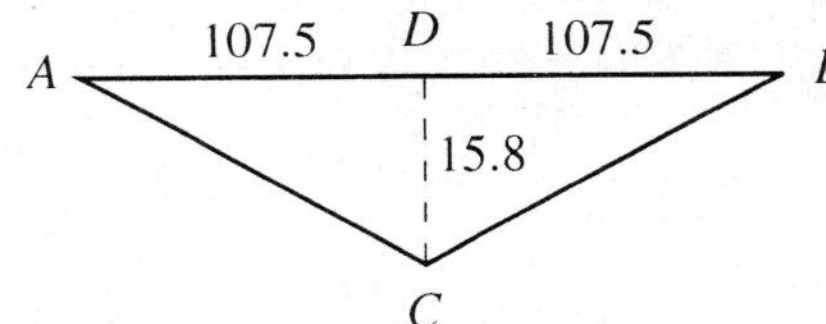

Finding $\angle DCB$ using the right triangle:

$$\tan\angle DCB = \frac{107.5}{15.8}$$

$$\angle DCB = \tan^{-1}\left(\frac{107.5}{15.8}\right) \approx 81.64°$$

Now re-drawing the tension vectors:

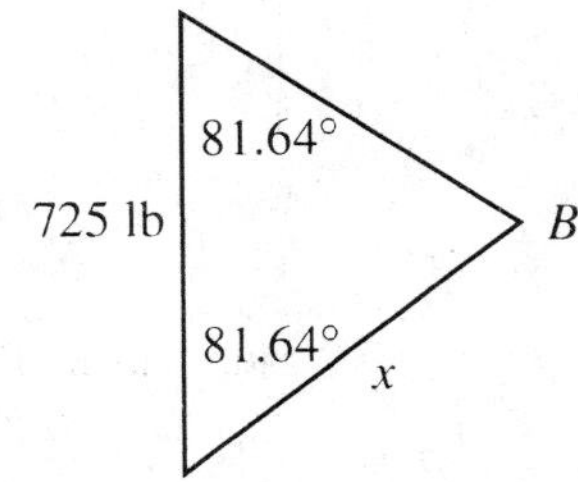

Note that $B = 180° - (81.64° + 81.64°) = 16.72°$. Using the law of sines:

$$\frac{x}{\sin 81.64°} = \frac{725}{\sin 16.72°}$$

$$x = \frac{725 \sin 81.64°}{\sin 16.72°} \approx 2490 \text{ pounds}$$

The tension is approximately 2,490 pounds.

40. Solving the equation:
$$5\cos\theta - 3 = 0$$
$$5\cos\theta = 3$$
$$\cos\theta = \tfrac{3}{5}$$
$$\theta \approx 53.1°, 306.9°$$

42. Solving by factoring:
$$3\cos\theta - 2\sin\theta\cos\theta = 0$$
$$\cos\theta(3 - 2\sin\theta) = 0$$

Either $\cos\theta = 0$ or $\sin\theta = \tfrac{3}{2}$, which is impossible. The solutions are $\theta = 90°, 270°$.

44. Solving by factoring:
$$10\cos^2\theta + \cos\theta - 3 = 0$$
$$(5\cos\theta + 3)(2\cos\theta - 1) = 0$$
$$\cos\theta = -\tfrac{3}{5}, \tfrac{1}{2}$$

If $\cos\theta = -\tfrac{3}{5}$, $\theta = 180° - 53.1° = 126.9°$ or $\theta = 180° + 53.1° = 233.1°$. If $\cos\theta = \tfrac{1}{2}$, $\theta = 60°$ or $\theta = 300°$. The solutions are $\theta = 60°, 126.9°, 233.1°, 300°$.

46. Using the quadratic formula with $a = 2$, $b = -6$, and $c = 3$:

$$\sin\theta = \frac{-(-6) \pm \sqrt{(-6)^2 - 4(2)(3)}}{2(2)} = \frac{6 \pm \sqrt{36 - 24}}{4} = \frac{6 \pm 2\sqrt{3}}{4} = \frac{3 \pm \sqrt{3}}{2}$$

Thus either $\sin\theta = \dfrac{3 + \sqrt{3}}{2} \approx 2.37$, which is impossible, or $\sin\theta = \dfrac{3 - \sqrt{3}}{2} \approx 0.63$, which occurs when $\theta \approx 39.3°, 140.7°$.

48. Solving by factoring:
$$2\sin x\cos x - \sqrt{3}\sin x = 0$$
$$\sin x\left(2\cos x - \sqrt{3}\right) = 0$$

Thus either $\sin x = 0$ or $\cos x = \dfrac{\sqrt{3}}{2}$. For $\sin x = 0$, $x = k\pi$. For $\cos x = \dfrac{\sqrt{3}}{2}$, $x = \dfrac{\pi}{6} + 2k\pi$ or $x = \dfrac{11\pi}{6} + 2k\pi$, where k represents any integer.

50. The reference angle is $\hat{\theta} = \sin^{-1}(0.7965) \approx 52.8°$. Thus either $\theta = 52.8°$ or $\theta = 180° - 52.8° = 127.2°$.

52. The reference angle is $\hat{\theta} = \sin^{-1}(0.2351) \approx 13.6°$. Thus either $\theta = 13.6°$ or $\theta = 180° - 13.6° = 166.4°$.

7.2 The Ambiguous Case

2.   Using the law of sines:
$$\frac{\sin B}{b} = \frac{\sin A}{a}$$
$$\frac{\sin B}{30} = \frac{\sin 150^\circ}{10}$$
$$\sin B = \frac{30\sin 150^\circ}{10} = 3\sin 150^\circ = 3 \cdot \tfrac{1}{2} = 1.5$$
Since $\sin B > 1$ is impossible, no such triangle exists.

4.   Using the law of sines:
$$\frac{\sin B}{b} = \frac{\sin A}{a}$$
$$\frac{\sin B}{12} = \frac{\sin 30^\circ}{6}$$
$$\sin B = 2\sin 30^\circ = 2 \cdot \tfrac{1}{2} = 1$$
$$B = 90^\circ$$
Since there is only one solution ($B = 90^\circ$), one triangle exists.

6.   Using the law of sines:
$$\frac{\sin B}{b} = \frac{\sin A}{a}$$
$$\frac{\sin B}{40} = \frac{\sin 20^\circ}{30}$$
$$\sin B = \frac{40\sin 20^\circ}{30} \approx 0.4560$$
$$B \approx 27^\circ, 153^\circ$$
Since $153^\circ + 20^\circ = 173^\circ < 180^\circ$, the third angle C is possible. Two triangles exist.

8.   Use the law of sines to find B:
$$\frac{\sin B}{b} = \frac{\sin A}{a}$$
$$\frac{\sin B}{37} = \frac{\sin 43^\circ}{31}$$
$$\sin B = \frac{37\sin 43^\circ}{31} \approx 0.8140$$
$$B \approx 54^\circ, 126^\circ$$
If $B = 54^\circ$, $C = 180^\circ - (43^\circ + 54^\circ) = 83^\circ$. Use the law of sines to find c:
$$\frac{\sin C}{c} = \frac{\sin A}{a}$$
$$\frac{\sin 83^\circ}{c} = \frac{\sin 43^\circ}{31}$$
$$c = \frac{31\sin 83^\circ}{\sin 43^\circ} \approx 45 \text{ ft}$$
If $B = 126^\circ$, $C = 180^\circ - (43^\circ + 126^\circ) = 11^\circ$. Use the law of sines to find c:
$$\frac{\sin C}{c} = \frac{\sin A}{a}$$
$$\frac{\sin 11^\circ}{c} = \frac{\sin 43^\circ}{31}$$
$$c = \frac{31\sin 11^\circ}{\sin 43^\circ} \approx 8.7 \text{ ft}$$

10.   Use the law of sines to find B:

$$\frac{\sin B}{b} = \frac{\sin A}{a}$$

$$\frac{\sin B}{50.2} = \frac{\sin 124.3°}{27.3}$$

$$\sin B = \frac{50.2 \sin 124.3°}{27.3} \approx 1.52$$

Since $\sin B > 1$ is impossible, no such triangle exists.

12.   Use the law of sines to find B:

$$\frac{\sin B}{b} = \frac{\sin C}{c}$$

$$\frac{\sin B}{821} = \frac{\sin 51°30'}{707}$$

$$\sin B = \frac{821 \sin 51°30'}{707} \approx 0.9088$$

$$B \approx 65.34° \approx 65°20'$$

$$B \approx 114.66° \approx 114°40'$$

If $B = 65°20'$, $A = 180° - (65°20' + 51°30') = 63°10'$. Use the law of sines to find a:

$$\frac{\sin A}{a} = \frac{\sin C}{c}$$

$$\frac{\sin 63°10'}{a} = \frac{\sin 51°30'}{707}$$

$$a = \frac{707 \sin 63°10'}{\sin 51°30'} \approx 806 \text{ m}$$

If $B = 114°40'$, $A = 180° - (114°40' + 51°30') = 13°50'$. Use the law of sines to find a:

$$\frac{\sin A}{a} = \frac{\sin C}{c}$$

$$\frac{\sin 13°50'}{a} = \frac{\sin 51°30'}{707}$$

$$a = \frac{707 \sin 13°50'}{\sin 51°30'} \approx 216 \text{ m}$$

14.   Use the law of sines to find C:

$$\frac{\sin C}{c} = \frac{\sin B}{b}$$

$$\frac{\sin C}{3.48} = \frac{\sin 62°40'}{6.78}$$

$$\sin C = \frac{3.48 \sin 62°40'}{6.78} \approx 0.456$$

$$C \approx 27.13° \approx 27°10'$$

$$C \approx 152.87° \approx 152°50'$$

Note this second value of C is impossible, since $B + C \approx 216° > 180°$.

If $C = 27°10'$, $A = 180° - (62°40' + 27°10') = 90°10'$. Use the law of sines to find a:

$$\frac{\sin A}{a} = \frac{\sin B}{b}$$

$$\frac{\sin 90°10'}{a} = \frac{\sin 62°40'}{6.78}$$

$$a = \frac{6.78 \sin 90°10'}{\sin 62°40'} \approx 7.63 \text{ inches}$$

16.   Use the law of sines to find A:

$$\frac{\sin A}{a} = \frac{\sin B}{b}$$

$$\frac{\sin A}{8.4} = \frac{\sin 30^\circ}{4.2}$$

$$\sin A = \frac{8.4 \sin 30^\circ}{4.2} = 1$$

$$A = 90^\circ$$

If $A = 90^\circ$, $C = 180^\circ - (30^\circ + 90^\circ) = 60^\circ$. Use the law of sines to find c:

$$\frac{\sin C}{c} = \frac{\sin B}{b}$$

$$\frac{\sin 60^\circ}{c} = \frac{\sin 30^\circ}{4.2}$$

$$c = \frac{4.2 \sin 60^\circ}{\sin 30^\circ} \approx 7.3 \text{ cm}$$

18.   Use the law of sines to find B:

$$\frac{\sin B}{b} = \frac{\sin A}{a}$$

$$\frac{\sin B}{7.6} = \frac{\sin 65^\circ}{7.1}$$

$$\sin B = \frac{7.6 \sin 65^\circ}{7.1} \approx 0.9701$$

$$B \approx 76^\circ, 104^\circ$$

If $B = 76^\circ$, $C = 180^\circ - (65^\circ + 76^\circ) = 39^\circ$. Use the law of sines to find c:

$$\frac{\sin C}{c} = \frac{\sin A}{a}$$

$$\frac{\sin 39^\circ}{c} = \frac{\sin 65^\circ}{7.1}$$

$$c = \frac{7.1 \sin 39^\circ}{\sin 65^\circ} \approx 4.9 \text{ yd}$$

If $B = 104^\circ$, $C = 180^\circ - (65^\circ + 104^\circ) = 11^\circ$. Use the law of sines to find c:

$$\frac{\sin C}{c} = \frac{\sin A}{a}$$

$$\frac{\sin 11^\circ}{c} = \frac{\sin 65^\circ}{7.1}$$

$$c = \frac{7.1 \sin 11^\circ}{\sin 65^\circ} \approx 1.5 \text{ yd}$$

20.   Use the law of sines to find B:

$$\frac{\sin B}{b} = \frac{\sin C}{c}$$

$$\frac{\sin B}{92.4} = \frac{\sin 73.4^\circ}{51.1}$$

$$\sin B = \frac{92.4 \sin 73.4^\circ}{51.1} \approx 1.73$$

Since $\sin B > 1$ is impossible, no such triangle exists.

22.   Drawing the figure, where b represents the desired distance:

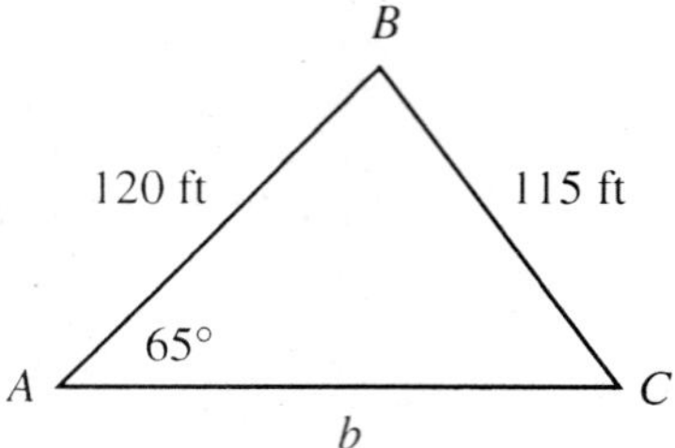

First find C using the law of sines:
$$\frac{\sin C}{120} = \frac{\sin 65°}{115}$$
$$\sin C = \frac{120 \sin 65°}{115} \approx 0.9457$$
$$C \approx 71.0°, 109.0°$$

Thus $B = 180° - (65° + 71°) = 44°$ or $B = 180° - (65° + 109°) = 6°$. Now find b using the law of sines:

$$\frac{\sin 44°}{b} = \frac{\sin 65°}{115} \qquad\qquad \frac{\sin 6°}{b} = \frac{\sin 65°}{115}$$

$$b = \frac{115 \sin 44°}{\sin 65°} \approx 88 \text{ ft} \qquad\qquad b = \frac{115 \sin 6°}{\sin 65°} \approx 13 \text{ ft}$$

The distance between the anchor points is either 88 feet or 13 feet.

24.   Drawing the vector:

26.   Drawing the vector:

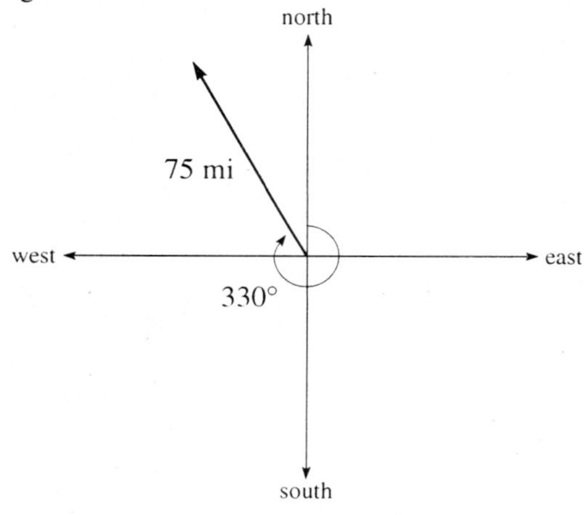

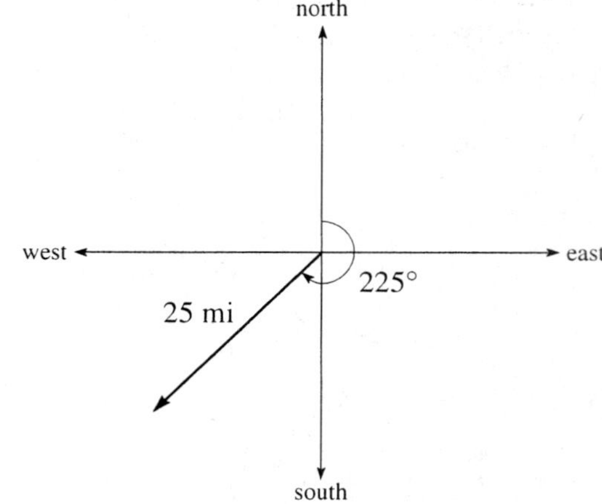

28. Begin by drawing a figure, where **S** represents the ship's heading and **T** represents the true course:

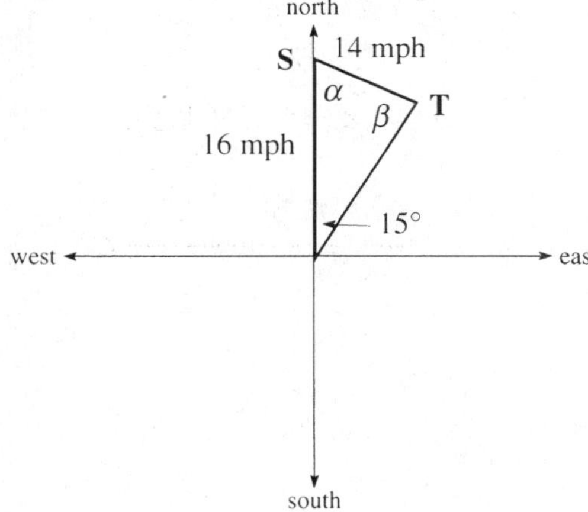

First find β using the law of sines:

$$\frac{\sin\beta}{16} = \frac{\sin 15°}{14}$$

$$\sin\beta = \frac{16\sin 15°}{14} \approx 0.2958$$

$$\beta \approx 17°, 163°$$

If $\beta = 17°$, $\alpha = 180° - (15° + 17°) = 148°$, in which case the current's direction is 32° from due north. If $\beta = 163°$, $\alpha = 180° - (15° + 163°) = 2°$, in which case the current's direction is 178° from due north.

30. Drawing the figure:

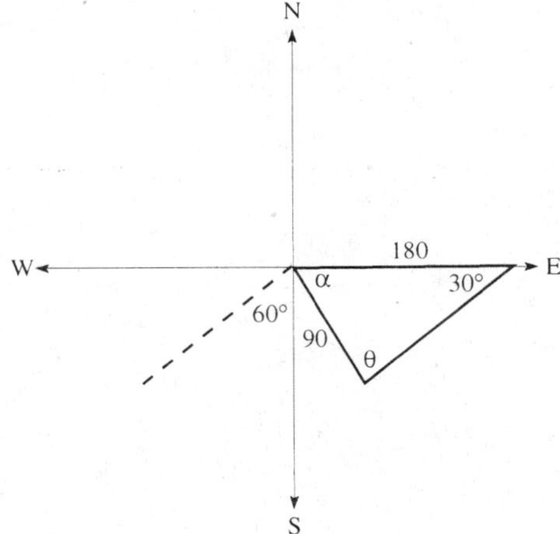

Using the law of sines:

$$\frac{\sin\theta}{180} = \frac{\sin 30°}{90}$$

$$\sin\theta = 2\sin 30°$$

$$\sin\theta = 1$$

$$\theta = 90°$$

Then $\alpha = 60°$, so the true course is 150° from due north.

32. Drawing a figure, where x represents the required distance:

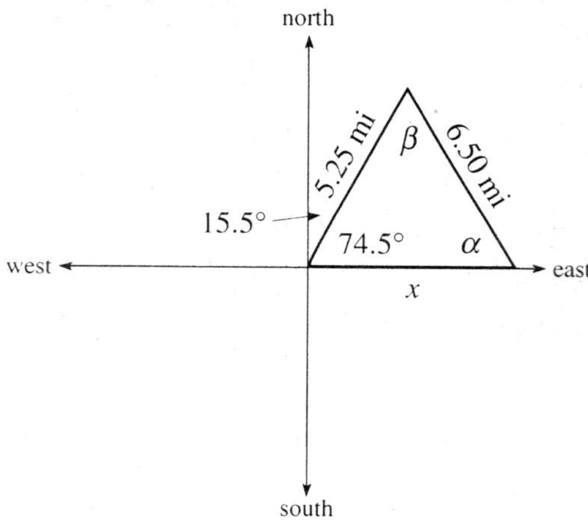

First find α using the law of sines:
$$\frac{\sin \alpha}{5.25} = \frac{\sin 74.5°}{6.50}$$
$$\sin \alpha = \frac{5.25 \sin 74.5°}{6.50} \approx 0.7783$$
$$\alpha \approx 51.1°$$
Thus $\beta = 180° - (74.5° + 51.1°) = 54.4°$. Now find x using the law of sines:
$$\frac{\sin 54.4°}{x} = \frac{\sin 74.5°}{6.50}$$
$$x = \frac{6.50 \sin 54.4°}{\sin 74.5°} \approx 5.48 \text{ miles}$$
He will have to walk 5.48 miles back to the service station.

34. Since $\csc \theta = \dfrac{1}{\sin \theta}$, the equation becomes:
$$2 \sin \theta - 1 = \frac{1}{\sin \theta}$$
$$2 \sin^2 \theta - \sin \theta = 1$$
$$2 \sin^2 \theta - \sin \theta - 1 = 0$$
$$(2 \sin \theta + 1)(\sin \theta - 1) = 0$$
$$\sin \theta = -\tfrac{1}{2}, 1$$
$$\theta = 90°, 210°, 330°$$

36. Using the double-angle formula for cosine:
$$\cos 2\theta + 3 \cos \theta - 2 = 0$$
$$2 \cos^2 \theta - 1 + 3 \cos \theta - 2 = 0$$
$$2 \cos^2 \theta + 3 \cos \theta - 3 = 0$$
Using the quadratic formula with $a = 2$, $b = 3$, and $c = -3$:
$$\cos \theta = \frac{-3 \pm \sqrt{3^2 - 4(2)(-3)}}{2(2)} = \frac{-3 \pm \sqrt{9 + 24}}{4} = \frac{-3 \pm \sqrt{33}}{4}$$
So either $\cos \theta = \dfrac{-3 - \sqrt{33}}{4} \approx -2.186$, which is impossible, or $\cos \theta = \dfrac{-3 + \sqrt{33}}{4} \approx 0.686$, which occurs when $\theta \approx 46.7°, 313.3°$.

38. Using the double-angle formula for cosine:

$$7\sin^2\theta - 9\cos 2\theta = 0$$

$$7\sin^2\theta - 9\left(1 - 2\sin^2\theta\right) = 0$$

$$7\sin^2\theta - 9 + 18\sin^2\theta = 0$$

$$25\sin^2\theta = 9$$

$$\sin^2\theta = \frac{9}{25}$$

$$\sin\theta = \frac{3}{5}, -\frac{3}{5}$$

So either $\sin\theta = \frac{3}{5}$, which occurs when $\theta \approx 36.9°, 143.1°$, or $\sin\theta = -\frac{3}{5}$, which occurs when $\theta \approx 216.9°, 323.1°$.

40. Substituting $\cos^2 x = 1 - \sin^2 x$:

$$2\left(1 - \sin^2 x\right) - \sin x = 1$$

$$2 - 2\sin^2 x - \sin x = 1$$

$$-2\sin^2 x - \sin x + 1 = 0$$

$$2\sin^2 x + \sin x - 1 = 0$$

$$(2\sin x - 1)(\sin x + 1) = 0$$

$$\sin x = \frac{1}{2}, -1$$

For $\sin x = \frac{1}{2}$, $x = \dfrac{\pi}{6} + 2k\pi$ or $x = \dfrac{5\pi}{6} + 2k\pi$. For $\sin x = -1$, $x = \dfrac{3\pi}{2} + 2k\pi$.

42. Squaring each side:

$$(\sin x - \cos x)^2 = 1$$

$$\sin^2 x - 2\sin x \cos x + \cos^2 x = 1$$

$$1 - \sin 2x = 1$$

$$\sin 2x = 0$$

$$2x = 0, \pi, 2\pi, 3\pi$$

$$x = 0, \frac{\pi}{2}, \pi, \frac{3\pi}{2}$$

Since this equation was solved by squaring, these possible solutions must be checked:

$$x = 0: \quad \sin 0 - \cos 0 = 0 - 1 = -1 \quad \text{(doesn't check)}$$

$$x = \frac{\pi}{2}: \quad \sin\frac{\pi}{2} - \cos\frac{\pi}{2} = 1 - 0 = 1 \quad \text{(checks)}$$

$$x = \pi: \quad \sin\pi - \cos\pi = 0 - (-1) = 1 \quad \text{(checks)}$$

$$x = \frac{3\pi}{2}: \quad \sin\frac{3\pi}{2} - \cos\frac{3\pi}{2} = -1 - 0 = -1 \quad \text{(doesn't check)}$$

All solutions can be written in the form $x = \dfrac{\pi}{2} + 2k\pi$ or $x = \pi + 2k\pi$, where k is any integer.

7.3 The Law of Cosines

2. Using the law of cosines to find c:
$$c^2 = a^2 + b^2 - 2ab\cos C = 120^2 + 66^2 - 2(120)(66)\cos 120° = 26676$$
$$c = \sqrt{26676} \approx 160 \text{ inches}$$

4. Using the law of cosines to find C:
$$c^2 = a^2 + b^2 - 2ab\cos C$$
$$15^2 = 13^2 + 14^2 - 2(13)(14)\cos C$$
$$225 = 365 - 364\cos C$$
$$-140 = -364\cos C$$
$$\cos C \approx 0.3846$$
$$C \approx 67°$$

6. Using the law of cosines to find b:
$$b^2 = a^2 + c^2 - 2ac\cos B = 3.7^2 + 6.4^2 - 2(3.7)(6.4)\cos 23° \approx 11.0549$$
$$b = \sqrt{11.0549} \approx 3.3 \text{ m}$$

8. Using the law of cosines to find A:
$$a^2 = b^2 + c^2 - 2bc\cos A$$
$$51^2 = 24^2 + 31^2 - 2(24)(31)\cos A$$
$$2601 = 1537 - 1488\cos A$$
$$1064 = -1488\cos A$$
$$\cos A \approx -0.7151$$
$$A \approx 136°$$

10. First find b using the law of cosines:
$$b^2 = a^2 + c^2 - 2ac\cos B = 76.3^2 + 42.8^2 - 2(76.3)(42.8)\cos 16.3° \approx 1384.77$$
$$b = \sqrt{1384.77} \approx 37.2 \text{ m}$$
Now find C using the law of sines:
$$\frac{\sin C}{c} = \frac{\sin B}{b}$$
$$\frac{\sin C}{42.8} = \frac{\sin 16.3°}{37.2}$$
$$\sin C = \frac{42.8\sin 16.3°}{37.2} \approx 0.3229$$
$$C \approx 18.8°, 161.2°$$
Note that $C \approx 161.2°$ is impossible, since a is the largest side, thus A must be the largest angle. Finally, $A = 180° - (16.3° + 18.8°) = 144.9°$.

12. First find B using the law of cosines:
$$b^2 = a^2 + c^2 - 2ac\cos B$$
$$75^2 = 48^2 + 63^2 - 2(48)(63)\cos B$$
$$5625 = 6273 - 6048\cos B$$
$$-648 = -6048\cos B$$
$$\cos B \approx 0.1071$$
$$B \approx 84°$$
Now find A using the law of sines:
$$\frac{\sin A}{a} = \frac{\sin B}{b}$$
$$\frac{\sin A}{48} = \frac{\sin 84°}{75}$$
$$\sin A = \frac{48\sin 84°}{75} \approx 0.6365$$
$$A \approx 39°$$
Finally $C = 180° - (39° + 84°) = 57°$.

14. First find a using the law of cosines:
$$a^2 = b^2 + c^2 - 2bc\cos A = 63.4^2 + 75.2^2 - 2(63.4)(75.2)\cos 124°40' \approx 15098.32$$
$$a = \sqrt{15098.32} \approx 123 \text{ km}$$
Now find C using the law of sines:
$$\frac{\sin C}{c} = \frac{\sin A}{a}$$
$$\frac{\sin C}{75.2} = \frac{\sin 124°40'}{123}$$
$$\sin C = \frac{75.2\sin 124°40'}{123} \approx 0.5028$$
$$C \approx 30.19° = 30°10'$$
Finally $B = 180° - (124°40' + 30°10') = 25°10'$.

16. First find A using the law of cosines:
$$a^2 = b^2 + c^2 - 2bc\cos A$$
$$832^2 = 623^2 + 345^2 - 2(623)(345)\cos A$$
$$692224 = 507154 - 429870\cos A$$
$$185070 = -429870\cos A$$
$$\cos A \approx -0.4305$$
$$A \approx 115.5°$$
Now find B using the law of sines:
$$\frac{\sin B}{b} = \frac{\sin A}{a}$$
$$\frac{\sin B}{623} = \frac{\sin 115.5°}{832}$$
$$\sin B = \frac{623\sin 115.5°}{832} \approx 0.6759$$
$$B \approx 42.5°$$
Finally $C = 180° - (115.5° + 42.5°) = 22.0°$.

18. Using the law of cosines and substituting $b^2 + c^2$ for a^2:
$$a^2 = b^2 + c^2 - 2bc\cos A$$
$$b^2 + c^2 = b^2 + c^2 - 2bc\cos A$$
$$0 = -2bc\cos A$$
$$\cos A = 0$$
$$A = 90°$$

20. Draw the figure, where x represents the length of the longer side:

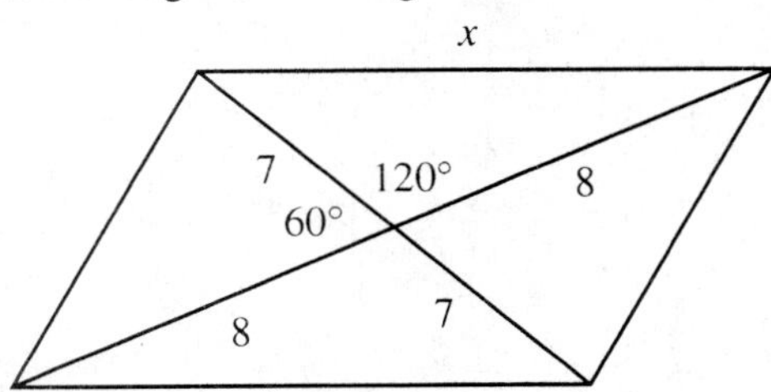

Using the law of cosines:
$$x^2 = 7^2 + 8^2 - 2(7)(8)\cos 120° = 169$$
$$x = \sqrt{169} = 13 \text{ m}$$
The length of the longer side is 13 m.

22. After 2 hours the ships have traveled 36 mi and 44 mi, respectively. Draw the figure, where x represents their distance apart after 2 hours:

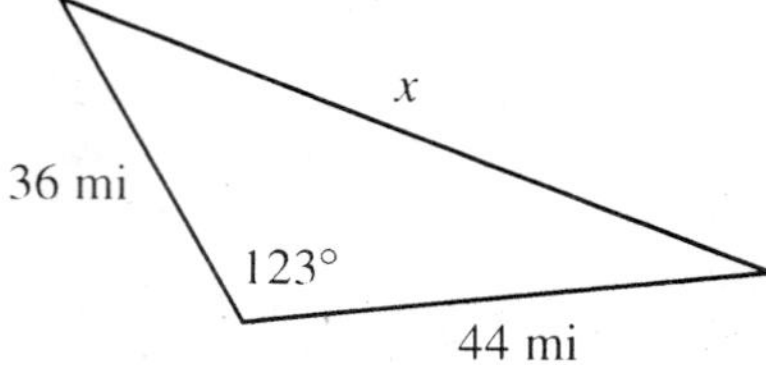

Using the law of cosines:

$$x^2 = 36^2 + 44^2 - 2(36)(44)\cos 123° \approx 4957.42$$

$$x = \sqrt{4957.42} \approx 70 \text{ mi}$$

The two ships are approximately 70 miles apart.

24. Draw the figure (using 42 mi and 36 mi for the ship distances), where x represents the distance between the two ships after 3 hours:

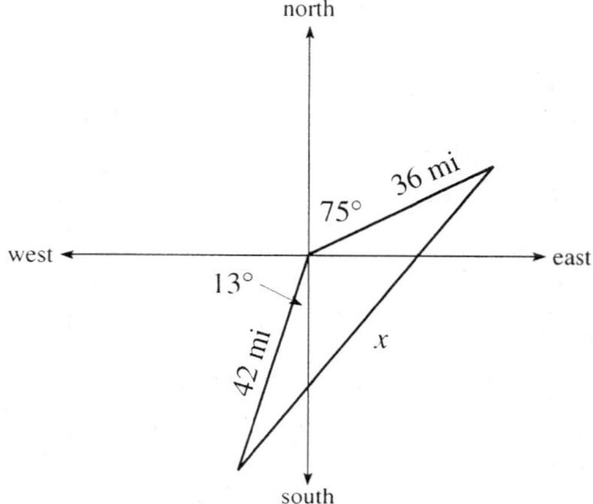

Note that the angle opposite x is $13° + 90° + 15° = 118°$. Using the law of cosines:

$$x^2 = 42^2 + 36^2 - 2(42)(36)\cos 118° \approx 4479.68$$

$$x = \sqrt{4479.68} \approx 67 \text{ mi}$$

26.   Draw the figure, where **P** represents the plane, **W** represents the wind, and **T** represents the true course:

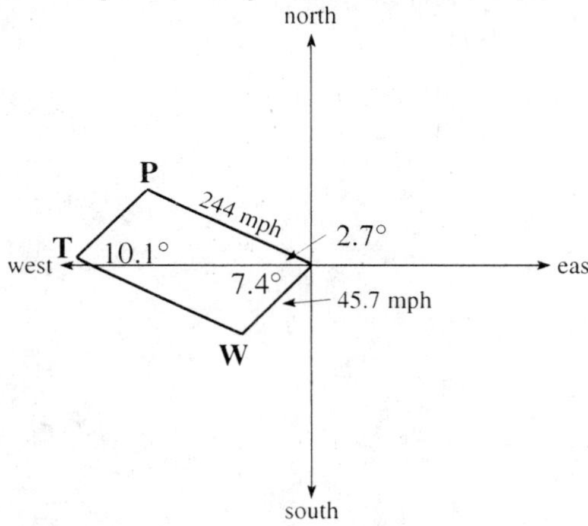

Note the angle at **P** is $180° - 10.1° = 169.9°$, so we have the triangle:

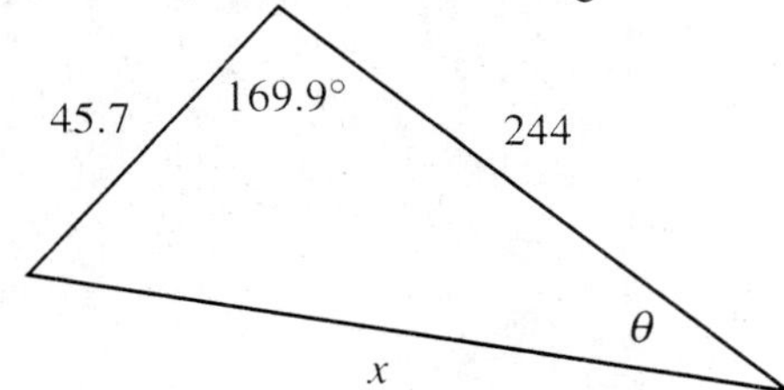

Find x using the law of cosines:
$$x^2 = 45.7^2 + 244^2 - 2(45.7)(244)\cos 169.9° \approx 83580.49$$
$$x = \sqrt{83580.49} \approx 289 \text{ mph}$$
Find θ using the law of sines:
$$\frac{\sin\theta}{45.7} = \frac{\sin 169.9°}{289}$$
$$\sin\theta = \frac{45.7\sin 169.9°}{289} \approx 0.0277$$
$$\theta \approx 1.6°$$
The ground speed is approximately 289 mph, and the true course is $272.7° - 1.6° = 271.1°$ from due north.

28. Draw the figure, where **P** represents the plane, **W** represents the wind, and **T** represents the true course:

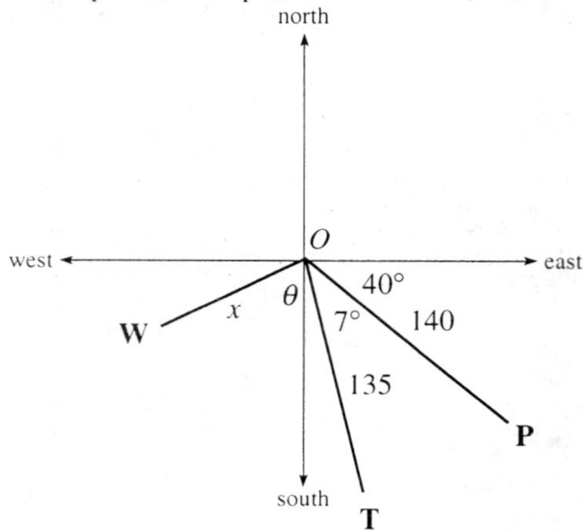

We have the triangle:

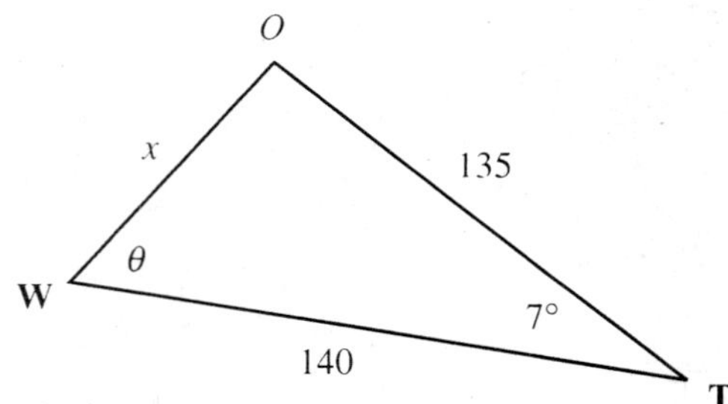

Find x using the law of cosines:

$$x^2 = 135^2 + 140^2 - 2(135)(140)\cos 7° \approx 306.76$$

$$x = \sqrt{306.76} \approx 17.5 \text{ mph}$$

Find θ using the law of sines:

$$\frac{\sin\theta}{135} = \frac{\sin 7°}{17.5}$$

$$\sin\theta = \frac{135\sin 7°}{17.5} \approx 0.9401$$

$$\theta \approx 70°$$

The wind speed is approximately 18 mph, and the wind direction is $270° - 70° + 40° = 240°$ from due north.

30. Let x represent the length of the down tube. Using the law of cosines:

$$x^2 = 55.0^2 + 53.5^2 - 2(55.0)(53.5)\cos 78.0° \approx 4663.69$$

$$x = \sqrt{4663.69} \approx 68.3$$

The down tube is approximately 68.3 cm long.

32. Let x represent the length of the down tube. Using the law of cosines:

$$x^2 = 56.9^2 + 57.0^2 - 2(56.9)(57.0)\cos 73.0° \approx 4590.11$$

$$x = \sqrt{4590.11} \approx 67.8$$

The down tube is approximately 67.8 cm long. Let θ represent the angle between the seat tube and the down tube. Using the law of sines:

$$\frac{\sin\theta}{56.9} = \frac{\sin 73.0°}{67.8}$$

$$\sin\theta = \frac{56.9\sin 73.0°}{67.8} \approx 0.8026$$

$$\theta = \sin^{-1}(0.8026) \approx 53.4°$$

The angle between the seat tube and the down tube is approximately 53.4°.

34. The radian solutions are:

$$4x = \frac{\pi}{2} + 2k\pi \qquad\qquad 4x = \frac{3\pi}{2} + 2k\pi$$

$$x = \frac{\pi}{8} + \frac{k\pi}{2} \qquad\qquad x = \frac{3\pi}{8} + \frac{k\pi}{2}$$

Note that these solutions combine as $x = \dfrac{3\pi}{8} + \dfrac{k\pi}{4}$, where k is any integer.

36. Since $\tan^2 4x = 1$, $\tan 4x = \pm 1$. The radian solutions are:

$$4x = \frac{\pi}{4} + k\pi \qquad\qquad 4x = \frac{3\pi}{4} + k\pi$$

$$x = \frac{\pi}{16} + \frac{k\pi}{4} \qquad\qquad x = \frac{3\pi}{16} + \frac{k\pi}{4}$$

Note that these solutions combine as $x = \dfrac{\pi}{16} + \dfrac{k\pi}{8}$, where k is any integer.

38. Factoring the equation:

$$3\sin^2 2\theta - 2\sin 2\theta - 5 = 0$$
$$(3\sin 2\theta - 5)(\sin 2\theta + 1) = 0$$
$$\sin 2\theta = \tfrac{5}{3}, -1$$

Note that $\sin 2\theta = \frac{5}{3}$ has no solutions. For $\sin 2\theta = -1$:

$$2\theta = 270° + 360°k$$
$$\theta = 135° + 180°k$$

40. Using the addition formula for cosine:

$$\cos 3\theta \cos 2\theta - \sin 3\theta \sin 2\theta = -1$$
$$\cos(3\theta + 2\theta) = -1$$
$$\cos 5\theta = -1$$
$$5\theta = 180° + 360°k$$
$$\theta = 36° + 72°k$$

42. If $\sin\theta - \cos\theta = 0$, $\sin\theta = \cos\theta$ and thus $\tan\theta = 1$. Thus $\theta = 45°, 225°$.

44. Draw the figure:

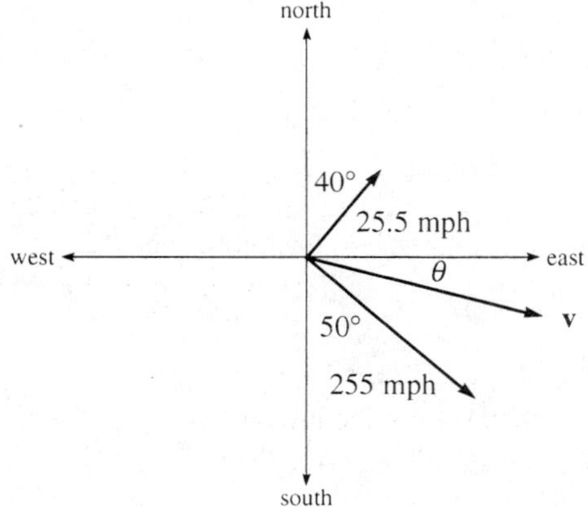

The resultant velocity vector has components:

$$\mathbf{v} = \langle 25.5\sin 40° + 255\sin 50°, 25.5\cos 40° - 255\cos 50°\rangle \approx \langle 211.73, -144.38\rangle$$

The ground speed is given by: $|\mathbf{v}| = \sqrt{(211.73)^2 + (-144.38)^2} = \sqrt{65675.18} \approx 256$ mph

To find the true course, first find θ:

$$\tan\theta = \frac{144.38}{211.73} \approx 0.6819$$
$$\theta = \tan^{-1}(0.6819) \approx 34°$$

The true course of the plane is $90° + 34° = 124°$ from due north.

46.   Sketching the vectors:

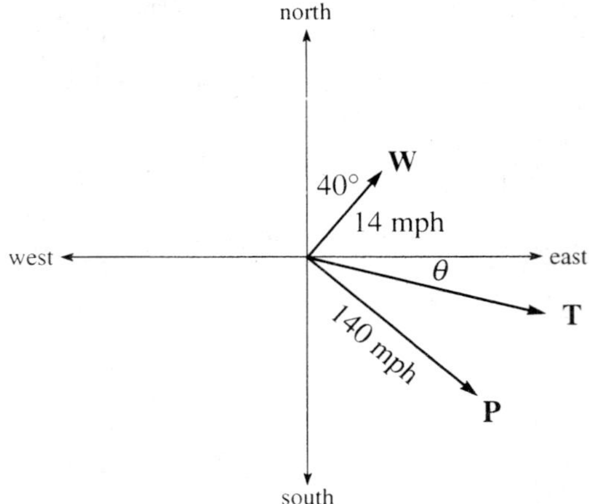

Computing the horizontal and vertical components:

$\mathbf{W}_x = 14\cos 50° \approx 9.00$ $\qquad\qquad$ $\mathbf{W}_y = 14\sin 50° \approx 10.72$

$\mathbf{P}_x = 140\cos 40° \approx 107.25$ $\qquad\qquad$ $\mathbf{P}_y = -140\sin 40° \approx -89.99$

So the true course vector has components:

$\mathbf{T}_x = 9.00 + 107.25 = 116.25$ $\qquad\qquad$ $\mathbf{T}_y = 10.72 - 89.99 = -79.27$

The ground speed is therefore given by: $|\mathbf{T}| = \sqrt{116.25^2 + (-79.27)^2} \approx 141$ mi / hr. Now find angle θ:

$$\tan\theta = \frac{79.27}{116.25} \approx 0.6819$$

$$\theta = \tan^{-1}(0.6865) \approx 34.3°$$

$$90° - \theta = 55.7°$$

The true course of the plane is S 55.7° E, or direction 124.3° from due north.

7.4 The Area of a Triangle

2.   Using the area formula: $S = \frac{1}{2}ab\sin C = \frac{1}{2}(10)(12)\sin 120° \approx 52.0$ cm^2

4.   Using the area formula: $S = \frac{1}{2}ac\sin B = \frac{1}{2}(76.3)(42.8)\sin 16.3° \approx 458$ m^2

6.   Using the area formula: $S = \frac{1}{2}bc\sin A = \frac{1}{2}(63.4)(75.2)\sin 124°40' \approx 1960$ km^2

8.   Note that $A = 180° - (57° + 31°) = 92°$. Using the area formula: $S = \dfrac{a^2 \sin B \sin C}{2\sin A} = \dfrac{7.3^2 \sin 57° \sin 31°}{2\sin 92°} \approx 11.5$ m^2

10.   Note that $B = 180° - (110.4° + 21.8°) = 47.8°$. Using the area formula:

$$S = \frac{c^2 \sin A \sin B}{2\sin C} = \frac{240^2 \sin 110.4° \sin 47.8°}{2\sin 21.8°} \approx 53,800 \text{ in.}^2$$

12.   Note that $A = 180° - (14°20' + 75°40') = 90°$. Using the area formula:

$$S = \frac{b^2 \sin A \sin C}{2\sin B} = \frac{2.72^2 \sin 90° \sin 75°40'}{2\sin 14°20'} \approx 14.5 \text{ ft}^2$$

14.   Compute $s = \frac{1}{2}(a + b + c) = \frac{1}{2}(23 + 34 + 45) = 51$. Using Heron's formula:

$$S = \sqrt{s(s-a)(s-b)(s-c)} = \sqrt{51(51-23)(51-34)(51-45)} \approx 382 \text{ in.}^2$$

16.   Compute $s = \frac{1}{2}(a + b + c) = \frac{1}{2}(48 + 75 + 63) = 93$. Using Heron's formula:

$$S = \sqrt{s(s-a)(s-b)(s-c)} = \sqrt{93(93-48)(93-75)(93-63)} \approx 1500 \text{ yd}^2$$

18.   Compute $s = \frac{1}{2}(a + b + c) = \frac{1}{2}(8.32 + 6.23 + 3.45) = 9$. Using Heron's formula:

$$S = \sqrt{s(s-a)(s-b)(s-c)} = \sqrt{9(9-8.32)(9-6.23)(9-3.45)} \approx 9.70 \text{ ft}^2$$

20. Use $a = 24.1$, $b = 31.4$, and $c = 32.4$. Compute $s = \frac{1}{2}(a+b+c) = \frac{1}{2}(24.1+31.4+32.4) = 43.95$. Using Heron's formula: $S = \sqrt{s(s-a)(s-b)(s-c)} = \sqrt{43.95(43.95-24.1)(43.95-31.4)(43.95-32.4)} \approx 355.609$ in.^{2}

Since the parallelogram consists of two of these triangles, its area is approximately $2(355.609) \approx 711$ in.^{2}.

22. Note that $B = 180° - (25° + 110°) = 45°$. Using the area formula:

$$S = \frac{b^2 \sin A \sin C}{2 \sin B}$$

$$80 = \frac{b^2 \sin 25° \sin 110°}{2 \sin 45°}$$

$$160 \sin 45° = b^2 \sin 25° \sin 110°$$

$$b^2 = \frac{160 \sin 45°}{\sin 25° \sin 110°} \approx 284.89$$

$$b \approx \sqrt{284.89} \approx 16.9 \text{ in.}$$

24. Since $\sin t = -y$ and $\cos t = x$:

$$\sin^2 t + \cos^2 t = 1$$

$$(-y)^2 + x^2 = 1$$

$$x^2 + y^2 = 1$$

The graph is a circle with center $= (0,0)$ and radius $= 1$:

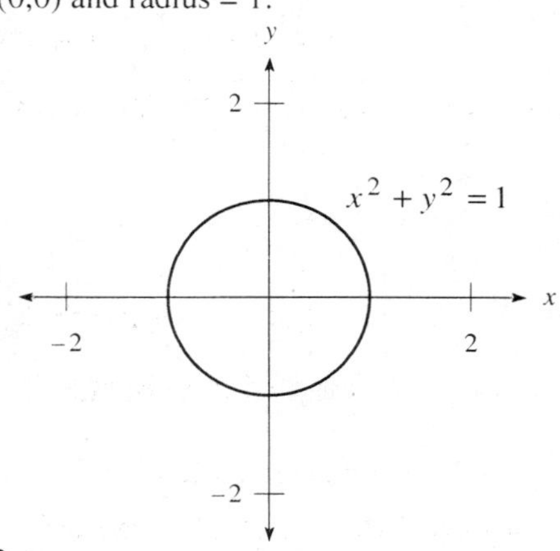

26. Since $\cos t = x + 3$ and $\sin t = y - 2$:

$$\sin^2 t + \cos^2 t = 1$$

$$(y-2)^2 + (x+3)^2 = 1$$

$$(x+3)^2 + (y-2)^2 = 1$$

The graph is a circle with center $= (-3,2)$ and radius $= 1$:

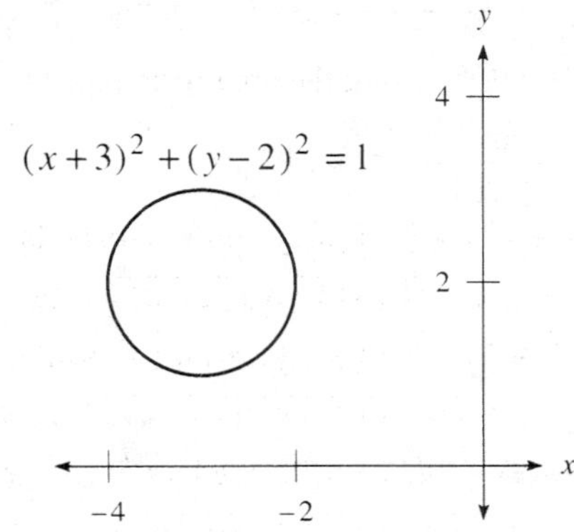

28. Since $x = 4\cot t$, $\cot t = \dfrac{x}{4}$. Since $y = 2\csc t$, $\csc t = \dfrac{y}{2}$. Therefore:

$$\cot^2 t + 1 = \csc^2 t$$

$$\left(\frac{x}{4}\right)^2 + 1 = \left(\frac{y}{2}\right)^2$$

$$\frac{x^2}{16} + 1 = \frac{y^2}{4}$$

$$\frac{y^2}{4} - \frac{x^2}{16} = 1$$

The graph would be a hyperbola.

30. Using the double-angle formula for cosine: $y = \cos 2t = 2\cos^2 t - 1$. Since $x = \cos t$, the equation is $y = 2x^2 - 1$.
The graph would be a portion of a parabola.

7.5 Vectors: An Algebraic Approach

2. Write $\mathbf{V} = \langle 1, 4 \rangle$. Drawing the vector:

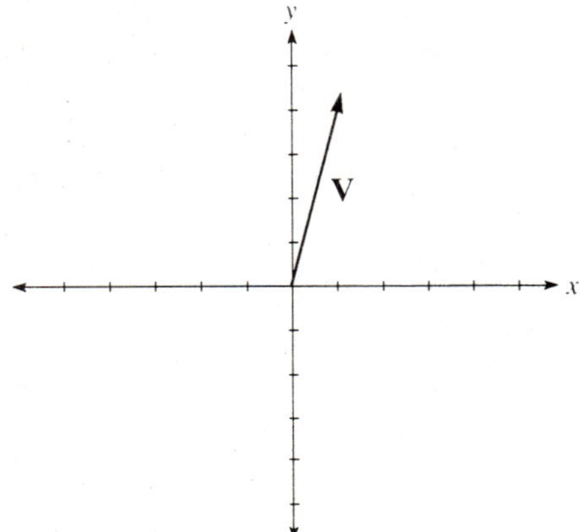

4. Write $\mathbf{V} = \langle -2, 5 \rangle$. Drawing the vector:

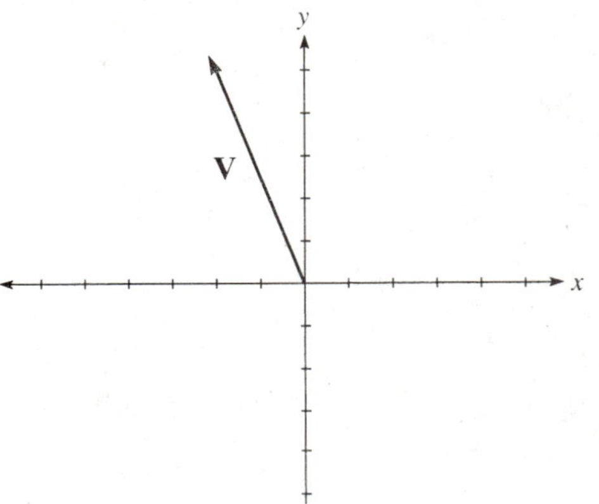

6. Write $\mathbf{V} = \langle 5, -5 \rangle$. Drawing the vector:

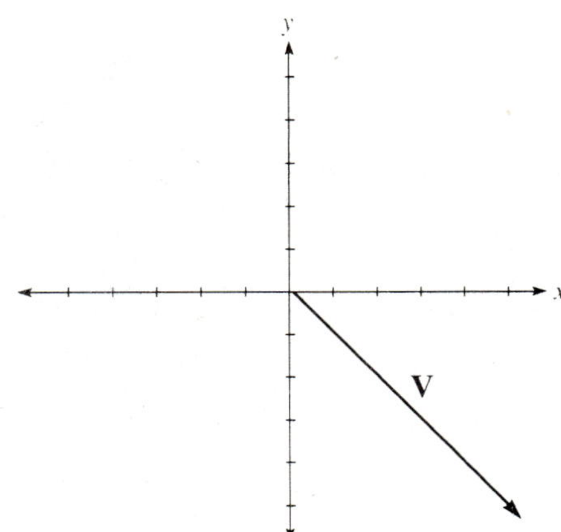

8. Write $\mathbf{V} = \langle -4, -6 \rangle$. Drawing the vector:

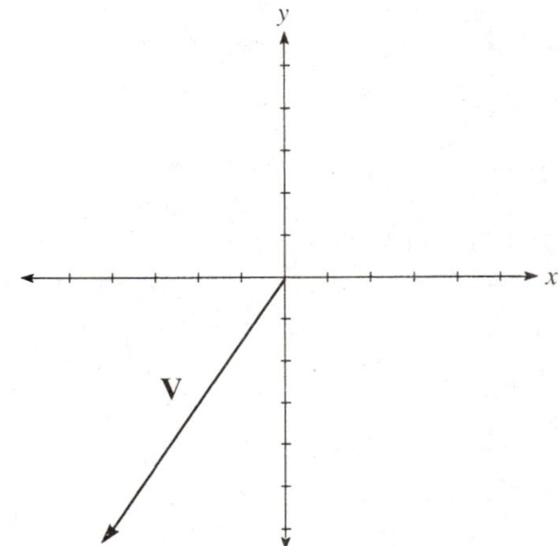

10. Write $V = 5\mathbf{i} + 2\mathbf{j}$. Drawing the vector:

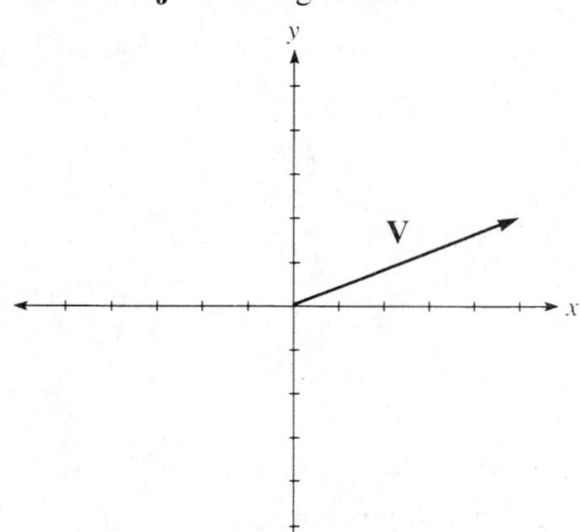

12. Write $V = -6\mathbf{i} + 3\mathbf{j}$. Drawing the vector:

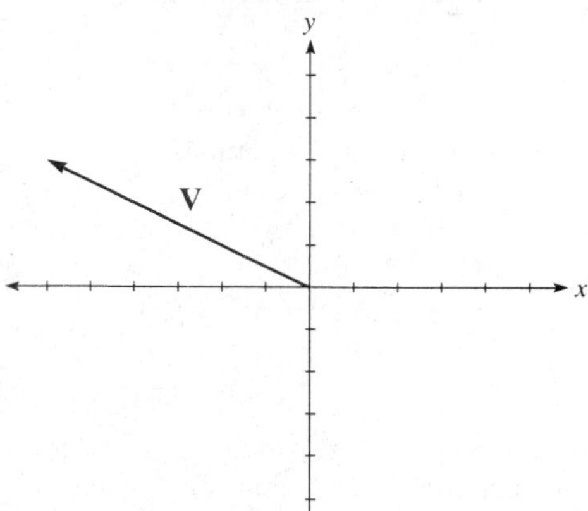

14. Write $V = 5\mathbf{i} - 4\mathbf{j}$. Drawing the vector:

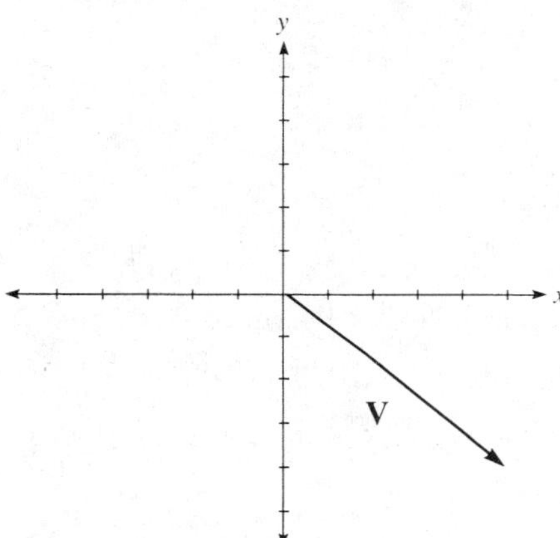

16. Write $V = -5\mathbf{i} - \mathbf{j}$. Drawing the vector:

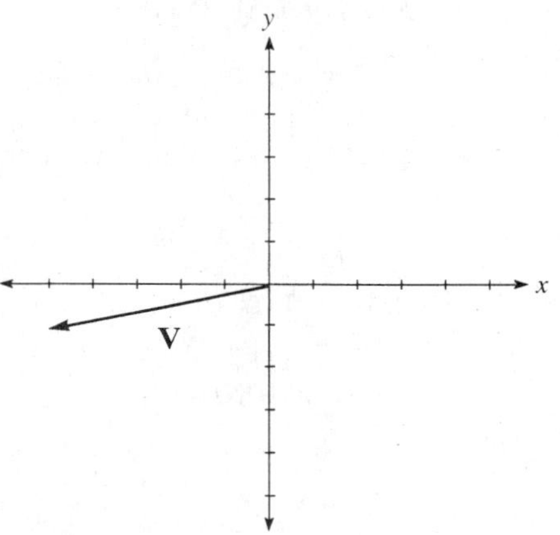

18. Let $V = \langle -3, 7 \rangle$. The magnitude is given by: $|V| = \sqrt{(-3)^2 + 7^2} = \sqrt{9 + 49} = \sqrt{58}$

20. Let $V = \langle 0, 5 \rangle$. The magnitude is given by: $|V| = \sqrt{0^2 + 5^2} = \sqrt{0 + 25} = \sqrt{25} = 5$

22. Let $V = \langle -8, -3 \rangle$. The magnitude is given by: $|V| = \sqrt{(-8)^2 + (-3)^2} = \sqrt{64 + 9} = \sqrt{73}$

24. The magnitude is given by: $|V| = \sqrt{6^2 + 8^2} = \sqrt{36 + 64} = \sqrt{100} = 10$

26. The magnitude is given by: $|U| = \sqrt{20^2 + (-21)^2} = \sqrt{400 + 441} = \sqrt{841} = 29$

28. The magnitude is given by: $|W| = \sqrt{3^2 + 1^2} = \sqrt{9 + 1} = \sqrt{10}$

30. Finding the indicated quantities:
$$U + V = \langle -4, 4 \rangle + \langle 4, 4 \rangle = \langle 0, 8 \rangle$$
$$U - V = \langle -4, 4 \rangle - \langle 4, 4 \rangle = \langle -8, 0 \rangle$$
$$2U - 3V = 2\langle -4, 4 \rangle - 3\langle 4, 4 \rangle = \langle -8, 8 \rangle - \langle 12, 12 \rangle = \langle -20, -4 \rangle$$

32. Finding the indicated quantities:
$$U + V = \langle -5, 0 \rangle + \langle 0, 1 \rangle = \langle -5, 1 \rangle$$
$$U - V = \langle -5, 0 \rangle - \langle 0, 1 \rangle = \langle -5, -1 \rangle$$
$$2U - 3V = 2\langle -5, 0 \rangle - 3\langle 0, 1 \rangle = \langle -10, 0 \rangle - \langle 0, 3 \rangle = \langle -10, -3 \rangle$$

34. Finding the indicated quantities:
$$U + V = \langle 1, 4 \rangle + \langle -2, 5 \rangle = \langle -1, 9 \rangle$$
$$U - V = \langle 1, 4 \rangle - \langle -2, 5 \rangle = \langle 3, -1 \rangle$$
$$2U - 3V = 2\langle 1, 4 \rangle - 3\langle -2, 5 \rangle = \langle 2, 8 \rangle - \langle -6, 15 \rangle = \langle 8, -7 \rangle$$

36. Finding the indicated quantities:
$$\mathbf{U} + \mathbf{V} = (-\mathbf{i} + \mathbf{j}) + (\mathbf{i} + \mathbf{j}) = 2\mathbf{j}$$
$$\mathbf{U} - \mathbf{V} = (-\mathbf{i} + \mathbf{j}) - (\mathbf{i} + \mathbf{j}) = -\mathbf{i} + \mathbf{j} - \mathbf{i} - \mathbf{j} = -2\mathbf{i}$$
$$3\mathbf{U} + 2\mathbf{V} = 3(-\mathbf{i} + \mathbf{j}) + 2(\mathbf{i} + \mathbf{j}) = -3\mathbf{i} + 3\mathbf{j} + 2\mathbf{i} + 2\mathbf{j} = -\mathbf{i} + 5\mathbf{j}$$

38. Finding the indicated quantities:
$$\mathbf{U} + \mathbf{V} = -3\mathbf{i} + 5\mathbf{j}$$
$$\mathbf{U} - \mathbf{V} = -3\mathbf{i} - 5\mathbf{j}$$
$$3\mathbf{U} + 2\mathbf{V} = 3(-3\mathbf{i}) + 2(5\mathbf{j}) = -9\mathbf{i} + 10\mathbf{j}$$

40. Finding the indicated quantities:
$$\mathbf{U} + \mathbf{V} = (5\mathbf{i} + 3\mathbf{j}) + (3\mathbf{i} + 5\mathbf{j}) = 8\mathbf{i} + 8\mathbf{j}$$
$$\mathbf{U} - \mathbf{V} = (5\mathbf{i} + 3\mathbf{j}) - (3\mathbf{i} + 5\mathbf{j}) = 5\mathbf{i} + 3\mathbf{j} - 3\mathbf{i} - 5\mathbf{j} = 2\mathbf{i} - 2\mathbf{j}$$
$$3\mathbf{U} + 2\mathbf{V} = 3(5\mathbf{i} + 3\mathbf{j}) + 2(3\mathbf{i} + 5\mathbf{j}) = 15\mathbf{i} + 9\mathbf{j} + 6\mathbf{i} + 10\mathbf{j} = 21\mathbf{i} + 19\mathbf{j}$$

42. The vector is: $\mathbf{U} = \langle 25\cos 110°, 25\sin 110° \rangle \approx \langle -8.6, 23 \rangle = -8.6\mathbf{i} + 23\mathbf{j}$ (two significant figures)

44. The vector is: $\mathbf{F} = \langle 30\cos 285°, 30\sin 285° \rangle \approx \langle 7.8, -29 \rangle = 7.8\mathbf{i} - 29\mathbf{j}$ (two significant figures)

46. The magnitude is given by: $|\mathbf{V}| = \sqrt{5^2 + (-5)^2} = \sqrt{25 + 25} = \sqrt{50} = 5\sqrt{2}$

Since $\tan\theta = \dfrac{-5}{5} = -1$ and θ lies in the fourth quadrant, $\theta = 315°$.

48. The magnitude is given by: $|\mathbf{F}| = \sqrt{\left(-2\sqrt{3}\right)^2 + (2)^2} = \sqrt{12 + 4} = \sqrt{16} = 4$

Since $\tan\theta = \dfrac{2}{-2\sqrt{3}} = -\dfrac{1}{\sqrt{3}}$ and θ lies in the second quadrant, $\theta = 150°$.

50. Write $\mathbf{W}$ as $\langle 0, -25 \rangle$, $\mathbf{F}$ as $\langle |\mathbf{F}|\cos 10°, |\mathbf{F}|\sin 10° \rangle$, and $\mathbf{N}$ as $\langle -|\mathbf{N}|\cos 80°, |\mathbf{N}|\sin 80° \rangle$. Since $\mathbf{W} + \mathbf{F} + \mathbf{N} = \mathbf{0}$, we have:
$$0 + |\mathbf{F}|\cos 10° - |\mathbf{N}|\cos 80° = 0$$
$$-25 + |\mathbf{F}|\sin 10° + |\mathbf{N}|\sin 80° = 0$$
These simplify to the equations:
$$|\mathbf{F}|\cos 10° - |\mathbf{N}|\cos 80° = 0$$
$$|\mathbf{F}|\sin 10° + |\mathbf{N}|\sin 80° = 25$$

Solving the first equation for $|\mathbf{F}|$ results in $|\mathbf{F}| = \dfrac{|\mathbf{N}|\cos 80°}{\cos 10°}$. Substituting into the second equation:
$$\frac{|\mathbf{N}|\cos 80°}{\cos 10°} \cdot \sin 10° + |\mathbf{N}|\sin 80° = 25$$
$$|\mathbf{N}|\cos 80°\sin 10° + |\mathbf{N}|\sin 80°\cos 10° = 25\cos 10°$$
$$|\mathbf{N}|(\sin 80°\cos 10° + \cos 80°\sin 10°) = 25\cos 10°$$
$$|\mathbf{N}|\sin(80° + 10°) = 25\cos 10°$$
$$|\mathbf{N}| = \frac{25\cos 10°}{\sin 90°} = 25\cos 10° \approx 25 \text{ lb}$$

Therefore: $|\mathbf{F}| = \dfrac{|\mathbf{N}|\cos 80°}{\cos 10°} = \dfrac{25\cos 10°\cos 80°}{\cos 10°} = 25\cos 80° \approx 4.3 \text{ lb}$

52.   Write $\mathbf{W}$ as $\langle 0, -58 \rangle$, $\mathbf{F}$ as $\langle |\mathbf{F}|\cos 8.5°, |\mathbf{F}|\sin 8.5° \rangle$, and $\mathbf{N}$ as $\langle -|\mathbf{N}|\cos 81.5°, |\mathbf{N}|\sin 81.5° \rangle$. Since $\mathbf{W} + \mathbf{F} + \mathbf{N} = \mathbf{0}$, we have:

$$0 + |\mathbf{F}|\cos 8.5° - |\mathbf{N}|\cos 81.5° = 0$$
$$-58 + |\mathbf{F}|\sin 8.5° + |\mathbf{N}|\sin 81.5° = 0$$

These simplify to the equations:

$$|\mathbf{F}|\cos 8.5° - |\mathbf{N}|\cos 81.5° = 0$$
$$|\mathbf{F}|\sin 8.5° + |\mathbf{N}|\sin 81.5° = 58$$

Solving the first equation for $|\mathbf{N}|$ results in $|\mathbf{N}| = \dfrac{|\mathbf{F}|\cos 8.5°}{\cos 81.5°}$. Substituting into the second equation:

$$|\mathbf{F}|\sin 8.5° + \frac{|\mathbf{F}|\cos 8.5°}{\cos 81.5°}\sin 81.5° = 58$$
$$|\mathbf{F}|\sin 8.5°\cos 81.5° + |\mathbf{F}|\cos 8.5°\sin 81.5°° = 58\cos 81.5°$$
$$|\mathbf{F}|(\sin 8.5°\cos 81.5° + \cos 81.5°\sin 8.5°) = 58\cos 81.5°$$
$$|\mathbf{F}|\sin(8.5° + 81.5°) = 58\cos 81.5°$$
$$|\mathbf{F}| = \frac{58\cos 81.5°}{\sin 90°} = 58\cos 81.5° \approx 8.57 \text{ lb}$$

Tyler must push with a force of approximately 8.57 lb.

54.   Re-draw the figure as follows:

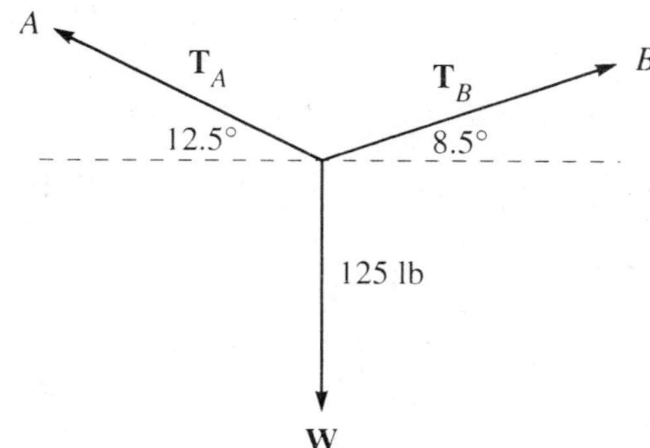

Write $\mathbf{W}$ as $\langle 0, -125 \rangle$, $\mathbf{T}_A$ as $\langle -|\mathbf{T}_A|\cos 12.5°, |\mathbf{T}_A|\sin 12.5° \rangle$, and $\mathbf{T}_B$ as $\langle |\mathbf{T}_B|\cos 8.5°, |\mathbf{T}_B|\sin 8.5° \rangle$. Since $\mathbf{W} + \mathbf{T}_A + \mathbf{T}_B = \mathbf{0}$, we have:

$$0 - |\mathbf{T}_A|\cos 12.5° + |\mathbf{T}_B|\cos 8.5° = 0$$
$$-125 + |\mathbf{T}_A|\sin 12.5° + |\mathbf{T}_B|\sin 8.5° = 0$$

These simplify to the equations:

$$|\mathbf{T}_B|\cos 8.5° - |\mathbf{T}_A|\cos 12.5° = 0$$
$$|\mathbf{T}_B|\sin 8.5° + |\mathbf{T}_A|\sin 12.5° = 125$$

Solving the first equation for $\left|\mathbf{T}_B\right|$ results in $\left|\mathbf{T}_B\right| = \dfrac{\left|\mathbf{T}_A\right|\cos 12.5°}{\cos 8.5°}$. Substituting into the second equation:

$$\frac{\left|\mathbf{T}_A\right|\cos 12.5°}{\cos 8.5°} \cdot \sin 8.5° + \left|\mathbf{T}_A\right|\sin 12.5° = 125$$

$$\left|\mathbf{T}_A\right|\cos 12.5°\sin 8.5° + \left|\mathbf{T}_A\right|\sin 12.5°\cos 8.5° = 125\cos 8.5°$$

$$\left|\mathbf{T}_A\right|(\cos 12.5°\sin 8.5° + \sin 12.5°\cos 8.5°) = 125\cos 8.5°$$

$$\left|\mathbf{T}_A\right|\sin(12.5° + 8.5°) = 125\cos 8.5°$$

$$\left|\mathbf{T}_A\right|\sin 21° = 125\cos 8.5°$$

$$\left|\mathbf{T}_A\right| = \frac{125\cos 8.5°}{\sin 21°} \approx 345 \text{ lb}$$

Therefore: $\left|\mathbf{T}_B\right| = \dfrac{125\cos 8.5°\cos 12.5°}{\sin 21°\cos 8.5°} = \dfrac{125\cos 12.5°}{\sin 21°} \approx 341 \text{ lb}$

The tensions at A and B are 345 lb and 341 lb, respectively.

7.6 Vectors: The Dot Product

2.   Finding the dot product: $\langle 3,4\rangle \bullet \langle 5,5\rangle = (3)(5) + (4)(5) = 15 + 20 = 35$

4.   Finding the dot product: $\langle 11,-8\rangle \bullet \langle 4,-7\rangle = (11)(4) + (-8)(-7) = 44 + 56 = 100$

6.   Finding the dot product: $(-\mathbf{i} + \mathbf{j}) \bullet (\mathbf{i} + \mathbf{j}) = (-1)(1) + (1)(1) = -1 + 1 = 0$

8.   Finding the dot product: $(-3\mathbf{i}) \bullet (5\mathbf{j}) = (-3)(0) + (0)(5) = 0 + 0 = 0$

10.   Finding the dot product: $(5\mathbf{i} + 3\mathbf{j}) \bullet (3\mathbf{i} + 5\mathbf{j}) = (5)(3) + (3)(5) = 15 + 15 = 30$

12.   Using $\cos\theta = \dfrac{\mathbf{U} \bullet \mathbf{V}}{|\mathbf{U}||\mathbf{V}|}$ to find the angle:

$$\cos\theta = \frac{(-4)(0) + (0)(17)}{\sqrt{(-4)^2 + 0^2}\sqrt{0^2 + 17^2}} = \frac{0+0}{\sqrt{16}\sqrt{289}} = 0$$

$$\theta = 90°$$

14.   Using $\cos\theta = \dfrac{\mathbf{U} \bullet \mathbf{V}}{|\mathbf{U}||\mathbf{V}|}$ to find the angle:

$$\cos\theta = \frac{(-3)(6) + (5)(3)}{\sqrt{(-3)^2 + 5^2}\sqrt{6^2 + 3^2}} = \frac{-18+15}{\sqrt{34}\sqrt{45}} = \frac{-3}{\sqrt{1580}} \approx -0.0767$$

$$\theta = 94.4°$$

16.   Using $\cos\theta = \dfrac{\mathbf{U} \bullet \mathbf{V}}{|\mathbf{U}||\mathbf{V}|}$ to find the angle:

$$\cos\theta = \frac{(11)(-4) + (7)(6)}{\sqrt{11^2 + 7^2}\sqrt{(-4)^2 + 6^2}} = \frac{-44+42}{\sqrt{170}\sqrt{52}} = \frac{-2}{\sqrt{8840}} \approx -0.0213$$

$$\theta = 91.2°$$

18.   Compute their dot product: $\langle 1,1\rangle \bullet \langle 1,-1\rangle = (1)(1) + (1)(-1) = 1 - 1 = 0$
Thus $\cos\theta = 0$, so $\theta = 90°$. So $\mathbf{i} + \mathbf{j}$ and $\mathbf{i} - \mathbf{j}$ are perpendicular.

20.   Compute their dot product: $\langle 2,1\rangle \bullet \langle 1,-2\rangle = (2)(1) + (1)(-2) = 2 - 2 = 0$
Thus $\cos\theta = 0$, so $\theta = 90°$. So $2\mathbf{i} + \mathbf{j}$ and $\mathbf{i} - 2\mathbf{j}$ are perpendicular.

22.   Since the dot product must equal 0:

$$(a\mathbf{i} + 6\mathbf{j}) \bullet (9\mathbf{i} + 12\mathbf{j}) = 0$$

$$9a + 72 = 0$$

$$9a = -72$$

$$a = -8$$

24.   The work is: $\mathbf{F} \cdot \mathbf{d} = (45\mathbf{i} - 12\mathbf{j}) \cdot (70\mathbf{i} + 15\mathbf{j}) = (45)(70) + (-12)(15) = 2970$ ft - lb

26.   The work is: $\mathbf{F} \cdot \mathbf{d} = (-6\mathbf{i} + 19\mathbf{j}) \cdot (8\mathbf{i} + 55\mathbf{j}) = (-6)(8) + (19)(55) = 997$ ft - lb

28.   The work is: $\mathbf{F} \cdot \mathbf{d} = (54\mathbf{i} + 0\mathbf{j}) \cdot (20\mathbf{i} + 0\mathbf{j}) = (54)(20) + (0)(0) = 1080$ ft - lb

30.   The work is: $\mathbf{F} \cdot \mathbf{d} = (0\mathbf{i} + 13\mathbf{j}) \cdot (44\mathbf{i} + 0\mathbf{j}) = (0)(44) + (13)(0) = 0$ ft - lb

32.   Assume $\mathbf{U} \cdot \mathbf{V} = 0$. Since $\mathbf{U} \cdot \mathbf{V} = |\mathbf{U}||\mathbf{V}|\cos\theta$, $|\mathbf{U}||\mathbf{V}|\cos\theta = 0$. Since $\mathbf{U}$ and $\mathbf{V}$ are nonzero vectors, $\cos\theta = 0$ and thus $\theta = 90°$. Thus $\mathbf{U}$ is perpendicular to $\mathbf{V}$. Assume $\mathbf{U}$ is perpendicular to $\mathbf{V}$. Then $\theta = 90°$, so $\cos\theta = 0$. So $|\mathbf{U}||\mathbf{V}|\cos\theta = |\mathbf{U}||\mathbf{V}| \cdot 0 = 0$, so $\mathbf{U} \cdot \mathbf{V} = 0$. This proves both directions of the theorem.

34.   The force vector is $\mathbf{F} = \langle 15\cos 25°, -15\sin 25° \rangle$ and the direction vector is $\mathbf{d} = \langle 50, 0 \rangle$.
The work is: $\mathbf{F} \cdot \mathbf{d} = \langle 15\cos 25°, -15\sin 25° \rangle \cdot \langle 50, 0 \rangle = 750\cos 25° \approx 680$ ft - lb

36.   The force vector is $\mathbf{F} = \langle 25\cos 30°, 25\sin 30° \rangle$ and the direction vector is $\mathbf{d} = \langle 300, 0 \rangle$.
The work is: $\mathbf{F} \cdot \mathbf{d} = \langle 25\cos 30°, 25\sin 30° \rangle \cdot \langle 300, 0 \rangle = 7500\cos 30° \approx 6500$ ft - lb

Chapter 7 Test

1.   Using the law of sines to find b:
$$\frac{b}{\sin B} = \frac{a}{\sin A}$$
$$\frac{b}{\sin 70°} = \frac{3.8}{\sin 32°}$$
$$b = \frac{3.8\sin 70°}{\sin 32°} \approx 6.7 \text{ in.}$$

2.   Using the law of sines to find b:
$$\frac{b}{\sin B} = \frac{c}{\sin C}$$
$$\frac{b}{\sin 118°} = \frac{2.9}{\sin 37°}$$
$$b = \frac{2.9\sin 118°}{\sin 37°} \approx 4.3 \text{ in.}$$

3.   First note that $C = 180° - (38.2° + 63.4°) = 78.4°$. Using the law of sines to find a:
$$\frac{a}{\sin A} = \frac{c}{\sin C}$$
$$\frac{a}{\sin 38.2°} = \frac{42.0}{\sin 78.4°}$$
$$a = \frac{42\sin 38.2°}{\sin 78.4°} \approx 26.5 \text{ cm}$$
Using the law of sines to find b:
$$\frac{b}{\sin B} = \frac{c}{\sin C}$$
$$\frac{b}{\sin 63.4°} = \frac{42.0}{\sin 78.4°}$$
$$b = \frac{42\sin 63.4°}{\sin 78.4°} \approx 38.3 \text{ cm}$$

4.   First note that $B = 180° - (24.7° + 106.1°) = 49.2°$. Using the law of sines to find a:
$$\frac{a}{\sin A} = \frac{b}{\sin B}$$
$$\frac{a}{\sin 24.7°} = \frac{34.0}{\sin 49.2°}$$
$$a = \frac{34\sin 24.7°}{\sin 49.2°} \approx 18.8 \text{ cm}$$
Using the law of sines to find c:
$$\frac{c}{\sin C} = \frac{b}{\sin B}$$
$$\frac{c}{\sin 106.1°} = \frac{34.0}{\sin 49.2°}$$
$$c = \frac{34\sin 106.1°}{\sin 49.2°} \approx 43.2 \text{ cm}$$

5.   Finding B using the law of sines:

$$\frac{a}{\sin A} = \frac{b}{\sin B}$$

$$\frac{12}{\sin 60°} = \frac{42}{\sin B}$$

$$\sin B = \frac{42 \sin 60°}{12} \approx 3.031$$

Since $-1 \le \sin B \le 1$, this is impossible. No such triangle exists.

6.   Findng B using the law of sines:

$$\frac{a}{\sin A} = \frac{b}{\sin B}$$

$$\frac{29}{\sin 42°} = \frac{21}{\sin B}$$

$$\sin B = \frac{21 \sin 42°}{29} \approx 0.4845$$

$$B \approx 29°, 151°$$

But $B = 151°$ is impossible, since $151° + 42° = 193° > 180°$. So $B = 29°$ is the only possible value of B, thus exactly one triangle exists meeting the specified conditions.

7.   Findng B using the law of sines:

$$\frac{b}{\sin B} = \frac{a}{\sin A}$$

$$\frac{7.9}{\sin B} = \frac{6.5}{\sin 51°}$$

$$\sin B = \frac{7.9 \sin 51°}{6.5} \approx 0.9445$$

$$B = 71°, 109°$$

If $B = 71°$, $C = 180° - (51° + 71°) = 58°$. Finding c using the law of sines:

$$\frac{c}{\sin C} = \frac{a}{\sin A}$$

$$\frac{c}{\sin 58°} = \frac{6.5}{\sin 51°}$$

$$c = \frac{6.5 \sin 58°}{\sin 51°} \approx 7.1 \text{ ft}$$

If $B = 109°$, $C = 180° - (51° + 109°) = 20°$. Finding c using the law of sines:

$$\frac{c}{\sin C} = \frac{a}{\sin A}$$

$$\frac{c}{\sin 20°} = \frac{6.5}{\sin 51°}$$

$$c = \frac{6.5 \sin 20°}{\sin 51°} \approx 2.9 \text{ ft}$$

8. Finding B using the law of sines:

$$\frac{b}{\sin B} = \frac{a}{\sin A}$$

$$\frac{9.4}{\sin B} = \frac{4.8}{\sin 26°}$$

$$\sin B = \frac{9.4 \sin 26°}{4.8} \approx 0.8585$$

$$B = 59°, 121°$$

If $B = 59°$, $C = 180° - (26° + 59°) = 95°$. Finding c using the law of sines:

$$\frac{c}{\sin C} = \frac{a}{\sin A}$$

$$\frac{c}{\sin 95°} = \frac{4.8}{\sin 26°}$$

$$c = \frac{4.8 \sin 95°}{\sin 26°} \approx 11 \text{ ft}$$

If $B = 121°$, $C = 180° - (26° + 121°) = 33°$. Finding c using the law of sines:

$$\frac{c}{\sin C} = \frac{a}{\sin A}$$

$$\frac{c}{\sin 33°} = \frac{4.8}{\sin 26°}$$

$$c = \frac{4.8 \sin 33°}{\sin 26°} \approx 6.0 \text{ ft}$$

9. Finding c using the law of cosines:

$$c^2 = a^2 + b^2 - 2ab \cos C = 10^2 + 12^2 - 2(10)(12) \cos 60° = 124$$

$$c = \sqrt{124} \approx 11 \text{ cm}$$

10. Finding c using the law of cosines:

$$c^2 = a^2 + b^2 - 2ab \cos C = 10^2 + 12^2 - 2(10)(12) \cos 120° = 364$$

$$c = \sqrt{364} \approx 19 \text{ cm}$$

11. Finding C using the law of cosines:

$$c^2 = a^2 + b^2 - 2ab \cos C$$
$$9^2 = 5^2 + 7^2 - 2(5)(7) \cos C$$
$$81 = 74 - 70 \cos C$$
$$7 = -70 \cos C$$
$$\cos C = -0.1$$
$$C \approx 95.7°$$

12. Finding B using the law of cosines:

$$b^2 = a^2 + c^2 - 2ac \cos B$$
$$12^2 = 10^2 + 11^2 - 2(10)(11) \cos B$$
$$144 = 221 - 220 \cos B$$
$$-77 = -220 \cos B$$
$$\cos B \approx 0.35$$
$$B \approx 69.5°$$

13. Finding c using the law of cosines:

$$c^2 = a^2 + b^2 - 2ab \cos C = 6.4^2 + 2.8^2 - 2(6.4)(2.8) \cos 119° \approx 66.176$$

$$c = \sqrt{66.176} \approx 8.1 \text{ m}$$

Finding A using the law of sines:

$$\frac{a}{\sin A} = \frac{c}{\sin C}$$

$$\frac{6.4}{\sin A} = \frac{8.1}{\sin 119°}$$

$$\sin A = \frac{6.4 \sin 119°}{8.1} \approx 0.6911$$

$$A \approx 44°$$

Finally $B = 180° - (44° + 119°) = 17°$.

14. Finding a using the law of cosines:
$$a^2 = b^2 + c^2 - 2bc\cos A = 3.7^2 + 6.2^2 - 2(3.7)(6.2)\cos 35° \approx 14.547$$
$$a = \sqrt{14.547} \approx 3.8 \text{ m}$$
Finding B using the law of sines:
$$\frac{b}{\sin B} = \frac{a}{\sin A}$$
$$\frac{3.7}{\sin B} = \frac{3.8}{\sin 35°}$$
$$\sin B = \frac{3.7\sin 35°}{3.8} \approx 0.5585$$
$$B \approx 34°$$
Finally $C = 180° - (35° + 34°) = 111°$.

15. Using the area formula: $S = \frac{1}{2}ab\sin C = \frac{1}{2}(26.5)(38.3)\sin 78.4° \approx 497 \text{ cm}^2$

16. Using the area formula: $S = \frac{1}{2}ab\sin C = \frac{1}{2}(18.8)(34.0)\sin 106.1° \approx 307 \text{ cm}^2$

17. Using the area formula: $S = \frac{1}{2}ab\sin C = \frac{1}{2}(10)(12)\sin 60° \approx 52 \text{ cm}^2$

18. Using the area formula: $S = \frac{1}{2}ab\sin C = \frac{1}{2}(10)(12)\sin 120° \approx 52 \text{ cm}^2$

19. Compute $s = \frac{1}{2}(a+b+c) = \frac{1}{2}(5+7+9) = 10.5$. Using Heron's formula:
$$S = \sqrt{s(s-a)(s-b)(s-c)} = \sqrt{10.5(10.5-5)(10.5-7)(10.5-9)} \approx 17 \text{ km}^2$$

20. Compute $s = \frac{1}{2}(a+b+c) = \frac{1}{2}(10+12+11) = 16.5$. Using Heron's formula:
$$S = \sqrt{s(s-a)(s-b)(s-c)} = \sqrt{16.5(16.5-10)(16.5-12)(16.5-11)} \approx 52 \text{ km}^2$$

21. Drawing the figure:

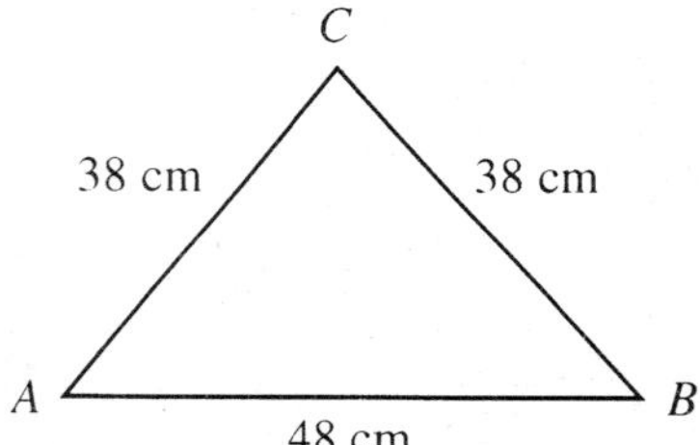

Finding A using the law of cosines:
$$a^2 = b^2 + c^2 - 2bc\cos A$$
$$38^2 = 38^2 + 48^2 - 2(38)(48)\cos A$$
$$1444 = 3748 - 3648\cos A$$
$$-2304 = -3648\cos A$$
$$\cos A \approx 0.6316$$
$$A \approx 51°$$
The two equal angles are approximately $51°$.

22.   Drawing the figure, where h represents the height of the lamp pole:

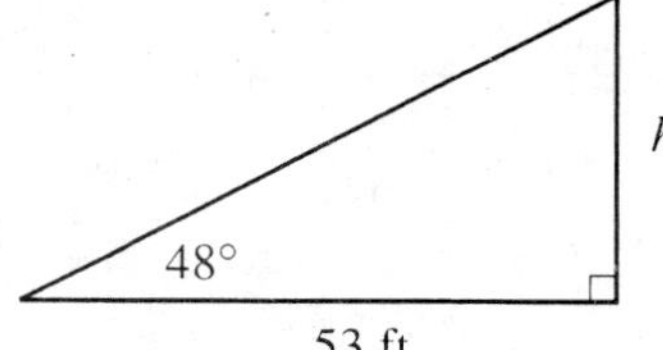

Since this is a right triangle:
$$\tan 48° = \frac{h}{53}$$
$$h = 53\tan 48° \approx 59 \text{ ft}$$
The lamp pole is approximately 59 feet tall.

23.   Drawing the figure:

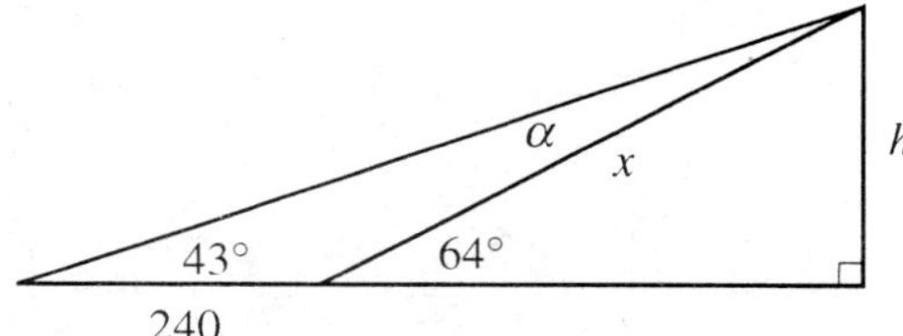

Note that $\alpha = 180° - (43° + 116°) = 21°$. Find x using the law of sines:
$$\frac{x}{\sin 43°} = \frac{240}{\sin 21°}$$
$$x = \frac{240\sin 43°}{\sin 21°} \approx 456.74 \text{ ft}$$
Therefore:
$$\sin 64° = \frac{h}{456.74}$$
$$h = 456.74\sin 64° \approx 410 \text{ ft}$$
The building is approximately 410 feet tall.

24.   Draw the figure, where x represents the length of the shorter side:

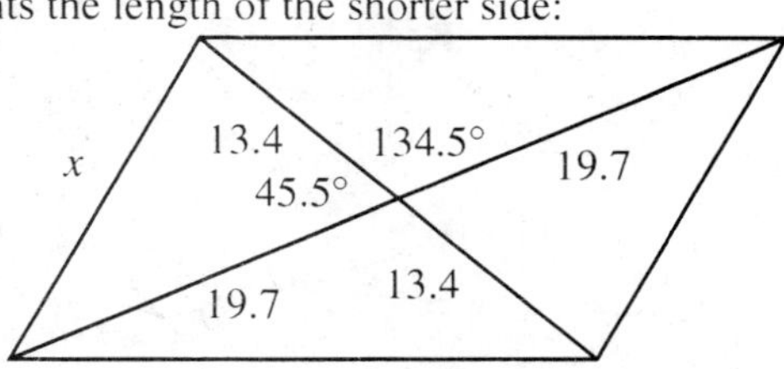

Find x using the law of cosines:
$$x^2 = 13.4^2 + 19.7^2 - 2(13.4)(19.7)\cos 45.5° \approx 197.60$$
$$x = \sqrt{197.60} \approx 14.1 \text{ km}$$
The length of the shorter side is approximately 14.1 km.

25. Re-draw the triangle to find angle C:

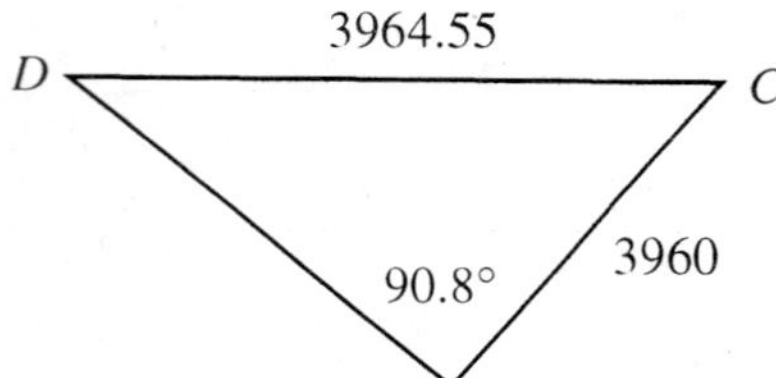

Finding D using the law of sines:
$$\frac{3960}{\sin D} = \frac{3964.55}{\sin 90.8°}$$
$$\sin D = \frac{3960 \sin 90.8°}{3964.55} \approx 0.9988$$
$$D \approx 87.14°$$

So $C = 180° - (90.8° + 87.14°) \approx 2.06°$. Converting to radians: $C = 2.06° \cdot \dfrac{\pi}{180°} = \dfrac{2.06\pi}{180}$ radians

The arc length is given by: $s = rC = 3960 \cdot \dfrac{2.06\pi}{180} \approx 142$ miles

26. Drawing the figure, where x represents his distance from the starting point:

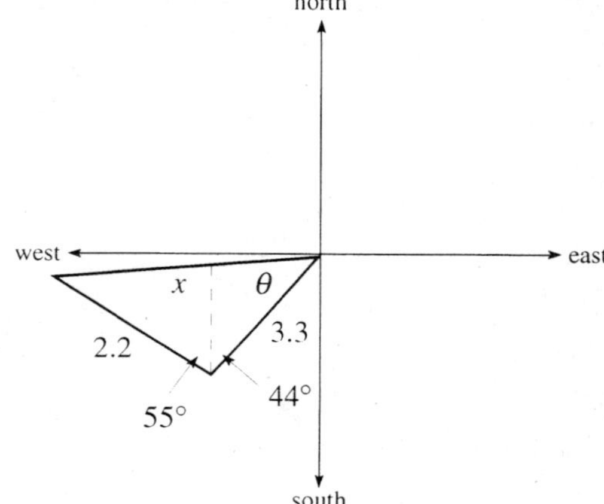

Re-draw the triangle:

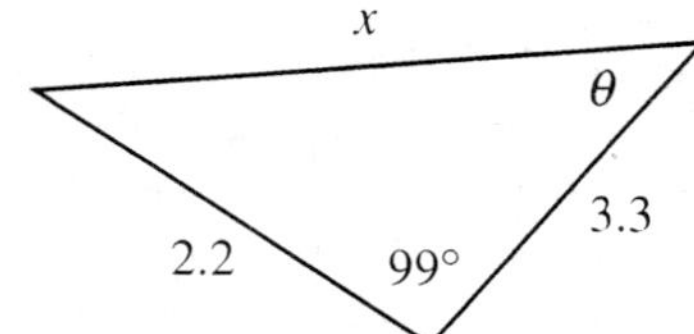

Use the law of cosines to find x:
$$x^2 = 2.2^2 + 3.3^2 - 2(2.2)(3.3)\cos 99° \approx 18.00$$
$$x = \sqrt{18.00} \approx 4.24 \text{ mi}$$
Use the law of sines to find θ:
$$\frac{2.2}{\sin \theta} = \frac{4.24}{\sin 99°}$$
$$\sin \theta = \frac{2.2 \sin 99°}{4.24} \approx 0.5125$$
$$\theta \approx 31°$$
He is approximately 4.2 miles from his starting point, with bearing S 75° W.

27.   Drawing the figure, where d represents the required distance:

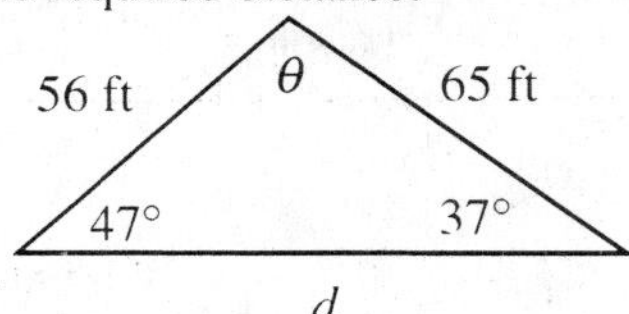

Note that $\theta = 180° - (37° + 47°) = 96°$. Finding d using the law of cosines:
$$d^2 = 56^2 + 65^2 - 2(56)(65)\cos 96° \approx 8121.97$$
$$d = \sqrt{8121.97} \approx 90 \text{ ft}$$
The stakes are approximately 90 feet apart.

28.   Drawing the figure, where P represents the plane, T represents the true course, and x represents the ground speed:

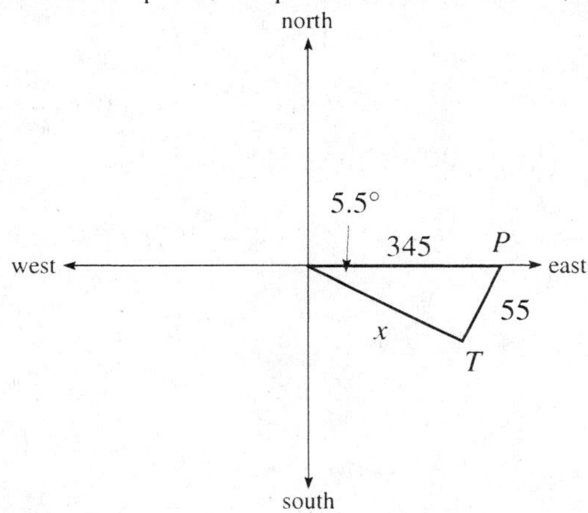

Using the law of sines to find the angle at T:
$$\frac{345}{\sin T} = \frac{55}{\sin 5.5°}$$
$$\sin T = \frac{345 \sin 5.5°}{55} \approx 0.6012$$
$$T \approx 36.96°, 143.04°$$
If $T \approx 36.96°$, $P = 180° - (36.96° + 5.5°) \approx 137.54°$. Finding x using the law of cosines:
$$x^2 = 345^2 + 55^2 - 2(345)(55)\cos 137.54° \approx 150047$$
$$x = \sqrt{150047} \approx 387 \text{ mph}$$
If $T \approx 143.04°$, $P = 180° - (143.04° + 5.5°) \approx 31.46°$. Finding x using the law of cosines:
$$x^2 = 345^2 + 55^2 - 2(345)(55)\cos 31.46° \approx 89678$$
$$x = \sqrt{89678} \approx 299 \text{ mph}$$
The possibilities for the ground speed of the plane are 299 mph or 387 mph.

29. Drawing the figure, where x represents the distance from the first person to the tree:

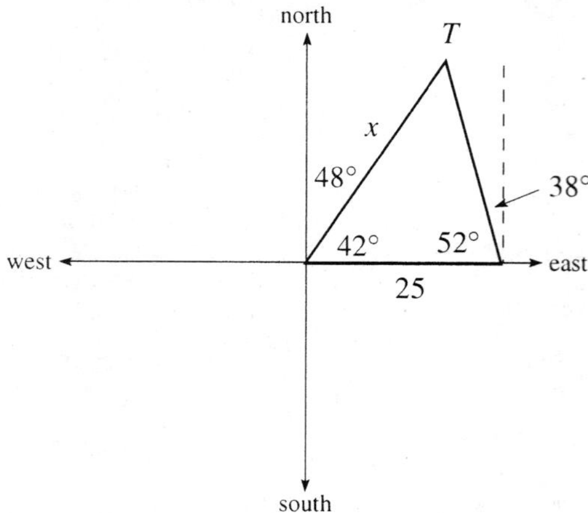

Note that $T = 180° - (42° + 52°) \approx 86°$. Finding x using the law of sines:

$$\frac{x}{\sin 52°} = \frac{25}{\sin 86°}$$

$$x = \frac{25 \sin 52°}{\sin 86°} \approx 19.75 \text{ ft}$$

Now consider the triangle where h represents the height of the tree:

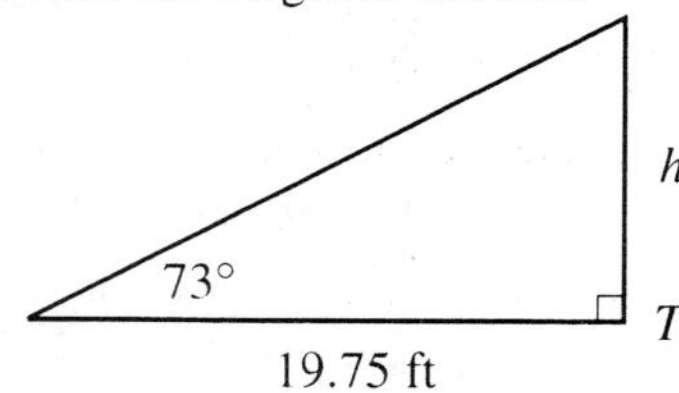

Since this is a right triangle (we are assuming the tree is vertical):

$$\tan 73° = \frac{h}{19.75}$$

$$h = 19.75 \tan 73° \approx 65 \text{ ft}$$

The tree is approximately 65 feet tall.

30. Drawing the figure, where P represents the plane and W represents the wind:

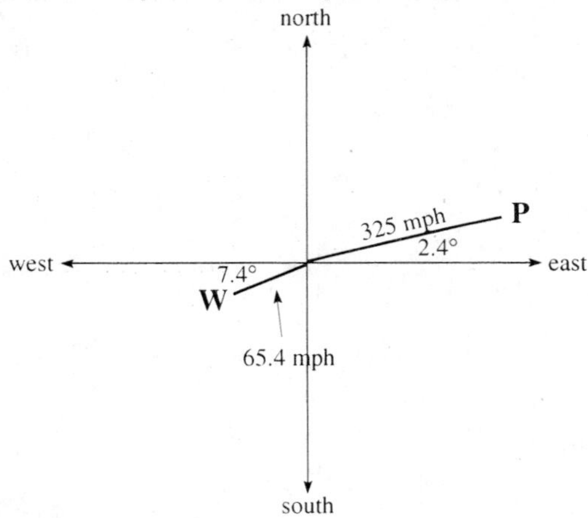

The plane and wind vectors are given by:
$$\mathbf{P} = \langle 325\cos 2.4°, 325\sin 2.4° \rangle$$
$$\mathbf{W} = \langle -65.4\cos 7.4°, -65.4\sin 7.4° \rangle$$

Adding results in: $\mathbf{P} + \mathbf{W} = \langle 259.86, 5.19 \rangle$

The magnitude is given by: $|\mathbf{P} + \mathbf{W}| = \sqrt{259.86^2 + 5.19^2} \approx 260$ mph

The direction is given by: $\theta = \tan^{-1}\left(\dfrac{5.19}{259.86}\right) \approx 1.1°$

The ground speed is approximately 260 mph and the true course is 88.9° from due north.

31. Finding the magnitude: $|\mathbf{U}| = \sqrt{5^2 + 12^2} = \sqrt{25 + 144} = \sqrt{169} = 13$

32. Finding the vector: $3\mathbf{U} + 5\mathbf{V} = 3(5\mathbf{i} + 12\mathbf{j}) + 5(-4\mathbf{i} + \mathbf{j}) = 15\mathbf{i} + 36\mathbf{j} - 20\mathbf{i} + 5\mathbf{j} = -5\mathbf{i} + 41\mathbf{j}$

33. Finding the vector: $3\mathbf{U} - 5\mathbf{V} = 3(5\mathbf{i} + 12\mathbf{j}) - 5(-4\mathbf{i} + \mathbf{j}) = 15\mathbf{i} + 36\mathbf{j} + 20\mathbf{i} - 5\mathbf{j} = 35\mathbf{i} + 31\mathbf{j}$

34. First find the vector: $2\mathbf{V} - \mathbf{W} = 2(-4\mathbf{i} + \mathbf{j}) - (\mathbf{i} - 4\mathbf{j}) = -8\mathbf{i} + 2\mathbf{j} - \mathbf{i} + 4\mathbf{j} = -9\mathbf{i} + 6\mathbf{j}$

Therefore: $|2\mathbf{V} - \mathbf{W}| = \sqrt{(-9)^2 + 6^2} = \sqrt{81 + 36} = \sqrt{117}$

35. Finding the dot product: $\mathbf{V} \cdot \mathbf{W} = (-4)(1) + (1)(-4) = -4 - 4 = -8$

36. Since $\mathbf{U} \cdot \mathbf{V} = |\mathbf{U}||\mathbf{V}|\cos\theta$, we have:
$$\mathbf{U} \cdot \mathbf{V} = |\mathbf{U}||\mathbf{V}|\cos\theta$$
$$(5)(-4) + (12)(1) = \sqrt{5^2 + 12^2}\,\sqrt{(-4)^2 + 1^2}\,\cos\theta$$
$$-20 + 12 = \sqrt{169}\,\sqrt{17}\,\cos\theta$$
$$-8 = \sqrt{2873}\,\cos\theta$$
$$\cos\theta = \frac{-8}{\sqrt{2873}} \approx -0.1493$$
$$\theta \approx 98.6°$$

37. $\mathbf{V} = a\mathbf{i} + b\mathbf{j}$ can be represented as a vector between the points $(0,0)$ and (a,b). Thus the slope is $m = \dfrac{b-0}{a-0} = \dfrac{b}{a}$.

38. Finding the dot product: $(3\mathbf{i} + 6\mathbf{j}) \cdot (-8\mathbf{i} + 4\mathbf{j}) = (3)(-8) + (6)(4) = -24 + 24 = 0$

Since their dot product is 0, the two vectors must be perpendicular.

39.   Since the dot product must be 0:
$$(5\mathbf{i}+12\mathbf{j})\bullet(4\mathbf{i}+b\mathbf{j})=0$$
$$20+12b=0$$
$$12b=-20$$
$$b=-\frac{5}{3}$$

40.   The work is: $\mathbf{F}\bullet\mathbf{d}=(33\mathbf{i}-4\mathbf{j})\bullet(56\mathbf{i}+10\mathbf{j})=(33)(56)+(-4)(10)=1808$

Chapter 8
Complex Numbers and Polar Coordinates

8.1 Complex Numbers

2. Writing in terms of i: $\sqrt{-49} = i\sqrt{49} = 7i$

4. Writing in terms of i: $\sqrt{-400} = i\sqrt{400} = 20i$

6. Writing in terms of i: $\sqrt{-45} = i\sqrt{45} = 3i\sqrt{5}$

8. Writing in terms of i: $\sqrt{-20} = i\sqrt{20} = 2i\sqrt{5}$

10. Simplifying: $\sqrt{-25} \cdot \sqrt{-1} = i\sqrt{25} \cdot i\sqrt{1} = 5i \cdot i = 5i^2 = -5$

12. Simplifying: $\sqrt{-16} \cdot \sqrt{-4} = i\sqrt{16} \cdot i\sqrt{4} = 4i \cdot 2i = 8i^2 = -8$

14. Setting the real parts equal:
$$-x = 2$$
$$x = -2$$

Setting the imaginary parts equal:
$$10y = -5$$
$$y = -\tfrac{1}{2}$$

16. Setting the real parts equal:
$$7x - 1 = 2$$
$$7x = 3$$
$$x = \tfrac{3}{7}$$

Setting the imaginary parts equal:
$$5y + 2 = 4$$
$$5y = 2$$
$$y = \tfrac{2}{5}$$

18. Setting the real parts equal:
$$x^2 - 2x = 8$$
$$x^2 - 2x - 8 = 0$$
$$(x - 4)(x + 2) = 0$$
$$x = -2, 4$$

Setting the imaginary parts equal:
$$y^2 = 2y - 1$$
$$y^2 - 2y + 1 = 0$$
$$(y - 1)^2 = 0$$
$$y = 1$$

20. Setting the real parts equal:
$$\sin x = -\cos x$$
$$\tan x = -1$$
$$x = \frac{3\pi}{4}, \frac{7\pi}{4}$$

Setting the imaginary parts equal:
$$\cos y = -1$$
$$y = \pi$$

22. Setting the real parts equal:
$$\cos^2 x + 1 = 2\cos x$$
$$\cos^2 x - 2\cos x + 1 = 0$$
$$(\cos x - 1)^2 = 0$$
$$\cos x = 1$$
$$x = 0$$

Setting the imaginary parts equal:
$$\tan y = -1$$
$$y = \frac{3\pi}{4}, \frac{7\pi}{4}$$

24.   Combining the numbers: $(3-5i)+(2+4i)=5-i$

26.   Combining the numbers: $(5+2i)-(3+6i)=5+2i-3-6i=2-4i$

28.   Combining the numbers: $(11-6i)-(2-4i)=11-6i-2+4i=9-2i$

30.   Combining the numbers: $(2\cos x-3i\sin y)+(3\cos x-2i\sin y)=5\cos x-5i\sin y$

32.   Combining the numbers: $\left[(4-5i)-(2+i)\right]+(2+5i)=4-5i-2-i+2+5i=4-i$

34.   Combining the numbers:
$$(10-2i)-\left[(2+i)-(3-i)\right]=(10-2i)-(2+i-3+i)=(10-2i)-(-1+2i)=10-2i+1-2i=11-4i$$

36.   Computing the power: $i^{13}=i^{12}\cdot i=\left(i^4\right)^3\cdot i=1\cdot i=i$

38.   Computing the power: $i^{15}=i^{12}\cdot i^3=\left(i^4\right)^3\cdot i^2\cdot i=1\cdot(-1)\cdot i=-i$

40.   Computing the power: $i^{34}=i^{32}\cdot i^2=\left(i^4\right)^8\cdot i^2=1\cdot(-1)=-1$

42.   Computing the power: $i^{35}=i^{32}\cdot i^3=\left(i^4\right)^8\cdot i^2\cdot i=1\cdot(-1)\cdot i=-i$

44.   Computing the product: $6i(3+8i)=18i+48i^2=18i-48=-48+18i$

46.   Computing the product: $(2+4i)(3-i)=6+12i-2i-4i^2=6+10i+4=10+10i$

48.   Computing the product: $(3-2i)^2=(3-2i)(3-2i)=9-6i-6i+4i^2=9-12i-4=5-12i$

50.   Computing the product: $(4+5i)(4-5i)=16+20i-20i-25i^2=16+25=41$

52.   Computing the product: $(2+7i)(2-7i)=4+14i-14i-49i^2=4+49=53$

54.   Computing the product: $3i(1+2i)(3+i)=3i\left(3+6i+i+2i^2\right)=3i(3+7i-2)=3i(1+7i)=3i+21i^2=-21+3i$

56.   Computing the product: $4i(1-i)^2=4i(1-i)(1-i)=4i\left(1-i-i+i^2\right)=4i(1-2i-1)=4i(-2i)=-8i^2=8$

58.   Finding the quotient: $\dfrac{3i}{2+i}\cdot\dfrac{2-i}{2-i}=\dfrac{6i-3i^2}{4-i^2}=\dfrac{6i+3}{4+1}=\dfrac{3}{5}+\dfrac{6}{5}i$

60.   Finding the quotient: $\dfrac{3+2i}{3-2i}\cdot\dfrac{3+2i}{3+2i}=\dfrac{9+12i+4i^2}{9-4i^2}=\dfrac{9+12i-4}{9+4}=\dfrac{5+12i}{13}=\dfrac{5}{13}+\dfrac{12}{13}i$

62.   Finding the quotient: $\dfrac{5-2i}{-i}\cdot\dfrac{i}{i}=\dfrac{5i-2i^2}{-i^2}=\dfrac{5i+2}{1}=2+5i$

64.   Finding the quotient: $\dfrac{5+4i}{3+6i}\cdot\dfrac{3-6i}{3-6i}=\dfrac{15-18i-24i^2}{9-36i^2}=\dfrac{15-18i+24}{9+36}=\dfrac{39-18i}{45}=\dfrac{13}{15}-\dfrac{2}{5}i$

66.   Computing the value: $z_2 z_1=(2-3i)(2+3i)=4-9i^2=4+9=13$

68.   Computing the value: $z_3 z_1=(4+5i)(2+3i)=8+22i+15i^2=8+22i-15=-7+22i$

70.   Computing the value: $3z_1+2z_2=3(2+3i)+2(2-3i)=6+9i+4-6i=10+3i$

72.   Computing the value:
$$z_3\left(z_1-z_2\right)=(4+5i)\left[(2+3i)-(2-3i)\right]=(4+5i)(2+3i-2+3i)=(4+5i)(6i)=24i+30i^2=-30+24i$$

74.   Computing the product: $(x-4i)(x+4i)=x^2-16i^2=x^2+16$

76.   Substituting $x=3+2i$ into the equation:
$$x^2-6x+13=(3+2i)^2-6(3+2i)+13$$
$$=9+12i+4i^2-18-12i+13$$
$$=9+12i-4-18-12i+13$$
$$=(9-4-18+13)+(12-12)i$$
$$=0$$

78.   Substituting $x = a + bi$ into the equation:

$$x^2 - 2ax + \left(a^2 + b^2\right) = (a - bi)^2 - 2a(a - bi) + \left(a^2 + b^2\right)$$
$$= a^2 - 2abi + b^2 i^2 - 2a^2 + 2abi + a^2 + b^2$$
$$= a^2 - 2abi - b^2 - 2a^2 + 2abi + a^2 + b^2$$
$$= \left(a^2 - b^2 - 2a^2 + a^2 + b^2\right) + (-2ab + 2ab)i$$
$$= 0$$

80.   Solving $x - y = 10$ for x yields $x = y + 10$. Substituting into the second equation:

$$xy = -40$$
$$(y + 10)y = -40$$
$$y^2 + 10y = -40$$
$$y^2 + 10y + 40 = 0$$

Using $a = 1$, $b = 10$, and $c = 40$ in the quadratic formula:

$$y = \frac{-10 \pm \sqrt{10^2 - 4(1)(40)}}{2(1)} = \frac{-10 \pm \sqrt{100 - 160}}{2} = \frac{-10 \pm \sqrt{-60}}{2} = \frac{-10 \pm 2i\sqrt{15}}{2} = -5 \pm i\sqrt{15}$$

If $y = -5 + i\sqrt{15}$, $x = \left(-5 + i\sqrt{15}\right) + 10 = 5 + i\sqrt{15}$. If $y = -5 - i\sqrt{15}$, $x = \left(-5 - i\sqrt{15}\right) + 10 = 5 - i\sqrt{15}$. The solutions are $\left(5 + i\sqrt{15}, -5 + i\sqrt{15}\right)$ and $\left(5 - i\sqrt{15}, -5 - i\sqrt{15}\right)$.

82.   Solving $3x + y = 6$ for y yields $y = 6 - 3x$. Substituting into the second equation:

$$xy = 9$$
$$x(6 - 3x) = 9$$
$$6x - 3x^2 = 9$$
$$3x^2 - 6x + 9 = 0$$
$$x^2 - 2x + 3 = 0$$

Using $a = 1$, $b = -2$, and $c = 3$ in the quadratic formula:

$$y = \frac{-(-2) \pm \sqrt{(-2)^2 - 4(1)(3)}}{2(1)} = \frac{2 \pm \sqrt{4 - 12}}{2} = \frac{2 \pm \sqrt{-8}}{2} = \frac{2 \pm 2i\sqrt{2}}{2} = 1 \pm i\sqrt{2}$$

Substituting each value to find y:

$$x = 1 + i\sqrt{2}: \qquad y = 6 - 3\left(1 + i\sqrt{2}\right) = 6 - 3 - 3i\sqrt{2} = 3 - 3i\sqrt{2}$$
$$x = 1 - i\sqrt{2}: \qquad y = 6 - 3\left(1 - i\sqrt{2}\right) = 6 - 3 + 3i\sqrt{2} = 3 + 3i\sqrt{2}$$

The solutions are $\left(1 + i\sqrt{2}, 3 - 3i\sqrt{2}\right)$ and $\left(1 - i\sqrt{2}, 3 + 3i\sqrt{2}\right)$.

84.   Let $z = a + bi$, so the conjugate is $\bar{z} = a - bi$. The sum is given by: $z + \bar{z} = (a + bi) + (a - bi) = 2a$
Thus the sum $z + \bar{z}$ is a real number.

86.   Let $z_1 = a + bi$ and $z_2 = c + di$. Then:

$$z_1 - z_2 = (a + bi) - (c + di) = (a - c) + (b - d)i$$
$$z_2 - z_1 = (c + di) - (a + bi) = (c - a) + (d - b)i$$

Since $z_1 - z_2 \neq z_2 - z_1$, subtraction of two complex numbers is not a commutative operation.

88.   If $x = -5$ and $y = 12$, $r = \sqrt{(-5)^2 + 12^2} = \sqrt{25 + 144} = \sqrt{169} = 13$. Therefore $\sin\theta = \frac{y}{r} = \frac{12}{13}$ and $\cos\theta = \frac{x}{r} = -\frac{5}{13}$.

90.   If $x = 1$ and $y = -1$, $r = \sqrt{1^2 + (-1)^2} = \sqrt{1 + 1} = \sqrt{2}$. Therefore $\sin\theta = \frac{y}{r} = -\frac{1}{\sqrt{2}}$ and $\cos\theta = \frac{x}{r} = \frac{1}{\sqrt{2}}$.

92.   If $\sin\theta = \frac{1}{2}$ and θ terminates in quadrant II, $\theta = 150°$.

94.   First note $A = 180° - (24.2° + 63.8°) = 92.0°$. Find a using the law of sines:

$$\frac{a}{\sin A} = \frac{b}{\sin B}$$

$$\frac{a}{\sin 92°} = \frac{5.92}{\sin 24.2°}$$

$$a = \frac{5.92 \sin 92°}{\sin 24.2°} \approx 14.4 \text{ inches}$$

Find c using the law of sines:

$$\frac{c}{\sin C} = \frac{b}{\sin B}$$

$$\frac{c}{\sin 63.8°} = \frac{5.92}{\sin 24.2°}$$

$$c = \frac{5.92 \sin 63.8°}{\sin 24.2°} \approx 13.0 \text{ inches}$$

96.   Find A using the law of sines:

$$\frac{a}{\sin A} = \frac{b}{\sin B}$$

$$\frac{625}{\sin A} = \frac{521}{\sin 32.8°}$$

$$\sin A = \frac{625 \sin 32.8°}{521} \approx 0.6498$$

$$A \approx 40.5°, 139.5°$$

If $A = 40.5°$, $C = 180° - (40.5° + 32.8°) = 106.7°$. Find c using the law of sines:

$$\frac{c}{\sin C} = \frac{b}{\sin B}$$

$$\frac{c}{\sin 106.7°} = \frac{521}{\sin 32.8°}$$

$$c = \frac{521 \sin 106.7°}{\sin 32.8°} \approx 921 \text{ ft}$$

If $A = 139.5°$, $C = 180° - (139.5° + 32.8°) = 7.7°$. Find c using the law of sines:

$$\frac{c}{\sin C} = \frac{b}{\sin B}$$

$$\frac{c}{\sin 7.7°} = \frac{521}{\sin 32.8°}$$

$$c = \frac{521 \sin 7.7°}{\sin 32.8°} \approx 129 \text{ ft}$$

8.2 Trigonometric Form for Complex Numbers

2. The absolute value is given by: $\left|3-4i\right| = \sqrt{3^2+(-4)^2} = \sqrt{9+16} = \sqrt{25} = 5$. Graphing the complex number:

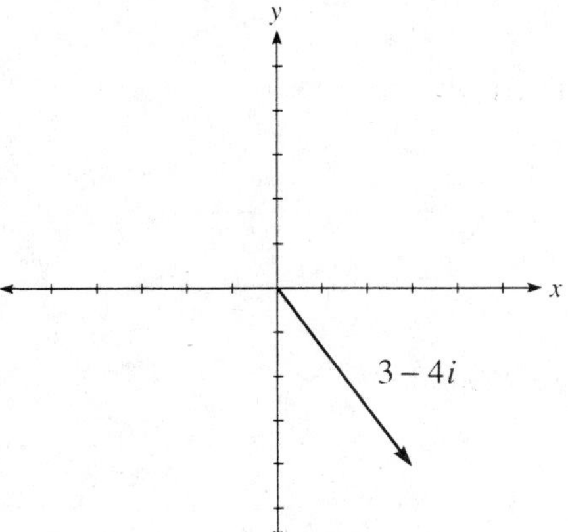

4. The absolute value is given by: $\left|1-i\right| = \sqrt{1^2+(-1)^2} = \sqrt{1+1} = \sqrt{2}$. Graphing the complex number:

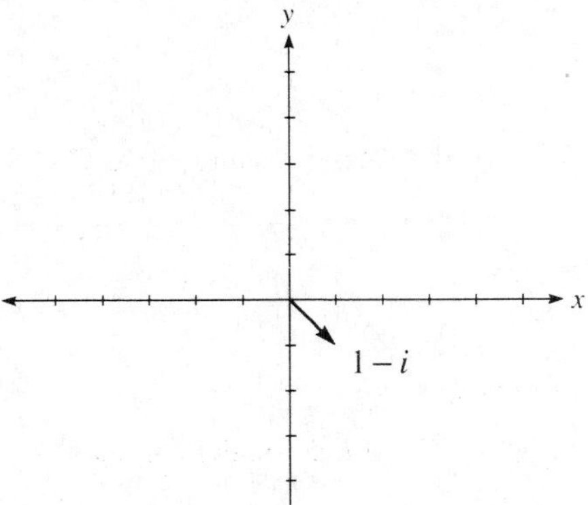

6. The absolute value is given by: $\left|4i\right| = \sqrt{0^2+4^2} = \sqrt{0+16} = \sqrt{16} = 4$. Graphing the complex number:

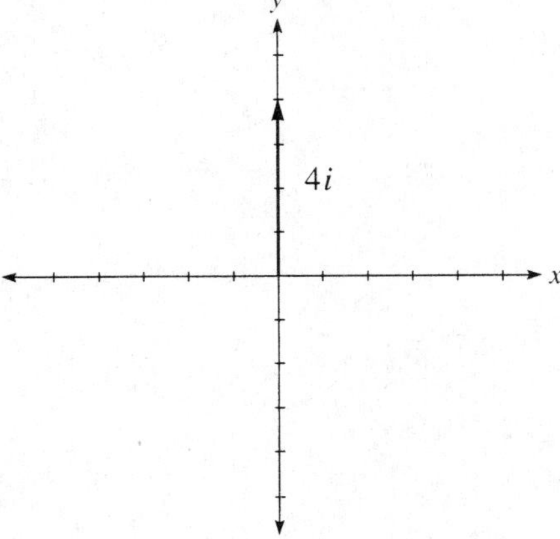

8.   The absolute value is given by: $\left|-4\right| = \sqrt{(-4)^2 + 0^2} = \sqrt{16+0} = \sqrt{16} = 4$. Graphing the complex number:

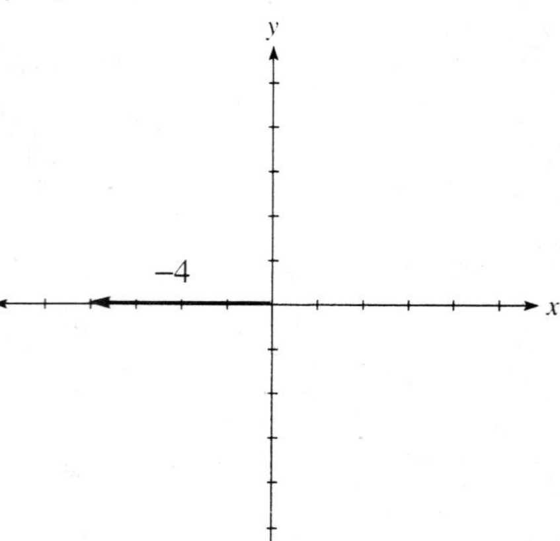

10.   The absolute value is given by: $\left|-3-4i\right| = \sqrt{(-3)^2 + (-4)^2} = \sqrt{9+16} = \sqrt{25} = 5$. Graphing the complex number:

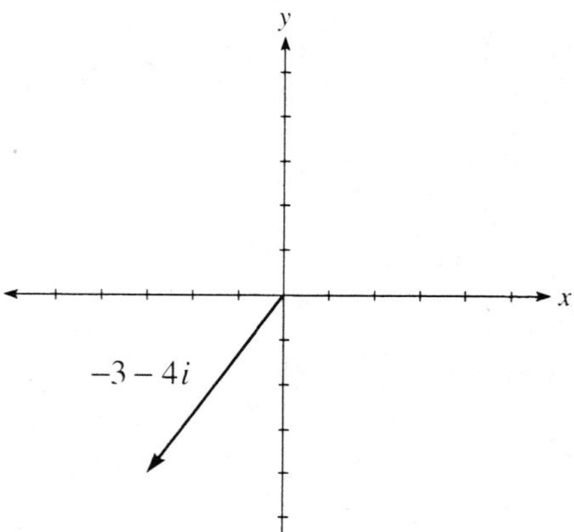

12.   The opposite is $-2-i$ and the conjugate is $2-i$:

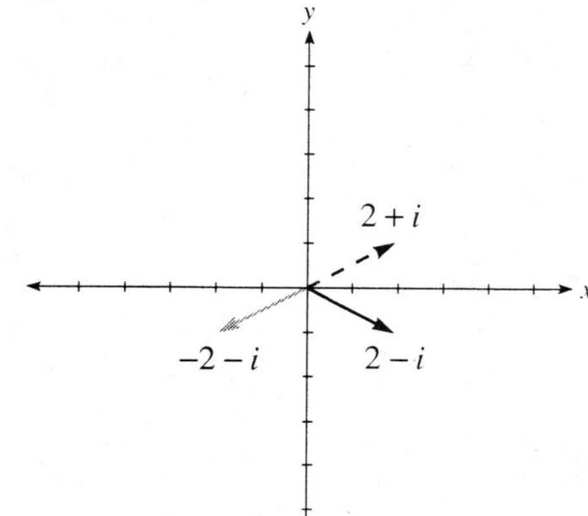

14.   The opposite is $3i$ and the conjugate is $3i$:

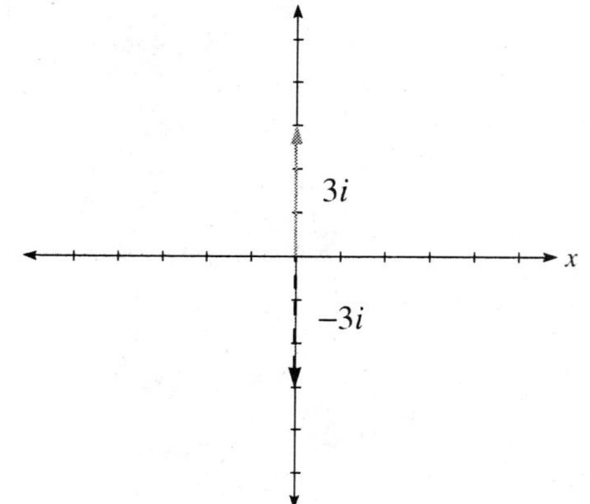

16.   The opposite is –5 and the conjugate is 5:

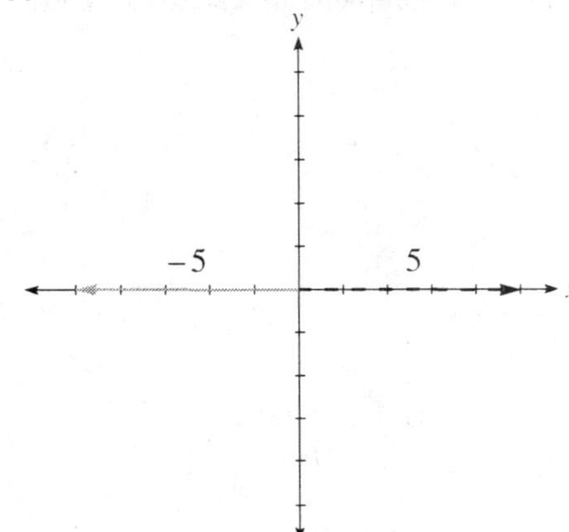

18.   The opposite is $2+5i$ and the conjugate is $-2+5i$:

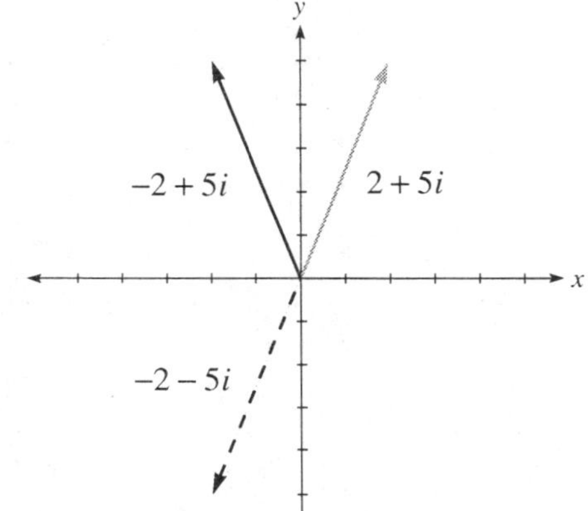

20.   Writing in standard form: $4\left(\cos 30^\circ + i\sin 30^\circ\right) = 4\left(\dfrac{\sqrt{3}}{2} + \dfrac{1}{2}i\right) = 2\sqrt{3} + 2i$

22.   Writing in standard form: $8\left(\cos 120^\circ + i\sin 120^\circ\right) = 8\left(-\dfrac{1}{2} + \dfrac{\sqrt{3}}{2}i\right) = -4 + 4i\sqrt{3}$

24.   Writing in standard form: $1\operatorname{cis}240^\circ = \cos 240^\circ + i\sin 240^\circ = -\dfrac{1}{2} - \dfrac{\sqrt{3}}{2}i$

26.   Writing in standard form: $\sqrt{2}\operatorname{cis}315^\circ = \sqrt{2}\left(\cos 315^\circ + i\sin 315^\circ\right) = \sqrt{2}\left(\dfrac{1}{\sqrt{2}} - \dfrac{1}{\sqrt{2}}i\right) = 1 - i$

28.   Writing in standard form: $100\left(\cos 70^\circ + i\sin 70^\circ\right) = \left(100\cos 70^\circ\right) + \left(100\sin 70^\circ\right)i \approx 34.20 + 93.97i$

30.   Writing in standard form: $100\left(\cos 171^\circ + i\sin 171^\circ\right) = \left(100\cos 171^\circ\right) + \left(100\sin 171^\circ\right)i \approx -98.77 + 15.64i$

32.   Writing in standard form: $1\operatorname{cis}261^\circ = \cos 261^\circ + i\sin 261^\circ \approx -0.16 - 0.99i$

34.   Writing in standard form: $10\operatorname{cis}318^\circ = 10\left(\cos 318^\circ + i\sin 318^\circ\right) = \left(10\cos 318^\circ\right) + \left(10\sin 318^\circ\right)i \approx 7.43 - 6.69i$

36.   Sketching the graph:

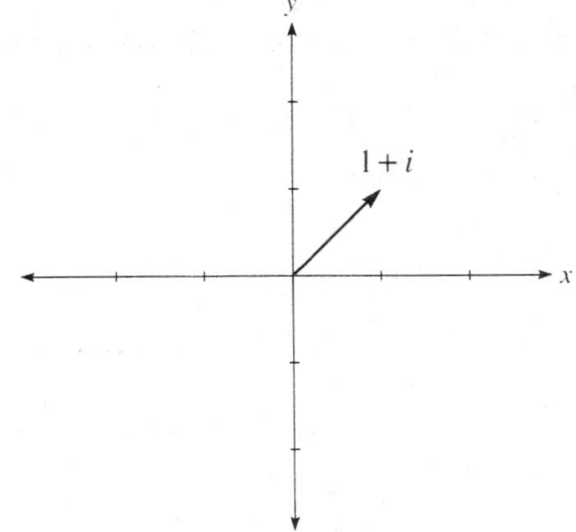

Here $x = 1$, $y = 1$, and $r = \sqrt{1^2 + 1^2} = \sqrt{1+1} = \sqrt{2}$. Since $\tan\theta = \dfrac{y}{x} = 1$, $\theta = 45^\circ$. So the trigonometric form is

$1 + i = \sqrt{2}\left(\cos 45^\circ + i\sin 45^\circ\right) = \sqrt{2}\operatorname{cis}45^\circ$.

38.   Sketching the graph:

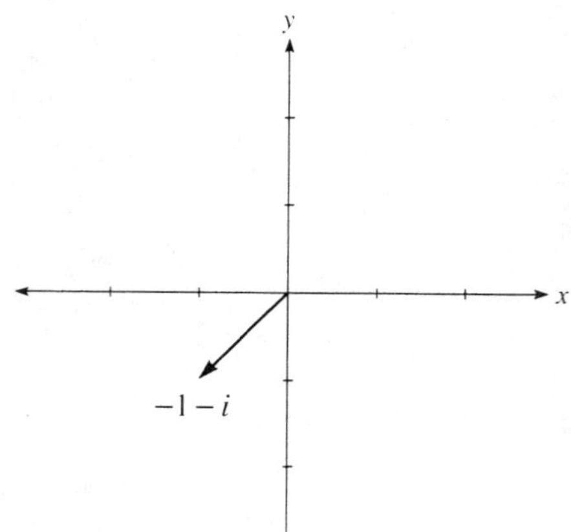

Here $x = -1$, $y = -1$, and $r = \sqrt{(-1)^2 + (-1)^2} = \sqrt{1+1} = \sqrt{2}$. Since $\tan\theta = \dfrac{y}{x} = \dfrac{-1}{-1} = 1$, $\theta = 225°$. So the trigonometric form is $-1 - i = \sqrt{2}\left(\cos 225° + i\sin 225°\right) = \sqrt{2}\,\text{cis}\,225°$.

40.   Sketching the graph:

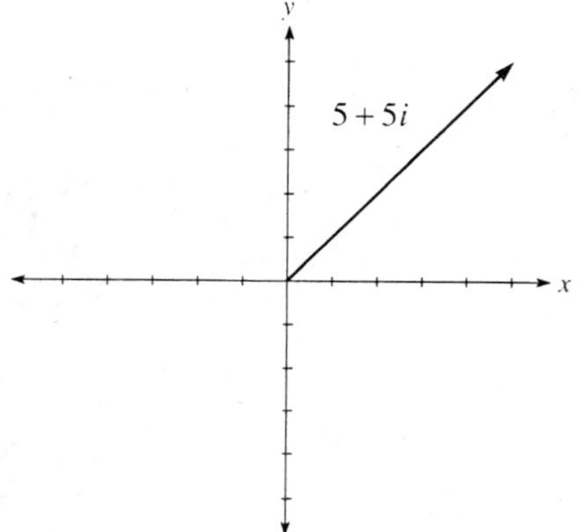

Here $x = 5$, $y = 5$, and $r = \sqrt{5^2 + 5^2} = \sqrt{25 + 25} = \sqrt{50} = 5\sqrt{2}$. Since $\tan\theta = \dfrac{y}{x} = \dfrac{5}{5} = 1$, $\theta = 45°$. So the trigonometric form is $5 + 5i = 5\sqrt{2}\left(\cos 45° + i\sin 45°\right) = 5\sqrt{2}\,\text{cis}\,45°$.

42.   Sketching the graph:

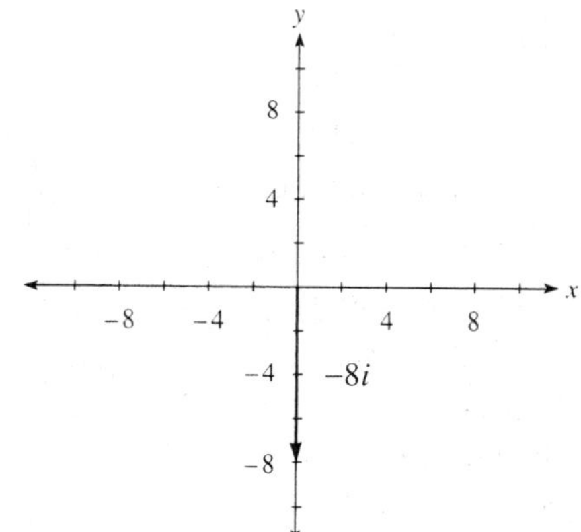

Here $x = 0$, $y = -8$, and $r = \sqrt{0^2 + (-8)^2} = \sqrt{0 + 64} = \sqrt{64} = 8$. Since $\tan\theta$ is undefined, $\theta = 270°$. So the trigonometric form is $-8i = 8\left(\cos 270° + i\sin 270°\right) = 8\,\text{cis}\,270°$.

44.   Sketching the graph:

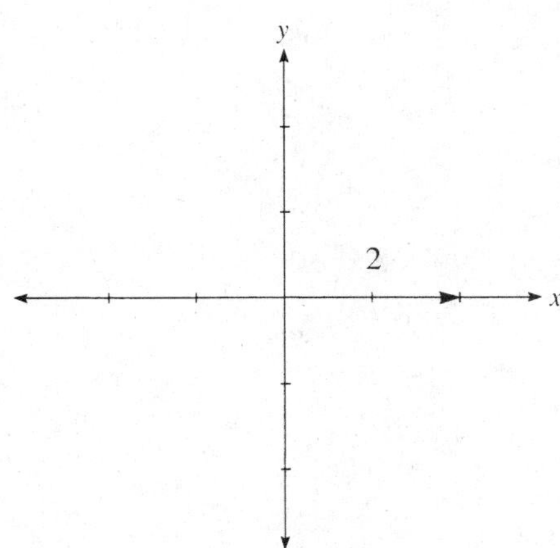

Here $x = 2$, $y = 0$, and $r = \sqrt{2^2 + 0^2} = \sqrt{4 + 0} = \sqrt{4} = 2$. Since $\tan\theta = \dfrac{0}{2} = 0$, $\theta = 0°$. So the trigonometric form is

$2 = 2(\cos 0° + i\sin 0°) = 2\operatorname{cis}0°$.

46.   Sketching the graph:

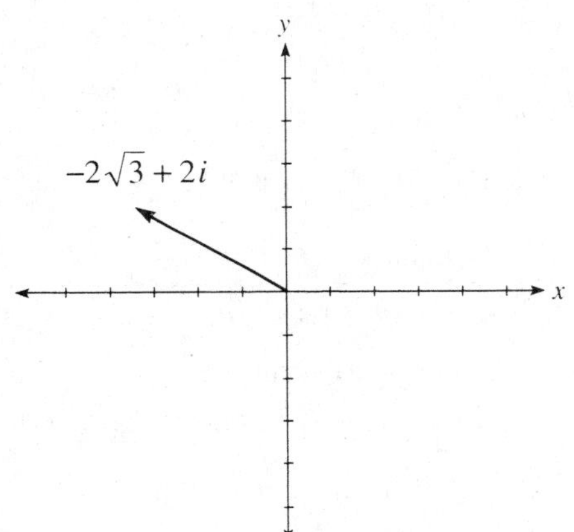

Here $x = -2\sqrt{3}$, $y = 2$, and $r = \sqrt{\left(-2\sqrt{3}\right)^2 + 2^2} = \sqrt{12 + 4} = \sqrt{16} = 4$. Since $\tan\theta = \dfrac{2}{-2\sqrt{3}} = -\dfrac{1}{\sqrt{3}}$, $\theta = 150°$. So the

trigonometric form is $-2\sqrt{3} + 2i = 4(\cos 150° + i\sin 150°) = 4\operatorname{cis}150°$.

48.   Here $x = 3$, $y = -4$, and $r = \sqrt{3^2 + (-4)^2} = \sqrt{9 + 16} = 5$. Since $\tan\theta = -\dfrac{4}{3}$ in quadrant IV,

$\theta = 360° - 53.13° = 306.87°$. So the trigonometric form is $3 - 4i = 5(\cos 306.87° + i\sin 306.87°) = 5\operatorname{cis}306.87°$.

50.   Here $x = 21$, $y = -20$, and $r = \sqrt{21^2 + (-20)^2} = \sqrt{441 + 400} = 29$. Since $\tan\theta = -\dfrac{20}{21}$ in quadrant IV,

$\theta = 360° - 43.60° = 316.40°$. So the trigonometric form is $20 - 21i = 29(\cos 316.40° + i\sin 316.40°) = 29\operatorname{cis}316.40°$.

52.   Here $x = 8$, $y = -15$, and $r = \sqrt{8^2 + (-15)^2} = \sqrt{64 + 225} = 17$. Since $\tan\theta = -\dfrac{15}{8}$ in quadrant IV,

$\theta = 360° - 61.93° = 298.07°$. So the trigonometric form is $8 - 15i = 17(\cos 298.07° + i\sin 298.07°) = 17\operatorname{cis}298.07°$.

54.   Here $x = 11$, $y = -2$, and $r = \sqrt{11^2 + (-2)^2} = \sqrt{121 + 4} = 5\sqrt{5}$. Since $\tan\theta = -\dfrac{2}{11}$ in quadrant IV,

$\theta = 360° - 10.30° = 349.70°$. So the trigonometric form is
$11 - 2i = 5\sqrt{5}(\cos 349.70° + i\sin 349.70°) = 5\sqrt{5}\operatorname{cis}349.70°$.

56.   The trigonometric form is $3 - 4i = 5(\cos 306.87° + i\sin 306.87°)$.

58.   The trigonometric form is $20 - 21i = 29(\cos 316.40° + i\sin 316.40°)$.

60.   The trigonometric form is $8 - 15i = 17(\cos 298.07° + i\sin 298.07°)$.

62.   The trigonometric form is $11 - 2i = 5\sqrt{5}(\cos 349.70° + i\sin 349.70°)$.

64.   The trigonometric forms are:
$$4i = 4(\cos 90° + i\sin 90°)$$
$$2 = 2(\cos 0° + i\sin 0°)$$
Therefore:
$$4i \cdot 2 = 4(\cos 90° + i\sin 90°) \cdot 2(\cos 0° + i\sin 0°)$$
$$= 8\left(\cos 90° \cos 0° + i\sin 90° \sin 0° + i\cos 90° \sin 0° + i^2 \sin 90° \sin 0°\right)$$
$$= 8(0 \cdot 1 + i \cdot 1 \cdot 1 + i \cdot 0 \cdot 0 + (-1) \cdot 1 \cdot 0)$$
$$= 8(0 + i + 0 + 0)$$
$$= 8i$$

66.   Writing each complex number in standard form:
$$2(\cos 60° + i\sin 60°) = 2\left(\frac{1}{2} + \frac{\sqrt{3}}{2}i\right) = 1 + i\sqrt{3}$$

$$2[\cos(-60°) + i\sin(-60°)] = 2\left(\frac{1}{2} - \frac{\sqrt{3}}{2}i\right) = 1 - i\sqrt{3}$$

68.   Finding the magnitude: $|z| = \sqrt{\cos^2 \theta + (-\sin \theta)^2} = \sqrt{\cos^2 \theta + \sin^2 \theta} = \sqrt{1} = 1$

70.   Using the addition formula for sine:
$$\sin 75° = \sin(30° + 45°) = \sin 30° \cos 45° + \cos 30° \sin 45° = \frac{1}{2} \cdot \frac{\sqrt{2}}{2} + \frac{\sqrt{3}}{2} \cdot \frac{\sqrt{2}}{2} = \frac{\sqrt{2} + \sqrt{6}}{4}$$

72.   Since A terminates in quadrant I, $\cos A > 0$ and thus: $\cos A = \sqrt{1 - \sin^2 A} = \sqrt{1 - \left(\frac{3}{5}\right)^2} = \sqrt{1 - \frac{9}{25}} = \sqrt{\frac{16}{25}} = \frac{4}{5}$

Since B terminates in quadrant I, $\cos B > 0$ and thus: $\cos B = \sqrt{1 - \sin^2 B} = \sqrt{1 - \left(\frac{5}{13}\right)^2} = \sqrt{1 - \frac{25}{169}} = \sqrt{\frac{144}{169}} = \frac{12}{13}$

Using the addition formula for cosine: $\cos(A + B) = \cos A \cos B - \sin A \sin B = \frac{4}{5} \cdot \frac{12}{13} - \frac{3}{5} \cdot \frac{5}{13} = \frac{48}{65} - \frac{15}{65} = \frac{33}{65}$

74.   Using the addition formula for cosine: $\cos 30° \cos 90° - \sin 30° \sin 90° = \cos(30° + 90°) = \cos 120° = -\frac{1}{2}$

76.   Using the addition formula for sine: $\sin 18° \cos 32° + \cos 18° \sin 32° = \sin(18° + 32°) = \sin 50°$

78.   Using the law of sines:
$$\frac{\sin B}{b} = \frac{\sin A}{a}$$
$$\frac{\sin B}{567} = \frac{\sin 45.6°}{678}$$
$$\sin B = \frac{567 \sin 45.6°}{678} \approx 0.5975$$
$$B \approx 36.7°, 143.3°$$
Since $143.3° + 45.6° = 188.9° > 180°$, only $B \approx 36.7°$ is valid.

80.   Using the law of sines:
$$\frac{\sin B}{b} = \frac{\sin A}{a}$$
$$\frac{\sin B}{567} = \frac{\sin 45.6°}{789}$$
$$\sin B = \frac{567 \sin 45.6°}{789} \approx 0.5134$$
$$B \approx 30.9°, 149.1°$$
Since $149.1° + 45.6° = 194.7° > 180°$, only $B \approx 30.9°$ is valid.

8.3 Products and Quotients in Trigonometric Form

2. Multiplying in trigonometric form:
$$5(\cos 15° + i \sin 15°) \cdot 2(\cos 25° + i \sin 25°) = 10\left[\cos(15° + 25°) + i \sin(15° + 25°)\right] = 10(\cos 40° + i \sin 40°)$$

4. Multiplying in trigonometric form:
$$9(\cos 115° + i \sin 115°) \cdot 4(\cos 51° + i \sin 51°) = 36\left[\cos(115° + 51°) + i \sin(115° + 51°)\right]$$
$$= 36(\cos 166° + i \sin 166°)$$

6. Multiplying in trigonometric form:
$$2\operatorname{cis}120° \cdot 4\operatorname{cis}30° = 2(\cos 120° + i \sin 120°) \cdot 4(\cos 30° + i \sin 30°)$$
$$= 8\left[\cos(120° + 30°) + i \sin(120° + 30°)\right]$$
$$= 8(\cos 150° + i \sin 150°)$$
$$= 8\operatorname{cis}150°$$

8. Finding the product in standard form: $z_1 z_2 = (1+i)(2+2i) = 2 + 4i + 2i^2 = 2 + 4i - 2 = 4i$

Writing each number in trigonometric form:
$$z_1 = 1 + i = \sqrt{2}\left(\cos 45° + i \sin 45°\right)$$
$$z_2 = 2 + 2i = 2\sqrt{2}\left(\cos 45° + i \sin 45°\right)$$

Multiplying in trigonometric form:
$$z_1 z_2 = \sqrt{2}\left(\cos 45° + i \sin 45°\right) \cdot 2\sqrt{2}\left(\cos 45° + i \sin 45°\right) = 4(\cos 90° + i \sin 90°) = 4(0 + 1i) = 4i$$

Note that the products are equal.

10. Finding the product in standard form: $z_1 z_2 = \left(-1 + i\sqrt{3}\right)\left(\sqrt{3} + i\right) = -\sqrt{3} + 2i + i^2\sqrt{3} = -\sqrt{3} + 2i - \sqrt{3} = -2\sqrt{3} + 2i$

Writing each number in trigonometric form:
$$z_1 = -1 + i\sqrt{3} = 2(\cos 120° + i \sin 120°)$$
$$z_2 = \sqrt{3} + i = 2(\cos 30° + i \sin 30°)$$

Multiplying in trigonometric form:
$$z_1 z_2 = 2(\cos 120° + i \sin 120°) \cdot 2(\cos 30° + i \sin 30°) = 4(\cos 150° + i \sin 150°) = 4\left(-\frac{\sqrt{3}}{2} + \frac{1}{2}i\right) = -2\sqrt{3} + 2i$$

Note that the products are equal.

12. Finding the product in standard form: $z_1 z_2 = (2i)(-5i) = -10i^2 = 10$

Writing each number in trigonometric form:
$$z_1 = 2i = 2(\cos 90° + i \sin 90°)$$
$$z_2 = -5i = 5(\cos 270° + i \sin 270°)$$

Multiplying in trigonometric form:
$$z_1 z_2 = 2(\cos 90° + i \sin 90°) \cdot 5(\cos 270° + i \sin 270°) = 10(\cos 360° + i \sin 360°) = 10(1 + 0i) = 10$$

Note that the products are equal.

14. Finding the product in standard form: $z_1 z_2 = (1+i)(3i) = 3i + 3i^2 = -3 + 3i$

Writing each number in trigonometric form:
$$z_1 = 1 + i = \sqrt{2}\left(\cos 45° + i \sin 45°\right)$$
$$z_2 = 3i = 3(\cos 90° + i \sin 90°)$$

Multiplying in trigonometric form:
$$z_1 z_2 = \sqrt{2}\left(\cos 45° + i \sin 45°\right) \cdot 3(\cos 90° + i \sin 90°)$$
$$= 3\sqrt{2}\left(\cos 135° + i \sin 135°\right)$$
$$= 3\sqrt{2}\left(-\frac{1}{\sqrt{2}} + \frac{1}{\sqrt{2}}i\right)$$
$$= -3 + 3i$$

Note that the products are equal.

16.   Finding the product in standard form: $z_1 z_2 = -3\left(\sqrt{3} + i\right) = -3\sqrt{3} - 3i$

Writing each number in trigonometric form:
$$z_1 = -3 = 3(\cos 180° + i \sin 180°)$$
$$z_2 = \sqrt{3} + i = 2(\cos 30° + i \sin 30°)$$

Multiplying in trigonometric form:

$$z_1 z_2 = 3(\cos 180° + i \sin 180°) \cdot 2(\cos 30° + i \sin 30°) = 6(\cos 210° + i \sin 210°) = 6\left(-\frac{\sqrt{3}}{2} - \frac{1}{2}i\right) = -3\sqrt{3} - 3i$$

Note that the products are equal.

18.   Using DeMoivre's Theorem:
$$\left[4(\cos 15° + i \sin 15°)\right]^3 = 4^3 \left[\cos(3 \cdot 15°) + i \sin(3 \cdot 15°)\right]$$
$$= 64(\cos 45° + i \sin 45°)$$
$$= 64\left(\frac{\sqrt{2}}{2} + \frac{\sqrt{2}}{2} i\right)$$
$$= 32\sqrt{2} + 32i\sqrt{2}$$

20.   Using DeMoivre's Theorem: $(\cos 18° + i \sin 18°)^{10} = \cos(10 \cdot 18°) + i \sin(10 \cdot 18°) = \cos 180° + i \sin 180° = -1 + 0i = -1$

22.   Using DeMoivre's Theorem:
$$(3 \operatorname{cis} 30°)^4 = \left[3(\cos 30° + i \sin 30°)\right]^4$$
$$= 3^4 \left[\cos(4 \cdot 30°) + i \sin(4 \cdot 30°)\right]$$
$$= 81(\cos 120° + i \sin 120°)$$
$$= 81\left(-\frac{1}{2} + \frac{\sqrt{3}}{2} i\right)$$
$$= -\frac{81}{2} + \frac{81\sqrt{3}}{2} i$$

24.   Using DeMoivre's Theorem:
$$\left(\sqrt{2} \operatorname{cis} 70°\right)^6 = \left[\sqrt{2}(\cos 70° + i \sin 70°)\right]^6$$
$$= \left(\sqrt{2}\right)^6 \left[\cos(6 \cdot 70°) + i \sin(6 \cdot 70°)\right]$$
$$= 8(\cos 420° + i \sin 420°)$$
$$= 8\left(\frac{1}{2} + \frac{\sqrt{3}}{2} i\right)$$
$$= 4 + 4i\sqrt{3}$$

26.   First write $1 + i$ in trigonometric form: $1 + i = \sqrt{2}(\cos 45° + i \sin 45°)$. Using DeMoivre's Theorem:
$$(1 + i)^5 = \left[\sqrt{2}(\cos 45° + i \sin 45°)\right]^5$$
$$= \left(\sqrt{2}\right)^5 \left[\cos(5 \cdot 45°) + i \sin(5 \cdot 45°)\right]$$
$$= 4\sqrt{2}(\cos 225° + i \sin 225°)$$
$$= 4\sqrt{2}\left(-\frac{1}{\sqrt{2}} - \frac{1}{\sqrt{2}} i\right)$$
$$= -4 - 4i$$

28. First write $\sqrt{3}+i$ in trigonometric form: $\sqrt{3}+i=2(\cos 30°+i\sin 30°)$. Using DeMoivre's Theorem:

$$\left(\sqrt{3}+i\right)^4=\left[2(\cos 30°+i\sin 30°)\right]^4$$
$$=2^4\left[\cos(4\bullet 30°)+i\sin(4\bullet 30°)\right]$$
$$=16(\cos 120°+i\sin 120°)$$
$$=16\left(-\frac{1}{2}+\frac{\sqrt{3}}{2}i\right)$$
$$=-8+8i\sqrt{3}$$

30. First write $-1+i$ in trigonometric form: $-1+i=\sqrt{2}\left(\cos 135°+i\sin 135°\right)$. Using DeMoivre's Theorem:

$$(-1+i)^8=\left[\sqrt{2}\left(\cos 135°+i\sin 135°\right)\right]^8$$
$$=\left(\sqrt{2}\right)^8\left[\cos(8\bullet 135°)+i\sin(8\bullet 135°)\right]$$
$$=16(\cos 1080°+i\sin 1080°)$$
$$=16(1+0i)$$
$$=16$$

32. First write $-2-2i$ in trigonometric form: $-2-2i=2\sqrt{2}\left(\cos 225°+i\sin 225°\right)$. Using DeMoivre's Theorem:

$$(-2-2i)^3=\left[2\sqrt{2}\left(\cos 225°+i\sin 225°\right)\right]^3$$
$$=\left(2\sqrt{2}\right)^3\left[\cos(3\bullet 225°)+i\sin(3\bullet 225°)\right]$$
$$=16\sqrt{2}\left(\cos 675°+i\sin 675°\right)$$
$$=16\sqrt{2}\left(\frac{1}{\sqrt{2}}-\frac{1}{\sqrt{2}}i\right)$$
$$=16-16i$$

34. Dividing in trigonometric form: $\dfrac{30(\cos 80°+i\sin 80°)}{10(\cos 30°+i\sin 30°)}=\dfrac{30}{10}\left[\cos(80°-30°)+i\sin(80°-30°)\right]=3(\cos 50°+i\sin 50°)$

36. Dividing in trigonometric form:
$$\frac{21(\cos 63°+i\sin 63°)}{14(\cos 44°+i\sin 44°)}=\frac{21}{14}\left[\cos(63°-44°)+i\sin(63°-44°)\right]=1.5(\cos 19°+i\sin 19°)$$

38. Dividing in trigonometric form:
$$\frac{6\operatorname{cis}120°}{8\operatorname{cis}90°}=\frac{6(\cos 120°+i\sin 120°)}{8(\cos 90°+i\sin 90°)}$$
$$=\frac{6}{8}\left[\cos(120°-90°)+i\sin(120°-90°)\right]$$
$$=\tfrac{3}{4}(\cos 30°+i\sin 30°)$$
$$=0.75\operatorname{cis}330°$$

40. Finding the quotient in standard form: $\dfrac{z_1}{z_2}=\dfrac{2-2i}{1-i}=\dfrac{2(1-i)}{1-i}=2$

Writing each number in trigonometric form:
$$z_1=2-2i=2\sqrt{2}\left(\cos 315°+i\sin 315°\right)$$
$$z_2=1-i=\sqrt{2}\left(\cos 315°+i\sin 315°\right)$$

Dividing in trigonometric form: $\dfrac{z_1}{z_2}=\dfrac{2\sqrt{2}\left(\cos 315°+i\sin 315°\right)}{\sqrt{2}\left(\cos 315°+i\sin 315°\right)}=2(\cos 0°+i\sin 0°)=2(1+0i)=2$

Note that the quotients are equal.

42. Finding the quotient in standard form: $\dfrac{z_1}{z_2} = \dfrac{1+i\sqrt{3}}{2i} \cdot \dfrac{i}{i} = \dfrac{i+i^2\sqrt{3}}{2i^2} = \dfrac{i-\sqrt{3}}{-2} = \dfrac{\sqrt{3}}{2} - \dfrac{1}{2}i$

Writing each number in trigonometric form:
$$z_1 = 1 + i\sqrt{3} = 2(\cos 60° + i\sin 60°)$$
$$z_2 = 2i = 2(\cos 90° + i\sin 90°)$$

Dividing in trigonometric form: $\dfrac{z_1}{z_2} = \dfrac{2(\cos 60° + i\sin 60°)}{2(\cos 90° + i\sin 90°)} = 1(\cos(-30°) + i\sin(-30°)) = \dfrac{\sqrt{3}}{2} - \dfrac{1}{2}i$

Note that the quotients are equal.

44. Finding the quotient in standard form: $\dfrac{z_1}{z_2} = \dfrac{6+6i}{-3-3i} = \dfrac{6(1+i)}{-3(1+i)} = \dfrac{6}{-3} = -2$

Writing each number in trigonometric form:
$$z_1 = 6 + 6i = 6\sqrt{2}(\cos 45° + i\sin 45°)$$
$$z_2 = -3 - 3i = 3\sqrt{2}(\cos 225° + i\sin 225°)$$

Dividing in trigonometric form: $\dfrac{z_1}{z_2} = \dfrac{6\sqrt{2}(\cos 45° + i\sin 45°)}{3\sqrt{2}(\cos 225° + i\sin 225°)} = 2(\cos(-180°) + i\sin(-180°)) = 2(-1 + 0i) = -2$

Note that the quotients are equal.

46. Finding the quotient in standard form: $\dfrac{z_1}{z_2} = \dfrac{-6}{3} = -2$

Writing each number in trigonometric form:
$$z_1 = -6 = 6(\cos 180° + i\sin 180°)$$
$$z_2 = 3 = 3(\cos 0° + i\sin 0°)$$

Dividing in trigonometric form: $\dfrac{z_1}{z_2} = \dfrac{6(\cos 180° + i\sin 180°)}{3(\cos 0° + i\sin 0°)} = 2(\cos 180° + i\sin 180°) = 2(-1 + 0i) = -2$

Note that the quotients are equal.

48. First write each complex number in trigonometric form:
$$\sqrt{3} + i = 2(\cos 30° + i\sin 30°)$$
$$2i = 2(\cos 90° + i\sin 90°)$$
$$1 + i = \sqrt{2}(\cos 45° + i\sin 45°)$$

Now simplify the expression:

$$\dfrac{\left(\sqrt{3}+i\right)^4 (2i)^5}{(1+i)^{10}} = \dfrac{\left[2(\cos 30° + i\sin 30°)\right]^4 \left[2(\cos 90° + i\sin 90°)\right]^5}{\left[\sqrt{2}(\cos 45° + i\sin 45°)\right]^{10}}$$

$$= \dfrac{16(\cos 120° + i\sin 120°) \cdot 32(\cos 450° + i\sin 450°)}{32(\cos 450° + i\sin 450°)}$$

$$= \dfrac{512(\cos 570° + i\sin 570°)}{32(\cos 450° + i\sin 450°)}$$

$$= 16(\cos 120° + i\sin 120°)$$

$$= 16\left(-\dfrac{1}{2} + \dfrac{\sqrt{3}}{2}i\right)$$

$$= -8 + 8i\sqrt{3}$$

50. First write each complex number in trigonometric form:

$$2 + 2i = 2\sqrt{2}\left(\cos 45° + i\sin 45°\right)$$
$$-3 + 3i = 3\sqrt{2}\left(\cos 135° + i\sin 135°\right)$$
$$\sqrt{3} + i = 2\left(\cos 30° + i\sin 30°\right)$$

Now simplify the expression:

$$\frac{(2+2i)^5(-3+3i)^3}{\left(\sqrt{3}+i\right)^{10}} = \frac{\left[2\sqrt{2}\left(\cos 45° + i\sin 45°\right)\right]^5\left[3\sqrt{2}\left(\cos 135° + i\sin 135°\right)\right]^3}{\left[2\left(\cos 30° + i\sin 30°\right)\right]^{10}}$$

$$= \frac{128\sqrt{2}\left(\cos 225° + i\sin 225°\right) \bullet 54\sqrt{2}\left(\cos 405° + i\sin 405°\right)}{1024\left(\cos 300° + i\sin 300°\right)}$$

$$= \frac{13824\left(\cos 630° + i\sin 630°\right)}{1024\left(\cos 300° + i\sin 300°\right)}$$

$$= \tfrac{27}{2}\left(\cos 330° + i\sin 330°\right)$$

$$= \tfrac{27}{2}\left(\frac{\sqrt{3}}{2} - \tfrac{1}{2}i\right)$$

$$= \frac{27\sqrt{3}}{4} - \tfrac{27}{4}i$$

52. Substituting into the equation:

$$x^2 - 2x + 4 = \left[2\left(\cos 300° + i\sin 300°\right)\right]^2 - 2 \bullet 2\left(\cos 300° + i\sin 300°\right) + 4$$

$$= 4\left(\cos 600° + i\sin 600°\right) - 4\left(\cos 300° + i\sin 300°\right) + 4$$

$$= 4\left(-\tfrac{1}{2} - \frac{\sqrt{3}}{2}i\right) - 4\left(\tfrac{1}{2} - \frac{\sqrt{3}}{2}i\right) + 4$$

$$= -2 - 2i\sqrt{3} - 2 + 2i\sqrt{3} + 4$$

$$= 0$$

Thus $x = 2\left(\cos 300° + i\sin 300°\right)$ is a solution to $x^2 - 2x + 4 = 0$.

54. To show x is a cube root of -1, we must show $x^3 = -1$. First write the number in trigonometric form:

$$x = \tfrac{1}{2} + \frac{\sqrt{3}}{2}i = \cos 60° + i\sin 60°$$

Therefore: $x^3 = \left(\cos 60° + i\sin 60°\right)^3 = \cos 180° + i\sin 180° = -1 + 0i = -1$

So $x = \tfrac{1}{2} + \frac{\sqrt{3}}{2}i$ is a cube root of -1.

56. Write $1 - i$ in trigonometric form: $1 - i = \sqrt{2}\left(\cos 315° + i\sin 315°\right)$

Therefore: $(1-i)^{-1} = \left[\sqrt{2}\left(\cos 315° + i\sin 315°\right)\right]^{-1} = \frac{1}{\sqrt{2}}\left[\cos(-315°) + i\sin(-315°)\right] = \frac{1}{\sqrt{2}}\left(\frac{1}{\sqrt{2}} + \frac{1}{\sqrt{2}}i\right) = \tfrac{1}{2} + \tfrac{1}{2}i$

58. Write $\sqrt{3} + i$ in trigonometric form: $\sqrt{3} + i = 2\left(\cos 30° + i\sin 30°\right)$

Therefore: $\left(\sqrt{3}+i\right)^{-1} = \left[2\left(\cos 30° + i\sin 30°\right)\right]^{-1} = \tfrac{1}{2}\left[\cos(-30°) + i\sin(-30°)\right] = \tfrac{1}{2}\left(\frac{\sqrt{3}}{2} - \tfrac{1}{2}i\right) = \frac{\sqrt{3}}{4} - \tfrac{1}{4}i$

60. Since $90° < A < 180°$, $\sin A > 0$ and thus: $\sin A = \sqrt{1 - \cos^2 A} = \sqrt{1 - \left(-\tfrac{1}{3}\right)^2} = \sqrt{1 - \tfrac{1}{9}} = \sqrt{\tfrac{8}{9}} = \frac{2\sqrt{2}}{3}$

Using the double-angle formula for sine: $\sin 2A = 2\sin A\cos A = 2\left(\frac{2\sqrt{2}}{3}\right)\left(-\tfrac{1}{3}\right) = -\frac{4\sqrt{2}}{9}$

62. Since $90° < A < 180°$, $45° < \dfrac{A}{2} < 90°$, thus $\cos\dfrac{A}{2} > 0$. Using the half-angle formula for cosine:

$$\cos\frac{A}{2} = \sqrt{\frac{1+\cos A}{2}} = \sqrt{\frac{1-\frac{1}{3}}{2}} = \sqrt{\frac{1}{3}} = \frac{1}{\sqrt{3}}$$

64. Using the result from Problem 62: $\sec\dfrac{A}{2} = \dfrac{1}{\cos\dfrac{A}{2}} = \dfrac{1}{\frac{1}{\sqrt{3}}} = \sqrt{3}$

66. Since $\sin\dfrac{A}{2} > 0$, using the half-angle formula for sine: $\sin\dfrac{A}{2} = \sqrt{\dfrac{1-\cos A}{2}} = \sqrt{\dfrac{1+\frac{1}{3}}{2}} = \sqrt{\dfrac{2}{3}} = \dfrac{\sqrt{2}}{\sqrt{3}}$

Therefore: $\tan\dfrac{A}{2} = \dfrac{\sin\dfrac{A}{2}}{\cos\dfrac{A}{2}} = \dfrac{\frac{\sqrt{2}}{\sqrt{3}}}{\frac{1}{\sqrt{3}}} = \sqrt{2}$

68. Drawing the figure, where x represents the length of the other guy wire:

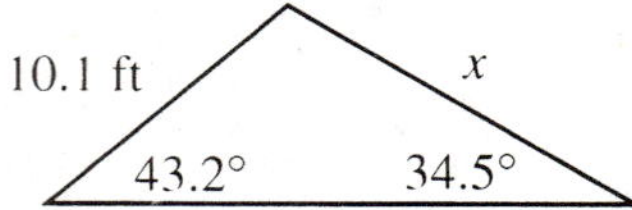

Using the law of sines:
$$\frac{x}{\sin 43.2°} = \frac{10.1}{\sin 34.5°}$$
$$x = \frac{10.1\sin 43.2°}{\sin 34.5°} \approx 12.2 \text{ ft}$$
The other guy wire is approximately 12.2 ft long.

70. Drawing the figure, where x represents the length of the longer diagonal:

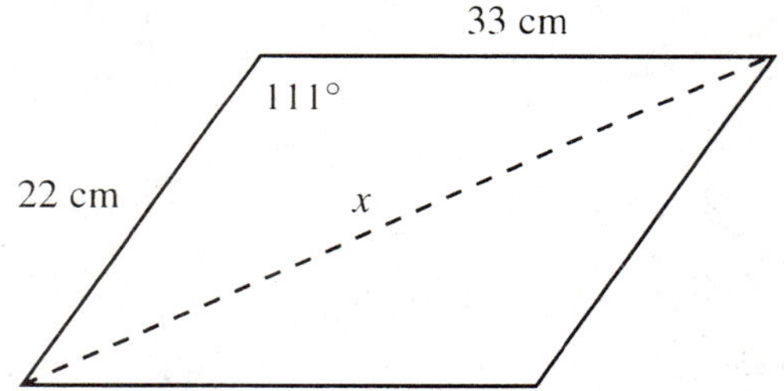

Using the law of cosines:
$$x^2 = 22^2 + 33^2 - 2(22)(33)\cos 111° \approx 2093$$
$$x = \sqrt{2093} \approx 46 \text{ cm}$$
The longer diagonal is approximately 46 cm long.

8.4 Roots of a Complex Number

2. The square roots are given by:
$$w_k = 16^{1/2}\left[\cos\frac{30° + 360°k}{2} + i\sin\frac{30° + 360°k}{2}\right] = 4\left[\cos(15° + 180°k) + i\sin(15° + 180°k)\right]$$
Substituting $k = 0,1$:
$$w_0 = 4(\cos 15° + i\sin 15°) \qquad\qquad w_1 = 4(\cos 195° + i\sin 195°)$$
Graphing these solutions:

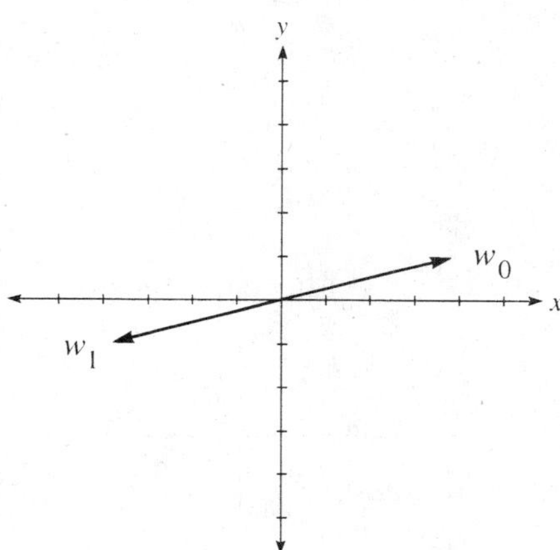

4. The square roots are given by:
$$w_k = 9^{1/2}\left[\cos\frac{310° + 360°k}{2} + i\sin\frac{310° + 360°k}{2}\right] = 3\left[\cos(155° + 180°k) + i\sin(155° + 180°k)\right]$$
Substituting $k = 0,1$:
$$w_0 = 3(\cos 155° + i\sin 155°) \qquad\qquad w_1 = 3(\cos 335° + i\sin 335°)$$
Graphing these solutions:

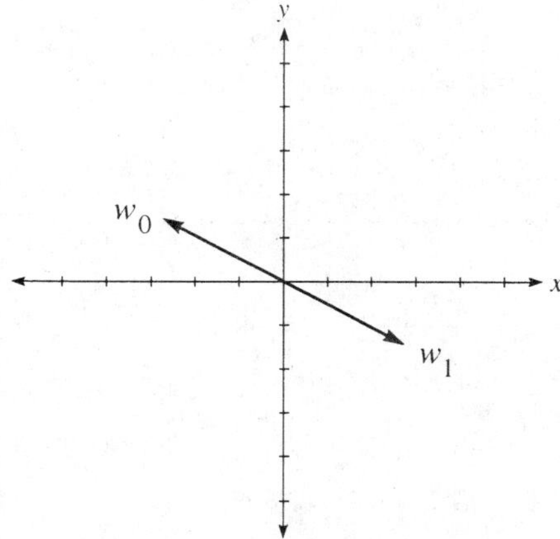

6.   The square roots are given by:
$$w_k = 81^{1/2}\left[\cos\frac{150° + 360°k}{2} + i\sin\frac{150° + 360°k}{2}\right] = 9\left[\cos(75° + 180°k) + i\sin(75° + 180°k)\right]$$

Substituting $k = 0,1$:
$$w_0 = 9(\cos 75° + i\sin 75°) = 9\operatorname{cis}75° \qquad w_1 = 9(\cos 255° + i\sin 255°) = 9\operatorname{cis}255°$$

Graphing these solutions:

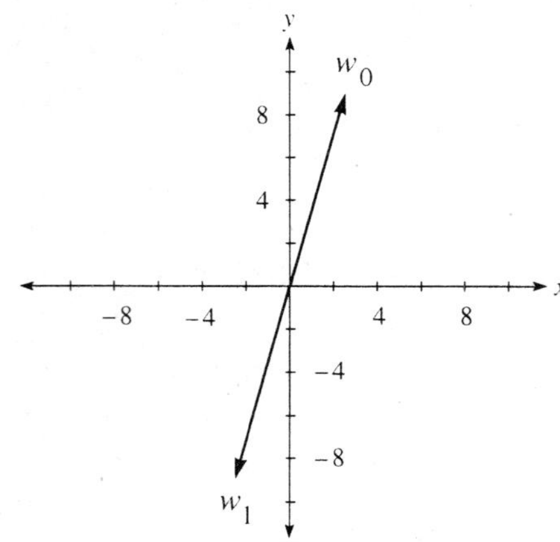

8.   First write the number in trigonometric form: $-2 + 2i\sqrt{3} = 4(\cos 120° + i\sin 120°)$
The square roots are given by:
$$w_k = 4^{1/2}\left[\cos\frac{120° + 360°k}{2} + i\sin\frac{120° + 360°k}{2}\right] = 2\left[\cos(60° + 180°k) + i\sin(60° + 180°k)\right]$$

Substituting $k = 0,1$:
$$w_0 = 2(\cos 60° + i\sin 60°) = 2\left(\frac{1}{2} + \frac{\sqrt{3}}{2}i\right) = 1 + i\sqrt{3}$$

$$w_1 = 2(\cos 240° + i\sin 240°) = 2\left(-\frac{1}{2} - \frac{\sqrt{3}}{2}i\right) = -1 - i\sqrt{3}$$

10.   First write the number in trigonometric form: $-4i = 4(\cos 270° + i\sin 270°)$
The square roots are given by:
$$w_k = 4^{1/2}\left[\cos\frac{270° + 360°k}{2} + i\sin\frac{270° + 360°k}{2}\right] = 2\left[\cos(135° + 180°k) + i\sin(135° + 180°k)\right]$$

Substituting $k = 0,1$:
$$w_0 = 2(\cos 135° + i\sin 135°) = 2\left(-\frac{\sqrt{2}}{2} + \frac{\sqrt{2}}{2}i\right) = -\sqrt{2} + i\sqrt{2}$$

$$w_1 = 2(\cos 315° + i\sin 315°) = 2\left(\frac{\sqrt{2}}{2} - \frac{\sqrt{2}}{2}i\right) = \sqrt{2} - i\sqrt{2}$$

12.   First write the number in trigonometric form: $25 = 25(\cos 0° + i\sin 0°)$

The square roots are given by: $w_k = 25^{1/2}\left[\cos\frac{0° + 360°k}{2} + i\sin\frac{0° + 360°k}{2}\right] = 5\left[\cos(180°k) + i\sin(180°k)\right]$

Substituting $k = 0,1$:
$$w_0 = 5(\cos 0° + i\sin 0°) = 5(1 + 0i) = 5 \qquad w_1 = 5(\cos 180° + i\sin 180°) = 5(-1 + 0i) = -5$$

14. First write the number in trigonometric form: $1 - i\sqrt{3} = 2(\cos 300° + i \sin 300°)$

The square roots are given by:

$$w_k = 2^{1/2}\left[\cos\frac{300° + 360°k}{2} + i \sin\frac{300° + 360°k}{2}\right] = \sqrt{2}\left[\cos(150° + 180°k) + i \sin(150° + 180°k)\right]$$

Substituting $k = 0,1$:

$$w_0 = \sqrt{2}(\cos 150° + i \sin 150°) = \sqrt{2}\left(-\frac{\sqrt{3}}{2} + \frac{1}{2}i\right) = -\frac{\sqrt{6}}{2} + \frac{\sqrt{2}}{2}i$$

$$w_1 = \sqrt{2}(\cos 330° + i \sin 330°) = \sqrt{2}\left(\frac{\sqrt{3}}{2} - \frac{1}{2}i\right) = \frac{\sqrt{6}}{2} - \frac{\sqrt{2}}{2}i$$

16. The cube roots are given by:

$$w_k = 27^{1/3}\left[\cos\frac{303° + 360°k}{3} + i \sin\frac{303° + 360°k}{3}\right] = 3\left[\cos(101° + 120°k) + i \sin(101° + 120°k)\right]$$

Substituting $k = 0,1,2$:

$$w_0 = 3(\cos 101° + i \sin 101°) \qquad w_1 = 3(\cos 221° + i \sin 221°)$$

$$w_2 = 3(\cos 341° + i \sin 341°)$$

18. First write the number in trigonometric form: $-4\sqrt{3} + 4i = 8(\cos 150° + i \sin 150°)$

The cube roots are given by:

$$w_k = 8^{1/3}\left[\cos\frac{150° + 360°k}{3} + i \sin\frac{150° + 360°k}{3}\right] = 2\left[\cos(50° + 120°k) + i \sin(50° + 120°k)\right]$$

Substituting $k = 0,1,2$:

$$w_0 = 2(\cos 50° + i \sin 50°) \qquad w_1 = 2(\cos 170° + i \sin 170°)$$

$$w_2 = 2(\cos 290° + i \sin 290°)$$

20. First write the number in trigonometric form: $8 = 8(\cos 0° + i \sin 0°)$

The cube roots are given by:

$$w_k = 8^{1/3}\left[\cos\frac{0° + 360°k}{3} + i \sin\frac{0° + 360°k}{3}\right] = 2\left[\cos(120°k) + i \sin(120°k)\right]$$

Substituting $k = 0,1,2$:

$$w_0 = 2(\cos 0° + i \sin 0°) \qquad w_1 = 2(\cos 120° + i \sin 120°)$$

$$w_2 = 2(\cos 240° + i \sin 240°)$$

22. First write the number in trigonometric form: $-64i = 64(\cos 270° + i \sin 270°)$

The cube roots are given by:

$$w_k = 64^{1/3}\left[\cos\frac{270° + 360°k}{3} + i \sin\frac{270° + 360°k}{3}\right] = 4\left[\cos(90° + 120°k) + i \sin(90° + 120°k)\right]$$

Substituting $k = 0,1,2$:

$$w_0 = 4(\cos 90° + i \sin 90°) \qquad w_1 = 4(\cos 210° + i \sin 210°)$$

$$w_2 = 4(\cos 330° + i \sin 330°)$$

24. Since $x^3 + 8 = 0$, $x^3 = -8$. We wish to find the three cube roots of -8. Writing -8 in trigonometric form:

$$-8 = 8(\cos 180° + i \sin 180°)$$

The cube roots are given by:

$$w_k = 8^{1/3}\left[\cos\frac{180° + 360°k}{3} + i\sin\frac{180° + 360°k}{3}\right] = 2\left[\cos(60° + 120°k) + i\sin(60° + 120°k)\right]$$

Substituting $k = 0, 1, 2$:

$$w_0 = 2(\cos 60° + i\sin 60°) = 2\left(\frac{1}{2} + \frac{\sqrt{3}}{2}i\right) = 1 + i\sqrt{3}$$

$$w_1 = 2(\cos 180° + i\sin 180°) = 2(-1 + 0i) = -2$$

$$w_2 = 2(\cos 300° + i\sin 300°) = 2\left(\frac{1}{2} - \frac{\sqrt{3}}{2}i\right) = 1 - i\sqrt{3}$$

26. Since $x^4 + 81 = 0$, $x^4 = -81$. We wish to find the four fourth roots of -81. Writing -81 in trigonometric form:

$$-81 = 81(\cos 180° + i\sin 180°)$$

The fourth roots are given by:

$$w_k = 81^{1/4}\left[\cos\frac{180° + 360°k}{4} + i\sin\frac{180° + 360°k}{4}\right] = 3\left[\cos(45° + 90°k) + i\sin(45° + 90°k)\right]$$

Substituting $k = 0, 1, 2, 3$:

$$w_0 = 3(\cos 45° + i\sin 45°) = 3\left(\frac{\sqrt{2}}{2} + \frac{\sqrt{2}}{2}i\right) = \frac{3\sqrt{2}}{2} + \frac{3\sqrt{2}}{2}i$$

$$w_1 = 3(\cos 135° + i\sin 135°) = 3\left(-\frac{\sqrt{2}}{2} + \frac{\sqrt{2}}{2}i\right) = -\frac{3\sqrt{2}}{2} + \frac{3\sqrt{2}}{2}i$$

$$w_2 = 3(\cos 225° + i\sin 225°) = 3\left(-\frac{\sqrt{2}}{2} - \frac{\sqrt{2}}{2}i\right) = -\frac{3\sqrt{2}}{2} - \frac{3\sqrt{2}}{2}i$$

$$w_3 = 3(\cos 315° + i\sin 315°) = 3\left(\frac{\sqrt{2}}{2} - \frac{\sqrt{2}}{2}i\right) = \frac{3\sqrt{2}}{2} - \frac{3\sqrt{2}}{2}i$$

28. The fourth roots are given by:

$$w_k = 1^{1/4}\left[\cos\frac{240° + 360°k}{4} + i\sin\frac{240° + 360°k}{4}\right] = \cos(60° + 90°k) + i\sin(60° + 90°k)$$

Substituting $k = 0, 1, 2, 3$:

$$w_0 = \cos 60° + i\sin 60° \qquad\qquad w_1 = \cos 150° + i\sin 150°$$

$$w_2 = \cos 240° + i\sin 240° \qquad\qquad w_3 = \cos 330° + i\sin 330°$$

30. The fifth roots are given by:

$$w_k = \left(10^{10}\right)^{1/5}\left[\cos\frac{75° + 360°k}{5} + i\sin\frac{75° + 360°k}{5}\right] = 100\left[\cos(15° + 72°k) + i\sin(15° + 72°k)\right]$$

Substituting $k = 0, 1, 2, 3, 4$:

$$w_0 = 100(\cos 15° + i\sin 15°) \approx 96.59 + 25.88i$$

$$w_1 = 100(\cos 87° + i\sin 87°) \approx 5.23 + 99.86i$$

$$w_2 = 100(\cos 159° + i\sin 159°) \approx -93.36 + 35.84i$$

$$w_3 = 100(\cos 231° + i\sin 231°) \approx -62.93 - 77.71i$$

$$w_4 = 100(\cos 303° + i\sin 303°) \approx 54.46 - 83.87i$$

32. Write 1 in trigonometric form: $1 = \cos 0° + i \sin 0°$

The sixth roots are given by: $w_k = 1^{1/6}\left[\cos\dfrac{0° + 360°k}{6} + i\sin\dfrac{0° + 360°k}{6}\right] = \cos(60°k) + i\sin(60°k)$

Substituting $k = 0,1,2,3,4,5$:

$$w_0 = \cos 0° + i\sin 0° \qquad\qquad w_1 = \cos 60° + i\sin 60°$$
$$w_2 = \cos 120° + i\sin 120° \qquad\qquad w_3 = \cos 180° + i\sin 180°$$
$$w_4 = \cos 240° + i\sin 240° \qquad\qquad w_5 = \cos 300° + i\sin 300°$$

Graphing these roots:

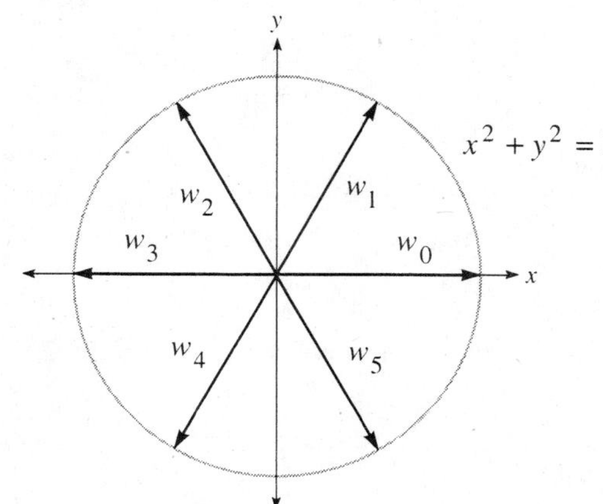

34. Using the quadratic formula with $a = 1$, $b = 2$, and $c = 4$:

$$x^2 = \frac{-2 \pm \sqrt{2^2 - 4(1)(4)}}{2(1)} = \frac{-2 \pm \sqrt{-12}}{2} = \frac{-2 \pm 2i\sqrt{3}}{2} = -1 \pm i\sqrt{3}$$

Thus $x^2 = -1 + i\sqrt{3}$ or $x^2 = -1 - i\sqrt{3}$. Writing $-1 + i\sqrt{3}$ in trigonometric form: $-1 + i\sqrt{3} = 2(\cos 120° + i\sin 120°)$

Its square roots are given by:

$$w_k = 2^{1/2}\left[\cos\frac{120° + 360°k}{2} + i\sin\frac{120° + 360°k}{2}\right] = \sqrt{2}\left[\cos(60° + 180°k) + i\sin(60° + 180°k)\right]$$

Substituting $k = 0,1$:

$$w_0 = \sqrt{2}(\cos 60° + i\sin 60°) \qquad\qquad w_1 = \sqrt{2}(\cos 240° + i\sin 240°)$$

Writing $-1 - i\sqrt{3}$ in trigonometric form: $-1 - i\sqrt{3} = 2(\cos 240° + i\sin 240°)$

Its square roots are given by:

$$w_k = 2^{1/2}\left[\cos\frac{240° + 360°k}{2} + i\sin\frac{240° + 360°k}{2}\right] = \sqrt{2}\left[\cos(120° + 180°k) + i\sin(120° + 180°k)\right]$$

Substituting $k = 0,1$:

$$w_0 = \sqrt{2}(\cos 120° + i\sin 120°) \qquad\qquad w_1 = \sqrt{2}(\cos 300° + i\sin 300°)$$

Summarizing, the solutions are $\sqrt{2}(\cos\theta + i\sin\theta)$, where $\theta = 60°, 120°, 240°, 300°$.

36. Using the quadratic formula with $a = 1$, $b = -2$, and $c = 2$:

$$x^2 = \frac{-(-2) \pm \sqrt{(-2)^2 - 4(1)(2)}}{2(1)} = \frac{2 \pm \sqrt{-4}}{2} = \frac{2 \pm 2i}{2} = 1 \pm i$$

Thus $x^2 = 1 + i$ or $x^2 = 1 - i$. Writing $1 + i$ in trigonometric form: $1 + i = \sqrt{2}\left(\cos 45° + i \sin 45°\right)$

Its square roots are given by:

$$w_k = \left(\sqrt{2}\right)^{1/2}\left[\cos\frac{45° + 360°k}{2} + i\sin\frac{45° + 360°k}{2}\right] = \sqrt[4]{2}\left[\cos(22.5° + 180°k) + i\sin(22.5° + 180°k)\right]$$

Substituting $k = 0, 1$:

$$w_0 = \sqrt[4]{2}\left(\cos 22.5° + i \sin 22.5°\right) \qquad w_1 = \sqrt[4]{2}\left(\cos 202.5° + i \sin 202.5°\right)$$

Writing $1 - i$ in trigonometric form: $1 - i = \sqrt{2}\left(\cos 315° + i \sin 315°\right)$

Its square roots are given by:

$$w_k = \left(\sqrt{2}\right)^{1/2}\left[\cos\frac{315° + 360°k}{2} + i\sin\frac{315° + 360°k}{2}\right] = \sqrt[4]{2}\left[\cos(157.5° + 180°k) + i\sin(157.5° + 180°k)\right]$$

Substituting $k = 0, 1$:

$$w_0 = \sqrt[4]{2}\left(\cos 157.5° + i \sin 157.5°\right) \qquad w_1 = \sqrt[4]{2}\left(\cos 337.5° + i \sin 337.5°\right)$$

Summarizing, the solutions are $\sqrt[4]{2}\left(\cos\theta + i\sin\theta\right)$, where $\theta = 22.5°, 157.5°, 202.5°, 337.5°$.

38. Note that $y = -2\cos(-3x) = -2\cos 3x$. The amplitude is 2, the period is $\dfrac{2\pi}{3}$, and the graph is a reflection of $y = 2\cos 3x$ across the x-axis:

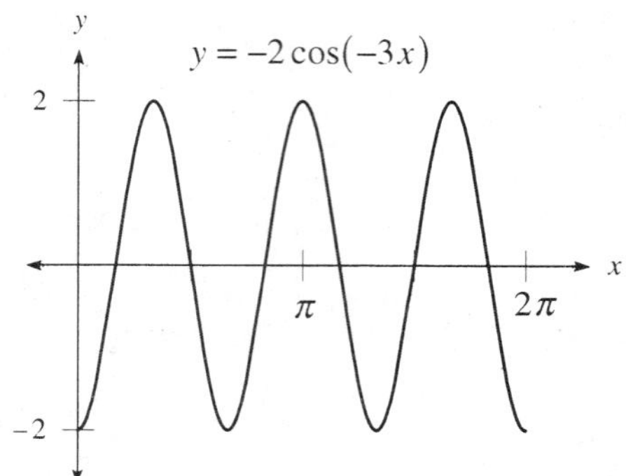

40. The amplitude is 1, the period is $\dfrac{2\pi}{2} = \pi$, and the phase shift is $\dfrac{\pi/2}{2} = \dfrac{\pi}{4}$. Note this graph is a reflection of $y = \cos\left(2x - \dfrac{\pi}{2}\right)$ across the x-axis:

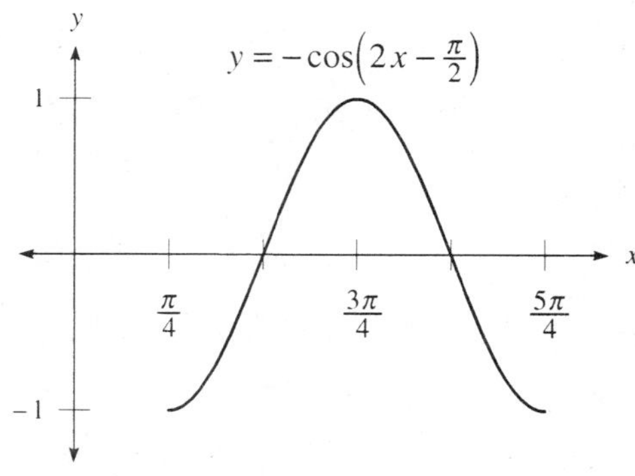

42. The amplitude is 3, the period is $\dfrac{2\pi}{\pi/3} = 6$, and the phase shift is $\dfrac{\pi/3}{\pi/3} = 1$. Sketching the graph:

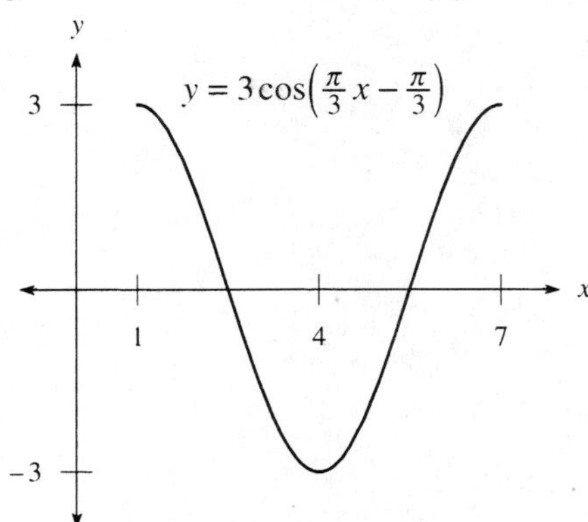

44. Using the area formula: $S = \frac{1}{2}ac\sin B = \frac{1}{2}(44.4)(22.2)\sin 21.8° \approx 183 \text{ cm}^2$

46. First compute $s = \frac{1}{2}(a+b+c) = \frac{1}{2}(5.4 + 4.3 + 3.2) = 6.45$. Using Heron's formula:

$$S = \sqrt{s(s-a)(s-b)(s-c)} = \sqrt{6.45(6.45-5.4)(6.45-4.3)(6.45-3.2)} \approx 6.9 \text{ ft}^2$$

8.5 Polar Coordinates

2. Graphing the ordered pair:

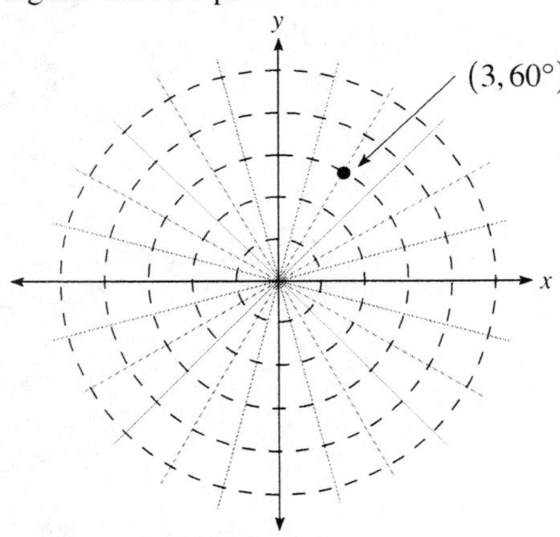

4. Graphing the ordered pair:

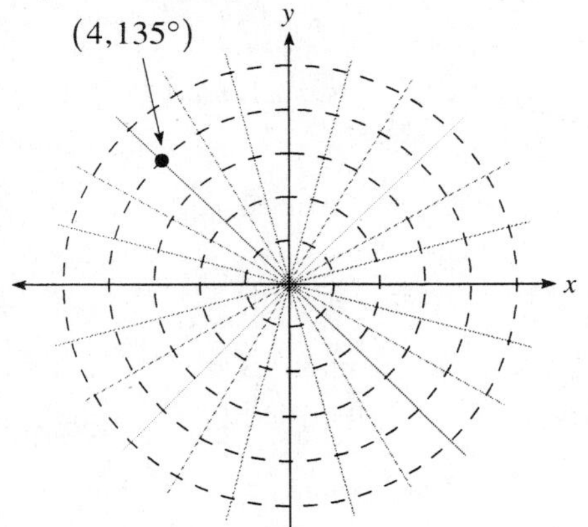

6. Graphing the ordered pair:

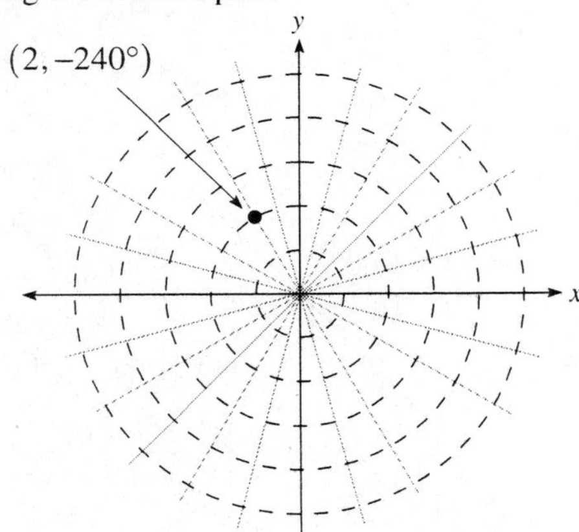

8. Graphing the ordered pair:

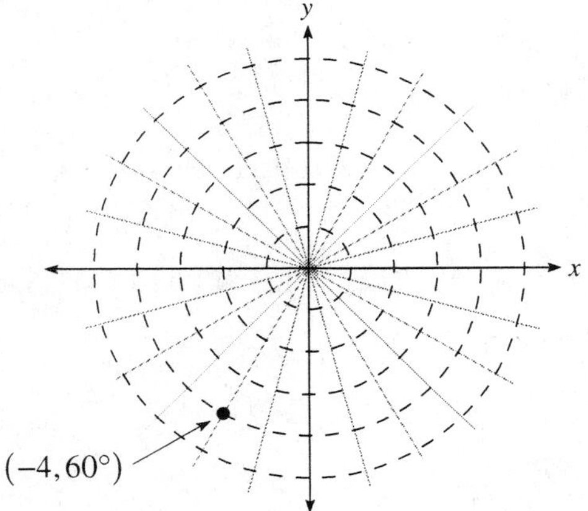

10.   Graphing the ordered pair:

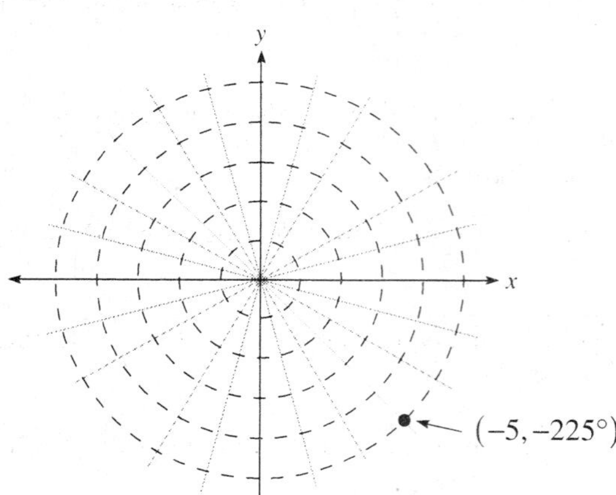

12.   Graphing the ordered pair:

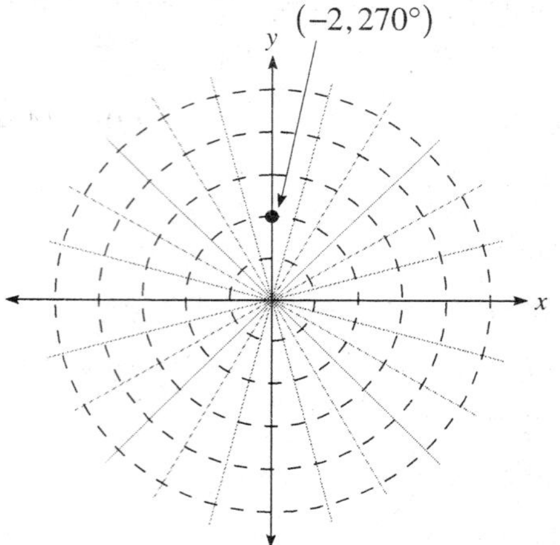

14.   Three other ordered pairs are $(1, -330°), (-1, 210°),$ and $(-1, -150°)$.

16.   Three other ordered pairs are $(3, -240°), (-3, -60°),$ and $(-3, 300°)$.

18.   Three other ordered pairs are $(-2, -315°), (2, 225°),$ and $(2, -135°)$.

20.   Here $r = -2$ and $\theta = 60°$. Converting to rectangular coordinates:

$$x = r\cos\theta = -2\cos 60° = -2 \cdot \tfrac{1}{2} = -1 \qquad y = r\sin\theta = -2\sin 60° = -2 \cdot \frac{\sqrt{3}}{2} = -\sqrt{3}$$

The rectangular coordinates are $\left(-1, -\sqrt{3}\right)$.

22.   Here $r = 1$ and $\theta = 180°$. Converting to rectangular coordinates:
$$x = r\cos\theta = 1\cos 180° = 1 \cdot (-1) = -1 \qquad y = r\sin\theta = 1\sin 180° = 1 \cdot 0 = 0$$
The rectangular coordinates are $(-1, 0)$.

24.   Here $r = \sqrt{2}$ and $\theta = -225°$. Converting to rectangular coordinates:

$$x = r\cos\theta = \sqrt{2}\cos(-225°) = \sqrt{2} \cdot \left(-\frac{1}{\sqrt{2}}\right) = -1$$

$$y = r\sin\theta = \sqrt{2}\sin(-225°) = \sqrt{2} \cdot \frac{1}{\sqrt{2}} = 1$$

The rectangular coordinates are $(-1, 1)$.

26.   Here $r = 4\sqrt{3}$ and $\theta = -30°$. Converting to rectangular coordinates:

$$x = r\cos\theta = 4\sqrt{3}\cos(-30°) = 4\sqrt{3} \cdot \left(\frac{\sqrt{3}}{2}\right) = 6$$

$$y = r\sin\theta = 4\sqrt{3}\sin(-30°) = 4\sqrt{3} \cdot \left(-\tfrac{1}{2}\right) = -2\sqrt{3}$$

The rectangular coordinates are $\left(6, -2\sqrt{3}\right)$.

28.   Here $r = 3$ and $\theta = 124°$. Converting to rectangular coordinates:
$$x = r\cos\theta = 3\cos 124° \approx -1.678 \qquad y = r\sin\theta = 3\sin 124° \approx 2.487$$
The rectangular coordinates are $(-1.678, 2.487)$.

30.   Here $r = -4$ and $\theta = 261°$. Converting to rectangular coordinates:
$$x = r\cos\theta = -4\cos 261° \approx 0.6257 \qquad y = r\sin\theta = -4\sin 261° \approx 3.951$$
The rectangular coordinates are $(0.6257, 3.951)$.

32.   Compute $r = \sqrt{(-3)^2 + (-3)^2} = \sqrt{9+9} = \sqrt{18} = 3\sqrt{2}$. Also $\tan\theta = \dfrac{-3}{-3} = 1$ with θ in quadrant III, so $\theta = 225°$. The polar coordinates are $\left(3\sqrt{2}, 225°\right)$.

34. Compute $r = \sqrt{2^2 + \left(-2\sqrt{3}\right)^2} = \sqrt{4+12} = \sqrt{16} = 4$. Also $\tan\theta = \dfrac{-2\sqrt{3}}{2} = -\sqrt{3}$ with θ in quadrant IV, so $\theta = 300°$. The polar coordinates are $(4, 300°)$.

36. Compute $r = \sqrt{(-2)^2 + 0^2} = \sqrt{4+0} = \sqrt{4} = 2$. Also $\tan\theta = \dfrac{0}{-2} = 0$, so $\theta = 180°$. The polar coordinates are $(2, 180°)$.

38. Compute $r = \sqrt{(-1)^2 + \left(-\sqrt{3}\right)^2} = \sqrt{1+3} = \sqrt{4} = 2$. Also $\tan\theta = \dfrac{-\sqrt{3}}{-1} = \sqrt{3}$ with θ in quadrant III, so $\theta = 240°$. The polar coordinates are $(2, 240°)$.

40. Compute $r = \sqrt{4^2 + 3^2} = \sqrt{16+9} = \sqrt{25} = 5$. Also $\tan\theta = \frac{3}{4}$ with θ in quadrant I, so $\theta \approx 36.9°$. The polar coordinates are $(5, 36.9°)$.

42. Compute $r = \sqrt{1^2 + (-2)^2} = \sqrt{1+4} = \sqrt{5}$. Also $\tan\theta = \dfrac{-2}{1} = -2$ with θ in quadrant IV, so $\theta \approx 296.6°$. The polar coordinates are $\left(\sqrt{5}, 296.6°\right)$.

44. Compute $r = \sqrt{(-3)^2 + (-2)^2} = \sqrt{9+4} = \sqrt{13}$. Also $\tan\theta = \dfrac{-2}{-3} = \frac{2}{3}$ with θ in quadrant III, so $\theta \approx 213.7°$. The polar coordinates are $\left(\sqrt{13}, 213.7°\right)$.

46. The polar coordinates are $(9.220, 102.5°)$.

48. The polar coordinates are $(7.616, 336.8°)$.

50. Since $r^2 = x^2 + y^2$, the equation is $x^2 + y^2 = 4$.

52. Since $x = r\cos\theta$, $\cos\theta = \dfrac{x}{r}$. Therefore:

$$r = 6\cos\theta$$
$$r = \frac{6x}{r}$$
$$r^2 = 6x$$
$$x^2 + y^2 = 6x$$

54. Using the double-angle formula for cosine:

$$r^2 = 4\cos 2\theta$$
$$r^2 = 4\left(\cos^2\theta - \sin^2\theta\right)$$
$$r^2 = 4\left(\frac{x^2}{r^2} - \frac{y^2}{r^2}\right)$$
$$r^4 = 4\left(x^2 - y^2\right)$$
$$\left(x^2 + y^2\right)^2 = 4\left(x^2 - y^2\right)$$

56. Since $x = r\cos\theta$ and $y = r\sin\theta$:
$$r(\cos\theta - \sin\theta) = 2$$
$$r\cos\theta - r\sin\theta = 2$$
$$x - y = 2$$

58. Substituting $x = r\cos\theta$ and $y = r\sin\theta$:
$$x + y = 5$$
$$r\cos\theta + r\sin\theta = 5$$
$$r(\cos\theta + \sin\theta) = 5$$

60. Since $r^2 = x^2 + y^2$, the equation is $r^2 = 9$, or $r = 3$.

62. Since $x^2 + y^2 = r^2$ and $x = r\cos\theta$:
$$x^2 + y^2 = 4x$$
$$r^2 = 4r\cos\theta$$
$$r = 4\cos\theta$$

64. Since $x = r\cos\theta$ and $y = r\sin\theta$:
$$y = -x$$
$$r\sin\theta = -r\cos\theta$$
$$\sin\theta = -\cos\theta$$
Note that $\theta = 135°$ is another way to write this equation.

66.   The amplitude is 6 and the period is 2π:

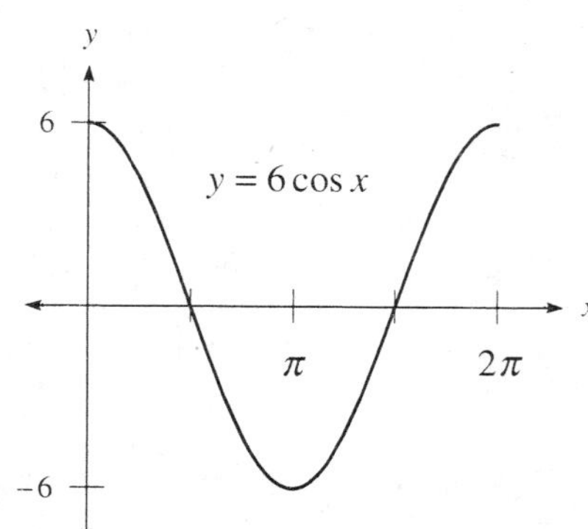

68.   The amplitude is 2 and the period is $\dfrac{2\pi}{4} = \dfrac{\pi}{2}$:

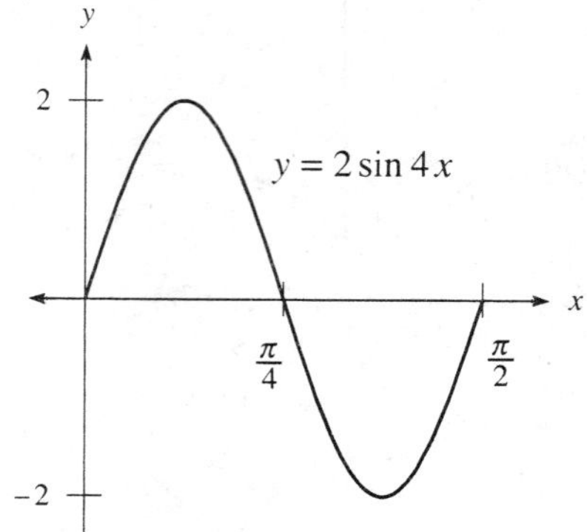

70.   The amplitude is 2 and the period is 2π. Note this is a vertical translation of $y = 2\cos x$ by 4 units:

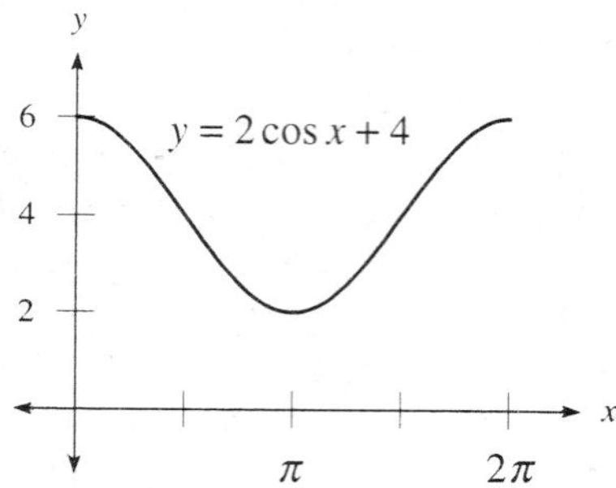

8.6 Equations in Polar Coordinates and Their Graphs

2.   Making a table of values:

θ	0°	45°	90°	135°	180°	225°	270°	315°	360°
$r = 4\sin\theta$	0	$2\sqrt{2}$	4	$2\sqrt{2}$	0	$-2\sqrt{2}$	-4	$-2\sqrt{2}$	0

Sketching the graph:

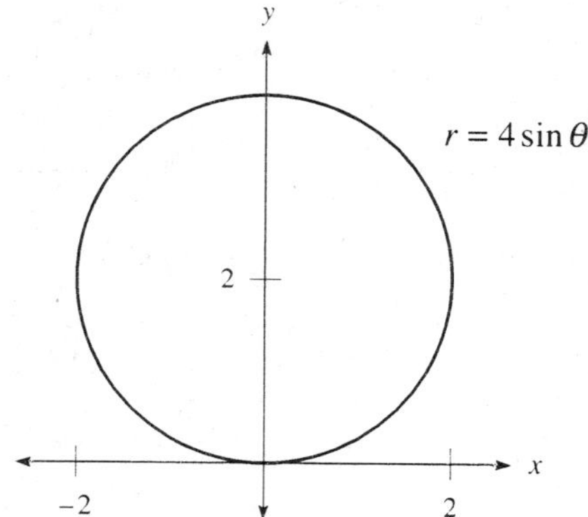

4.   Making a table of values:

θ	0°	45°	90°	135°	180°	225°	270°	315°	360°
$r = \cos 3\theta$	1	$-\dfrac{\sqrt{2}}{2}$	0	$\dfrac{\sqrt{2}}{2}$	-1	$\dfrac{\sqrt{2}}{2}$	0	$-\dfrac{\sqrt{2}}{2}$	1

Sketching the graph:

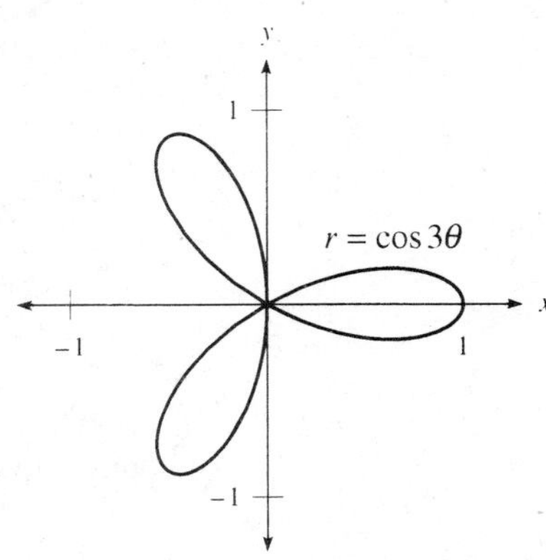

6.   Making a table of values:

θ	0°	15°	30°	45°	60°	75°	90°	105°	120°	135°	150°	165°	180°
$r = 2\sin 2\theta$	0	1	$\sqrt{3}$	2	$\sqrt{3}$	1	0	-1	$-\sqrt{3}$	-2	$-\sqrt{3}$	-1	0

Sketching the graph:

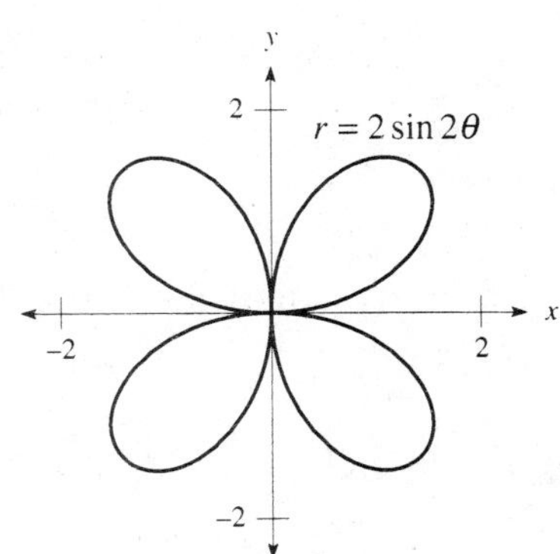

8.   Making a table of values:

θ	0°	15°	30°	45°	60°	75°	90°	105°	120°	135°	150°	165°	180°
$r = 3 + 3\cos\theta$	6	5.9	5.6	5.1	4.5	3.8	3	2.2	1.5	0.9	0.4	0.1	0

θ	180°	195°	210°	225°	240°	255°	270°	285°	300°	315°	330°	345°	360°
$r = 3 + 3\cos\theta$	0	0.1	0.4	0.9	1.5	2.2	3	3.8	4.5	5.1	5.6	5.9	6

Sketching the graph:

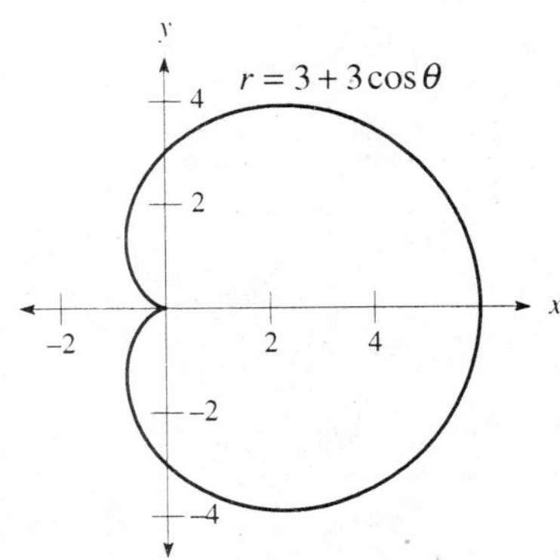

10.   Graphing the equation:

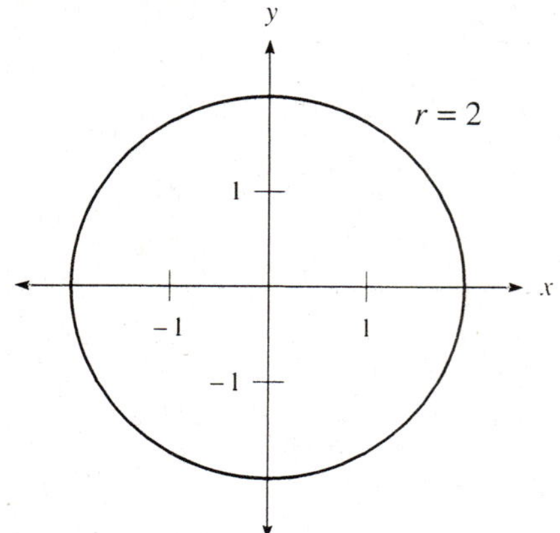

12.   Graphing the equation:

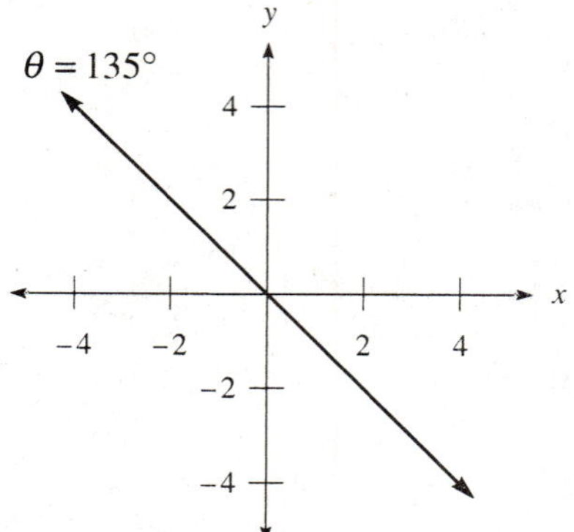

14.   Graphing the equation:

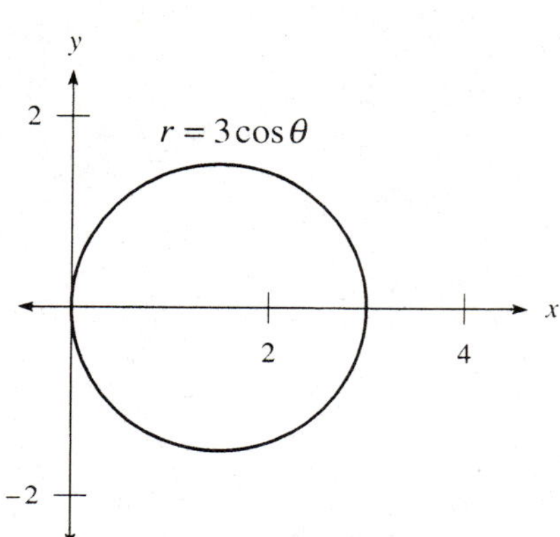

16.   Graphing the equation:

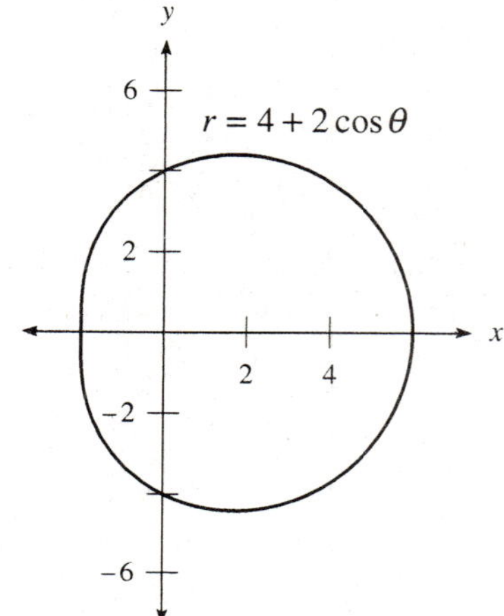

18.   Graphing the equation:

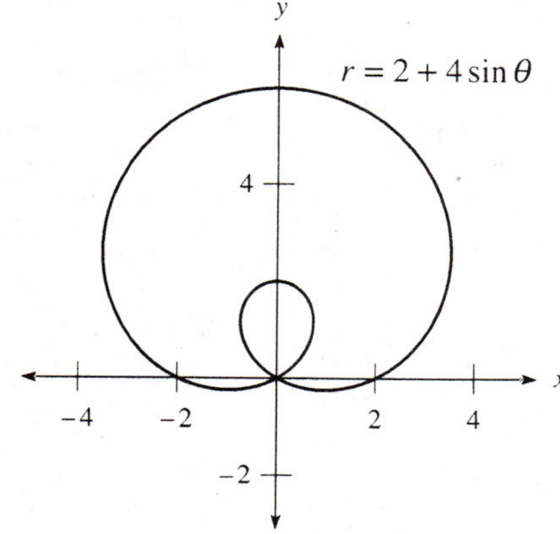

20.   Graphing the equation:

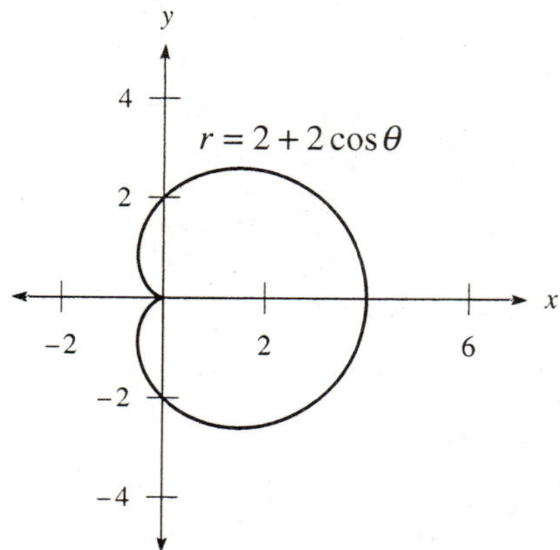

22.   Graphing the equation:

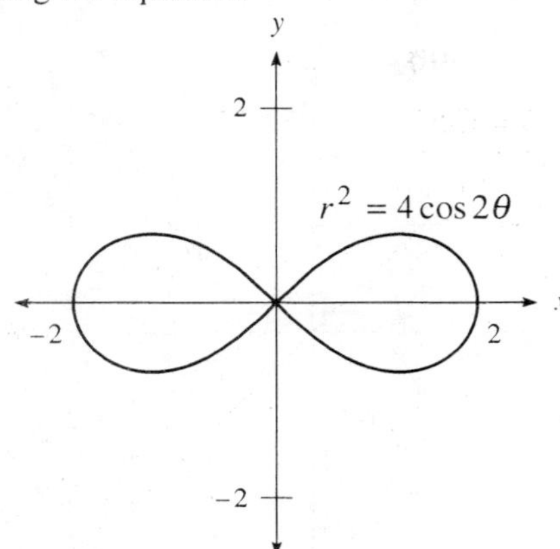

24.   Graphing the equation:

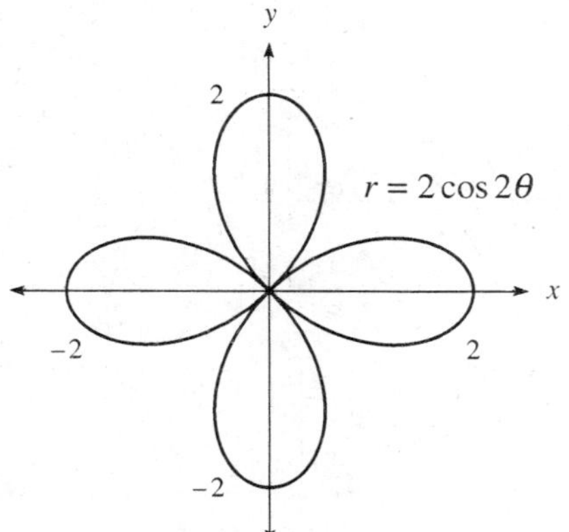

26.   Graphing the equation:

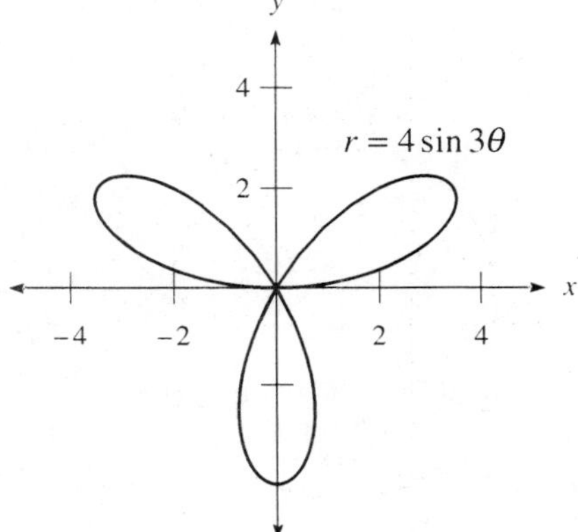

28.   Graphing the equation:

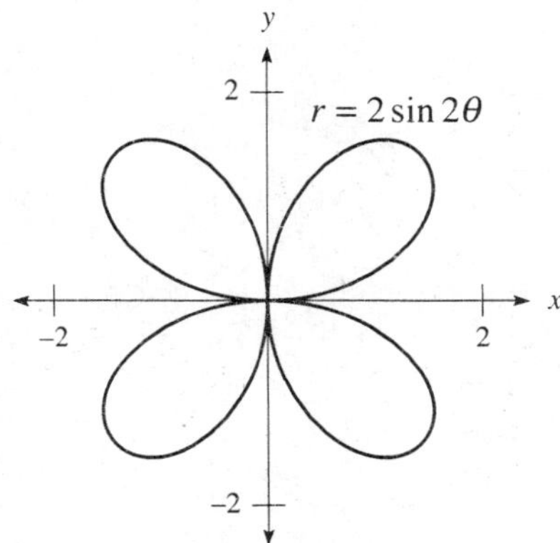

30.   Graphing the equation:

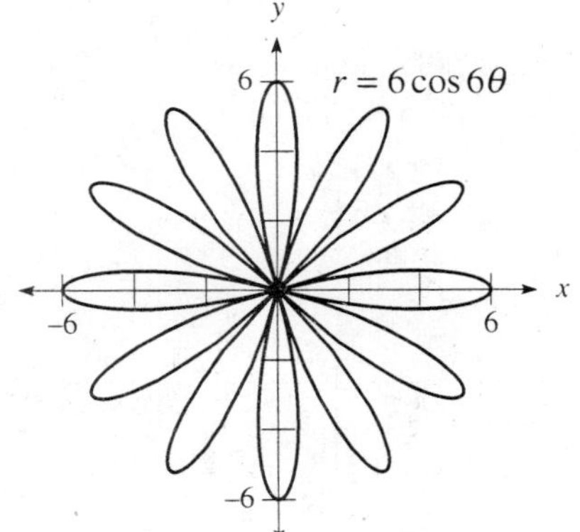

32.   Graphing the equation:

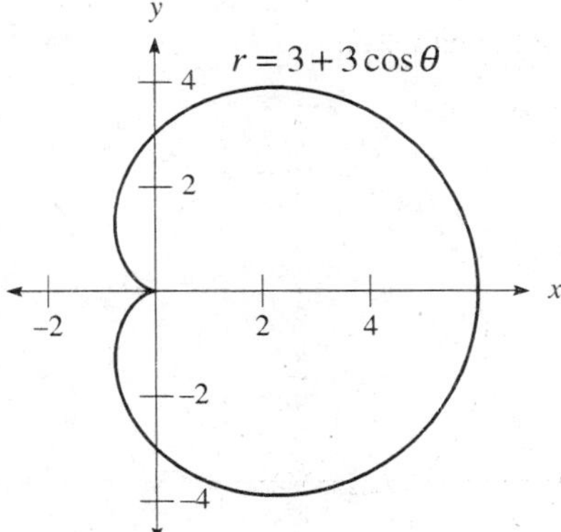

34.   Graphing the equation:

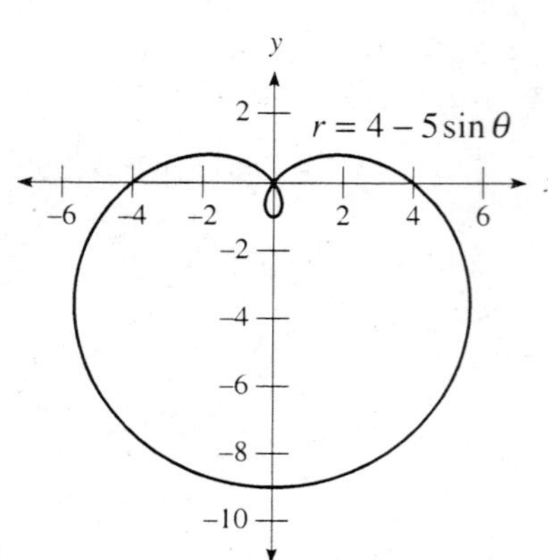

36.   Graphing the equation:

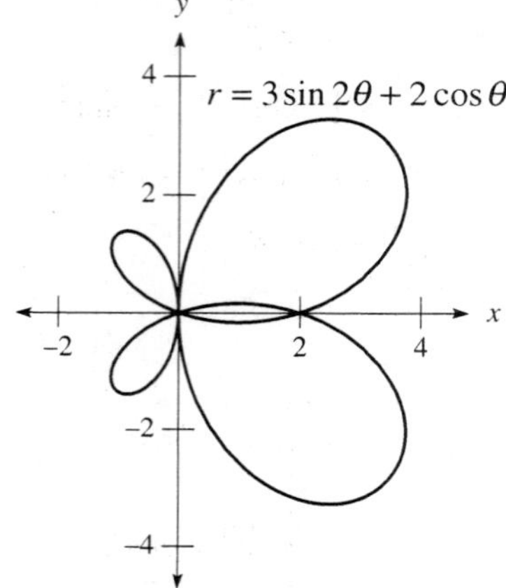

38.   Graphing the equation:

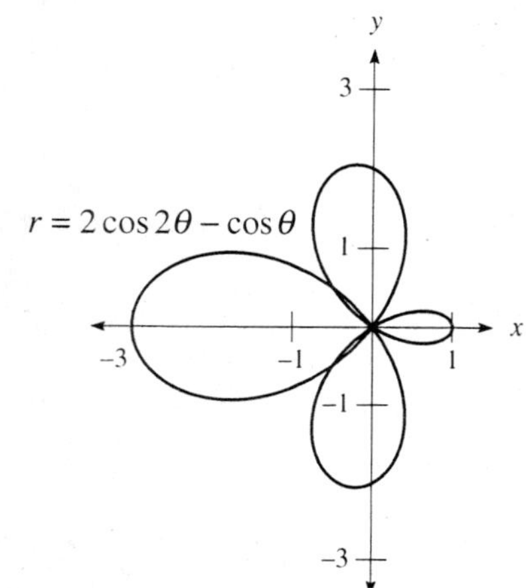

40.   Converting to polar coordinates:

$$x^2 + y^2 = 25$$
$$r^2 = 25$$
$$r = 5$$

Sketching the graph:

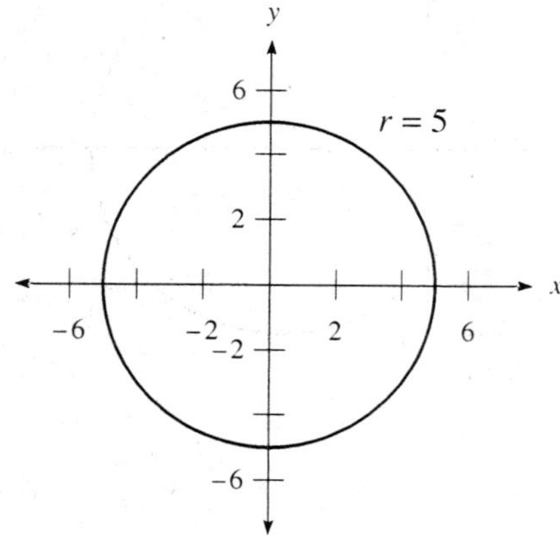

42.   Converting to polar coordinates:

$$x^2 + y^2 = 6y$$
$$r^2 = 6r\sin\theta$$
$$r = 6\sin\theta$$

Sketching the graph:

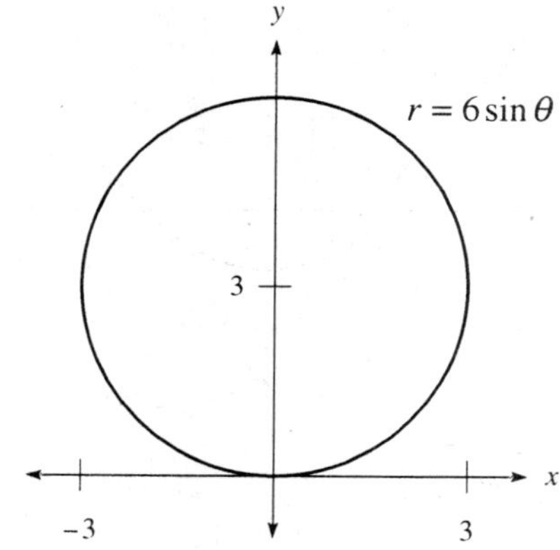

44.   Converting to polar coordinates:
$$\left(x^2 + y^2\right)^2 = x^2 - y^2$$
$$\left(r^2\right)^2 = r^2\cos^2\theta - r^2\sin^2\theta$$
$$r^4 = r^2\left(\cos^2\theta - \sin^2\theta\right)$$
$$r^2 = \cos 2\theta$$

Sketching the graph:

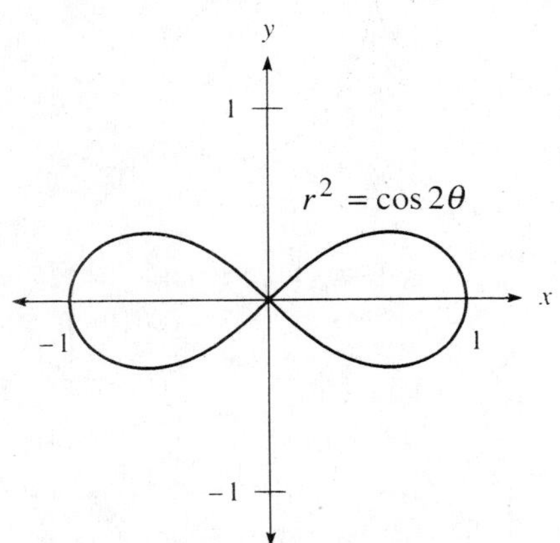

46.   Converting to rectangular coordinates:
$$r(3\cos\theta - 2\sin\theta) = 6$$
$$3r\cos\theta - 2r\sin\theta = 6$$
$$3x - 2y = 6$$

Sketching the graph:

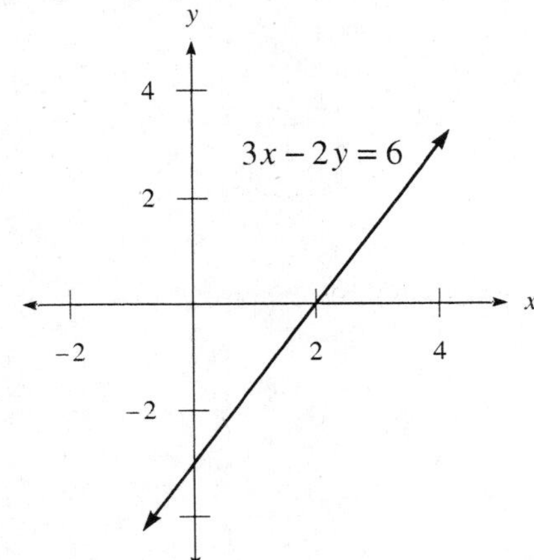

48.   Converting to rectangular coordinates:
$$r(1 - \sin\theta) = 1$$
$$r - r\sin\theta = 1$$
$$r = 1 + r\sin\theta$$
$$\sqrt{x^2 + y^2} = 1 + y$$
$$x^2 + y^2 = (1 + y)^2$$
$$x^2 + y^2 = 1 + 2y + y^2$$
$$x^2 = 2y + 1$$

Sketching the graph:

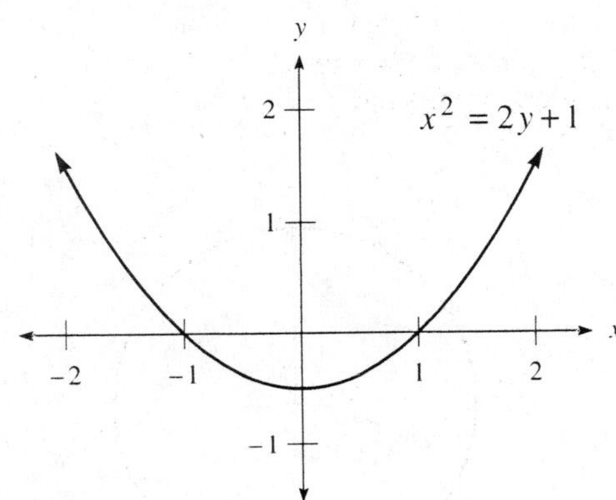

50.   Converting to rectangular coordinates:
$$r = 6\cos\theta$$
$$r^2 = 6r\cos\theta$$
$$x^2 + y^2 = 6x$$
$$x^2 - 6x + y^2 = 0$$
$$x^2 - 6x + 9 + y^2 = 9$$
$$(x - 3)^2 + y^2 = 9$$

Sketching the graph:

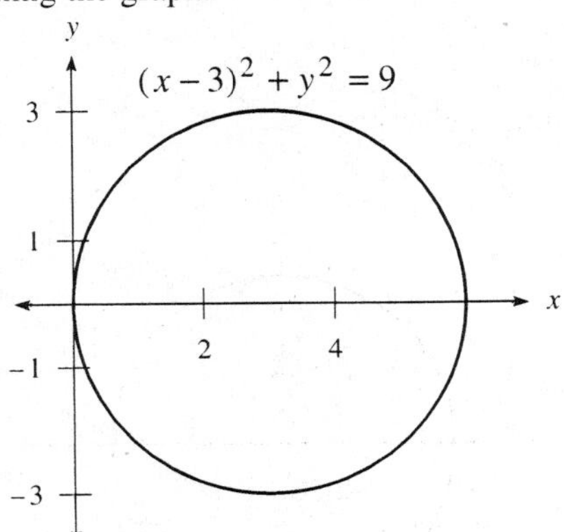

52. Graphing the two curves:

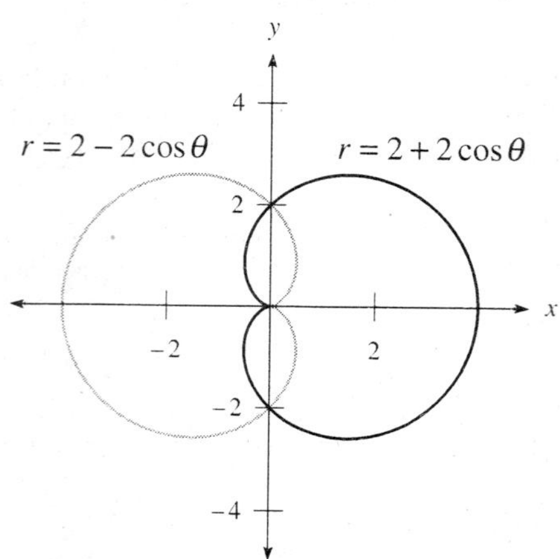

Two points they have in common are $(0,2) = (2,90°)$ and $(0,-2) = (2,270°)$. Note that although the rectangular point $(0,0)$ lies on both curves, and thus is a common point, the polar point $(0,0°)$ is not a point in common, since it does not satisfy both equations.

54. Graphing the equation:

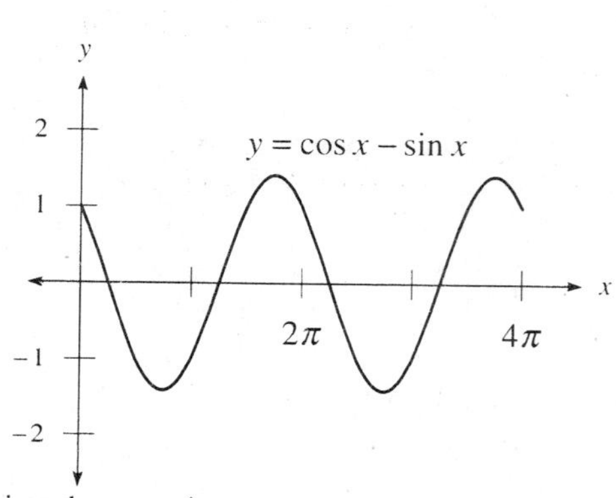

56. Graphing the equation:

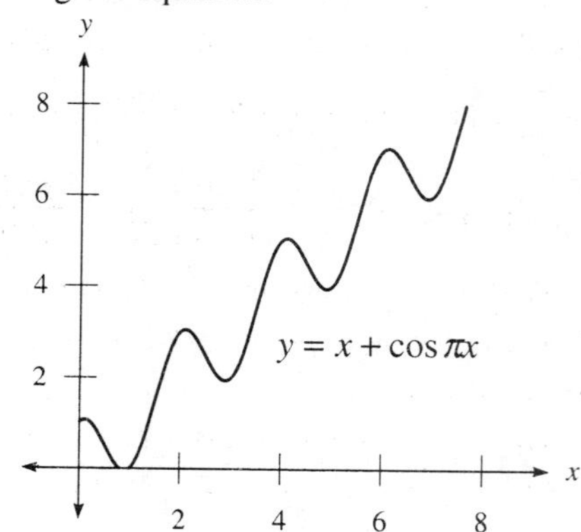

58. Graphing the equation:

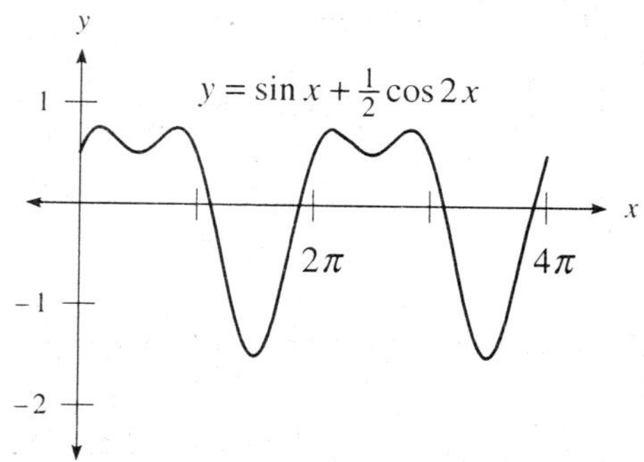

Chapter 8 Test

1. Writing in terms of i: $\sqrt{-25} = i\sqrt{25} = 5i$

2. Writing in terms of i: $\sqrt{-12} = i\sqrt{12} = 2i\sqrt{3}$

3. Setting the real parts equal:
$$7x = 14$$
$$x = 2$$

Setting the imaginary parts equal:
$$-6 = -3y$$
$$y = 2$$

4. Setting the real parts equal:
$$x^2 - 3x = 10$$
$$x^2 - 3x - 10 = 0$$
$$(x-5)(x+2) = 0$$
$$x = -2, 5$$

Setting the imaginary parts equal:
$$16 = 8y$$
$$y = 2$$

5. Combining the numbers: $(6-3i) + \left[(4-2i)-(3+i)\right] = (6-3i) + (4-2i-3-i) = 6-3i+1-3i = 7-6i$

6. Combining the numbers:
$$(7+3i) - \left[(2+i)-(3-4i)\right] = (7+3i) - (2+i-3+4i) = (7+3i) - (-1+5i) = 7+3i+1-5i = 8-2i$$

7. Computing the power: $i^{16} = \left(i^4\right)^4 = 1^4 = 1$

8. Computing the power: $i^{17} = i^{16} \cdot i = \left(i^4\right)^4 \cdot i = 1 \cdot i = i$

9. Computing the product: $(8+5i)(8-5i) = 64 + 40i - 40i - 25i^2 = 64 + 25 = 89$

10. Computing the product: $(3+5i)^2 = (3+5i)(3+5i) = 9 + 15i + 15i + 25i^2 = 9 + 30i - 25 = -16 + 30i$

11. Finding the quotient: $\dfrac{5-4i}{2i} \cdot \dfrac{i}{i} = \dfrac{5i-4i^2}{2i^2} = \dfrac{5i+4}{-2} = -2 - \frac{5}{2}i$

12. Finding the quotient: $\dfrac{6+5i}{6-5i} \cdot \dfrac{6+5i}{6+5i} = \dfrac{36+60i+25i^2}{36-25i^2} = \dfrac{36+60i-25}{36+25} = \dfrac{11+60i}{61} = \frac{11}{61} + \frac{60}{61}i$

13.
 a. The absolute value is: $|3+4i| = \sqrt{3^2 + 4^2} = \sqrt{9+16} = \sqrt{25} = 5$
 b. The opposite is: $-(3+4i) = -3-4i$
 c. The conjugate is: $3-4i$

14.
 a. The absolute value is: $|3-4i| = \sqrt{3^2 + (-4)^2} = \sqrt{9+16} = \sqrt{25} = 5$
 b. The opposite is: $-(3-4i) = -3+4i$
 c. The conjugate is: $3+4i$

15.
 a. The absolute value is: $|8i| = \sqrt{0^2 + 8^2} = \sqrt{0+64} = \sqrt{64} = 8$
 b. The opposite is: $-(8i) = -8i$
 c. The conjugate is: $-8i$

16.
 a. The absolute value is: $|-4| = \sqrt{(-4)^2 + 0^2} = \sqrt{16+0} = \sqrt{16} = 4$
 b. The opposite is: $-(-4) = 4$
 c. The conjugate is: -4

17. Writing in standard form: $8(\cos 330° + i\sin 330°) = 8\left(\dfrac{\sqrt{3}}{2} - \frac{1}{2}i\right) = 4\sqrt{3} - 4i$

18. Writing in standard form: $2\operatorname{cis}135° = 2(\cos 135° + i\sin 135°) = 2\left(-\dfrac{\sqrt{2}}{2} + \dfrac{\sqrt{2}}{2}i\right) = -\sqrt{2} + i\sqrt{2}$

19. Here $x = 2$, $y = 2$, and $r = \sqrt{2^2 + 2^2} = \sqrt{4+4} = \sqrt{8} = 2\sqrt{2}$. Since $\tan\theta = \dfrac{2}{2} = 1$ in quadrant I, $\theta = 45°$. So the trigonometric form is: $2 + 2i = 2\sqrt{2}(\cos 45° + i\sin 45°)$

20. Here $x = -\sqrt{3}$, $y = 1$, and $r = \sqrt{\left(-\sqrt{3}\right)^2 + 1^2} = \sqrt{3+1} = \sqrt{4} = 2$. Since $\tan\theta = -\dfrac{1}{\sqrt{3}}$ in quadrant II, $\theta = 150°$.

So the trigonometric form is: $-\sqrt{3} + i = 2(\cos 150° + i\sin 150°)$

21. Here $x = 0$, $y = 5$, and $r = \sqrt{0^2 + 5^2} = \sqrt{0 + 25} = \sqrt{25} = 5$. Since $\tan\theta$ is undefined, $\theta = 90°$. So the trigonometric form is: $5i = 5(\cos 90° + i\sin 90°)$

22. Here $x = -3$, $y = 0$, and $r = \sqrt{(-3)^2 + 0^2} = \sqrt{9 + 0} = \sqrt{9} = 3$. Since $\tan\theta = 0$, $\theta = 180°$. So the trigonometric form is: $-3 = 3(\cos 180° + i\sin 180°)$

23. Multiplying in trigonometric form:
$$5(\cos 25° + i\sin 25°) \cdot 3(\cos 40° + i\sin 40°) = 15[\cos(25° + 40°) + i\sin(25° + 40°)] = 15(\cos 65° + i\sin 65°)$$

24. Dividing in trigonometric form: $\dfrac{10(\cos 50° + i\sin 50°)}{2(\cos 20° + i\sin 20°)} = \dfrac{10}{2}[\cos(50° - 20°) + i\sin(50° - 20°)] = 5(\cos 30° + i\sin 30°)$

25. Using DeMoivre's Theorem: $[2(\cos 10° + i\sin 10°)]^5 = 2^5[\cos(5 \cdot 10°) + i\sin(5 \cdot 10°)] = 32(\cos 50° + i\sin 50°)$

26. Using DeMoivre's Theorem:
$$(3\text{cis}20°)^4 = [3(\cos 20° + i\sin 20°)]^4 = 3^4[\cos(4 \cdot 20°) + i\sin(4 \cdot 20°)] = 81(\cos 80° + i\sin 80°)$$

27. The square roots are given by:
$$w_k = 49^{1/2}\left[\cos\frac{50° + 360°k}{2} + i\sin\frac{50° + 360°k}{2}\right] = 7[\cos(25° + 180°k) + i\sin(25° + 180°k)]$$
Substituting $k = 0, 1$:
$$w_0 = 7(\cos 25° + i\sin 25°) \qquad\qquad w_1 = 7(\cos 205° + i\sin 205°)$$

28. First write the number in trigonometric form: $2 + 2i\sqrt{3} = 4(\cos 60° + i\sin 60°)$
The fourth roots are given by:
$$w_k = 4^{1/4}\left[\cos\frac{60° + 360°k}{4} + i\sin\frac{60° + 360°k}{4}\right] = \sqrt{2}[\cos(15° + 90°k) + i\sin(15° + 90°k)]$$
Substituting $k = 0, 1, 2, 3$:
$$w_0 = \sqrt{2}(\cos 15° + i\sin 15°) \qquad\qquad w_1 = \sqrt{2}(\cos 105° + i\sin 105°)$$
$$w_2 = \sqrt{2}(\cos 195° + i\sin 195°) \qquad\qquad w_3 = \sqrt{2}(\cos 285° + i\sin 285°)$$

29. Using the quadratic formula with $a = 1$, $b = -2\sqrt{3}$, and $c = 4$:
$$x^2 = \frac{-\left(-2\sqrt{3}\right) \pm \sqrt{\left(-2\sqrt{3}\right)^2 - 4(1)(4)}}{2(1)} = \frac{2\sqrt{3} \pm \sqrt{12 - 16}}{2} = \frac{2\sqrt{3} \pm \sqrt{-4}}{2} = \frac{2\sqrt{3} \pm 2i}{2} = \sqrt{3} \pm i$$
Thus either $x^2 = \sqrt{3} + i$ or $x^2 = \sqrt{3} - i$. Writing $\sqrt{3} + i$ in trigonometric form: $\sqrt{3} + i = 2(\cos 30° + i\sin 30°)$
So its square roots are given by:
$$w_k = 2^{1/2}\left[\cos\frac{30° + 360°k}{2} + i\sin\frac{30° + 360°k}{2}\right] = \sqrt{2}[\cos(15° + 180°k) + i\sin(15° + 180°k)]$$
Substituting $k = 0, 1$:
$$w_0 = \sqrt{2}(\cos 15° + i\sin 15°) \qquad\qquad w_1 = \sqrt{2}(\cos 195° + i\sin 195°)$$
Writing $\sqrt{3} + i$ in trigonometric form: $\sqrt{3} - i = 2(\cos 330° + i\sin 330°)$
So its square roots are given by:
$$w_k = 2^{1/2}\left[\cos\frac{330° + 360°k}{2} + i\sin\frac{330° + 360°k}{2}\right] = \sqrt{2}[\cos(165° + 180°k) + i\sin(165° + 180°k)]$$
Substituting $k = 0, 1$:
$$w_0 = \sqrt{2}(\cos 165° + i\sin 165°) \qquad\qquad w_1 = \sqrt{2}(\cos 345° + i\sin 345°)$$
Summarizing, the solutions are $\sqrt{2}(\cos\theta + i\sin\theta)$, where $\theta = 15°, 165°, 195°, 345°$.

30. The solutions will be the cube roots of -1. Since $-1 = \cos 180^\circ + i \sin 180^\circ$, the cube roots are given by:

$$w_k = 1^{1/3}\left[\cos\frac{180^\circ + 360^\circ k}{3} + i\sin\frac{180^\circ + 360^\circ k}{3}\right] = \cos(60^\circ + 120^\circ k) + i\sin(60^\circ + 120^\circ k)$$

Substituting $k = 0, 1, 2$:

$$w_0 = \cos 60^\circ + i\sin 60^\circ \qquad\qquad w_1 = \cos 180^\circ + i\sin 180^\circ$$
$$w_2 = \cos 300^\circ + i\sin 300^\circ$$

31. Two other points are $(4, -135^\circ)$ and $(-4, 45^\circ)$. The rectangular coordinates are given by:

$$x = r\cos\theta = 4\cos 225^\circ = 4\cdot\left(-\frac{\sqrt{2}}{2}\right) = -2\sqrt{2} \qquad y = r\sin\theta = 4\sin 225^\circ = 4\cdot\left(-\frac{\sqrt{2}}{2}\right) = -2\sqrt{2}$$

The rectangular coordinates are $\left(-2\sqrt{2}, -2\sqrt{2}\right)$.

32. Two other points are $(6, -120^\circ)$ and $(6, 240^\circ)$. The rectangular coordinates are given by:

$$x = r\cos\theta = -6\cos 60^\circ = -6\cdot\frac{1}{2} = -3 \qquad y = r\sin\theta = -6\sin 60^\circ = -6\cdot\frac{\sqrt{3}}{2} = -3\sqrt{3}$$

The rectangular coordinates are $\left(-3, -3\sqrt{3}\right)$.

33. Compute $r = \sqrt{(-3)^2 + 3^2} = \sqrt{9 + 9} = \sqrt{18} = 3\sqrt{2}$. Also $\tan\theta = \dfrac{3}{-3} = -1$ with θ in quadrant II, so $\theta = 135^\circ$. The polar coordinates are $\left(3\sqrt{2}, 135^\circ\right)$.

34. Compute $r = \sqrt{0^2 + 5^2} = \sqrt{0 + 25} = \sqrt{25} = 5$. Also $\tan\theta$ is undefined, so $\theta = 90^\circ$.
The polar coordinates are $(5, 90^\circ)$.

35. Converting to rectangular coordinates:

$$r = 6\sin\theta$$
$$r = 6\left(\frac{y}{r}\right)$$
$$r^2 = 6y$$
$$x^2 + y^2 = 6y$$

36. Using the double-angle formula for sine:

$$r = \sin 2\theta$$
$$r = 2\sin\theta\cos\theta$$
$$r = 2\cdot\frac{y}{r}\cdot\frac{x}{r}$$
$$r^3 = 2xy$$
$$\left(x^2 + y^2\right)^{3/2} = 2xy$$

37. Since $x = r\cos\theta$ and $y = r\sin\theta$:

$$x + y = 2$$
$$r\cos\theta + r\sin\theta = 2$$
$$r(\sin\theta + \cos\theta) = 2$$

38. Since $x^2 + y^2 = r^2$ and $y = r\sin\theta$:

$$x^2 + y^2 = 8y$$
$$r^2 = 8r\sin\theta$$
$$r = 8\sin\theta$$

39. Graphing the equation:

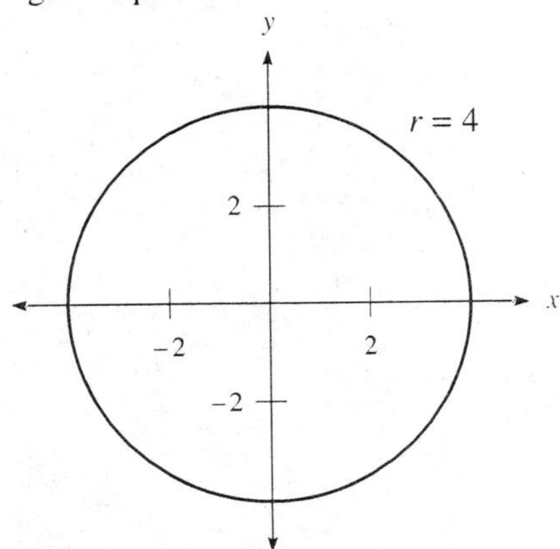

40. Graphing the equation:

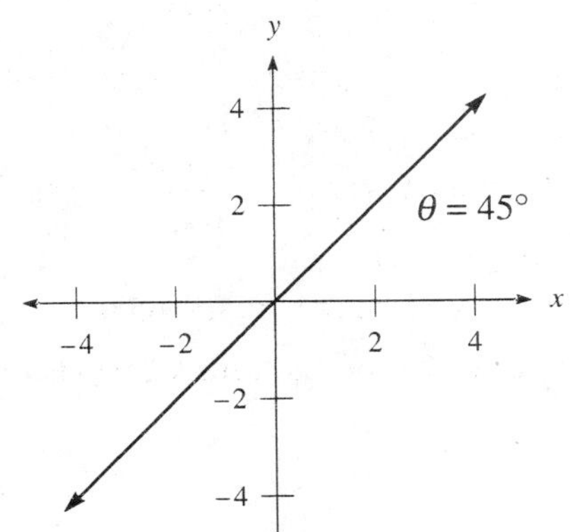

41. Graphing the equation:

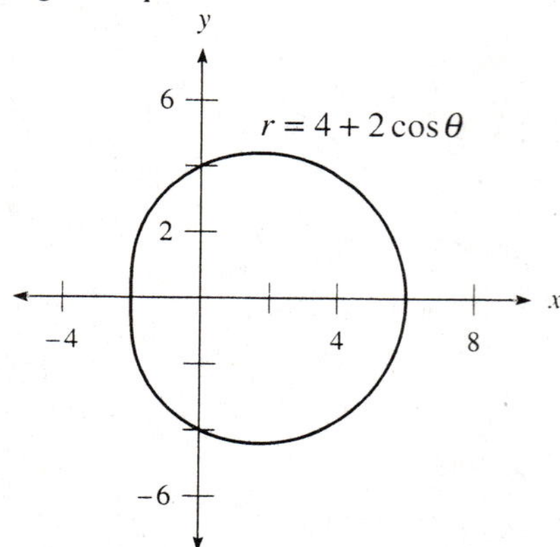

42. Graphing the equation:

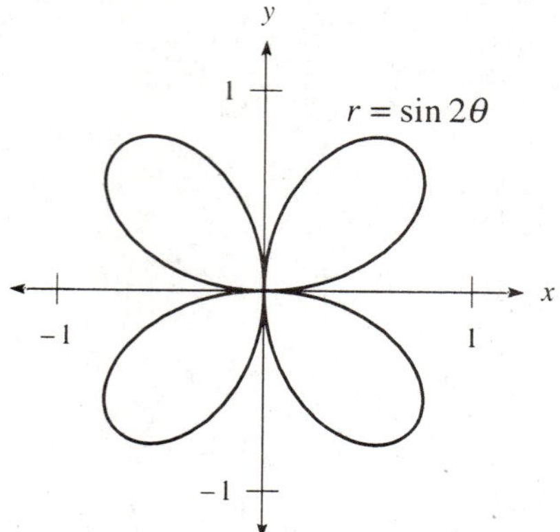

43. Graphing the equation:

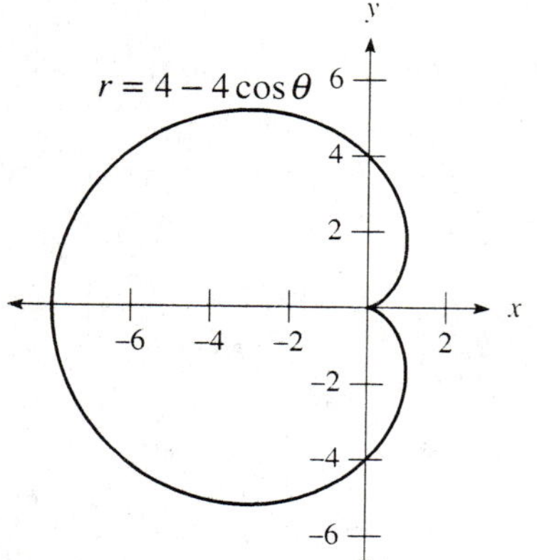

44. Graphing the equation:

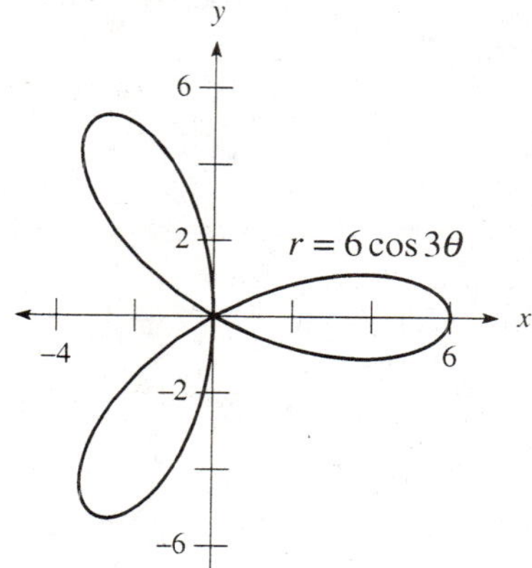

45. Graphing the equation:

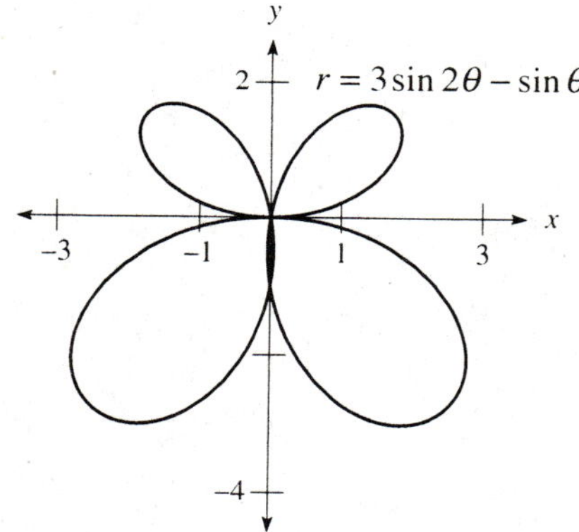

46. Graphing the equation:

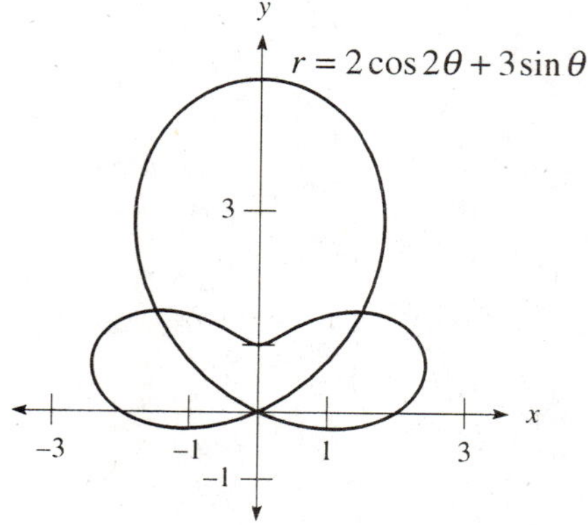

Appendix A
Review of Functions

A.1 Introduction to Functions

2. The domain is {2,3,5} and the range is {1,3,7}. This is a function.
4. The domain is {−1,3} and the range is {−4,2,5}. This is not a function.
6. The domain is {3,5} and the range is {−2,−1}. This is not a function.
8. Yes, since it passes the vertical line test. **10.** Yes, since it passes the vertical line test.
12. No, since it fails the vertical line test. **14.** No, since it fails the vertical line test.
16. The domain is all real numbers and the range is $\{y \mid y \geq 4\}$. This is a function.

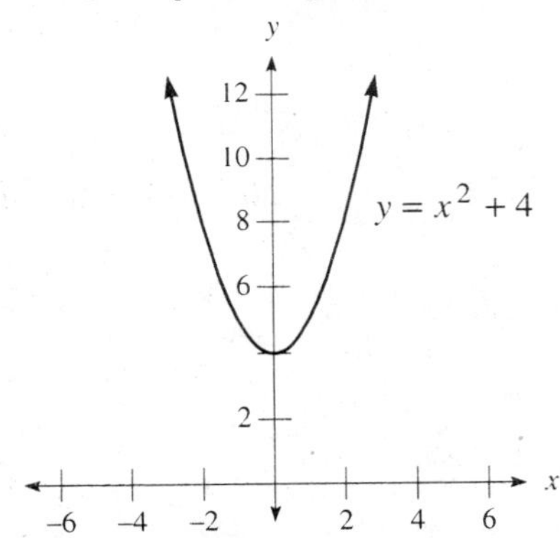

18. The domain is $\{x \mid x \geq -9\}$ and the range is all real numbers. This is not a function.

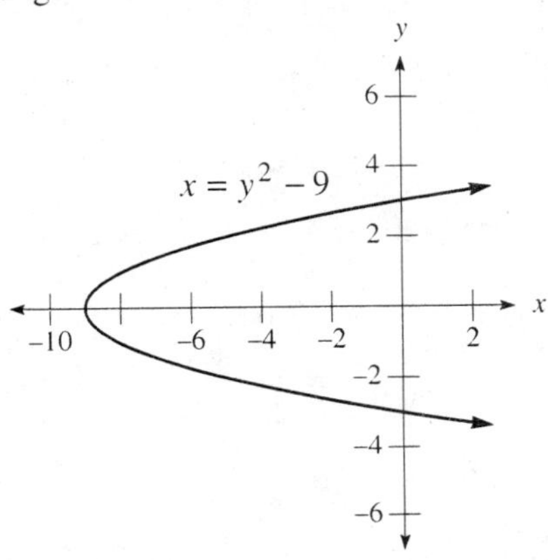

20.   The domain is all real numbers and the range is $\{y \mid y \geq 0\}$. This is a function.

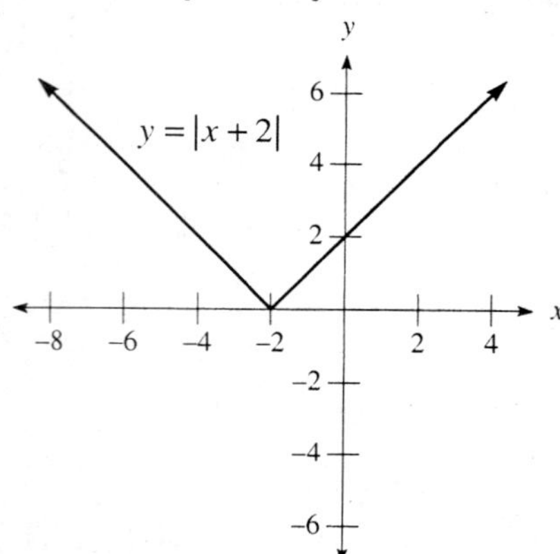

22.   The domain is all real numbers and the range is $\{y \mid y \geq 2\}$. This is a function.

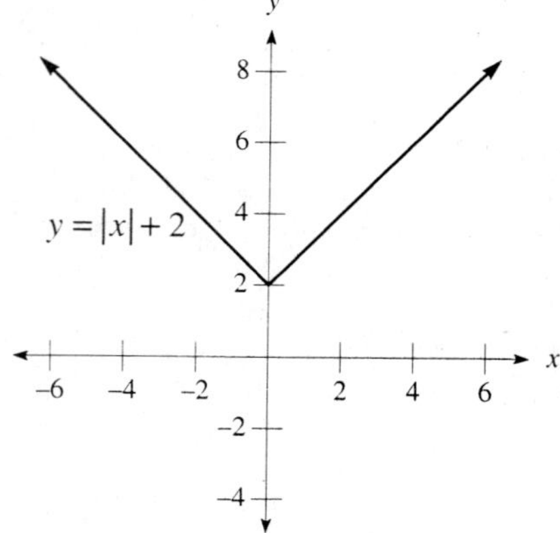

24. **a.** The equation is $y = 5.25x$ for $15 \le x \le 30$.

Hours Worked	Gross Pay ($)
x	y
15	78.75
20	105.00
25	131.25
30	157.50

b. Completing the table:

c. Constructing a line graph:

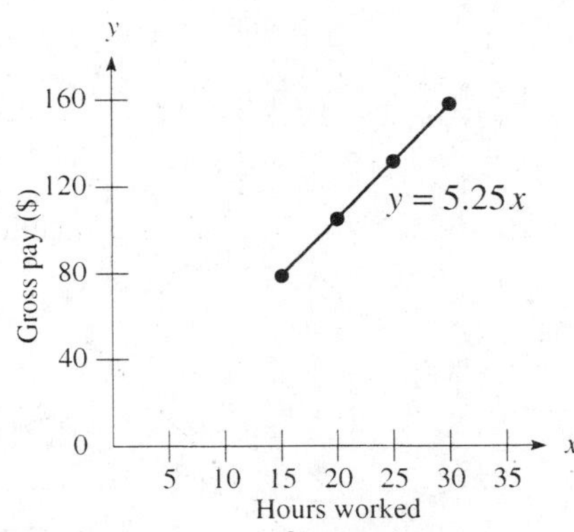

d. The domain is $\{x \mid 15 \le x \le 30\}$ and the range is $\{y \mid 78.75 \le y \le 157.50\}$.

e. The minimum is \$78.75 and the maximum is \$157.50.

26. **a.** Completing the table:

Distance (ft)	Intensity
d	I
1	120
2	30
3	13.3
4	7.5
5	4.8
6	3.3

b. Graphing the function:

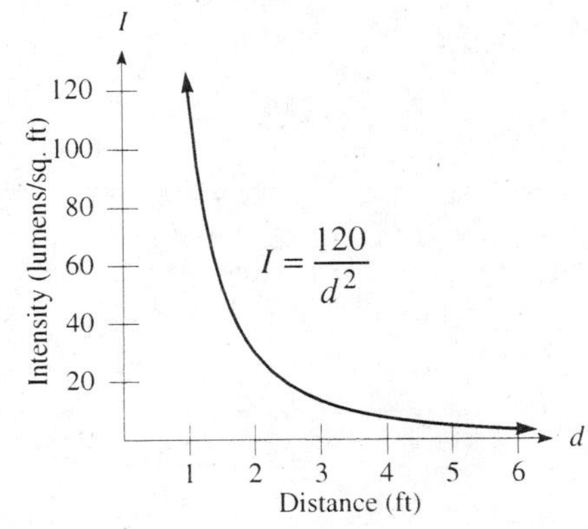

28. **a.** The equation is $P = 4x + 4$. The restriction on x is $x \geq 0$.

  b. Graphing the relationship:

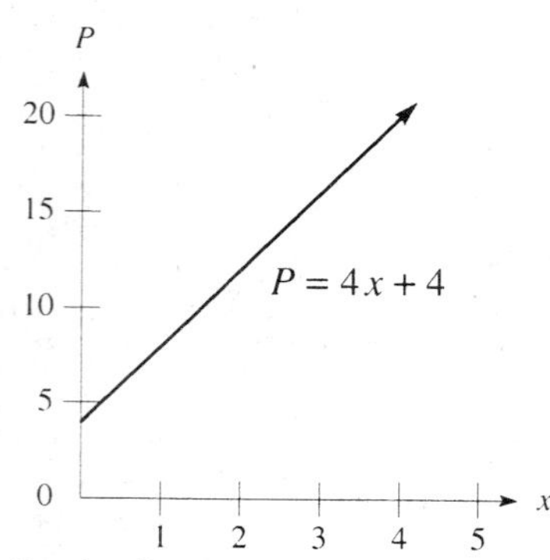

30. Evaluating the function: $f(3) = 2(3) - 5 = 6 - 5 = 1$

32. Evaluating the function: $g(-2) = (-2)^2 + 3(-2) + 4 = 4 - 6 + 4 = 2$

34. Evaluating the function: $f(-4) = 2(-4) - 5 = -8 - 5 = -13$

36. Evaluating the function: $g(2) = (2)^2 + 3(2) + 4 = 4 + 6 + 4 = 14$

38. First evaluate each function:

$$f(2) = 2(2) - 5 = 4 - 5 = -1 \qquad g(3) = (3)^2 + 3(3) + 4 = 9 + 9 + 4 = 22$$

Now evaluating: $f(2) - g(3) = -1 - 22 = -23$

40. First evaluate each function:

$$g(-1) = (-1)^2 + 3(-1) + 4 = 1 - 3 + 4 = 2 \qquad f(-1) = 2(-1) - 5 = -2 - 5 = -7$$

Now evaluating: $g(-1) + f(-1) = 2 - 7 = -5$

42. Evaluating the function: $g(0) = 2(0) - 1 = 0 - 1 = -1$

44. Evaluating the function: $f(1) = 3(1)^2 - 4(1) + 1 = 3 - 4 + 1 = 0$

46. Evaluating the function: $g(-1) = 2(-1) - 1 = -2 - 1 = -3$

48. Evaluating the function: $f(10) = 3(10)^2 - 4(10) + 1 = 300 - 40 + 1 = 261$

50. Evaluating the function: $g(3) = 2(3) - 1 = 6 - 1 = 5$ **52.** Evaluating the function: $g\left(\frac{1}{4}\right) = 2\left(\frac{1}{4}\right) - 1 = \frac{1}{2} - 1 = -\frac{1}{2}$

54. Evaluating the function: $g(b) = 2b - 1$ **56.** $g(1) = 1$

58. $f(3) = \frac{1}{2}$ **60.** $f(\pi) = 0$

62. Graphing the function:

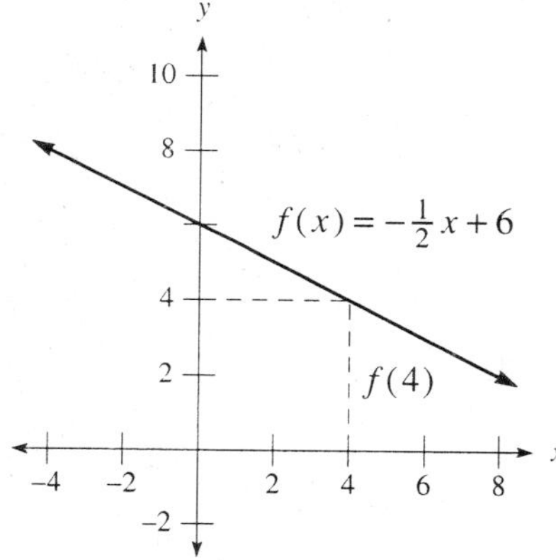

64. Evaluating: $s(4) = \frac{60}{4} = 15$; Minke's average speed is 15 miles per hour if it takes her 4 minutes to run a mile.

Evaluating: $s(5) = \frac{60}{5} = 12$; Minke's average speed is 12 miles per hour if it takes her 5 minutes to run a mile.

66. Since the radius cannot be negative, $A(-10)$ does not make sense.

68.   **a.** $g(2) - f(2) = 3 - 2 = 1$ **b.** $f(1) + g(1) = 1 + 2 = 3$

   c. $f[g(3)] = f(4) = 0$ **d.** $g[f(3)] = g(1) = 2$

70.   Finding where $f(x) = x$:

$$-\tfrac{1}{2}x + 6 = x$$

$$6 = \tfrac{3}{2}x$$

$$x = 4$$

A.2   The Inverse of a Function

2.   Let $y = f(x)$. Switch x and y and solve for y:

$$2y - 5 = x$$
$$2y = x + 5$$
$$y = \frac{x + 5}{2}$$

The inverse is $f^{-1}(x) = \dfrac{x + 5}{2}$.

4.   Let $y = f(x)$. Switch x and y and solve for y:

$$y^3 - 2 = x$$
$$y^3 = x + 2$$
$$y = \sqrt[3]{x + 2}$$

The inverse is $f^{-1}(x) = \sqrt[3]{x + 2}$.

6.   Let $y = f(x)$. Switch x and y and solve for y:

$$\frac{y - 2}{y - 3} = x$$
$$y - 2 = xy - 3x$$
$$y - xy = 2 - 3x$$
$$y(1 - x) = 2 - 3x$$
$$y = \frac{2 - 3x}{1 - x} = \frac{3x - 2}{x - 1}$$

The inverse is $f^{-1}(x) = \dfrac{3x - 2}{x - 1}$.

8.   Let $y = f(x)$. Switch x and y and solve for y:

$$\frac{y + 7}{2} = x$$
$$y + 7 = 2x$$
$$y = 2x - 7$$

The inverse is $f^{-1}(x) = 2x - 7$.

10.   Let $y = f(x)$. Switch x and y and solve for y:

$$\tfrac{1}{3}y + 1 = x$$
$$y + 3 = 3x$$
$$y = 3x - 3$$

The inverse is $f^{-1}(x) = 3x - 3$.

12.   Let $y = f(x)$. Switch x and y and solve for y:

$$\frac{3y + 2}{5y + 1} = x$$
$$3y + 2 = 5xy + x$$
$$3y - 5xy = x - 2$$
$$y(3 - 5x) = x - 2$$
$$y = \frac{x - 2}{3 - 5x} = \frac{2 - x}{5x - 3}$$

The inverse is $f^{-1}(x) = \dfrac{2 - x}{5x - 3}$.

14. Finding the inverse:

$$3y + 1 = x$$
$$3y = x - 1$$
$$y = \frac{x-1}{3}$$

The inverse is $y^{-1} = \dfrac{x-1}{3}$. Graphing each curve:

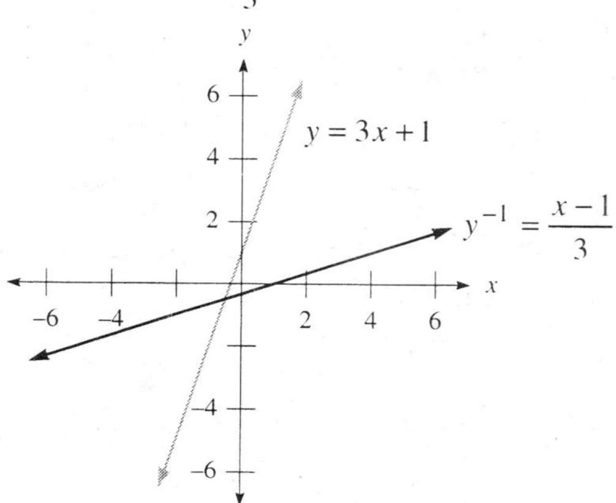

16. Finding the inverse:

$$y^2 + 1 = x$$
$$y^2 = x - 1$$
$$y = \pm\sqrt{x-1}$$

The inverse is $y^{-1} = \pm\sqrt{x-1}$. Graphing each curve:

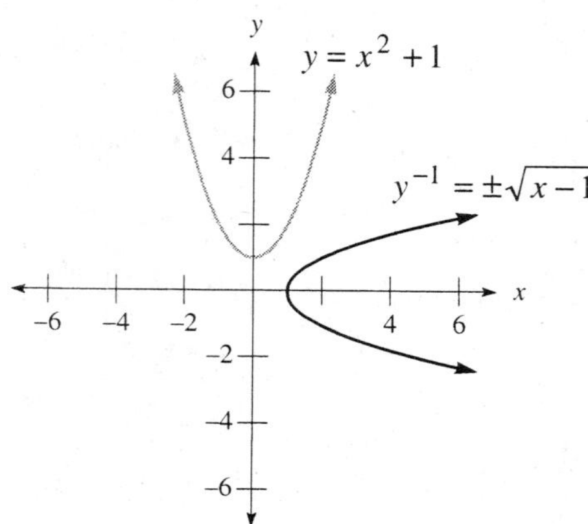

18. Finding the inverse:

$$y^2 + 2y - 3 = x$$
$$y^2 + 2y + 1 = x + 3 + 1$$
$$(y+1)^2 = x + 4$$
$$y + 1 = \pm\sqrt{x+4}$$
$$y = -1 \pm \sqrt{x+4}$$

The inverse is $y^{-1} = -1 \pm \sqrt{x+4}$. Graphing each curve:

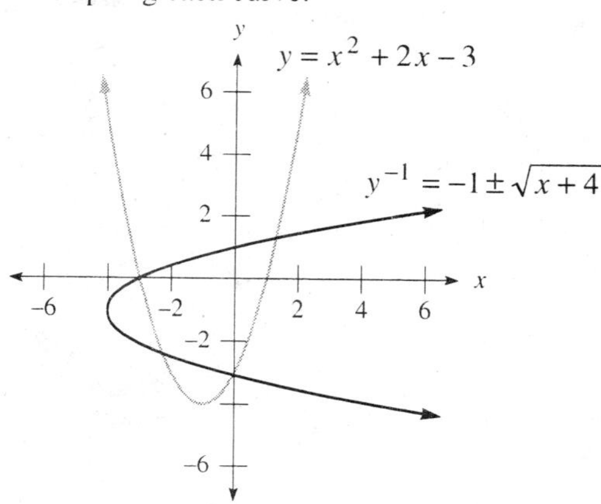

20.   The inverse is $x = -2$. Graphing each curve:

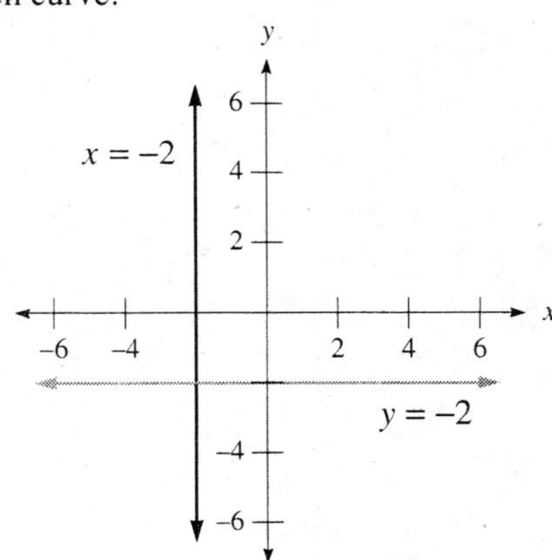

22.   Finding the inverse:

$$\tfrac{1}{3}y - 1 = x$$
$$y - 3 = 3x$$
$$y = 3x + 3$$

The inverse is $y^{-1} = 3x + 3$. Graphing each curve:

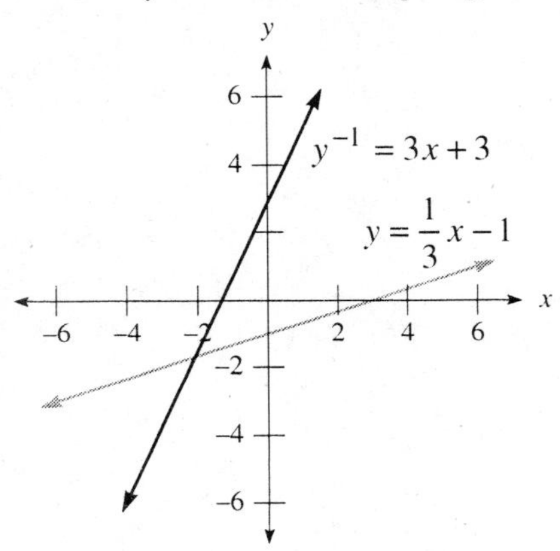

24.   Finding the inverse:

$$y^3 - 2 = x$$
$$y^3 = x + 2$$
$$y = \sqrt[3]{x + 2}$$

The inverse is $y^{-1} = \sqrt[3]{x + 2}$. Graphing each curve:

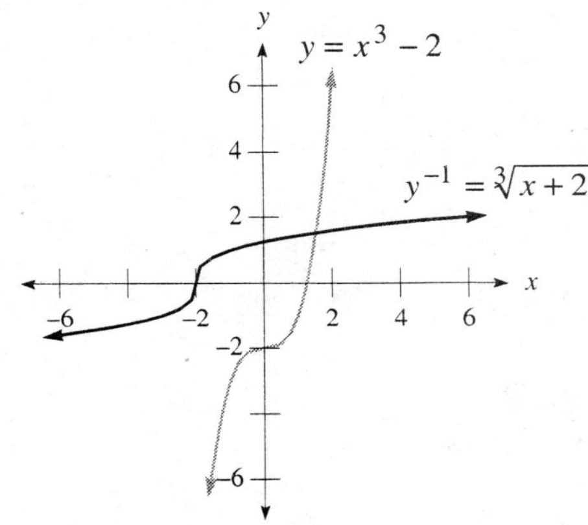

26.   Finding the inverse:
$$\sqrt{y} + 2 = x$$
$$\sqrt{y} = x - 2$$
$$y = (x-2)^2$$

The inverse is $y^{-1} = (x-2)^2$, $x \geq 2$. Graphing each curve:

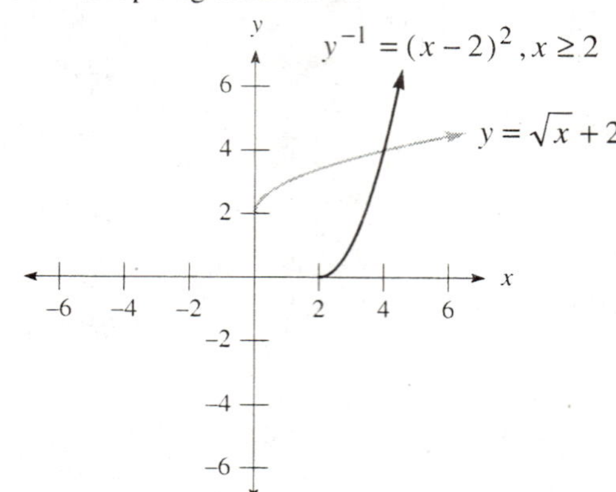

28.   **a.**   No, there are two x-values paired with the same y-value.
　　b.   Yes, each y-value is paired with only one x-value.

30.   **a.**   Evaluating the function: $f(-4) = \frac{1}{2}(-4) + 5 = -2 + 5 = 3$

　　b.   Evaluating the function: $f^{-1}(-4) = 2(-4) - 10 = -8 - 10 = -18$

　　c.   Evaluating the function: $f\left[f^{-1}(-4)\right] = f(-18) = \frac{1}{2}(-18) + 5 = -9 + 5 = -4$

　　d.   Evaluating the function: $f^{-1}\left[f(-4)\right] = f^{-1}(3) = 2(3) - 10 = 6 - 10 = -4$

32.   Let $y = f(x)$. Switch x and y and solve for y:
$$\frac{a}{y} = x$$
$$y = \frac{a}{x}$$

The inverse is $f^{-1}(x) = \frac{a}{x}$.

34.   **a.**   The domain of f is $D_f = \{-6, 2, 3, 6\}$, and the range of f is $R_f = \{-3, -2, 3, 4\}$.
　　b.   The domain of g is $D_g = \{-3, -2, 3, 4\}$, and the range of g is $R_g = \{-6, 2, 3, 6\}$.
　　c.   Note that $D_f = R_g$ and $R_f = D_g$.　　　　　　**d.**   Yes, f is a one-to-one function.
　　e.   Yes, g is a one-to-one function.

36.   Finding the inverse:
$$6 - 8y = x$$
$$8y = 6 - x$$
$$y = \frac{6-x}{8}$$

So $f^{-1}(x) = \frac{6-x}{8}$. Now verifying the inverse: $f\left[f^{-1}(x)\right] = f\left(\frac{6-x}{8}\right) = 6 - 8\left(\frac{6-x}{8}\right) = 6 - 6 + x = x$

38.   Finding the inverse:

$$y^3 - 8 = x$$
$$y^3 = x + 8$$
$$y = \sqrt[3]{x+8}$$

So $f^{-1}(x) = \sqrt[3]{x+8}$. Now verifying the inverse: $f\left[f^{-1}(x)\right] = f\left(\sqrt[3]{x+8}\right) = \left(\sqrt[3]{x+8}\right)^3 - 8 = x + 8 - 8 = x$

40.   **a.**   The inverse is $f^{-1}(x) = \dfrac{x-7}{2}$.   **b.**   The inverse is $f^{-1}(x) = (x+9)^2, x \geq -9$.

   c.   The inverse is $f^{-1}(x) = \sqrt[3]{x+4}$.   **d.**   The inverse is $f^{-1}(x) = \sqrt[3]{x^2+4}, x \geq 0$.

Appendix B
Exponential and Logarithmic Functions

B.1   Exponential Functions

2. Evaluating: $f(0) = 3^0 = 1$

4. Evaluating: $g(-4) = \left(\frac{1}{2}\right)^{-4} = 16$

6. Evaluating: $f(-1) = 3^{-1} = \frac{1}{3}$

8. Evaluating: $f(2) - g(-2) = 3^2 - \left(\frac{1}{2}\right)^{-2} = 9 - 4 = 5$

10. Graphing the function:

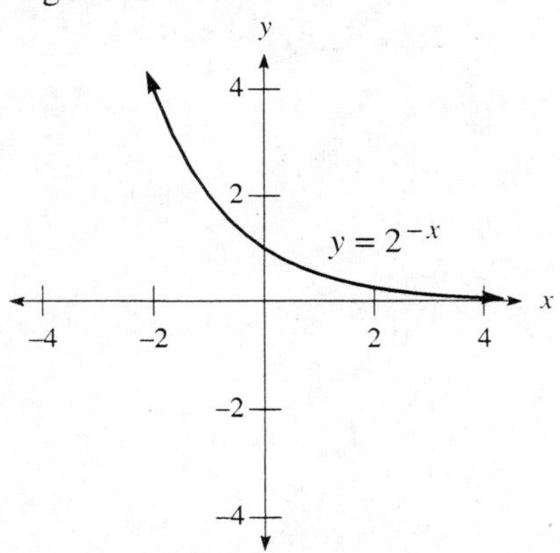

12. Graphing the function:

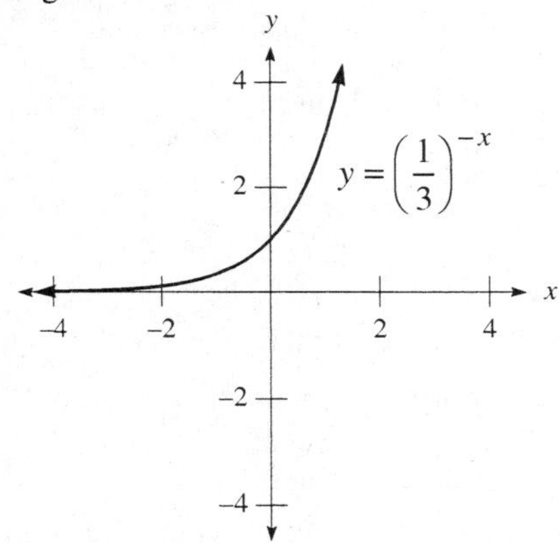

14. Graphing the function:

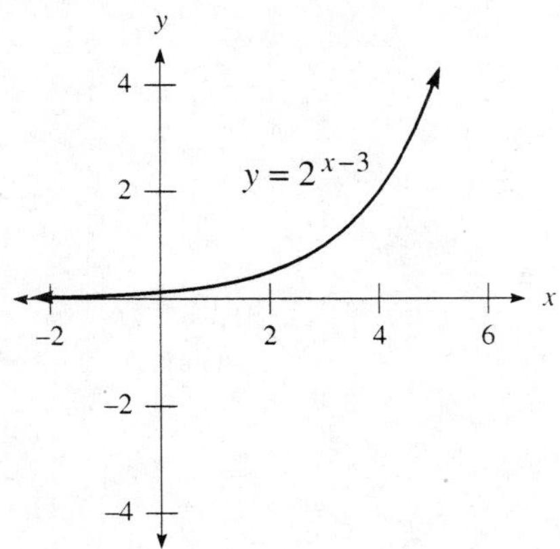

16. Graphing the function:

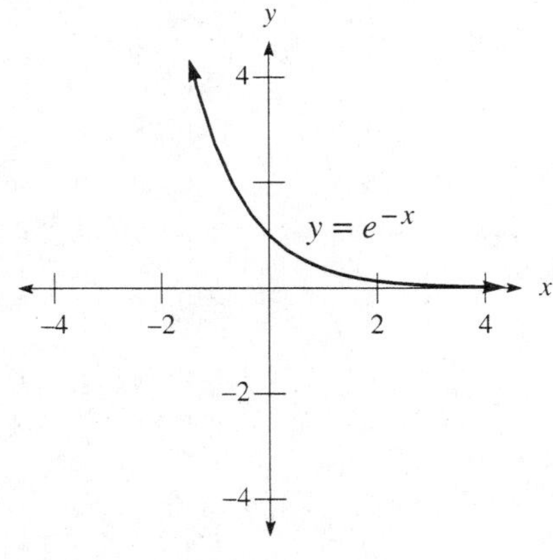

18.   Graphing the function:

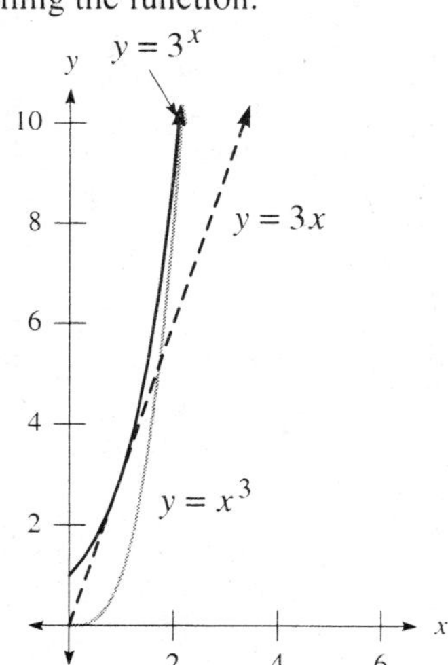

20.   Graphing the function:

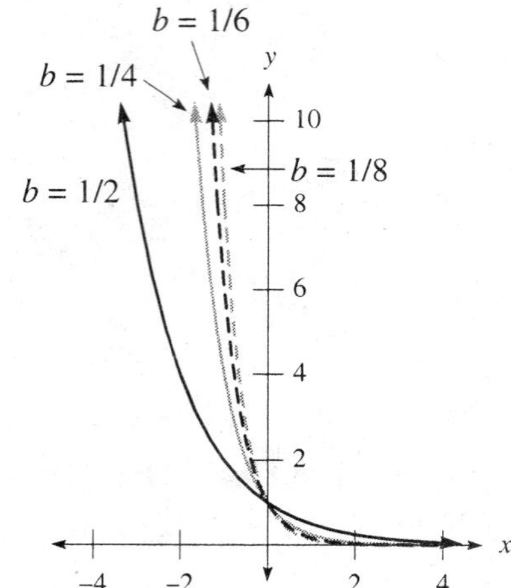

22.   The equation is: $h(n) = A\left(\frac{1}{2}\right)^{n}$

Substituting $n = 8$: $h(8) = 10\left(\frac{1}{2}\right)^{8} \approx 0.039$ feet

24.   For 2000, substitute $t = 10$: $p(10) = 1.89(1.25)^{10} \approx \$17.60$ per pound

For 2005, substitute $t = 15$: $p(15) = 1.89(1.25)^{15} \approx \$53.72$ per pound

26.   **a.**   The equation is $A(t) = 500\left(1 + \frac{0.08}{12}\right)^{12t}$.   **b.**   Substitute $t = 5$: $A(5) = 500\left(1 + \frac{0.08}{12}\right)^{60} \approx \$744.92$

c.   Substitute $t = 5$ into the compound interest formula: $A(5) = 500e^{0.08 \times 5} \approx \$745.91$

28.   The amount will be: $A = 200\left(1 + \frac{0.08}{2}\right)^{2 \cdot 10} = 200(1.04)^{20} \approx \$438.22$

30.   Substituting $x = 1$, $x = 2$, and $x = 3$:

$f(1) = 50 \cdot 4^{1} = 200$ bacteria

$f(2) = 50 \cdot 4^{2} = 800$ bacteria

$f(3) = 50 \cdot 4^{3} = 3200$ bacteria

32.   **a.**   Completing the table:

Years since 1960 t	Cost $C(t)$
0	\$0.10
15	\$0.24
40	\$1.00
50	\$1.78
90	\$17.84

b.   At the beginning of the year 2000, the cost will be \$1.00.

34.   Graphing the function:

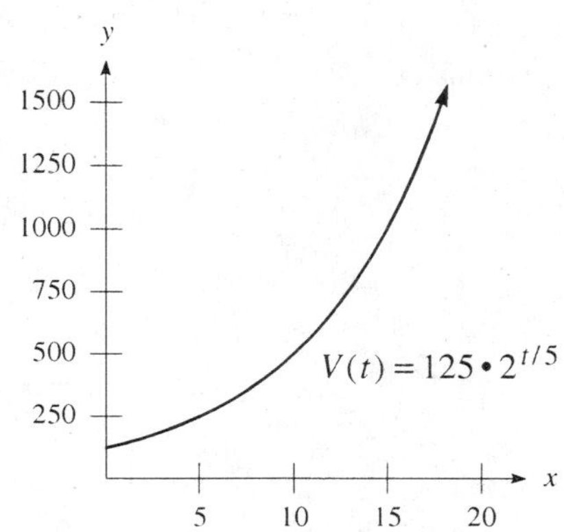

36.   **a.**   Substitute $t = 4.75$: $V(4.75) = 375{,}000(1 - 0.25)^{4.75} \approx \$95{,}625.18$

    b.   The domain is $\{t \mid 0 \leq t \leq 7\}$.

    c.   Sketching the graph:

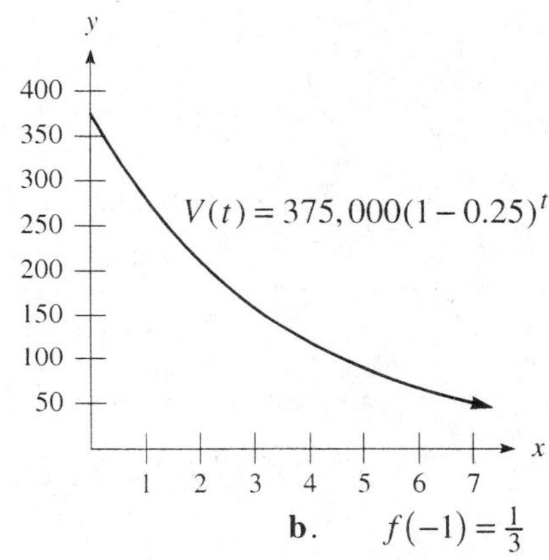

38.   **a.** $f(0) = 1$ **b.** $f(-1) = \frac{1}{3}$

    c. $f(1) = 3$ **d.** $g(0) = 1$

    e. $g(1) = \frac{1}{2}$ **f.** $g(-1) = 2$

B.2   Logarithms Are Exponents

2.   Writing in logarithmic form: $\log_3 9 = 2$

4.   Writing in logarithmic form: $\log_4 16 = 2$

6.   Writing in logarithmic form: $\log_{10} 0.001 = -3$

8.   Writing in logarithmic form: $\log_4 \frac{1}{16} = -2$

10.   Writing in logarithmic form: $\log_{1/3} 9 = -2$

12.   Writing in logarithmic form: $\log_3 81 = 4$

14.   Writing in exponential form: $2^3 = 8$

16.   Writing in exponential form: $2^5 = 32$

18.   Writing in exponential form: $9^1 = 9$

20.   Writing in exponential form: $10^{-4} = 0.0001$

22.   Writing in exponential form: $7^2 = 49$

24.   Writing in exponential form: $3^{-4} = \frac{1}{81}$

26.   Solving the equation:
$$\log_4 x = 3$$
$$x = 4^3 = 64$$

28.   Solving the equation:
$$\log_2 x = -4$$
$$x = 2^{-4} = \frac{1}{16}$$

30.   Solving the equation:
$$\log_3 27 = x$$
$$3^x = 27$$
$$x = 3$$

32.   Solving the equation:
$$\log_{25} 5 = x$$
$$25^x = 5$$
$$x = \frac{1}{2}$$

34. Solving the equation:
$$\log_x 16 = 4$$
$$x^4 = 16$$
$$x = 2$$

36. Solving the equation:
$$\log_x 8 = 2$$
$$x^2 = 8$$
$$x = \sqrt{8} = 2\sqrt{2}$$

38. Sketching the graph:

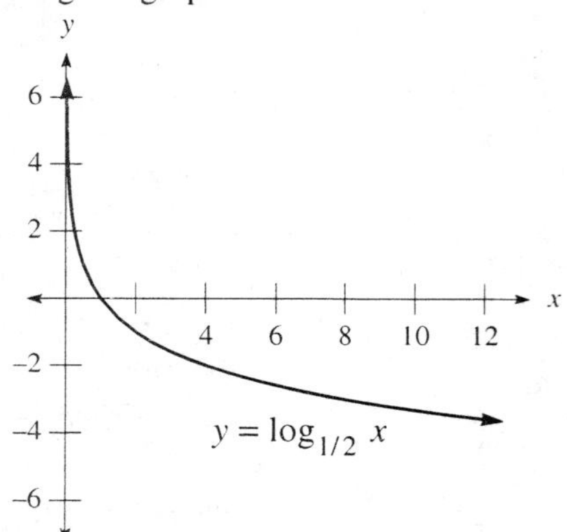

40. Sketching the graph:

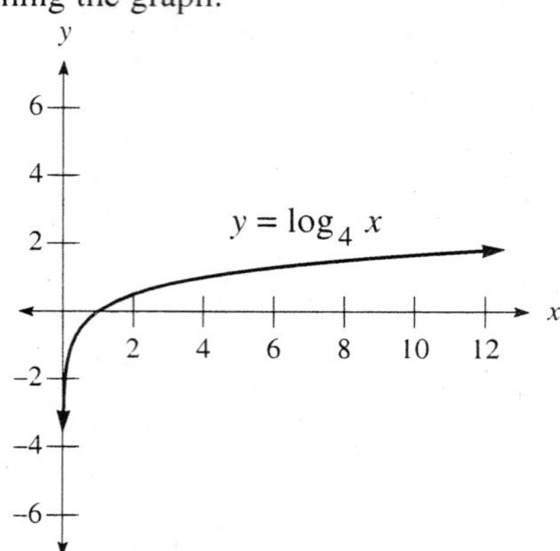

42. Sketching the graph:

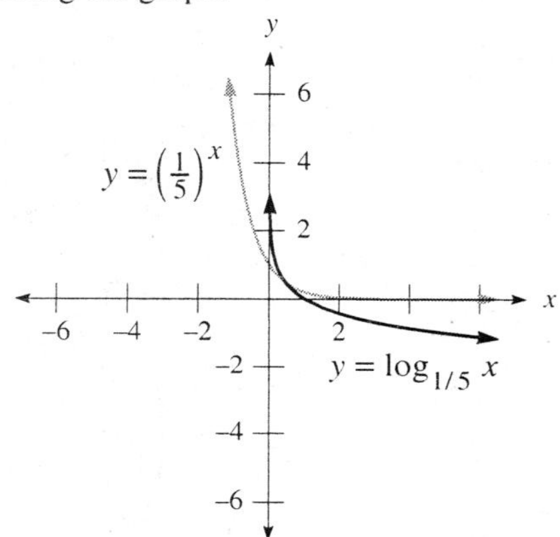

44. Sketching the graph:

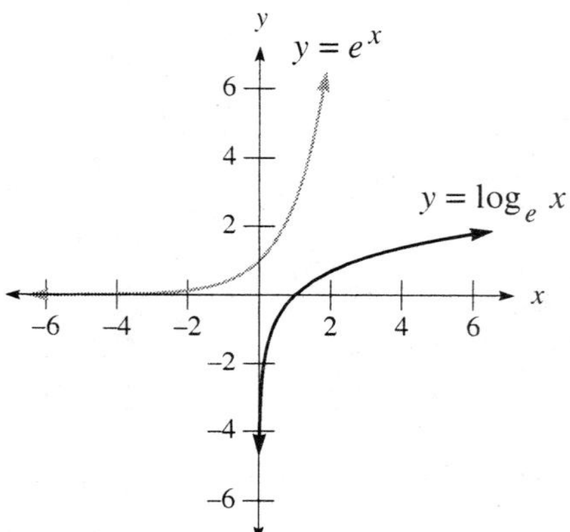

46. Simplifying the logarithm:
$$x = \log_3 9$$
$$3^x = 9$$
$$x = 2$$

48. Simplifying the logarithm:
$$x = \log_9 27$$
$$9^x = 27$$
$$3^{2x} = 3^3$$
$$2x = 3$$
$$x = \frac{3}{2}$$

50. Simplifying the logarithm:
$$x = \log_{10} 10000$$
$$10^x = 10000$$
$$x = 4$$

52. Simplifying the logarithm:
$$x = \log_4 4$$
$$4^x = 4$$
$$x = 1$$

54. Simplifying the logarithm:
$$x = \log_{10} 1$$
$$10^x = 1$$
$$x = 0$$

56. First find $\log_3 3$:
$$x = \log_3 3$$
$$3^x = 3$$
$$x = 1$$

Now find $\log_5 1$:
$$x = \log_5 1$$
$$5^x = 1$$
$$x = 0$$

58. First find $\log_2 8$:
$$x = \log_2 8$$
$$2^x = 8$$
$$x = 3$$

Now find $\log_3 3$:
$$x = \log_3 3$$
$$3^x = 3$$
$$x = 1$$

Now find $\log_4 1$:
$$x = \log_4 1$$
$$4^x = 1$$
$$x = 0$$

60. The pH is given by: $\text{pH} = -\log_{10}\left(10^{-3}\right) = -(-3) = 3$

62. The $\left[\text{H}^+\right]$ is given by:
$$-\log_{10}\left[\text{H}^+\right] = 4$$
$$\log_{10}\left[\text{H}^+\right] = -4$$
$$\left[\text{H}^+\right] = 10^{-4}$$

64. Using the relationship $M = \log_{10} T$:
$$M = \log_{10} 100,000$$
$$10^M = 100,000$$
$$M = 5$$

66. It is 10^6 times as large.

68. **a.** Completing the table:

x	-1	0	1	2
$f(x)$	5	1	$\frac{1}{5}$	$\frac{1}{25}$

b. Completing the table:

x	5	1	$\frac{1}{5}$	$\frac{1}{25}$
$f^{-1}(x)$	-1	0	1	2

c. The equation is $f(x) = \left(\frac{1}{5}\right)^x$.

d. The equation is $f^{-1}(x) = \log_{1/5} x$.

B.3   Properties of Logarithms

2. Using properties of logarithms: $\log_2 5x = \log_2 5 + \log_2 x$

4. Using properties of logarithms: $\log_3 \dfrac{x}{5} = \log_3 x - \log_3 5$

6. Using properties of logarithms: $\log_7 y^3 = 3\log_7 y$

8. Using properties of logarithms: $\log_8 \sqrt{z} = \log_8 z^{1/2} = \frac{1}{2}\log_8 z$

10. Using properties of logarithms: $\log_{10} x^2 y^4 = \log_{10} x^2 + \log_{10} y^4 = 2\log_{10} x + 4\log_{10} y$

12. Using properties of logarithms: $\log_8 \sqrt[3]{xy^6} = \log_8 x^{1/3} y^2 = \log_8 x^{1/3} + \log_8 y^2 = \frac{1}{3}\log_8 x + 2\log_8 y$

14. Using properties of logarithms: $\log_b \dfrac{3x}{y} = \log_b 3x - \log_b y = \log_b 3 + \log_b x - \log_b y$

16. Using properties of logarithms: $\log_{10} \dfrac{5}{4y} = \log_{10} 5 - \log_{10} 4y = \log_{10} 5 - \log_{10} 4 - \log_{10} y$

18. Using properties of logarithms: $\log_{10} \dfrac{\sqrt{x} \bullet y}{z^3} = \log_{10} x^{1/2} + \log_{10} y - \log_{10} z^3 = \frac{1}{2}\log_{10} x + \log_{10} y - 3\log_{10} z$

20. Using properties of logarithms: $\log_{10} \dfrac{x^4 \sqrt[3]{y}}{\sqrt{z}} = \log_{10} x^4 + \log_{10} y^{1/3} - \log_{10} z^{1/2} = 4\log_{10} x + \frac{1}{3}\log_{10} y - \frac{1}{2}\log_{10} z$

22. Using properties of logarithms:
$$\log_b \sqrt[4]{\dfrac{x^4 y^3}{z^5}} = \log_b \dfrac{xy^{3/4}}{z^{5/4}} = \log_b x + \log_b y^{3/4} - \log_b z^{5/4} = \log_b x + \frac{3}{4}\log_b y - \frac{5}{4}\log_b z$$

24. Writing as a single logarithm: $\log_b x - \log_b z = \log_b \dfrac{x}{z}$

26. Writing as a single logarithm: $4\log_2 x + 5\log_2 y = \log_2 x^4 + \log_2 y^5 = \log_2 x^4 y^5$

28. Writing as a single logarithm: $\frac{1}{3}\log_{10} x - \frac{1}{4}\log_{10} y = \log_{10} x^{1/3} - \log_{10} y^{1/4} = \log_{10} \dfrac{\sqrt[3]{x}}{\sqrt[4]{y}}$

30. Writing as a single logarithm: $2\log_3 x + 3\log_3 y - \log_3 z = \log_3 x^2 + \log_3 y^3 - \log_3 z = \log_3 \dfrac{x^2 y^3}{z}$

32. Writing as a single logarithm: $3\log_{10} x - \log_{10} y - \log_{10} z = \log_{10} x^3 - \log_{10} y - \log_{10} z = \log_{10} \dfrac{x^3}{yz}$

34. Writing as a single logarithm: $3\log_{10} x - \frac{4}{3}\log_{10} y - 5\log_{10} z = \log_{10} x^3 - \log_{10} y^{4/3} - \log_{10} z^5 = \log_{10} \dfrac{x^3}{y^{4/3} z^5}$

36. Solving the equation:
$$\log_3 x + \log_3 3 = 1$$
$$\log_3 3x = 1$$
$$3x = 3^1$$
$$3x = 3$$
$$x = 1$$

38. Solving the equation:
$$\log_3 x + \log_3 2 = 2$$
$$\log_3 2x = 2$$
$$2x = 3^2$$
$$2x = 9$$
$$x = \frac{9}{2}$$

40. Solving the equation:
$$\log_6 x + \log_6 (x-1) = 1$$
$$\log_6 \left(x^2 - x\right) = 1$$
$$x^2 - x = 6^1$$
$$x^2 - x - 6 = 0$$
$$(x-3)(x+2) = 0$$
$$x = 3, -2$$

The solution is 3 (–2 does not check).

42. Solving the equation:
$$\log_4 (x-2) - \log_4 (x+1) = 1$$
$$\log_4 \frac{x-2}{x+1} = 1$$
$$\frac{x-2}{x+1} = 4^1$$
$$x - 2 = 4x + 4$$
$$-3x = 6$$
$$x = -2$$

There is no solution (–2 does not check).

44. Solving the equation:
$$\log_4 x + \log_4 (x+6) = 2$$
$$\log_4 \left(x^2 + 6x\right) = 2$$
$$x^2 + 6x = 4^2$$
$$x^2 + 6x - 16 = 0$$
$$(x+8)(x-2) = 0$$
$$x = 2, -8$$

The solution is 2 (–8 does not check).

46. Solving the equation:
$$\log_{27} x + \log_{27} (x+8) = \frac{2}{3}$$
$$\log_{27} \left(x^2 + 8x\right) = \frac{2}{3}$$
$$x^2 + 8x = 27^{2/3}$$
$$x^2 + 8x - 9 = 0$$
$$(x+9)(x-1) = 0$$
$$x = 1, -9$$

The solution is 1 (–9 does not check).

48. Solving the equation:
$$\log_2 \sqrt{x} + \log_2 \sqrt{6x+5} = 1$$
$$\log_2 \sqrt{6x^2 + 5x} = 1$$
$$\tfrac{1}{2}\log_2\left(6x^2 + 5x\right) = 1$$
$$\log_2\left(6x^2 + 5x\right) = 2$$
$$6x^2 + 5x = 2^2$$
$$6x^2 + 5x - 4 = 0$$
$$(3x+4)(2x-1) = 0$$
$$x = \tfrac{1}{2}, -\tfrac{4}{3}$$

The solution is $\tfrac{1}{2}$ ($-\tfrac{4}{3}$ does not check).

50. Solving for N: $N = \log_{10}\dfrac{100}{1} = \log_{10} 10^2 = 2$

52. Solving the equation:
$$7.4 = 6.1 + \log_{10}\left(\frac{x}{y}\right)$$
$$\log_{10}\left(\frac{x}{y}\right) = 1.3$$
$$\frac{x}{y} = 10^{1.3} \approx 20.0$$

54. Substituting $I_0 = 10^{-12}$:
$$D = 10\left(\log_{10} I - \log_{10}\left(10^{-12}\right)\right)$$
$$D = 10\left(\log_{10} I + 12\right)$$

B.4 Common Logarithms and Natural Logarithms

2. Evaluating the logarithm: $\log 426 \approx 2.6294$

4. Evaluating the logarithm: $\log 42,600 \approx 4.6294$

6. Evaluating the logarithm: $\log 0.4260 \approx -0.3706$

8. Evaluating the logarithm: $\log 0.0426 \approx -1.3706$

10. Evaluating the logarithm: $\log 4,900 \approx 3.6902$

12. Evaluating the logarithm: $\log 900 \approx 2.9542$

14. Evaluating the logarithm: $\log 10,200 \approx 4.0086$

16. Evaluating the logarithm: $\log 0.0312 \approx -1.5058$

18. Evaluating the logarithm: $\log 0.00052 \approx -3.2840$

20. Evaluating the logarithm: $\log 0.111 \approx -0.9547$

22. Solving for x:
$$\log x = 4.8802$$
$$x = 10^{4.8802} \approx 75,893$$

24. Solving for x:
$$\log x = -3.1198$$
$$x = 10^{-3.1198} \approx 0.000759$$

26. Solving for x:
$$\log x = 5.5911$$
$$x = 10^{5.5911} \approx 390,032$$

28. Solving for x:
$$\log x = -1.5670$$
$$x = 10^{-1.5670} \approx 0.0271$$

30. Solving for x:
$$\log x = -4.2000$$
$$x = 10^{-4.2000} \approx 0.0000631$$

32. Solving for x:
$$\log x = -1$$
$$x = 10^{-1} = \tfrac{1}{10}$$

34. Solving for x:
$$\log x = 1$$
$$x = 10^1 = 10$$

36. Solving for x:
$$\log x = -20$$
$$x = 10^{-20}$$

38. Solving for x:
$$\log x = 4$$
$$x = 10^4 = 10,000$$

40. Solving for x:
$$\log x = \log_3 9$$
$$\log x = 2$$
$$x = 10^2 = 100$$

42. Simplifying the logarithm: $\ln 1 = \ln e^0 = 0$ **44.** Simplifying the logarithm: $\ln e^{-3} = -3$

46. Simplifying the logarithm: $\ln e^y = y$

48. Using properties of logarithms: $\ln 10 e^{4t} = \ln 10 + \ln e^{4t} = \ln 10 + 4t$

50. Using properties of logarithms: $\ln A e^{-3t} = \ln A + \ln e^{-3t} = \ln A - 3t$

52. Evaluating the logarithm: $\ln 10 = \ln(2 \cdot 5) = \ln 2 + \ln 5 = 0.6931 + 1.6094 = 2.3025$

54. Evaluating the logarithm: $\ln \frac{1}{5} = \ln 5^{-1} = -\ln 5 = -1.6094$

56. Evaluating the logarithm: $\ln 25 = \ln 5^2 = 2\ln 5 = 2(1.6094) = 3.2188$

58. Evaluating the logarithm: $\ln 81 = \ln 3^4 = 4\ln 3 = 4(1.0986) = 4.3944$

60. Computing the pH: $\text{pH} = -\log\left(1.88 \times 10^{-6}\right) \approx 5.73$

62. Finding the concentration:

$$5.75 = -\log\left[\text{H}^+\right]$$
$$-5.75 = \log\left[\text{H}^+\right]$$
$$\left[\text{H}^+\right] = 10^{-5.75} \approx 1.78 \times 10^{-6}$$

64. Finding the magnitude:

$$6.6 = \log T$$
$$T = 10^{6.6} \approx 3.98 \times 10^6$$

66. Finding the magnitude:
$$8.7 = \log T$$
$$T = 10^{8.7} \approx 5.01 \times 10^8$$

68. For the first earthquake:
$$\log T_1 = 8.5$$
$$T_1 = 10^{8.5}$$

For the second earthquake:
$$\log T_2 = 5.5$$
$$T_2 = 10^{5.5}$$

The ratio is $\dfrac{T_1}{T_2} = \dfrac{10^{8.5}}{10^{5.5}} = 10^3 = 1000$ times stronger.

70. Finding the magnitude:
$$7.7 = \log T$$
$$T = 10^{7.7} \approx 5.01 \times 10^7$$

The magnitude is 5.01×10^7 times larger.

72. Finding the rate of depreciation:
$$\log(1 - r) = \tfrac{1}{4}\log\frac{3000}{9000}$$
$$\log(1 - r) \approx -0.1193$$
$$1 - r \approx 10^{-0.1193}$$
$$r = 1 - 10^{-0.1193}$$
$$r \approx 0.240 = 24\%$$

74. Finding the rate of depreciation:
$$\log(1 - r) = \tfrac{1}{3}\log\frac{5750}{7550}$$
$$\log(1 - r) \approx -0.0394$$
$$1 - r \approx 10^{-0.0394}$$
$$r = 1 - 10^{-0.0394}$$
$$r \approx 0.087 = 8.7\%$$

76. It appears to approach e. Completing the table:

x	$\left(1 + \dfrac{1}{x}\right)^x$
1	2.0000
10	2.5937
50	2.6916
100	2.7048
500	2.7156
1,000	2.7169
10,000	2.7181
1,000,000	2.7183

B.5 Exponential Equations and Change of Base

2. Solving the equation:
$$4^x = 3$$
$$\ln 4^x = \ln 3$$
$$x \ln 4 = \ln 3$$
$$x = \frac{\ln 3}{\ln 4} \approx 0.7925$$

4. Solving the equation:
$$3^x = 4$$
$$\ln 3^x = \ln 4$$
$$x \ln 3 = \ln 4$$
$$x = \frac{\ln 4}{\ln 3} \approx 1.2619$$

6. Solving the equation:
$$7^{-x} = 8$$
$$\ln 7^{-x} = \ln 8$$
$$-x \ln 7 = \ln 8$$
$$x = -\frac{\ln 8}{\ln 7} \approx -1.0686$$

8. Solving the equation:
$$8^{-x} = 7$$
$$\ln 8^{-x} = \ln 7$$
$$-x \ln 8 = \ln 7$$
$$x = -\frac{\ln 7}{\ln 8} \approx -0.9358$$

10. Solving the equation:
$$9^{x+1} = 3$$
$$3^{2x+2} = 3^1$$
$$2x + 2 = 1$$
$$2x = -1$$
$$x = -\tfrac{1}{2}$$

12. Solving the equation:
$$3^{x-1} = 9$$
$$3^{x-1} = 3^2$$
$$x - 1 = 2$$
$$x = 3$$

14. Solving the equation:
$$2^{2x+1} = 3$$
$$\ln 2^{2x+1} = \ln 3$$
$$(2x+1)\ln 2 = \ln 3$$
$$2x + 1 = \frac{\ln 3}{\ln 2}$$
$$2x = \frac{\ln 3}{\ln 2} - 1$$
$$x = \tfrac{1}{2}\left(\frac{\ln 3}{\ln 2} - 1\right) \approx 0.2925$$

16. Solving the equation:
$$2^{1-2x} = 3$$
$$\ln 2^{1-2x} = \ln 3$$
$$(1-2x)\ln 2 = \ln 3$$
$$1 - 2x = \frac{\ln 3}{\ln 2}$$
$$-2x = \frac{\ln 3}{\ln 2} - 1$$
$$x = \tfrac{1}{2}\left(1 - \frac{\ln 3}{\ln 2}\right) \approx -0.2925$$

18. Solving the equation:
$$10^{3x-4} = 15$$
$$\ln 10^{3x-4} = \ln 15$$
$$(3x-4)\ln 10 = \ln 15$$
$$3x - 4 = \frac{\ln 15}{\ln 10}$$
$$3x = \frac{\ln 15}{\ln 10} + 4$$
$$x = \tfrac{1}{3}\left(\frac{\ln 15}{\ln 10} + 4\right) \approx 1.7254$$

20. Solving the equation:
$$9^{7-3x} = 5$$
$$\ln 9^{7-3x} = \ln 5$$
$$(7-3x)\ln 9 = \ln 5$$
$$7 - 3x = \frac{\ln 5}{\ln 9}$$
$$-3x = \frac{\ln 5}{\ln 9} - 7$$
$$x = \tfrac{1}{3}\left(7 - \frac{\ln 5}{\ln 9}\right) \approx 2.0892$$

22. Evaluating the logarithm: $\log_9 27 = \dfrac{\log 27}{\log 9} = 1.5000$

24. Evaluating the logarithm: $\log_{27} 9 = \dfrac{\log 9}{\log 27} \approx 0.6667$

26. Evaluating the logarithm: $\log_3 12 = \dfrac{\log 12}{\log 3} \approx 2.2619$

28. Evaluating the logarithm: $\log_{12} 3 = \dfrac{\log 3}{\log 12} \approx 0.4421$

30. Evaluating the logarithm: $\log_6 180 = \dfrac{\log 180}{\log 6} \approx 2.8982$

32. Evaluating the logarithm: $\log_5 462 = \dfrac{\log 462}{\log 5} \approx 3.8122$

34. Evaluating the logarithm: $\ln 3,450 \approx 8.1461$

36. Evaluating the logarithm: $\ln 0.0345 \approx -3.3668$

38. Evaluating the logarithm: $\ln 100 \approx 4.6052$

40. Evaluating the logarithm: $\ln 450,000 \approx 13.0170$

42. Using the compound interest formula:

$$500\left(1+\frac{0.06}{12}\right)^{12t}=1000$$

$$\left(1+\frac{0.06}{12}\right)^{12t}=2$$

$$\ln\left(1+\frac{0.06}{12}\right)^{12t}=\ln 2$$

$$12t\ln\left(1+\frac{0.06}{12}\right)=\ln 2$$

$$t=\frac{\ln 2}{12\ln\left(1+\frac{0.06}{12}\right)}\approx 11.6$$

It will take 11.6 years.

44. Using the compound interest formula:

$$1000\left(1+\frac{0.12}{6}\right)^{6t}=4000$$

$$\left(1+\frac{0.12}{6}\right)^{6t}=4$$

$$\ln\left(1+\frac{0.12}{6}\right)^{6t}=\ln 4$$

$$6t\ln\left(1+\frac{0.12}{6}\right)=\ln 4$$

$$t=\frac{\ln 4}{6\ln\left(1+\frac{0.12}{6}\right)}\approx 11.7$$

It will take 11.7 years.

46. Using the compound interest formula:

$$P\left(1+\frac{0.08}{4}\right)^{4t}=3P$$

$$\left(1+\frac{0.08}{4}\right)^{4t}=3$$

$$\ln\left(1+\frac{0.08}{4}\right)^{4t}=\ln 3$$

$$4t\ln\left(1+\frac{0.08}{4}\right)=\ln 3$$

$$t=\frac{\ln 3}{4\ln\left(1+\frac{0.08}{4}\right)}\approx 13.9$$

It will take 13.9 years.

48. Using the compound interest formula:

$$25\left(1+\frac{0.06}{2}\right)^{2t}=50$$

$$\left(1+\frac{0.06}{2}\right)^{2t}=2$$

$$\ln\left(1+\frac{0.06}{2}\right)^{2t}=\ln 2$$

$$2t\ln\left(1+\frac{0.06}{2}\right)=\ln 2$$

$$t=\frac{\ln 2}{2\ln\left(1+\frac{0.06}{2}\right)}\approx 11.7$$

It was invested 11.7 years ago.

50. Using the continuous interest formula:

$$1000e^{0.12t}=4000$$

$$e^{0.12t}=4$$

$$0.12t=\ln 4$$

$$t=\frac{\ln 4}{0.12}\approx 11.6$$

It will take 11.6 years.

52. Using the continuous interest formula:

$$500e^{0.12t}=1500$$

$$e^{0.12t}=3$$

$$0.12t=\ln 3$$

$$t=\frac{\ln 3}{0.12}\approx 9.16$$

It will take 9.16 years.

54. The current population is 100,000.

56. Finding when $P(t)=45,000$:

$$15,000e^{0.08t}=45,000$$

$$e^{0.08t}=3$$

$$0.08t=\ln 3$$

$$t=\frac{\ln 3}{0.08}\approx 13.7$$

It will take approximately 13.7 years. This value is verified with a graphing calculator.

58. Solving for t:

$$A = Pe^{-rt}$$
$$e^{-rt} = \frac{A}{P}$$
$$-rt = \ln\frac{A}{P}$$
$$t = \frac{\ln A - \ln P}{-r} = -\frac{1}{r}\ln\frac{A}{P}$$

60. Solving for t:

$$A = P2^{kt}$$
$$2^{kt} = \frac{A}{P}$$
$$\ln 2^{kt} = \ln\frac{A}{P}$$
$$kt \ln 2 = \ln A - \ln P$$
$$t = \frac{\ln A - \ln P}{k \ln 2}$$

62. Solving for t:

$$A = P(1+r)^{t}$$
$$(1+r)^{t} = \frac{A}{P}$$
$$\ln(1+r)^{t} = \ln\frac{A}{P}$$
$$t \ln(1+r) = \ln A - \ln P$$
$$t = \frac{\ln A - \ln P}{\ln(1+r)}$$